STEAM-PLANT
OPERATION

FOURTH EDITION

STEAM-PLANT OPERATION

EVERETT B. WOODRUFF

*Project Engineer, A. M. Kinney, Inc.; Former
Instructor in Steam Power Engineering, Ohio
College of Applied Science, University of Cincinnati;
Member, American Society of Mechanical Engineers,
Engineering Society of Cincinnati, National
Association of Power Engineers; Licensed, Professional
Engineer, Mechanical (Ohio and New York),
Stationary Engineer (Ohio)*

HERBERT B. LAMMERS

*Consultant; Former Instructor in Steam Power Engineering,
Ohio College of Applied Science, University of Cincinnati;
Member and Past President, Engineering
Society of Cincinnati;
Licensed, Professional Engineer,
Mechanical (Ohio), Stationary Engineer (Ohio)*

McGRAW-HILL BOOK COMPANY

*New York   St. Louis   San Francisco   Auckland   Bogotá
Dusseldorf   Johannesburg   London   Madrid
Mexico   Montreal   New Delhi   Panama
Paris   Sao Paulo   Singapore
Sydney   Tokyo   Toronto*

Library of Congress Cataloging in Publication Data

Woodruff, Everett Bowman.
  Steam-plant operation.

  Includes index.
  1. Steam power-plants. I. Lammers, Herbert B.,
joint author. II. Title.
TJ405.W6 1976   621.1   76-27829
ISBN 0-07-071731-1

    90 KPKP   865432

*The editors for this book were Tyler G. Hicks and Virginia Anne Fechtmann,
the designer was Naomi Auerbach, and the production supervisor
was Frank P. Bellantoni. It was set in Caledonia
by Bi-Comp, Incorporated.*

It was printed and bound by The Kingsport Press.

CONTENTS

Preface .... *ix*

1. BOILERS **1**

  1.1 *The Boiler* 1
  1.2 *Fire-Tube Boilers* 2
  1.3 *Water-Tube Boilers* 9
  1.4 *Steam Separators and Superheaters* 21
  1.5 *Heat-Recovery Equipment* 26
  1.6 *Furnace Construction* 32
  1.7 *Industrial and Utility Boilers* 39
       Questions and Problems 57

2. DESIGN AND CONSTRUCTION OF BOILERS **61**

  2.1 *Materials Used in Boiler Construction* 61
  2.2 *Stresses in Tubes, Boiler Shells, and Drums* 65
  2.3 *Drum and Shell Construction* 71
  2.4 *Riveted Joints* 73
  2.5 *Welded Construction* 76
  2.6 *Braces and Stays* 79
  2.7 *Manholes, Handholes, and Fittings* 81
  2.8 *Boiler Assembly* 82
  2.9 *Heating Surface and Capacity* 85
  2.10 *Boiler-Capacity Calculation* 89
       Questions and Problems 93

3. COMBUSTION OF FUELS **96**

  3.1 *The Combustion Process* 96
  3.2 *The Theory of Combustion* 100
  3.3 *The Air Supply* 111
  3.4 *Coal* 117
  3.5 *Fuel Oil* 126

3.6	*Gas*	130
3.7	*By-Product Fuels*	132
3.8	*Control of the Combustion Process*	133
	Questions and Problems	140

4. SETTINGS, COMBUSTION EQUIPMENT, AND HEATING SURFACES   **142**

4.1	*Boiler Settings*	142
4.2	*Hand Firing*	145
4.3	*Chain-and Traveling-Grate Stokers*	147
4.4	*Underfeed Stoker*	152
4.5	*Spreader Stoker*	162
4.6	*Pulverized Fuel*	176
4.7	*Fluidized-Bed Combustion*	194
4.8	*Fuel Oil*	195
4.9	*Gas*	206
4.10	*Automatic Operation of Boilers*	210
4.11	*Chimneys*	214
4.12	*Mechanical Draft*	216
4.13	*Steam and Air Jets*	222
4.14	*Air Pollution Control*	223
	Questions and Problems	230

5. BOILER ACCESSORIES   **232**

5.1	*Water Columns*	232
5.2	*Fusible Plugs*	237
5.3	*Steam Gauges*	239
5.4	*Feedwater Regulators*	241
5.5	*Safety Valves*	246
5.6	*Blowdown Apparatus*	254
5.7	*Nonreturn Valves*	259
5.8	*Steam Headers*	262
5.9	*Soot Blowers*	263
5.10	*Valves*	269
5.11	*Instruments and Automatic Combustion Control*	275
	Questions and Problems	280

6. OPERATION AND MAINTENANCE OF STEAM BOILERS   **283**

6.1	*Putting Boilers in Operation*	283
6.2	*Normal Operation*	286
6.3	*Operating in an Emergency*	299
6.4	*Idle Boilers*	303
6.5	*Maintenance—Causes and Reduction*	305
6.6	*Boilers Internal Inspection*	312
6.7	*Making Repairs*	314
	Questions and Problems	322

7. PUMPS **324**

7.1   *Pumps* 324
7.2   *Injectors* 325
7.3   *Duplex Pumps* 328
7.4   *Simplex Pumps* 338
7.5   *Power Pumps* 339
7.6   *Vacuum Pumps* 340
7.7   *Rotary Pumps* 343
7.8   *Centrifugal Pumps* 346
7.9   *Pump Installation and Operation* 360
7.10 *Pump Testing and Calculations* 364
7.11 *Pump Maintenance* 370
      Questions and Problems 376

8. RECIPROCATING STEAM ENGINES **380**

8.1   *Pistons and Cylinders* 381
8.2   *Pistons Rods, Crossheads, and Connecting Rods* 384
8.3   *Crank, Flywheel, and Valve Mechanism* 387
8.4   *Frames and Foundations* 390
8.5   *Governers* 390
8.6   *Engine Speed* 396
8.7   *Compound Engines* 398
8.8   *Condensing Engines* 400
8.9   *Uniflow Engines* 401
      Questions and Problems 402

9. VALVE-OPERATING MECHANISM **405**

9.1   *Cycle of Events* 405
9.2   *Function of Valves* 408
9.3   *The Slide Valve* 415
9.4   *Piston Valves* 417
9.5   *Corliss Valves* 419
9.6   *Poppet Valves* 425
9.7   *The Uniflow Engine* 428
      Questions and Problems 433

10. OPERATION AND MAINTENANCE OF STEAM ENGINES **435**

10.1 *Inspection of Engines* 435
10.2 *Lubrication of Steam Engines* 437
10.3 *Operating a Steam Engine* 444
10.4 *Maintenance of Steam Engines* 447
10.5 *Economy of Steam Engines* 451
10.6 *Steam-Engine Indicators* 453
10.7 *Engine Rating and Efficiency* 458
      Questions and Problems 463

11. STEAM TURBINES AND AUXILIARIES **466**

11.1   *Turbines: General* 466

11.2   *Condensers* 497

11.3   *Spray Ponds and Cooling Towers* 504

11.4   *Condenser Auxiliaries* 511

        Questions and Problems 515

12. OPERATING AND MAINTAINING TURBINES AND AUXILIARIES **517**

12.1   *Turbines* 517

12.2   *Condensers* 530

12.3   *Cooling, Towers and Spray Ponds* 536

12.4   *Auxiliaries* 541

        Questions and Problems 545

13. AUXILIARY STEAM-PLANT EQUIPMENT **547**

13.1   *Open Feedwater Heaters* 547

13.2   *Closed Feedwater Heaters* 554

13.3   *Ion-Exchange Water Conditioners* 557

13.4   *Evaporators* 560

13.5   *Boiler Blowdown* 562

13.6   *Piping System* 564

13.7   *Steam Traps* 573

13.8   *Pipeline Separators and Strainers* 578

13.9   *Lubricants and Lubricating Devices* 581

        Questions and Problems 590

APPENDIX A: USEFUL DATA **592**

APPENDIX B: ARITHMETIC **596**

APPENDIX C: DEFINITIONS **600**

APPENDIX D: STEAM TABLES AND CHARTS **604**

APPENDIX E: CALCULATION OF THE STAYED AREA OF THE HEADS OF HORIZONTAL RETURN TABULAR BOILERS **615**

APPENDIX F: CALCULATION OF THE MEAN EFFECTIVE PRESSURE OF A STEAM ENGINE BY APPLYING THE ORDINATE METHOD TO THE INDICATOR DIAGRAM **617**

Answers to Problems   619

Index   621

PREFACE

TO THE FOURTH EDITION

It is with pleasure and satisfaction that we have this opportunity of sharing our 50 years of experience in the design and operation of steam plants with you. In compiling this text and the subsequent editions we imagine ourselves by your side and helping you as you endeavor to increase your knowledge of this subject. We are gratified by the wide acceptance of the original text and subsequent editions. It is satisfying to know that many have benefited by the information contained in this text. We continue to present the practical man with a working knowledge of the fundamental principles and operating procedures involved in steam plants.

The fourth edition provides information on the latest designs and improved techniques for operating and maintaining all types of steam-plant equipment. The older types of equipment are retained to illustrate fundamentals, to provide information required to answer questions asked on operating license examinations, and to supply operators and maintenance men with information otherwise difficult to obtain.

The student, whether preparing for an examination or advancing his general knowledge of steam plants, is advised to study the text and then attempt the questions and problems at the end of the chapters. They serve as a measure of the grasp of the material presented. Much more material is covered in the text than is included in the questions and problems.

The descriptive material and illustrations have been revised to reflect the recent trends in steam plants. Automation and the automatic plants present new requirements in knowledgeable operation and maintenance procedures. Safety and economy may be "built in" the original equipment, but continued safe and economical operation depends upon thorough routine inspections and preventive maintenance. Improved water-conditioning equipment and techniques provide opportunities to secure trouble-free operating with low maintenance cost. High-speed machinery reduces initial costs and provides low-maintenance, trouble-free service only when such equipment is adequately lubricated, operated and maintained. Economics in steam generation apply not only in the large, high-pressure units but also in the small and medium sizes for which factory-assembled, complete packaged units are being supplied. Fuel economy is secured by a reduction in the amount of excess air and air infiltration. The use of two or more fuels in the same furnace results in complicated combustion equipment and controls.

We are grateful to the publishers and authors who gave us permission to use extracts from their material and to the manufacturers of equipment who provided information and illustrations.

Everett B. Woodruff
Herbert B. Lammers

CHAPTER 1

BOILERS

1.1   THE BOILER.   A boiler is a closed vessel in which water, under pressure, is transformed into steam by the application of heat; open vessels and those generating steam at atmospheric pressure are not considered to be boilers.   In the furnace, the chemical energy in the fuel is converted into heat; it is the function of the boiler to transfer this heat to the water in the most efficient manner.

The *ideal* boiler embodies

1. Simplicity in construction, excellent workmanship, and materials conducive to low maintenance cost
2. Design and construction to accommodate expansion and contraction properties of materials
3. Adequate steam and water space, delivery of clean steam, and good water circulation
4. A furnace setting conducive to efficient combustion and maximum rate of heat transfer
5. Responsiveness to sudden demands and overloads
6. Accessibility for cleaning and repair
7. A factor of safety of at least code requirement

A boiler should be designed to absorb the maximum amount of heat released in the process of combustion.   This heat is transmitted to the boiler by *radiation, conduction,* and *convection,* the percentage of each depending upon boiler design.

"Radiant" heat is heat radiated from a hot to a cold body and depends on the temperature difference and the color of the body which receives the heat.   Absorption of radiant heat increases with the furnace temperature; it depends on many factors but primarily on the area of the tubes exposed to the heat rays.

"Conduction" heat is heat which passes from the gas to the tube by physical contact. The heat passes from molecule of metal to molecule of metal with no displacement of the molecules. The amount of absorption depends on the conductivity or heat-absorption qualities of the material through which the heat must pass.

"Convection" heat is heat transmitted from the hot to the cold body by movement of the conveying substance. In this case, the hot body is the boiler gas; the cold body, the boiler tube containing water.

In designing a boiler, each form of heat transmission is given special consideration. In the operation of a boiler unit, all three forms of heat transmission occur simultaneously and cannot readily be divorced from each other.

Considerable progress has been made in boiler design from the standpoint both of safety and of efficiency of the fuel-burning equipment. More and more emphasis is being placed on efficiency and flexibility. Recent advances in boiler design include forced circulation of water, pressurized furnaces, hot-water boilers, extremely high steam pressures and temperatures, automation, and large-capacity units producing 5 to 6 million lb of steam per hr.

Boilers are built in a variety of sizes, shapes, and forms to fit conditions peculiar to the individual plant and to meet varying requirements. With increasing fuel cost, greater attention is being given to improvement of the furnace's combustion efficiency. Many boilers are designed to burn multiple fuels in order to take advantage of the fuel most economically available.

Increased boiler "availability" has made practical units of increased capacity, and this has resulted in lower installation and operating costs. For the small plant, all boilers should preferably be of the same type, size, and capacity since standardization of equipment makes possible uniform operating procedures, reduces spare-parts stock to a minimum, and contributes to lower overall costs.

The field of application is diversified. Boilers are used to produce steam for heating, process, and power generation; to operate steam engines, turbines, pumps, etc. This text is concerned with boilers employed in stationary practice.

1.2   FIRE-TUBE BOILERS.   Fire-tube boilers are so named because the products of combustion pass through tubes or flues, which are surrounded by water. They may be either *internally* fired (Fig. 1.1) or *externally* fired (Fig. 1.3). Internally fired boilers are those in which the grate and combustion chamber are enclosed within the boiler shell.

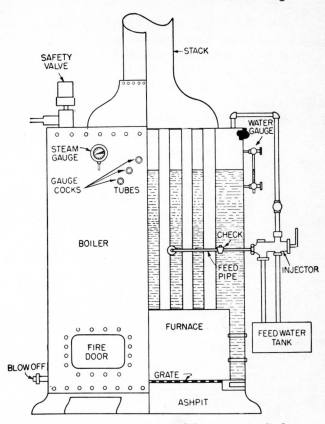

FIG. 1.1 *Sectional view of vertical boiler—exposed-tube type.*

Externally fired boilers are those in which the setting, including furnace and grates, is separate and distinct from boiler shell. Fire-tube boilers are classified as vertical or horizontal tubular.

The vertical fire-tube boiler consists of a cylindrical shell with enclosed firebox (Figs. 1.1, 1.2). Here tubes extend from the *crown sheet* (firebox) to the upper tube sheet. Holes are drilled in each sheet to receive the tubes, which are then rolled to produce a tight fit, and the ends are beaded over. Screwed stay bolts are used in the water leg (firebox to shell) with the ends riveted over.

In the vertical *exposed-tube* boiler (Fig. 1.1), the upper tube sheet and tube ends are above the normal water level, extending into the steam space. This type of construction reduces the moisture carry-over and superheats the steam leaving the boiler. However, the upper tube ends, not being protected by water, may become overheated and leak at the point where they are expanded into the tube sheet. The furnace is water-cooled and is formed by an extension of the outer and inner shell which is riveted to the lower tube sheet. The upper tube sheet is riveted directly to the shell. When the boiler is operated, water is carried some distance below the top of the tube sheet; the area above the water level is steam space.

In *submerged-tube* boilers (Fig. 1.2), the tubes are rolled into the upper tube sheet, which is below the water level; the outer shell extends above the top of the tube sheet. A conical-shaped section of the plate is riveted to the sheet so that the space above the tube sheet provides a smoke outlet. Space between the inner and outer sheets comprises the steam space. This design permits carrying the water level above the upper tube sheet, thus preventing overheating of the tube ends. The conical-shaped section is difficult to construct, requiring staying, and is subject to leaks.

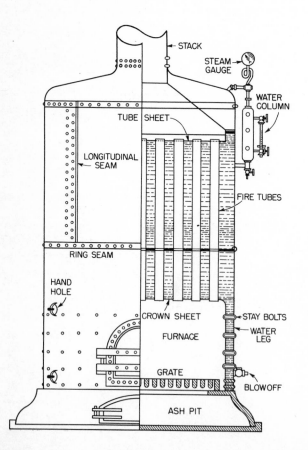

FIG. 1.2 *Sectional view of vertical boiler—submerged-tube type.*

Since vertical boilers are portable, they are used to power hoisting devices and operate fire engines and tractors as well as for stationary practice. They range in size from 5 to 75 hp; tube sizes range from 2 to 3 in. in diameter, pressures to 100 psi, diameters from 3 to 5 ft, and height from 5 to 10 ft. With the exposed-tube arrangement, super-heated steam 10 to 15°F may be secured.

Vertical fire-tube boilers are rapid steamers, their initial cost is low, and they occupy little floor space. Boilers of this type usually employ a standard base. Combustion efficiency is improved when the boiler is elevated and set on a refractory base to secure added furnace volume. This is especially important if bituminous coal is to be burned and smoke is to be reduced to a minimum. If the boiler is stoker-fired, either raise the boiler or pit the stoker for the required setting height.

Horizontal fire-tube boilers are of many varieties, the most common being the horizontal-return tubular (HRT) boiler (Fig. 1.3). This boiler has a long cylindrical shell supported by the furnace side walls and set on saddles equipped with rollers to permit movement of the boiler as it expands and contracts. It may also be suspended from hangers (Fig. 1.4) and supported by overhead beams; here the boiler is free to move independently of the setting. Expansion and contraction work no hardship on the brick setting, and thus maintenance is reduced.

The required boiler-shell length is secured by riveting (Fig. 1.3) several plates together. The seam running the length of the shell is called a "longitudinal" joint and is of butt-strap construction. Note that this

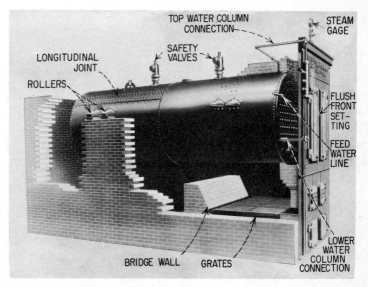

FIG. 1.3 *Horizontal return-tubular boiler and setting.* (*Erie City Iron Works.*)

joint is above the fire line to avoid overheating. The "circumferential" joint is a lap joint.

A return tubular boiler (Fig. 1.4) has its plates joined by fusion welding. This type of construction is superior to that of a riveted boiler since there are no joints to overheat. As a result, the life of the boiler is lengthened, maintenance is reduced, and at the same time higher rates of firing are permitted.

The boiler setting includes grates (or stoker), bridge wall, and combustion space. The products of combustion are made to pass from the grate, over the bridge wall (and under the shell), to the rear end of the boiler. Gases return through the tubes to the front end of the boiler, where they connect to the breeching or stack. The shell is bricked in slightly below the top row of tubes to prevent overheating of the longitudinal joint and to keep the hot gases from coming into contact with that portion of the boiler plate above the waterline.

The conventional HRT boiler is set to slope 1 to 3 in. from front to rear. A blowoff line is connected to the underside of the shell at the rear end of the boiler to permit drainage and removal of impurities; it is extended through the setting, where blowoff valves are attached. The line is protected from the heat by a brick lining or protective sleeve. Safety valves and water column are located as shown in Fig. 1.3. A *dry pipe* is frequently installed in the top of the drum to separate the moisture from the steam before the steam passes to the steam-nozzle intake.

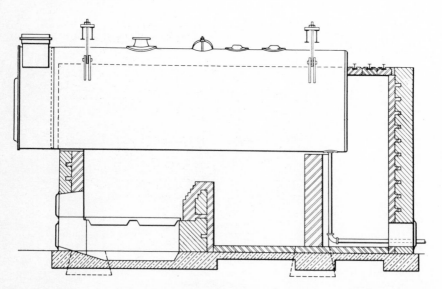

FIG. 1.4   *Horizontal-return tubular boiler and setting—overhanging front.*

Tubes and flues are used in the HRT boiler as a means of increasing the heating surface and steaming capacity; they provide additional strength for the tube sheets.   Sizes below 4 in. are usually referred to as "tubes"; those above 4 in., as "flues."   For tube size, we designate the external diameter; for flue size, the internal diameter.

Tubes are rolled into the tube sheet by means of an expander and then beaded over, whereas flues are riveted in place.   The size of the tube or flue depends upon the type of fuel used, the draft loss desired, etc.   Tube spacing and staggering are resorted to in an effort to improve the water circulation.

Still another type of HRT boiler is the horizontal four-pass forced-draft packaged unit (Fig. 1.5), which can be fired with gas or fuel oil.   In heavy oil-fired models, the burner has a retractable nozzle for ease in cleaning and replacing.

Gases from the combustion chamber reverse at the rear to pass downward to the tubes directly beneath the chamber.   Again they reverse

FIG. 1.5 *Horizontal four-pass forced-draft packaged boiler. (Cleaver Brooks Co.)*

to pass through the tube bank above the combustion chamber and reverse and pass through the top tube section to the stack.

Such units are available in sizes of 15 to 600 hp with pressures of 15 to 250 psi. They are a source of high- or low-pressure steam or hot water. These units are compact, requiring a minimum of space and headroom; are automatic in operation; have a low initial cost; and do not need a tall stack. As such they find application and acceptance in many locations. By reason of their compactness, however, they are not readily accessible for inspection and repairs.

The Scotch marine boiler (Fig. 1.6) is a fire-tube return tubular unit. It consists of a cylindrical shell containing the firebox and tubes; the tubes surround the upper portion of the firebox and are rolled into tube sheets on each end of the boiler. Combustion gases pass to the rear of the furnace, returning through the tubes to the front, where they are discharged to the stack.

The water-cooled furnace and limited furnace volume make smokeless combustion difficult when firing bituminous coal unless over-fire air or steam jets are employed.

Scotch marine boilers are self-contained, do not require a setting, and are internally fired. They are portable packaged units requiring a minimum of space and headroom. Units of this type are to be found in marine and stationary service burning coal, oil, and gas. For coal

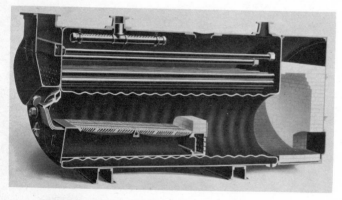

FIG. 1.6   *Scotch marine boiler.*

burning, the long grates make cleaning of fires very difficult. Scotch marine boilers are built in diameters 3 to 8 ft, lengths 4 to 18 ft, capacity to 350 hp, and pressures to 200 psi.

1.3 WATER-TUBE BOILERS. A water-tube boiler is one in which the products of combustion pass around tubes containing water. The tubes are interconnected to common water channels and to a steam outlet or outlets. For some boilers, baffles to direct the gas flow are not required. For others baffles are installed in the tube bank to direct the gases across the heating surfaces and to secure maximum heat absorption. The baffles may be of refractory or membrane construction, as discussed later. There are an endless variety of boilers designed to meet specific needs, so care must be exercised in the selection and based on plant requirements and space limitations. Water-tube boilers are classified as (1) vertical, (2) vertically inclined, (3) horizontal, and (4) a combination of the foregoing, to describe a few.

The steam generator (Fig. 1.7) is a self-contained packaged portable unit available in sizes from 16.5 to 175 hp (approximately 6,000 pph of steam) with pressures to 295 psig. This unit is automatic in operation and requires little floor space; it is fired with oil or gas or with a combination of these fuels. Units can be grouped together to provide a total output of approximately 70,000 pph of steam.

The heating surface consists of a single continuous coil, there being no drums or headers; hence hazards of explosion are eliminated. The unit is responsive to steam demands, and pressure can be generated quickly.

When a forced controlled circulation system is used, only a small quantity of water is required to produce the desired steam output. Heating coils are enclosed in a steel insulated jacket; a small quantity of air is circulated between the insulation and jacket for cooling.

In operation, water is pumped from a receiver to the accumulator, the level in the accumulator being maintained by a liquid-level controller. The bottom of the accumulator is equipped with a blowdown valve. A pump draws water from the accumulator and forces it under pressure through the heating and steam coils, counterflow to the products of combustion. Steam is generated in the lower portion of the coil and passes through a separator in the accumulator to the steam outlet; water separated from the steam is retained in the accumulator.

A feature of this unit is the ring thermostat tube as an integral part of the heating coil. This ring expands and makes contact with the thermostat fuel control to maintain constant steam temperature and acts as a safety device if overheating or water shortage occurs.

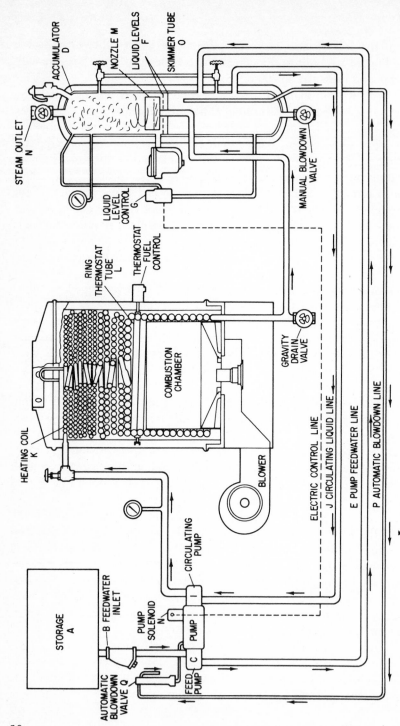

FIG. 1.7 Steam generator. (*Clayton Mfg. Co.*)

ACCUMULATOR D

NOZZLE M

LIQUID LEVELS F

SKIMMER TUBE O

STEAM OUTLET N

LIQUID LEVEL CONTROL G

RING THERMOSTAT TUBE L

THERMOSTAT FUEL CONTROL

HEATING COIL K

COMBUSTION CHAMBER

MANUAL BLOWDOWN VALVE

GRAVITY DRAIN VALVE

BLOWER

ELECTRIC CONTROL LINE

J CIRCULATING LIQUID LINE

E PUMP FEEDWATER LINE

P AUTOMATIC BLOWDOWN LINE

STORAGE A

B FEEDWATER INLET

PUMP SOLENOID N

CIRCULATING PUMP

PUMP C

FEED PUMP

AUTOMATIC BLOWDOWN VALVE Q

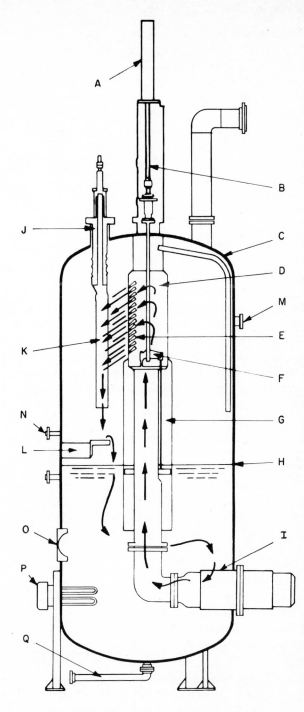

FIG. 1.8 Coates electrode high-voltage steam boiler; interior view. (A) Control cylinder. (B) Control cylinder rod. (C) Boiler shell. (D) Jet column. (E) Jets. (F) Control linkage. (G) Control sleeve. (H) Water level. (I) Circulating pump. (J) Insulator. (K) Electrode. (L) Counter electrode. (M) Safety valve. (N) Water level control. (O) Manhole. (P) Standby heater. (Q) Tank drain. (Cam Industries Inc.)

The electric steam boiler (Fig. 1.8) provides steam at high pressure; it is a packaged unit generating steam for heating and process. The small units (1,000 to 10,000 pph of steam) operate at low voltage, while the larger units (7,000 to 100,000 pph of steam) operate at 13,800 volts. Such units find application in educational institutions, commercial and office buildings, hospitals, processing plants, etc.

Operation is as follows: water from the lower part of the boiler shell (3) is pumped to the jet column (4) and flows through the jets (5) to strike the electrodes (11) thus creating a path for the electrical current. As the unevaporated portion of the water flows from the electrode to the counterelectrode (12), a second path for current is created. Regulation of the boiler output is accomplished by hydraulically lifting the control sleeve (7) to intercept and divert the streams of water from some or all of the jets (5), to prevent the water from striking the electrode. The control sleeve (7) is moved by the lift cylinder (1), which is positioned by the boiler pressure and load control system to hold the steam pressure constant or to limit the kw output to a desired level.

To shut the boiler off, it is only necessary to stop the pump. A proportioning-type feedwater regulator (not shown) is used to maintain a constant water level. Water failure simply causes the boiler to cease operation with no overheating or danger involved.

The advantages of the electric steam generator are: compactness; safety of operation; absence of storage tanks; the use of electric power during off-peak periods; responsiveness to demand. Its disadvantages lie in the use of high voltage; power costs; and availability, if needed to be used during other than off-peak periods. So when such an installation is contemplated, all factors, including the initial cost, need to be considered. This is an interior view which also describes how the unit operates; the unit arrives as a package enclosed in the casing.

The vertical-inclined water-tube boiler (Fig. 1.9) has four drums. The upper drums are set on saddles fastened to horizontal beams; the center drum is suspended from an overhead beam by slings; the mud drum hangs free, suspended from the tubes.

Water enters the right-hand drum (top) and flows down the vertical bank of tubes to the mud drum; it then moves up the inclined bank of tubes, passing through the center drum, returning to its point of origin. Steam is made to pass around a baffle plate; in the process most of the entrained moisture is removed, before the steam enters the circulators and as the moisture is on its way to the steam drum (separator) and steam outlet. The steam in passing through the circulators receives a small degree of superheat, 10 to 15°F by the time it enters the "separator."

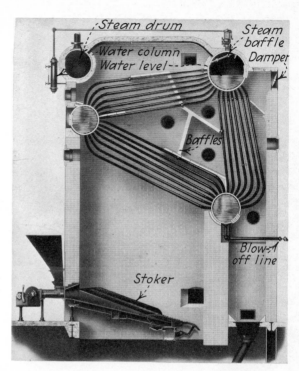

FIG. 1.9 *Water-tube boiler—four-drum type.* (*Riley Stoker Corp.*)

A cross-drum straight-tube sectional-header boiler is shown in Fig. 1.10, arranged for oil firing and equipped with an interdeck superheater. It is designed with a steep inclination of the main tube bank to provide rapid circulation. Boilers of this general type fit into locations where headroom may be a deciding factor and where efficient operating standards must be maintained. These boilers can be designed to use any of the fuels currently available.

The Stirling boiler (Fig. 1.11) is built in a number of designs to meet various space and headroom limitations Three drums are set transversely, interconnected to the lower (mud) drum by tubes slightly inclined. The tubes are shaped and curved at the ends so as to enter the drum radially.

The upper drums are interconnected by steam circulators (top) and by water circulators (bottom). Note that the center drum has tubes leaving, to enter the rear tube bank; this is done to improve the water circulation. The heating surface is then a combination of waterwall surface, boiler tubes, and a small amount of drum surface. The interdeck superheater and economizer likewise contain heating surface.

The furnace is water-cooled. Downcomers from the upper drum supply water to the side wall headers, with steam and water returning to

FIG. 1.10 *Cross-drum straight-tube boiler.* (*The Babcock & Wilcox Co.*)

the drum from the wall tubes. For the boiler proper, feedwater passing through the economizer enters the left (top) drum and flows down the rear bank of tubes to the mud drum. Steam generated in the first two banks of boiler tubes returns to the right and center drums; note the interconnection of drums, top and bottom. Finally all the steam generated in the boiler and waterwalls reaches the left-hand drum where it is made to pass through baffles or a steam scrubber or combination of the two, to reduce the moisture content of the steam before passing to the superheater. The superheater consists of a series of tube loops; the steam then passes to the outlet steam header.

The upper drums are supported at the ends, by lugs resting on steel columns; the lower drum is suspended from the tubes and is free to move, imposing no hardship on the setting, by expansion. The superheater headers are supported (at each end) from slings attached to steel columns overhead.

The products of combustion pass over the first bank of tubes; through the superheater; down across the second pass; reverse and up through

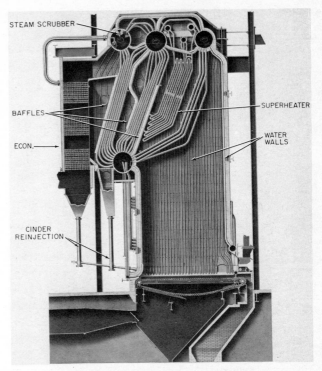

STEAM SCRUBBER

BAFFLES

ECON.

CINDER
REINJECTION

SUPERHEATER

WATER
WALLS

FIG. 1.11 *Stirling boiler; fired with Rotograte spreader stoker.* (*The Babcock & Wilcox Co.*)

the third pass. Upon leaving the boiler, the gas enters the economizer traveling down (counterflow to gas flow) through the tubes to the exit.

Most of the steam is generated in the waterwalls and the first bank of boiler tubes since this heating surface is exposed to radiant heat. This unit is fired with a Rotograte spreader stoker; it may also be gas-oil or combination firing, providing flexibility in operation. Fly ash collected at the bottom of the third pass and from the economizer is reinjected back onto the grate, through a series of nozzles located in the rear wall. Over-fire air is introduced through nozzles in the rear and side walls, to reduce smoke emission to a minimum.

Vertically inclined boilers are simple in construction, have independent expansion of component parts, are rapid steamers, and are accessible for cleaning and repair. They have excellent water circulation; they require added headroom for tube renewal. Boiler tubes are usually 3 to $3\frac{1}{4}$ in. in diameter, with spacing provided for making repairs and removals convenient.

There are many varieties of packaged boilers. The smaller units are completely factory-assembled and ready for shipment. The larger units are of modular construction with final assembly and erection done in the field. Shown in Fig. 1.12 is a unit in the process of shop assembly: a two-drum Keystone generator. The generator furnace is composed of 2-in. tangent tubes on 2-in. centers forming a water-cooled wall; the wall directs the flow of gases from the front of the unit, through the furnace, and around both sides at the rear into the convection zones, then toward the front of the unit with top vertical gas discharge. The Keystone generator is available in capacities ranging from 6,000 to 350,000 pph of steam, with pressures to 2,000 psi and temperatures to 950°F. These units are gas-oil–fired.

If desired the boiler can be equipped with a radiant-type superheater installed in the rear of the furnace (it is also removable). Likewise a fin-tube economizer can be installed, mounted on top of the steam generator; or an air preheater can be added, installed and mounted where space permits, either on top of the boiler or to one side. Burners are pre-assembled and ready for installation in the front wall.

The waterwalls have a reinforced, welded steel casing providing a gas-tight envelope; jacketed insulation next to the wall tubes backed with corrugated lagging provides exterior protection. Piping for the gen-

FIG. 1.12 *Keystone steam generator. (Zurn Industries; Erie City Energy Div.)*

FIG. 1.13 *Type FM integral-furnace boiler. (The Babcock & Wilcox Co.)*

erator is preengineered, to speed installation.   The unit has no refractory baffles or header handhole plates, that require maintenance.

Steam upon leaving the drum must first pass through a steam separator before going to the superheater.   The separator may consist of a series of baffle plates to throw out the water; when high-quality steam is desired, vortex steam separators are used in combination with chevron scrubbers.   Steam quality of 1 ppm can be attained.

The FM packaged boiler (Fig. 1.13) is available in capacities from 8,000 to 160,000 pph of steam, with pressures of 250 to 925 psi and temperatures to 825°F.   Superheaters, economizers, and air preheaters can be added, with operating and economic design consideration.

The packaged unit includes burners, soot blowers, pump and heating set, etc.   The steam drum is provided with scrubbers or cyclones to meet steam requirements.

Features of the FM boiler are: absence of headers, eliminating the need for handhole openings; complex inspection and gasket replacement are not required; tube-bank access ports with refractory knockout plugs provide for ease in inspection and tube cleaning; outer shell lagging permits use of outdoor installation.   A rugged steel base frame supports the entire boiler, ready to be placed in service after connections to header and water line have been made.

For units size 8,000 to 75,000 pph of steam, a welded casing is employed; from 75,000 to 160,000 pph of steam, membrane wall construction with partial welded casing.   Units to 75,000 pph of steam are shop-assembled and shipped as a package unit; over 75,000 pph, the forced-draft fan and connecting duct are shipped separately and assembled on the site.

The path of flue gases is clearly shown: from the burner to the rear of furnace (radiant section), reversing to pass through the superheater and convection passes. These units are oil-, gas-, or combination-fired. An instrument control panel accompanies the boiler installation.

The SD steam generator (Fig. 1.14) has been developed to meet demands of power and process steam producers in a wide range of sizes to a capacity in excess of 65 mw. This unit is available in capacities to 650,000 pph of steam; 1,800 psi; 960°F. It is a complete waterwall furnace construction, with a radiant and convection superheater and four-pass gas travel. If economics dictate, an air preheater or economizer (or combination) can be added. The unit shown is for a pressurized furnace designed for oil, gas, and low-heat-value fuel.

FIG. 1.14   *SD steam generator.*   (*Foster Wheeler Corp.*)

The combination radiant-convection superheaters provide a flat superheater temperature over the normal operating range. When superheat control is necessary, a spray-type desuperheater is installed. This is located in an intermediate steam temperature zone that insures mixing and rapid, complete evaporation of the injected water.

The rear drum and lower waterwall headers are bottom-supported (for coal, top-supported), permitting upward expansion. Waterwall tubes are fin-tube-welded; wall construction is such that the unit is pressure-tight for operating either with a balanced draft or when employing a pressurized furnace. The upper steam drum is equipped with chevron dryers and horizontal separators to provide dry steam to the superheater.

The prominent nose at the top of the furnace ensures gas turbulence and good distribution of gases as they flow through the boiler. The unit is front-fired and has a deep furnace, from which the gases pass through the radiant superheater; down through the convection superheater; up and through the first bank of boiler tubes; reverse and down through the rear bank of tubes to the boiler exit (to the right of the lower drum).

The outer walls are of the welded fin-tube-type; baffles are constructed of welded fin-tubes; the single-pass gas arrangement eliminates corrosion due to sharp turns in the gas stream.

HTW Boilers.[1] High-temperature water boilers are manufactured in a variety of sizes and pressures (100 to 300 psi) permitting a supply temperature for the system of 320 to 420°F. Many units are shop-assembled and shipped as packaged units. The large units are shipped in component assemblies; it is necessary to install refractory and insulation after the pressure parts are erected. The packaged units are usually oil- or gas-fired; for the larger units, oil, gas, or a spreader stoker is employed.

The high-pressure water can be converted into low-pressure steam for process, if desired. For example, with a system operating at a maximum temperature of 365°F, the unit is capable of providing steam at 100 psi.

A high-temperature water system is defined as a fluid system operating at temperatures above 212°F and requiring the application of pressure to keep the fluid in a fluid state. Whereas a steam boiler operates at a fixed temperature which is its saturation temperature, a water system, depending upon its use, can be varied from an extremely low to a relatively high temperature.

The average water temperature within a complete system will vary with load demand, and as a result an expansion tank is used to provide for expansion and contraction of the water volume as its average temperature varies. To maintain pressure in the system we can employ (*a*)

[1] Courtesy International Boiler Works.

steam pressurization or (*b*) gas pressurization in the expansion tank. For the latter we can use air or nitrogen.

The pressure is maintained independent of the heating load by means of automatic or manual control. Firing of each boiler is controlled by the water temperature leaving the boiler.

The hot-water system has much to recommend it by reason of its flexibility. For the normal hot-water system there are no blowdown losses and little or no makeup; installation costs are lower than for a steam-heating system; the system requires less attention and maintenance. The system can be smaller than for an equivalent steam system because of the huge water-storage capacity; peak loads and pickup are likewise minimized, with resulting uniform firing cycles and higher combustion efficiency.

The high-temperature water system is a closed system. When applied to heating systems, the largest advantage is for the heating of multiple buildings. For such applications, the simplicity of the system helps reduce the initial cost. Headers can follow the contour of the ground. Traps, condensate pumps, and receivers are eliminated together with the maintenance for these auxiliaries.

Only a small amount of makeup is required to replace the amount of water that leaks out of the system at valve stems, pump shafts, and similar packed points. Since there is little or no free oxygen in the system, return line corrosion is reduced or eliminated, as contrasted with wet returns from the steam system wherein excessive maintenance is frequently required. Feedwater treatment can be reduced to a minimum.

COMPARISON OF FIRE- AND WATER-TUBE BOILERS

Fire-tube boilers ranging to 500 hp and water-tube boilers with a capacity ranging to approximately 80,000 lb per hr are factory-assembled and shipped in one package. Elimination of field-assembly work, the compact design, and standardization result in a lower cost than that of comparable field-erected boilers. Factory-assembled units are superior to those erected in the field. Tube sections and other components, in sizes suitable for shipment, reduce overall cost and erection time.

Fire-tube boilers are preferred to water-tube boilers because of their lower initial cost, their compactness, and the fact that little or no setting is required. They occupy a minimum of floor space. However, they have the following inherent disadvantages: drums and joints are exposed to furnace heat, increasing the hazard of explosion; the water volume is large and the circulation poor, resulting in slow response to changes in steam demand; tube and shell are subjected to variations in temperature and subsequent stress owing to unequal expansion and contraction; and capacity, pressure, and steam temperature are limited.

Packaged boilers may be of the fire-tube or water-tube variety. They are usually oil- or gas-fired or employ a combination of fuels. Less time is required to procure packaged units, which hence are immediately available. Completely shop-assembled, they can be placed in service very quickly. The packaged units are automatic, requiring a minimum of attention, and hence reduce labor costs. In compacting, however, the furnace and heating surfaces are reduced to a minimum, resulting in high heat-transfer rates, with consequent overheating and increased maintenance and operating difficulties. So caution must be exercised in the selection of packaged units since the tendency toward compactness can be carried too far; such compactness also makes the units somewhat inaccessible for repairs. Because these units operate at high ratings and high heat transfer, it is important to provide optimum water conditioning at all times; otherwise overheating and damage to the boiler may result.

Water-tube boilers are available in various capacities for high-pressure and high-temperature steam. The use of tubes of small diameter results in rapid heat transmission, rapid response to steam demands, and high efficiency; moreover, failure of a small tube will not result in a disastrous explosion, as may be experienced with failure of the shell on a fire-tube boiler. Water-tube boilers require elaborate settings; cost per pound of steam is usually higher than that of packaged or fire-tube boilers in the range for which such units are most frequently designed. However, when we get beyond this capacity range, for high-pressure and high-temperature steam, only the water-tube boiler is available. Air infiltration, which plagued the earlier water-tube boiler, has now been largely eliminated in the design by means of improved expansion joints, casings completely enclosing the unit, pressurized furnaces, etc. Feedwater regulation is no longer a problem when the automatic feedwater regulator is employed. In addition, water-tube boilers are capable of burning any economically available fuel with excellent efficiency, whereas packaged units must use liquid or gaseous fuels to avoid fouling the heating surfaces.

So in selecting a boiler many factors other than first cost are to be considered; important are availability, maintenance costs, labor costs, space, and a host of other factors. Most important perhaps are fuel costs. During the life of the equipment, we can expect fuel costs to be two to four or more times the cost of boiler and firing equipment.

1.4 STEAM SEPARATORS AND SUPERHEATERS. In the past difficulty with "carry-over" and impurities in the steam was frequently encountered (carry-over is the passing of water and impurities to the steam outlet). Efforts were directed to reduce carry-over to a minimum by separating the water from the steam through the installation of baffles and the dry pipe; both measures met with some success. The dry pipe runs the length of the drum, the ends being closed; the upper side of the pipe is drilled with many small holes. The top center of this pipe is connected to the steam outlet. Steam entering through the series of holes is made to change direction before entering the steam outlet, and in the process the water is separated from the steam. The bottom of the dry pipe contains a drain, which is run below the normal water level in the drum. The dry pipe is installed near the top of the drum so as not to require removal for routine inspection and repairs inside the drum.

The dry pipe proved to be fairly effective for small boilers but unsuited for units operating at high rating and for steam drums with limited steam capacity. Placing a baffle ahead of the dry pipe offered some slight improvement in steam quality but was still not considered entirely satisfactory.

Modern practice requires clean steam for process, for the superheater, and for the turbine and engine. An important contribution to increased boiler capacity and high rating is the fact that the modern boiler is protected by clean feedwater. The application of both external and internal feedwater treatment is supplemented by the use of steam scrubbers and separators.

The cyclone separators illustrated in cross-sectional elevation in Figs. 1.15 and 1.16 overcome many of the shortcomings previously mentioned for the baffle and dry pipe. For one drum there is a single row of cyclone steam separators with scrubbers, running the entire length of

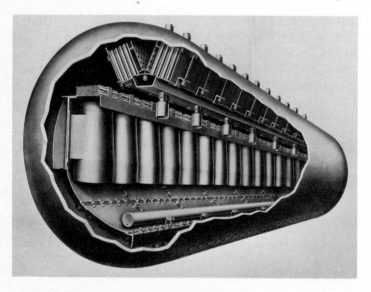

FIG. 1.15   *Single-row arrangement of cyclone steam separators with scrubbers.*   (*The Babcock & Wilcox Co.*)

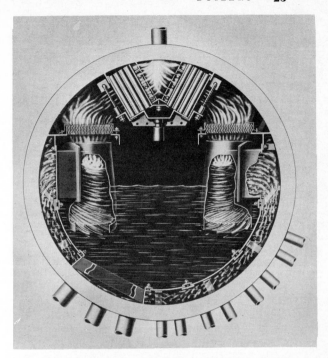

FIG. 1.16 *Double-row arrangement of cyclone-type primary steam separators, with scrubber elements at top of drum for secondary separation.* (*The Babcock & Wilcox Co.*)

the drum; for another, two rows. Baffle plates are located above each cyclone, and there is a series of corrugated scrubber elements at the entrance to the steam outlet. Water from the scrubber elements drains to a point below the normal water level.

In the installation shown in Fig. 1.16, operation is as follows: (*a*) Steam and water from the risers enter the drum from behind the baffle plate to mix with the water (washed) before entering the cyclone; the cyclone is open at top and bottom. (*b*) Water is thrown to the side of the cyclone by centrifugal force. (*c*) Additional separation of water and steam occurs in the passage of steam through the baffle plates. (*d*) On entering the scrubber elements, water is also removed with steam passing to outlet header. Separators of this type can reduce the solids' carry-over to a very low value, depending on the type of feed-water treatment employed, the rate of evaporation, and the water solids' concentration. The cyclone and scrubber elements are removable for cleaning and inspection.

Superheaters. Steam that has been heated above the temperature corresponding to its pressure is said to be "superheated." This steam contains more heat than does saturated steam at the same pressure (see Chap. 2), and the added heat provides more energy for the turbine for conversion to electric power.

Superheater surface is that surface which has steam on one side and hot gases on the other. The tubes are therefore dry except for the steam (gas) which circulates through them. Overheating of the tubes is prevented by designing the unit to accommodate the heat transfer required for a given steam velocity through the tubes, based on the desired exit temperature. To accomplish this, it is necessary to have the steam distributed uniformly to all the superheater tubes and at a velocity sufficient to provide a scrubbing action to avoid overheating of the metal. It is assumed that tubes are clean and carry-over is at a minimum.

Superheaters are referred to as "convection," "radiant," or "combination" types. The convection superheater is placed somewhere in the gas stream, where it receives most of its heat by convection. As the name implies, a radiant superheater is placed in or near the furnace, where it receives the bulk of its heat by radiation.

The conventional convection-type superheater employs two headers into which seamless tubes are rolled. The headers are baffled so that the steam is made to pass back and forth through the connecting tubes, which carry their proportioned amount of steam, the total emerging at the desired temperature. The drums are small, and access to the tubes is had by removal of handhole caps similar to those for boiler-tube access. Superheater tubes are frequently covered with fins to increase the heating surface.

With either the radiant or the convection-type superheater, it is difficult to maintain a uniform steam-outlet temperature; so a combination superheater is installed (Fig. 1.17). The *radiant* section is shown above the screen tubes in the furnace; the *convection* section lies between the first and second gas passages. Steam leaving the boiler drum first passes through the convection section and then to the radiant section and to the steam header. Even this arrangement may not produce the desired results in maintaining a constant steam temperature within the limits prescribed, and so a bypass damper, shown at the bottom of the second pass of the boiler, is sometimes employed. A damper of this type can be operated to bypass the gas or portion of the gas around the convection section, thus controlling the final steam-outlet temperature for various boiler ratings.

At times the superheat temperature may exceed desired limits even though the bypass dampers are employed. So a *desuperheater* is em-

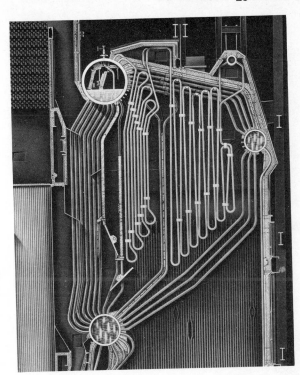

FIG. 1.17 *Combination radiant and convection-type super-heater.* (*Riley Stoker Corp.*)

ployed; this is a device for adding water to the steam to reduce its temperature before the steam enters the turbine.

The final exit-steam temperature is influenced by many factors, such as gas flow, gas velocity, gas temperature, steam flow and velocity, ash accumulation on furnace walls and heat-exchange surfaces, method of firing, burner arrangement, type of fuel fired, etc.

In the *convection* superheater, steam temperature increases with the rating, whereas in the *radiant* superheater steam temperature decreases with the rating. For maximum economy a constant superheat tempera-ture is desirable. Moreover, uniform steam temperatures are desired to avoid the problem of expansion and contraction in the steam header and turbine.

Overheating of the superheater tubes is a matter for the designer, who is required to produce uniform steam flow and velocity of steam through the tubes. This is accomplished in a variety of ways: by spac-ing the take-offs from the steam drum to the superheater, by installing baffles in the superheater header, by placing ferrules in the tubes at the steam entrance to the tubes, or by other means. Care must be exercised to secure uniform flow without an excessive pressure drop

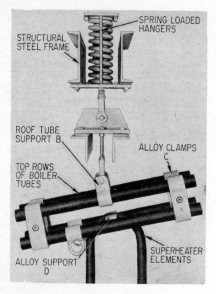

FIG. 1.18 *Superheater support and tube clamps; method of support. (Riley Stoker Corp.)*

through the superheater. The superheater must also be designed to permit drainage when the unit is placed in or taken out of service.

Modern boilers may have twin furnaces, one containing the superheater and the other the reheater section; the superheater is usually a combination radiant and convection section, with various types of arrangements. Constant steam temperature is possible if the necessary precautions are taken in designing the unit.

Various methods are employed to support the superheater tubes, as shown in Figs. 1.17 and 1.18. Superheater tubes vary in size from 1 to 2 in. in diameter; temperatures range to 1050°F.

Superheated steam has many advantages: it can be transmitted for long distances with little heat loss; condensation is reduced or eliminated; superheated steam contains more heat (energy), and hence less steam is required; and erosion of turbine blading is reduced to a minimum.

1.5 HEAT-RECOVERY EQUIPMENT. In the boiler heat balance, the greatest loss is that due to heat in the exit gases. So to operate a boiler unit at maximum efficiency it becomes necessary to reduce this loss to an absolute minimum. This goal is accomplished by installing economizers and air preheaters.

Theoretically it is possible to reduce the exit-gas temperature to that of the incoming air. Certain economic limitations prevent carrying the temperature reduction too far, since the fixed charges on the added investment to accomplish this aim may more than offset any savings obtained. Furthermore, if reduction in temperature is carried below the dew point (the temperature at which condensation occurs), corrosion difficulties may be experienced. So savings resulting from the installation of heat-recovery apparatus must be balanced against added investment and maintenance costs.

An *economizer* is a heat exchanger located somewhere in the gas passage between the boiler and the stack, designed to recover some of the waste heat from the products of combustion. It consists of a series of tubes through which water flows on its way to the boiler. Economizers may be "parallel-flow" or "counterflow" types or a combination of the two. In parallel-flow economizers the gas and water flow

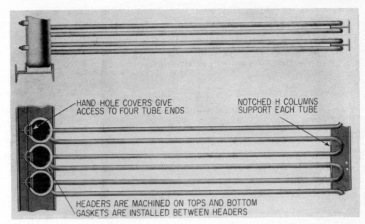

FIG. 1.19 *Construction of economizer.* (*Riley Stoker Corp.*)

in the same direction; in counterflow economizers, in opposite directions. For parallel flow, the hottest gases come into contact with the coldest feedwater; for counterflow, the reverse is true. Counterflow units are considered, usually, to be more efficient, resulting in increased heat absorption. The gas side of the economizer is usually of single-pass construction; gas baffling is sometimes used to increase the heat transfer. In operation, water enters at one end of the economizer and is directed through a system of return bends and headers until it enters the steam drum at a higher temperature. Economizers are referred to as "return tubular" because the water is made to pass back and forth through a series of return bends. Details of construction are shown in Figs. 1.19 and 1.20.

The original economizers were constructed of cast iron, whereas today steel is employed. Access to the tubes is secured through removal of forged-steel handhole covers, the tubes being rolled into the headers as shown. Economizers so constructed were used when the intention was to hold the pressure drop to a minimum and when feedwater conditions were such as to necessitate internal inspection and cleaning.

Instead of using headers, economizers using flanged joints similar to that shown in Fig. 1.20 are frequently constructed. Such units have the advantage of employing a minimum number of return-bend fittings, of the absence of handhole fittings and gaskets, and of freedom from expansion difficulties. A multiple number of take-offs to the steam drum provides uniform water distribution to the drum without disturbing the water level.

The modern economizer consists of a continuous coil of tubes rolled into headers at each end. This construction has the advantage of elimi-

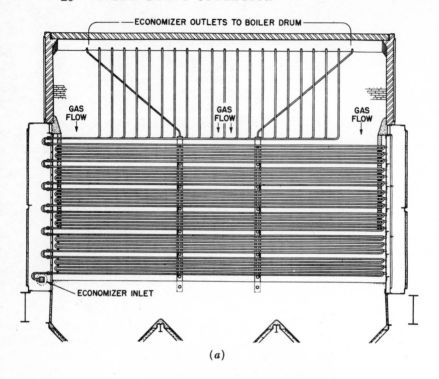

(*a*)

(*b*)

FIG. 1.20   *Return-bend economizers.   (a) Continuous-tube type.*
(*b*) *Loop-tube type.   (The Babcock & Wilcox Co.)*

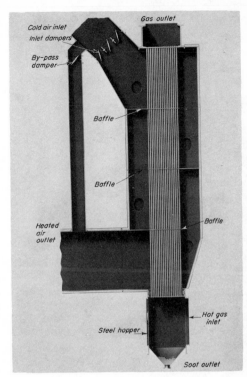

FIG. 1.21   *Tubular air pre-heater showing baffle arrange-ment.* (*The Babcock & Wilcox Co.*)

nating gaskets, handholes, etc.; it also permits acid cleaning of tubes, which is not possible with previous designs.

Tubes range in size from 1 to 2 in. in diameter; design usually provides a ratio of boiler to economizer heating surface of approximately 2:1.   In many cases, the economizer surface is made as large as that of the boiler heating surface.   The size of the economizer to be installed is influenced by many factors, such as cost, space availability, type of boiler units, nature of feedwater employed, and whether or not an air preheater is to be installed.   When both an economizer and an air pre-heater are to be installed, consideration must be given to preventing the exit-gas temperature from dropping below the dew point.

In large central power stations, economizers and air preheaters are both installed to secure maximum efficiency.   For the modern plant, typical efficiencies might be as follows: boiler efficiency, 74 per cent; boiler and economizer, 82 per cent; boiler, economizer, and air preheater, 88 per cent.

The *air preheater* consists of plates or tubes having hot gases on one side and air on the other.   The heat in the gas leaving the boiler or economizer is recovered by the incoming air, thereby reducing the temperature and increasing the efficiency. There are two types of air preheaters, "tubular" and "plate." The tubular type consists of a series of tubes (Fig. 1.21) through which the combustion gases pass, with air passing around the outside of the tube. In the illustration shown, baffles are arranged to make the preheater a three-pass unit. Tubes are expanded into tube sheets at top and bottom, the entire assembly being enclosed in a steel casing. Note the air-bypass dampers. The plate-type air preheater (Fig. 1.22) rotates slowly, exposing a section of the plates alternately to the exit gases and to the entering air. The plates comprise the heating surface.

In the regenerative air preheater (Fig. 1.23), the heating elements are stationary, with cold air hoods (top and bottom) rotating across the heating surfaces. The air preheater shown is mounted vertically but can be arranged horizontally or angularly to accommodate station installation design. The heating elements are assembled in bundles, ready-packed for ease in installation or removal.

To increase the service life of the elements, consideration is given to the following: (*a*) Excess temperature at the hot end—by the use of scale-resistant steel. (*b*) Corrosion at the cold end—by greater sheet

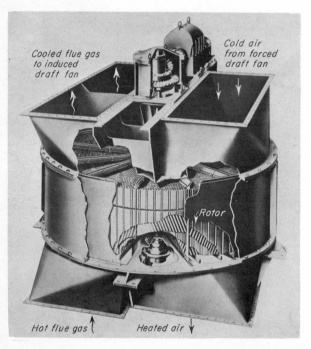

FIG. 1.22 *Air preheater-plate type. (The Air Preheater Corp.)*

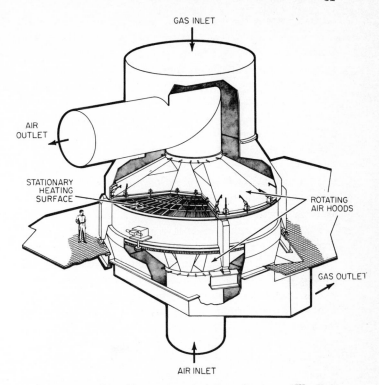

GAS INLET

AIR
OUTLET

STATIONARY
HEATING
SURFACE

ROTATING
AIR HOODS

GAS OUTLET

AIR INLET

FIG. 1.23 *Rothemuhle regenerative air preheater.* (*The Bab-cock & Wilcox Co.*)

thickness; low-alloyed steel; enameled sheets; glazed ceramics; honeycomb blocks made of ceramics. (*c*) Danger of clogging—by enlarged passage cross section; enameled sheets.

The unit is equipped with shifting soot blowers employing superheated steam or compressed air. Washing and fire-extinguishing devices consist of a series of spray nozzles mounted (vertical shaft) both in the flue duct (gas) at the upper collar seal level and within the top air hood. Washing can be carried out with the boiler remaining in operation and without reaching the gas side. Thermocouples are mounted at the cold end close to the heating surfaces and in the flue-gas and air ducts. They serve to monitor any falling below the acid dew point of the flue gases and also to give early warning of danger of fire.

Air heaters have been accepted as standard equipment in power-plant design, for which they are justified by increased economy in operation. The degree of preheat employed depends on many factors, such as furnace and boiler design, type of fuel-burning equipment, and fuel cost. Preheated air accelerates combustion by producing more rapid ignition and facilitates the burning of low-grade fuels; in the process it permits the use of low excess air, thereby increasing efficiency. When coal is burned in pulverized form, preheated air assists in drying the coal, increasing mill capacity and accelerating combustion.

For stoker firing, depending on the type of stoker and fuel burned, care must be taken not to operate with too high a preheated-air temperature. This high temperature may damage the grates; difficulty may also be experienced with matting of the fuel bed and clinkers. The degree of preheating is determined by the kind of fuel, the type of fuel-burning equipment, and the burning rate or grate-heat release. Preheated air at 300°F is usually considered the upper limit for stokers; for pulverized fuel, high-temperature preheated air is tempered when it enters and leaves the pulverizer.

For the air preheater, a low air-inlet or low exit-gas temperature or a combination of the two may result in corrosion when fuels containing sulfur are burned, should the metal temperature fall below the dew point. Two dew points need to be considered: the water dew point, which occurs at approximately 120°F; and the gas dew point, which varies with the quantity of sulfur trioxide in the gas and with other factors. The acid dew point occurs at a higher temperature than the water dew point. The metal temperature is considered to be approximately the average of the air-gas temperature at any given point. Corrosion may be prevented by preheating the air before it enters the preheater, by bypassing a portion of the air around the preheater, and by using alloys or corrosion-resistant metals.

When an air preheater is employed, the added draft loss requires the installation of an induced-draft fan. The use of an air preheater increases the overall unit efficiency from 2 to 10 per cent; the amount of increase depends on the unit location, the rating, and whether or not an economizer is also installed. While air preheaters increase the efficiency, this increase must be weighed against the added cost of installation, operation, and maintenance.

1.6 FURNACE CONSTRUCTION. Internally fired boilers, such as the firebox and packaged types, are self-contained and require no additional setting. Externally fired boilers require special consideration in terms of furnace construction, particularly since each installation is designed to meet specific plant requirements and space availability.

The horizontal-return tubular boiler (Fig. 1.3) is supported by the furnace walls; it is mounted on lugs set on rollers, permitting the boiler to move longitudinally. An improved method of installation is shown in Fig. 1.4; here the refractory walls need not carry the weight of the boiler.

Expansion and contraction for water-tube boilers are taken care of in a number of ways: (1) by suspending the drums and headers from slings attached to overhead columns; (2) by supporting the drum at the end, on columns or overhead beams; and (3) by anchoring the lower drum at the floor level, permitting expansion upward. For these installations the brickwork and casing can be placed in position and in close proximity to the drums, expansion joints being installed where necessary.

In the past, refractory arches were frequently installed in furnaces equipped with chain-grate stokers. Their primary purpose was to assist in maintaining stable ignition with a reduction in smoke emission. Such arches were difficult to maintain, resulting in frequent replacement that required outage of boiler units. These arches have largely been replaced by water-cooled arches or by small snub-nose refractory arches, also water-cooled. Over-fire air jets are provided for the elimination of smoke.

The low-head three-drum boiler (Fig. 1.24) is shown in the process of erection. The top drums are supported at the ends, resting on steel columns or beams; the lower drum hangs suspended from the tubes.

FIG. 1.24   *Low-head boiler in process of erection.   (The Bigelow Co.)*

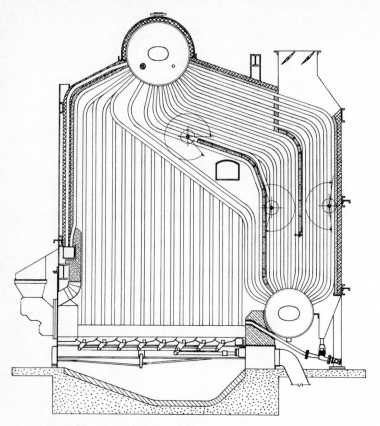

FIG. 1.25 *Vertical boiler with spreader stoker.* (*Union Iron Works Co.*)

Note that the boiler is set high to provide adequate furnace volume. This unit has a capacity of 20,000 pph of steam and is suitable for coal, oil, or gas firing.

The vertical boiler (Fig. 1.25) is a two-drum three-pass water-tube boiler, with side waterwalls. The steam drum is supported on steel beams while the mud drum is suspended from the inclined-vertical tubes. It is fired by a spreader stoker. Note the cinder reinjection at the rear of the furnace.

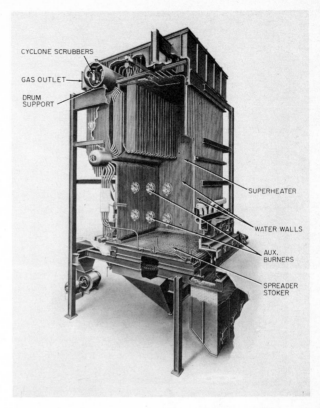

CYCLONE SCRUBBERS

GAS OUTLET

DRUM
SUPPORT

SUPERHEATER

WATER WALLS

AUX.
BURNERS

SPREADER
STOKER

FIG. 1.26 *The B. & W. Stirling SS boiler. (The Babcock & Wilcox Co.)*

The Stirling SS boiler (Fig. 1.26) is of waterwall construction with membrane walls. A radiant-type superheater is located in the furnace. It is fired by a travel-grate spreader stoker with auxiliary burners (gas or oil) located in the side wall providing flexibility for supplementary fuel firing.

The single-elevation top support assures an even downward expansion without differential stresses or binding. The drum rests on overhead steel beams; the superheater is hung from slings as shown. Boilers of this type are largely prefabricated; furnace walls are built in panel sections, in the shop. Later the panels are welded together to form the membrane wall furnace sections. So these units are carefully built under controlled factory conditions, for ease of erection, requiring a minimum amount of time to assemble.

This is a single-pass boiler and hence no baffling is required. There are no local areas of high-velocity products of combustion to cause tube erosion. Cinder return from the last pass of the boiler is by gravity to the rear of the grate. Over-fire air jets are provided to reduce air pollution. These units are available in capacities of 60,000 to 200,000 pph of steam; pressures of 160 to 1,050 psi; temperatures to 900°F.

Furnace heat release is expressed in Btu per cubic foot of furnace volume per hour. The permissible heat release varies with design, depending on whether the furnace is refractory-lined or water-cooled, the extent of water cooling, heat transfer, and the type of fuel burned. High furnace heat release is usually accompanied by high furnace temperatures. When coal low in ash-fusion temperature is being burned, the ash adheres to the refractory surface, causing erosion and spalling. The ash may also adhere to the heating surfaces, reducing the heat transfer and frequently fouling the gas passages with a loss in boiler capacity and efficiency. For the refractory-lined furnace, high furnace heat release is more critical than for waterwall installations.

Owing to the fact that refractory walls were unable to meet the severe service conditions to which they were subjected, waterwalls were introduced, even for some of the smaller boiler units. Excessive maintenance and outage of equipment are thus avoided; the addition of waterwalls increases the boiler capacity, for a given furnace size. Shown in Fig. 1.25 is a small vertical boiler equipped with waterwalls; it has a capacity of 30,000 pph of steam.

The first application of furnace water cooling was the installation of the water screen when burning pulverized fuel. This screen consisted of a series of tubes located above the ashpit and connected to the boiler water-circulating system. Its purpose was to reduce the temperature of the ash below its fusion point before the ash reached the ashpit; thus slagging was prevented.

The waterwall was added next. In replacing the refractory walls, the added heating surface increased the boiler output, and with the elimination of refractory maintenance, boiler availability was improved. The amount of water cooling which can be applied is determined in part by combustion conditions to be experienced at low rating, since excessive cooling reduces stability of ignition and combustion efficiency. So some furnaces are partially water-cooled, or the waterwall is partially insulated and based on design experience. Details of wall construction are illustrated in Figs. 1.27 and 1.28; tube construction with full and partial stud

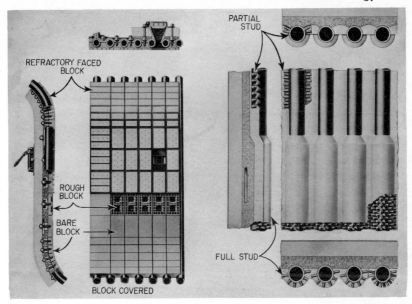

FIG. 1.27 *Water-cooled furnace-wall construction.* (*The Babcock & Wilcox Co.*)

tubes is shown. The studs are used to anchor the refractory in place while tie bars hold the tubes in line.

Various types of wall blocks are employed. The choice is determined by their individual capacity for heat conductivity and by the varying conditions to which they are exposed in different parts of the furnace. The blocks may be rough-faced or smooth, of bare metal or refractory-faced. Depending on known heat-transfer coefficients, blocks are applied to meet design specifications and to limit the heat input to the tubes in order to prevent overheating and other problems.

Special attention must be given to wall sections subjected to flame impingement, to tube bends, and to division walls and slag screens subject to the blast action of the flame. Special refractory materials provide protection against molten slag and erosion. The arrangement of studs and the extent of refractory covering are modified to meet the specific requirements of the individual furnace and the type of fuel burned. In operation any excess refractory is washed away until a state of thermal equilibrium is reached because of the cooling effect of the studs. Fully studded tubes are used to assist ignition and to promote complete combustion for sections of the furnace where maximum temperatures are desired. Partially studded tubes are usually used in cooler zones of the furnace and where more rapid heat absorption is advantageous.

Over the years efforts have been directed to the reduction of air in-

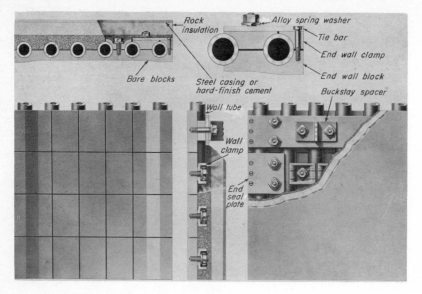

FIG. 1.28   *Block-covered wall, showing method of clamping blocks on tubes.* (*The Babcock & Wilcox Co.*)

filtration into the boiler setting and thereby to the improvement of unit efficiency while maintaining boiler capacity.   The use of waterwalls with welded outer casings has enabled this leakage to be reduced to a very considerable extent.   The pressurized furnace was the next step; it employs an all-welded casing behind the tube enclosure, the insulation being located behind the casing.

However, gas can still leak through the walls to cause overheating of the inner and outer casings.   Such leakage causes gas and fly ash to enter the casings; the gas may be saturated with sulfur, resulting in corrosion of the casings.   So a recent development is the membrane wall construction (Fig. 1.29).   Tightness is accomplished by welding a bar between the tubes, with insulation being placed behind the tubes, faced with outer casing or lagging.

For boilers such as those shown in Figs. 1.9 and 1.10, the bulk of the heat absorbed was the result of convection and conduction, only the lower rows of tubes receiving heat by radiation.   The square feet of heating surface was then employed to determine the capacity of the unit, approximately 10 sq ft of heating surface being considered capable of generating 34.5 pph of steam from and at 212°F feedwater temperature.   Where waterwalls comprise the greater portion of the heating surface, receiving most of the heat by radiation, the previous standard cannot be applied.   Hence for units of this type, steam capacity is provided by the designer, who bases his calculations on design performance data and experience with similar units in the field.

FIG. 1.29   *Membrane wall construction with block insulation and metal lagging. (The Babcock & Wilcox Co.)*

When pulverized fuel is being fired, difficulty may be experienced with deposits of furnace slag. This is especially troublesome when the coal contains ash having a low fusion temperature. The slag becomes very hard and difficult to remove, especially when it is attached to the brickwork. Furthermore, a portion of the refractory is frequently removed along with the slag, thus increasing the maintenance cost.

Furnaces can be designed to burn coal of any fusion range. If the ash is removed in the dry state, the unit is referred to as a "dry-bottom" furnace. Or for low-fusion-ash coal, the unit may be designed to remove the ash in liquid form; this unit is called a "wet-bottom" furnace. The liquid ash can be removed on a continuous basis; here the molten ash collects on the furnace floor, is made to flow over a weir located in the floor of the furnace, and drops into a bath of water below. Later the ash is removed from the hopper hydraulically. Or the molten ash may be permitted to remain and collect on the furnace floor to be tapped off at intervals. On being discharged, the molten ash encounters a jet of high-velocity water; the chilling of the ash causes it to break up into a fine granular form for ease of disposal.

1.7 INDUSTRIAL AND UTILITY BOILERS. Boilers are designed to meet every requirement of capacity and space limitations. So we have many packaged units, shop-assembled in various capacities and pressures. Others are of modular construction, adapted to space limitations. Many are field-erected yet have many of the component parts shop-assembled in modular form. Boilers with pressures of 3,650 psi and steam temperatures to 1050°F are available in many different designs. The large units are single-purpose boilers, providing steam to a single turbine having single- or double-reheat arrangements. Design includes twin furnaces for steam-and-reheat temperature control, forced water circulation, once-through boilers, pressurized furnaces, and many other innovations.

Boilers installed for the Tennessee Valley Authority (TVA) by the Babcock & Wilcox Co. have a capacity of 8 million pph of steam, operating at 3,650 psi with steam temperature to 1050°F. This is a universal-type steam generator, cyclone-fuel–fired and employing a pressurized furnace. It is approximately 225 ft high and 210 ft deep. The height is equivalent to that of a 25-story building. The steam generator employs 23 burners, consuming approximately 500 tons of coal per hr.

The PFI integral-furnace boiler (Fig. 1.30) is a standard bent-tube bottom-supported two-drum unit arranged with gas-tight membrane furnace and bare-tube boiler for pressurized or induced-draft operation, with completely water-cooled furnace and drum cyclones. Maximum shop subassembly facilitates field erection. Superheater units are equipped with an inverted loop, fully drainable. Gas flow is horizontal throughout the unit with multiple passes in the boiler bank. Capacities range from 80,000 to 700,000 pph of steam; pressures to 975 psi; steam saturation temperatures to 950°F. Fuel is usually oil or gas, singly or in combination; waste fuels can also be used.

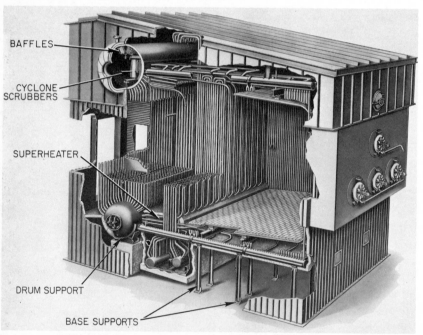

FIG. 1.30 *B. & W. integral-furnace boiler; type PFI.* (*The Babcock & Wilcox Co.*)

These units provide steam for heating, power, or process; they can be located outdoors if desired; they require a minimum of space; tube spacing provides relatively constant steam temperature over a wide load range; an air preheater or economizer can be readily adapted to the unit if economics so dictates.

For a pressurized furnace, no induced-draft fan or equipment is necessary, reducing operating and maintenance costs. All heat-absorbing tubes in the furnace, boiler, and superheater are the same diameter; hence investment in replacement parts is minimized. In case a tube repair is necessary, the membrane wall can be easily cut and rewelded to replace a damaged tube section.

In Fig. 1.31 is shown a unit with a capacity of 200,000 pph of steam, operating at 625 psi and temperatures to 750°F. It is equipped with waterwalls, a radiant superheater, an economizer, a tubular air preheater,

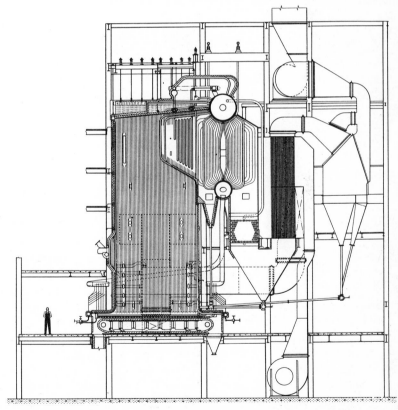

FIG. 1.31 *VU-40 boiler equipped with travel-grate spreader stoker; 200,000 lb per hr; 625 psig; 750°F. (Combustion Engr. Co. Inc.)*

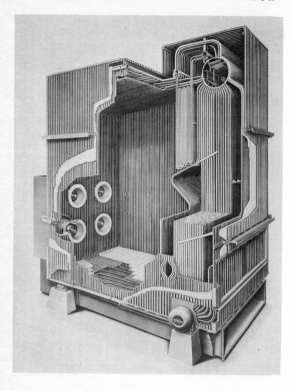

FIG. 1.32   *VU-60 boiler with horizontal burners; gas-oil-fired.* (*Combustion Engr. Co. Inc.*)

and a mechanical dust collector. It is fired by a continuous-ash-discharge spreader stoker and CE-type TT burners. It is designed to burn pine and gum bark, oil, natural gas, or coal. The boiler is supported from overhead steel columns, as shown.

The VU-60 boiler (Fig. 1.32) is a pressurized furnace unit, using bare waterwall tubes with floor tubes covered, as shown. A front-firing arrangement with tangential firing is available if desired. It is designed to burn oil, gas, and waste heating fuels. It is available in capacities from 100,000 to 400,000 pph of steam with temperatures to 900°F.

A preengineered steam generator, available for industrial and utility use in a variety of sizes, is shown in Fig. 1.33. It is a single-pass unit, thus eliminating turns, baffles, and pockets where concentrations of dust might accumulate. This type of construction also eliminates high velocities across the heating surfaces, which are conducive to tube erosion. The installation includes a superheater, a plate-type air preheater, an economizer, and a dust collector.

The prominent nose at the top of the furnace ensures gas turbulence, thus providing for good distribution of gases as they flow through the boiler and superheater. This unit is available in capacities to 500,000 pph of steam at 1,500 psi and temperatures to 960°F. Units such as this can be oil-, gas-, or pulverized-fuel–fired or combination-fired.

For the unit shown, primary air is introduced at the fuel inlet; secondary air, at the burner exit; and tertiary air, through the center of the burner, all three combining to produce a turbulent ignition with a minimum of excess air.

The waterwall tubes are fin-tube-welded. Wall construction is such that the unit is pressure-tight for operating either with a balanced draft or when employing a pressurized furnace. The drum is equipped with horizontal steam separators to provide dry steam to the superheater. Units such as this operate at high efficiency.

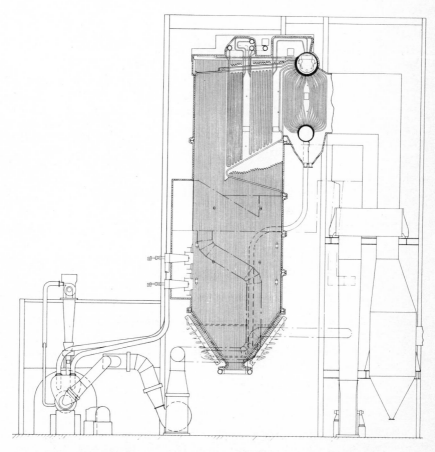

FIG. 1.33 *Pressurized steam generator, single-pass, with pulverized-fuel or combination firing; capacities to 500,000 lb per hr; pressure, 1,500 psig; 960°F. (Foster Wheeler Corp.)*

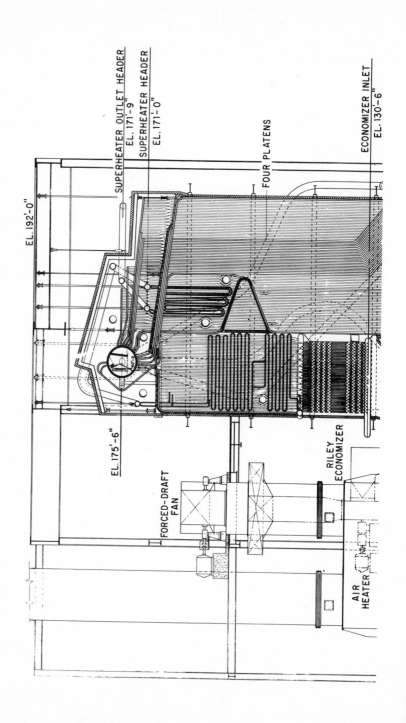

SUPERHEATER OUTLET HEADER
EL. 171'-9"

SUPERHEATER HEADER
EL. 171'-0"

FOUR PLATENS

ECONOMIZER INLET
EL. 130'-6"

EL. 192'-0"

EL. 175'-6"

FORCED-DRAFT
FAN

RILEY
ECONOMIZER

AIR
HEATER

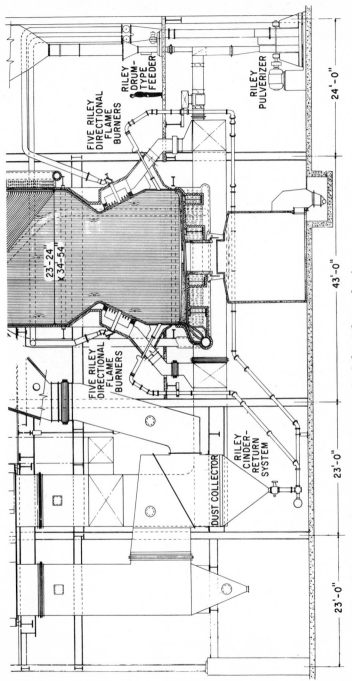

FIG. 1.34 *Riley Turbo furnace steam-generator unit fired by Riley pulverizers;
650,000 lb per hr; 1,250 psig; 900° F. (Riley Stoker Corp.)*

FIVE RILEY
DIRECTIONAL
FLAME
BURNERS

RILEY
DRUM-
TYPE
FEEDER

RILEY
PULVERIZER

23'-24"
X 34'-54"

FIVE RILEY
DIRECTIONAL
FLAME
BURNERS

RILEY
CINDER-
RETURN
SYSTEM

DUST COLLECTOR

24'-0"

43'-0"

23'-0"

23'-0"

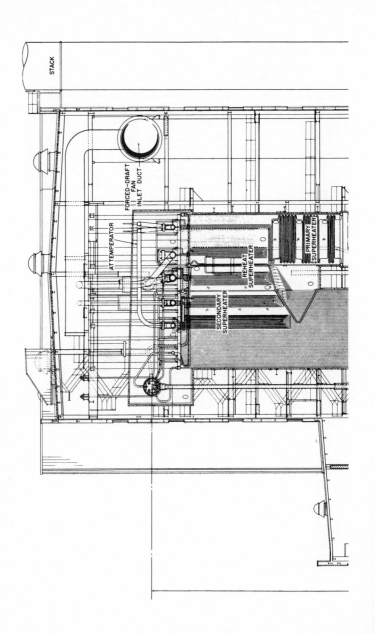

STACK

FORCED-DRAFT
FAN
INLET DUCT

ATTEMPERATOR

PRIMARY
SUPERHEATER

REHEAT
SUPERHEATER

SECONDARY
SUPERHEATER

46

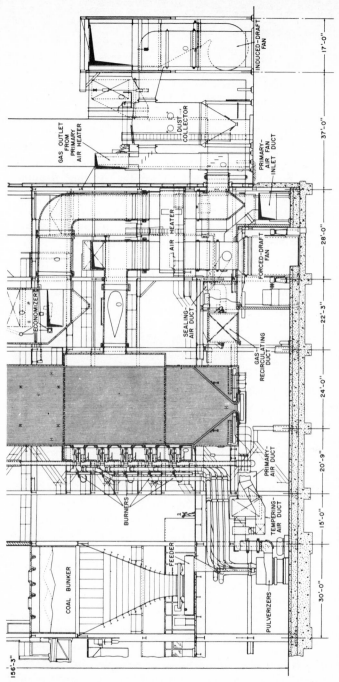

FIG. 1.35 A modern pulverized-coal-fired radiant boiler. (The Babock & Wilcox Co.)

A typical sectional elevation of a Riley Turbo furnace designed for pulverized-fuel firing is shown in Fig. 1.34. This is an open-pass furnace, water-cooled; installation includes a radiant superheater, an economizer, an air preheater, and a dust collector. Directional flame burners oppose each other and are arranged on one level; they produce turbulent combustion within the furnace hearth. A continuous-slag-discharge spout is mounted in the center of the furnace floor. The spout consists of a water-cooled coil using low-pressure water separate from the boiler water-pressure system. The spout forms the slag opening and weir; the molten ash passes over the weir and through the opening, dropping into a water bath for removal (wet-bottom furnace).

Combustion is completed substantially in the lowest furnace area, leaving the vertical furnace walls free from flame impingement and ash deposits. Fly ash from the various passes of the boiler and dust collector are returned to the furnace, where the remaining carbon is consumed and the ash converted to clinker for disposal.

The bottom of the furnace is constructed of tangential water-cooled tubes, seal-welded to each other and covered with chrome refractory to form a leakproof slag-pool floor. Units of this type are adaptable for any fuel use. When coal is burned, however, the fusion temperature of the ash must be in the 2450 to 2550°F range, or lower if the ash is to be removed in a liquid state. The Turbo furnace is applicable to industrial and utility stations and is available in varying capacities and arrangements to meet individual needs. The installation shown has a capacity of 650,000 pph of steam at 1,250 psig and 900°F.

Another pulverized-coal–fired radiant boiler is shown in Fig. 1.35. Proceeding from the water-cooled furnace, we have the gases passing over the secondary superheater (radiant type) to the reheater superheater, to the primary superheater, to the economizer and plate-type air preheater, to the dust collector, and finally to the stack. Steam from the boiler drum is made to pass through the primary superheater and then conveyed through the secondary superheater. At the outlet of the secondary superheater, the steam is made to pass through an attemperator (where water can be added) to the turbine. After passing through the turbine, the steam is returned to the boiler, passing through the reheat superheater and again returning to the turbine.

A gas-recirculating duct is connected at the base of the furnace for the introduction of waste gas and employed to control combustion conditions in the bottom of the furnace as well as the furnace-outlet gas temperature, if desired. High-preheated air (from the air preheater) is provided for drying the coal in the mill, with a tempering arrangement to control mill-outlet air temperature. Units such as this can be designed for various capacities, pressures, and steam temperatures.

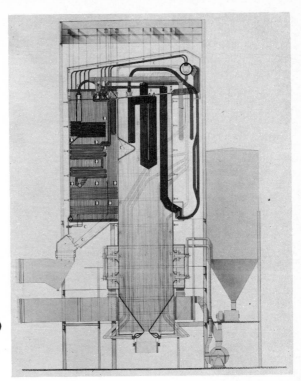

FIG. 1.36 *Modern coal-fired central-station unit; 1,620,000 lb per hr; 2,525 psig; 1005/1005°F. (Riley Stoker Corp.)*

This central-station unit (Fig. 1.36) has a capacity of 1,620,000 lb per hr at 2,525 psig and 1005/1005°F. It contains a combination radiant and convection superheater, a reheater section, an economizer, an air preheater, and a dust collector. Reheat-temperature control is secured by gas-flow–proportioning dampers. This boiler employs natural circulation of water. It also employs opposed firing, which creates extreme turbulence and is conducive to operating efficiently at low excess air.

A "natural-circulation" unit is one in which the pumping head is provided by the difference in density between the saturated liquid in the unbalanced downcomers and the steam-water mixture in the heated risers. A separating drum is required to provide the recirculated saturated liquid to the unheated downcomers and saturated steam to the superheater. For the "controlled-circulation" unit, the system employs a pump to ensure sufficient pumping head to secure the proper cooling of the furnace parts. A separating drum is provided, as for natural circulation.

A "once-through" boiler unit is considered to be one which does not employ recirculation at full load. Such a unit is shown in Fig. 1.37; it has a capacity of approximately 5 million pph of steam at 3,800 psi and 1010/1010°F. Eighteen Foster Wheeler mill pulverizers process coal necessary for the unit.

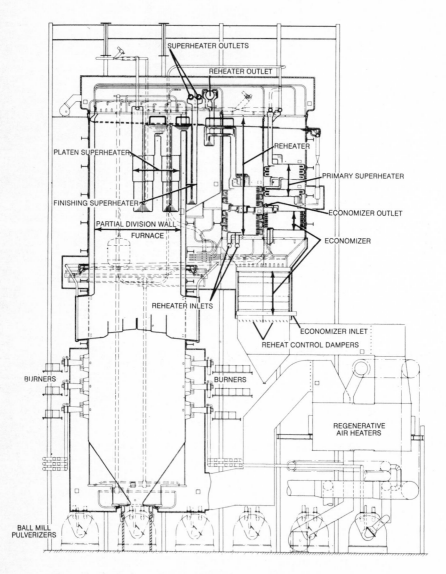

FIG. 1.37   *Once-through type of steam generator installed for the Allegheny Power System; capacity 5,000,000 pph; 3,800 psi; 1010/1010 °F.   (Foster Wheeler Corp.)*

In operation, feedwater flows through the economizer and upper partial division walls in the furnace. The heated fluid from this wall is routed through an external downcomer to supply the first enclosure pass in the furnace section. The furnace enclosure is cooled in this manner by several series-connected passes. Full mixing of the fluid between these passes is achieved as a means of lessening unbalance.

After leaving the furnace circuitry, the fluid is heated in the convection pass enclosure and the roof, with partial mixing of the fluid between each pass. From the roof circuit outlet, the fluid then is routed to the superheaters for final heating to full temperature.

The furnace is arranged for opposed firing. Pulverized fuel mixes with preheated air in the burner zone and is consumed. The flue gas flows upwards, through the platen and finishing superheaters. The primary superheater and reheater are installed in a parallel-pass arrangement with proportioning damper control of the flue gas flowing over the reheater. The economizer is partially installed in one of the parallel passes and in the section following the rejoining of the parallel passes. Flue gases are then directed through two regenerative air heaters to the electrostatic precipitators and to the stack. Three such units are installed in the Harrison station of the Allegheny Power System.

A cyclone furnace consists of a water-cooled horizontal cylinder into which fuel is fed and mixed with air; combustion is completed in a minimum of time with high temperatures generated in the burner (Figs.

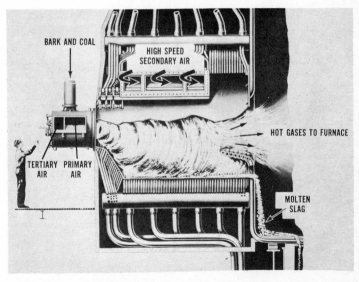

FIG. 1.38 *The cyclone furnace.* (*The Babcock & Wilcox Co.*)

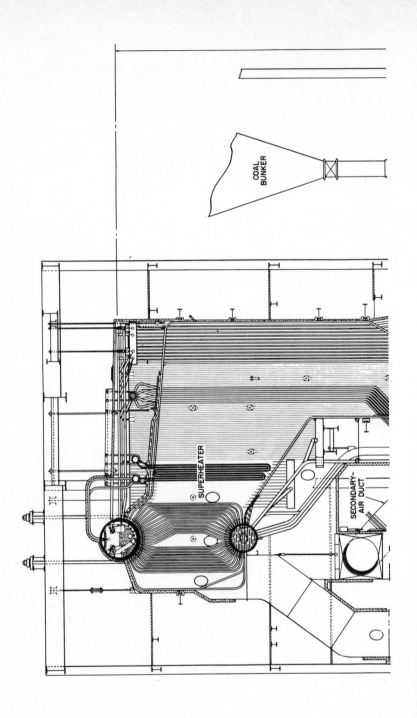

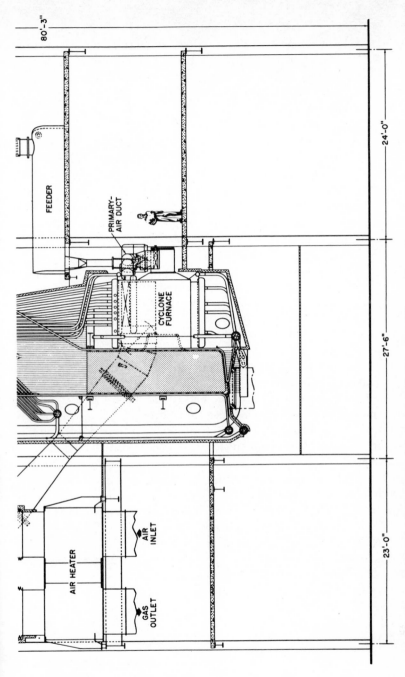

FIG. 1.39 Compact cyclone-furnace–fired boiler installed in the Clinton Corn Processing Co.) (The Babcock & Wilcox Co.)

1.38, 1.39). Coal is crushed to $\frac{1}{4}$-in. top size and drops into the cylinder on a tangent to the burner throat. Primary preheated air, approximately 15 per cent of the total requirements, enters with the coal, both tangentially to the burner. Secondary air also enters tangentially along the side of the burner.

Centrifugal force throws the ash and coal particles to the walls of the burner; secondary air entering tangentially sweeps past the coal-ash particles at high velocity to combine with the slag surface. The scrubbing action and intimate contact permit combustion with low excess air and high burner temperatures. The intense heat melts the ash trapped behind the lip of the burner; when in molten form, the ash discharges through the opening provided, draining into the adjacent boiler and to the slag tap. In some installations a small quantity of teritiary air is introduced at the end of the burner, mixing with the primary air-coal mixture at right angles to that flow.

Fly ash leaving the burner is reduced in temperature, being made to pass over a set of screen tubes. Fly ash collected in the dust-collector or boiler passages can be returned to the cyclone; here any carbon remaining can be consumed and the ash converted to slag for ease in removal.

One of the main advantages of the cyclone burner is its ability to handle low-grade low-fusion-ash coals. The use of a cyclone also reduces the amount of ash to the furnace and to the stack. As a comparison, the percentage of ash leaving the furnace on a cyclone might be 10 to 20 per cent, as compared with 40 to 60 per cent for the wet-bottom furnace and 85 per cent for the conventional pulverized-fuel–fired unit.

Fly ash is a nuisance since it lodges on the heating surface, reducing the heat transfer or causing slag buildup on the furnace and superheater tubes with consequent fouling. Soot blowers are required to remove this deposit, and at times water washing becomes necessary. Eventually much of the fly ash passes to the stack unless a dust collector is provided. For the cyclone, the fly-ash emission is frequently below permissible limits established by ordinance; hence a dust collector is not required, or the collector can be smaller and less expensive.

Since the combustion gases contain less fly ash, tube spacing can be reduced without danger of fouling. Since combustion is virtually completed in the cyclone, the furnace volume can be reduced; hence smaller furnaces and lower costs are possible. Since the coal is crushed rather than pulverized, we can eliminate pulverizers and their many auxiliaries, together with piping and air ducts as well as the maintenance of this auxiliary equipment. Power costs are likewise lower than those of the conventional pulverized-fuel–fired installation.

Cyclone burners are adaptable for burning bark, oil, gas, and coal or a combination of these fuels. From an operating standpoint the cyclone may be considered to be less critical in terms of safety (loss of ignition). Cyclone units are responsive to load demands; they operate with a high availability factor.

When cyclones are employed, the selection of coal having a maximum ash-softening temperature of 2450 to 2550°F is of importance. If the ash-softening temperature is too high, the ash will not flow from the cyclone or secondary furnace. The ash may actually solidify and make removal difficult, so that the unit must be taken out of service.

The recommended maximum ash-softening temperature (2450 to 2550°F) is not the complete index to coal's suitability for cyclone firing. Rather we must know the chemical constituents of the ash, from which we can determine the ash viscosity for varying furnace temperatures. The viscosity of the ash can be determined in the laboratory, where it is calculated from the chemical analysis of the ash, or by trial and error, which may prove to be very costly and time-consuming.

Nuclear Steam Generation.   The nuclear generator produces steam and hence is a boiler. The furnace for burning conventional organic fuels is replaced by a reactor which contains a core of nuclear fuel.

The heart of the reactor is the *core* (Fig. 1.40). The core employs uranium fuel, generating heat to produce steam. The core is encased in a pressure vessel, which in turn is enclosed in a heavy steel tank; shielding is provided to reduce nuclear radiation losses.

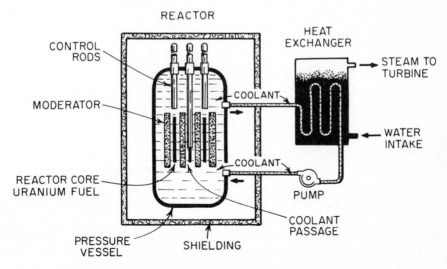

FIG. 1.40   *Atomic energy in use.   (U.S. Atomic Energy Commission.)*

The core containing the uranium fuel elements is assembled in plates or rods; these are required so a self-sustaining chain nuclear reaction takes place. The fuel for the most part is uranium enriched with the fissionable type of uranium, U-235.

When a neutron splits the nuclei of an atom, freeing other neutrons to keep up the fissioning process, the liberated neutrons travel at a high rate of speed. Since slow-moving neutrons are more effective in splitting nuclei than are fast-moving neutrons, we need to slow down the action. So to put the brakes on speeding neutrons, the slow-moving process is accomplished by the use of a moderator.

The *moderator* does the slowing down. Various materials are employed as moderators, such as graphite, ordinary water, and heavy water (water which contains heavy hydrogen instead of ordinary hydrogen). The moderator can put the brakes on speeding neutrons without absorbing them.

The *control rods* contain substances which absorb neutrons readily. They are arranged so that they may be inserted or withdrawn from within the fuel core as required. When the control rods are inserted into the reactor core, they absorb so many neutrons that the chain reaction is slowed or stopped. As the rods are withdrawn, the neutrons become active again and the nuclear chain reaction starts up again. So the control rods are used to raise or lower the power output of the reactor.

Another component of a reactor is the *coolant*. The function of the coolant is to remove heat; the intense heat developed in the core can be used to produce steam to generate power. The coolant may be ordinary water, heavy water, a gas, a liquid, or an organic material.

In one power reactor system, called a pressurized-water system (PWR), water is used as both the moderator and coolant. The water is kept under pressure in the reactor vessel. From the reactor vessel, the water is piped to a unit called a heat exchanger which turns the water to steam in a secondary piping system.

In another power reactor system, called the boiling-water system (BWR), water is again used as both moderator and coolant, but here the water is allowed to boil within the reactor vessel; the steam thus generated then passes directly to the turbine generator. Various kinds of moderators, coolants, and heat exchangers are used in other kinds of systems.

In the Babcock & Wilcox Co. PWR system (Fig. 1.41), water under high pressure (to prevent boiling) circulates past the fuel elements within the reactor vessel, passing to a heat exchanger (steam generator); the water (condensed steam) is returned by a pump in a closed system. Water is used both as a coolant and moderator.

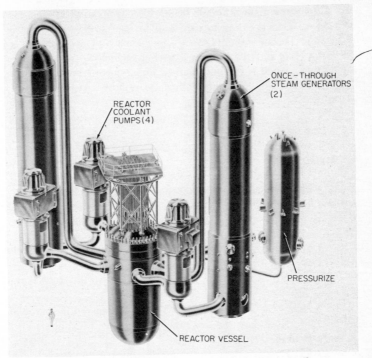

FIG. 1.41 *Nuclear steam system.* (*The Babcock & Wilcox Co.*)

Output from the reactor is controlled by positioning the control rods to vary the chain reaction and to satisfy steam demands. The heat exchanger is shell and tube, arranged vertically with straight tubes. For both the PWR or BWR systems we can employ a superheater or reheater, oil- or gas-fired, as economics dictate.

QUESTIONS AND PROBLEMS

1.1. What is a boiler?

1.2. What are the requirements of a good boiler?

1.3. What is meant by radiant heat? By conduction? By convection?

1.4. What is a fire-tube boiler? Describe a small vertical fire-tube boiler.

1.5. What is an exposed-tube-type unit? A submerged-tube-type unit? What are the advantages of each?

1.6. What are the advantages and disadvantages of a fire-tube boiler?

1.7. What is a crown sheet? What is its purpose?

1.8. What is a longitudinal joint? A circumferential joint?

1.9. How does a firebox boiler differ from an HRT boiler?

1.10. Make a sketch of an HRT boiler.

1.11. What is the difference between a tube and a flue?

1.12. Is an HRT boiler set level or inclined?   Why?

1.13. What is the best method of supporting an HRT boiler?

1.14. What is an internally fired boiler?   An externally fired unit?   Name two of each.

1.15. Describe a Scotch marine boiler.   Mention its advantages and disadvantages.

1.16. What are the advantages and disadvantages of some of the packaged boilers?

1.17. Describe water circulation in a Stirling boiler.

1.18. Why are tubes in a water-tube boiler inclined?

1.19. Describe the water circulation in a cross-drum water-tube boiler.

1.20. What are the main advantages and disadvantages of the water-tube boiler?

1.21. What is the best method of supporting a water-tube boiler?

1.22. Are all water-tube boilers suspended from overhead or set on columns to take care of expansion and contraction?   If not, how is this sometimes taken care of?

1.23. What is a dry pipe?   Where is it located in the boiler?

1.24. How does the dry pipe compare with scrubbers in the procurement of dry steam?

1.25. Describe how a cyclone steam separator works.

1.26. Some boilers have a combination of baffles and scrubbers in the steam drum.   What are the advantages of so doing?

1.27. Why are water-tube boilers considered to be less dangerous than fire-tube boilers?

1.28. What are the advantages of a hot-water boiler or generator?

1.29. What is a high-temperature water system?   What are its advantages over a steam-heating system?

1.30. What is a superheater?   What is the advantage of using a superheater? Where is the superheater located in the furnace?

1.31. What is a radiant-type superheater?   A convection-type superheater? What are their advantages?

1.32. What is an economizer?   What is the advantage of using an economizer?   Is it placed ahead or behind the air preheater, assuming a preheater is also installed?   Why?

1.33. What is an air preheater?   Mention two types and their advantages.

1.34. What is meant by the term "parallel flow" and "counterflow" in reference to economizers and air preheaters?   What are the advantages of each method?

1.35. What effect, if any, does the dew point have on corrosion of economizers and air pre-heaters?   Does the sulfur in the fuel alter the dew point? Why?

1.36. In furnace construction, what items must be given consideration to reduce maintenance to a minimum?

1.37. What is the purpose of the waterwall?   For water cooling of the floor of the furnace?

1.38. What is meant by membrane furnace-wall construction?   What is its advantage?

1.39. What is the difference between a BWR and a PWR nuclear reactor system?

CHAPTER 2

DESIGN AND CONSTRUCTION OF BOILERS

Boiler drums, shells, braces, stays, and tubes are all subject to continual stress and elevated temperatures when a boiler is in service. Boilers must be made adequately strong and of suitable materials to withstand these forces and temperatures. Boiler design requires a detailed study of the forces that are exerted on the various parts and of the temperatures to which they will be exposed.

2.1 MATERIALS USED IN BOILER CONSTRUCTION. When two forces of equal intensity are acting upon the ends of a bar or plate and each is acting in opposite directions, the bar or plate is said to be "in tension." If the forces are each acting toward the center, the material is said to be "in compression." When two forces are acting in the same plane, in very much the same way as the forces produced by a pair of shears, the material between these two forces is said to be "in shear" (Fig. 2.1). Boiler plate and stay bolts in boilers are subjected to a tensile stress. Rivets, in most cases, are subjected to a shearing stress. Stress is the internal resistance that the material offers to being deformed by the external force. As long as the material does not fail, the stress is equal to the external force. The external force is usually expressed in pounds per square inch. For example, the force on a stay bolt may be 4,000 lb. If the bolt is 1 in. in diameter, the area will be $1 \times 1 \times 0.7854 = 0.7854$ sq in. The force per square inch on the stay bolt will be

$$4,000 \div 0.7854 = 5,092 \text{ psi}$$

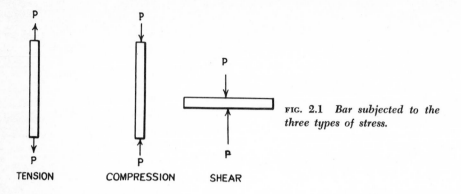

FIG. 2.1 *Bar subjected to the three types of stress.*

TENSION   COMPRESSION   SHEAR

When a material is loaded, it is deformed or changed in shape. Up to a certain load, for any given material, the amount of this deformation is proportional to the load. In the case of a tensile stress, the deformation is called "elongation." If a given bar will deflect 0.013 in. with a load of 15,000 lb, it will deflect 0.026 in. with a load of 30,000 lb. After a certain load has been reached, this proportion will no longer hold true; neither will the material return to its original shape when the load is removed. The load in pounds per square inch required to cause the material to be permanently deflected is called the "elastic limit." The load in pounds per square inch required to cause complete failure in a piece of material is called the "ultimate strength."

There is a close relation between the compressive and the tensile strength of steel. However, in boiler construction the tensile strength of steel is of foremost importance. The tensile strength of steel is determined by placing a test specimen in a testing machine and subjecting it to a measured "pull" until it fails. The elastic limit is determined by making two marks on the test specimen and noting the elongation as the force is applied. When the elongation is no longer proportional to the applied force, the elastic limit has been reached. Before the specimen actually fails, the cross-sectional area is usually much reduced as a result of the elongation. The ultimate strength is calculated from the original area. The tensile strength of steel used in boiler construction varies from 50,000 to 100,000 psi and the elastic limit from 25,000 to 55,000 psi. There is also a relation between the tensile and the shearing strengths of steel; that is, a piece of steel that has high tensile strength also has high shearing strength. The shearing strength of boiler sheets and rivets varies from 40,000 to 48,000 psi.

The cold-bending test is very valuable in determining the ductility of steel and is practical because only very simple equipment is necessary. The specimen is bent cold in a specified manner, and the outside of the bend is examined for cracks. Plates ¾ in. and under are bent back and hammered out flat upon themselves. Plates from ¾ to 1¼ in. thick are bent around a pin having a diameter equal to the thickness of the plate. For plates over 1¼ in. thick, the pin diameter is twice the thickness of the plates. Rivets are tested by bending them back upon themselves through a 180° angle. Plates and rivets suitable for use in boiler construction should withstand this test without showing cracks on the tension side.

The steel used in boiler construction must be of first quality with the constituents carefully controlled. Unlike the steel used in general construction, the steel in boilers must withstand the load at elevated temperatures. The temperature has a more serious effect upon the boiler than has the pressure. Excessive temperatures must be guarded against just as carefully as excessive pressures. Good grades of low-carbon steel maintain and even increase in strength up to a temperature of 700°F, but above this temperature the strength of carbon steel decreases very rapidly and it becomes necessary to resort to the use of steels containing alloys (an alloy is a mixture of two or more metals). Small percentages of molybdenum in the steel used to manufacture superheater tubes, piping, and valves increase the ability of these parts to withstand high temperatures.

When plain carbon steel is subjected to high temperatures, there is a gradual change in the internal structure. The combination of stress and temperature produces an extremely small deformation of the steel known as "creep." This is permanent, but normally it is too small to be measured even over the life of the boiler. Loads applied repeatedly to a piece of material in the form of vibration will eventually cause failure even when the load is below the determined elastic limit. When the correct materials are used and the boiler is not abused, these factors do not cause difficulty or limit the useful life of the boiler. They are mentioned here to emphasize the importance of correct design and operation.

The quality of steel is determined by chemical analysis for alloys and impurities. Carbon is reduced to approximately 0.25 per cent in the boiler plate. Too high a percentage of carbon will cause the steel to harden and crack under the influence of pressure and temperature. More than 0.25 per cent carbon is allowed in flanges and fittings. The common impurities in steel are sulfur, phosphorus, manganese, and silicon. Sulfur reduces the ductility of steel and makes it difficult to work, especially when hot. Phosphorus makes steel hard and to some degree strong; it is undesirable in boiler plate because it lessens its ability to withstand vibratory forces. Manganese makes steel hard and difficult to cut and work; it is considered desirable because it combines with sulfur and lessens the effect of the sulfur. Silicon has a tendency to make steel hard, but since it does not decrease its tensile strength, it is considered desirable.

The chemical analysis is also used to control the introduction of chromium, nickel, molybdenum, and other substances used in making alloy steel for high-temperature boiler tubes.

Steel used in boiler construction is manufactured by two different processes: the bessemer and the open-hearth. The bessemer process was invented in 1855 and for years was the principal method of making steel. The development and perfection of the open-hearth process led to its gradual and almost universal adoption.

The open-hearth process consists of heating pig iron and scrap iron by exposing them to the heat of a gas flame until the carbon and impurities have been reduced to the desired percentage. The open-hearth furnace consists of a shallow, dish-shaped cavity having a capacity of 15 to 200 tons of steel. An arched roof is built over the container. Burners for natural gas or blast-furnace gas enter the sides of the furnace. The heat from the flame is reflected down upon the steel in the bottom of the furnace. The furnace may be charged with either pig iron or scrap. Part of the heat is obtained from the oxidation of the impurities in the iron, and the remainder from the fuel burned. It requires from 10 to 12 hr to make steel by this process. The slowness with which the process progresses makes it possible to analyze samples, to continue the reduction of impurities, and to add the necessary materials to obtain the desired percentage of alloys.

After analysis shows the metal to have the proper percentage of carbon, etc., it is taken from the furnace. In some furnaces the metal is drawn through a spout by opening a taphole. Other furnaces are arranged so that they may be tilted and the metal poured out. In either case the hot metal flows from the furnace into ingot molds. After the ingots have cooled they are taken to the rolling mill, heated, and worked into plates of the desired thickness and shapes for boiler construction.

Cast steel has the same chemical and physical properties as the steel used in making boiler plates. The steel from the open-hearth furnaces is cast into the final parts. Steel castings are costly because of the difficulties encountered and the special techniques required in their manufacture. However, they are widely used in making fittings and headers.

Cast iron was at one time used in the manufacture of boilers. It was employed mainly for drumheads, fittings, and mud drums. Its use is restricted to low-pressure heating boilers. It is dangerous because it is impossible to know whether or not a casting is homogeneous. Cast iron, although having high compressive strength, has low tensile strength. It has the advantage of being less affected by corrosion than steel.

Malleable iron is made by annealing cast iron. The white cast iron, which has most of the carbon in a combined state, must be used in making malleable iron. The castings are packed in a decarbonizing substance such as iron oxide and held at a temperature of 1400°F from 60 to 100 hr. The carbon is taken out of the casting for a depth of ¼ in. Malleable iron in many respects resembles cast steel. It is used in making headers and boxes, having cross-section dimensions not exceeding 7 by 7 in. in boilers that do not exceed a pressure of 200 psig.

Ductile cast iron is a high-magnesium-treated ferrous product. It possesses superior mechanical properties with possibilities for use in making pressure castings, and with special casting procedures it can be made abrasion-resistant.

Copper is an unimportant material in boiler construction. This is chiefly because of its high cost and low tensile strength. Brass and other alloys are used in making many small boiler fittings, valves, etc. Their resistance to corrosion is their chief advantage.

2.2 STRESSES IN TUBES, BOILER SHELLS, AND DRUMS. In order to understand how boiler drums and shells are designed it is necessary to understand the location and magnitude of the stresses to which they are subjected. These stresses can best be understood by considering a seamless cylinder with heads in each end. When pressure is applied to this shell, stresses are set up which tend to cause failure at three different locations.

First, the total pressure on each head is equal to the pressure in the shell—in pounds per square inch multiplied by the area of the head in square inches. This total pressure will exert the same force on the sides of the tank as if two forces A and B (Fig. 2.2), each equal to

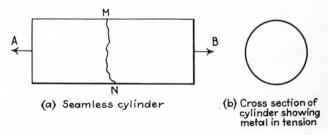

(a) Seamless cylinder

(b) Cross section of cylinder showing metal in tension

FIG. 2.2   *Tension in transverse section of a drum or shell.*

the total pressure on one end, were pulling on the ends of the shell. This force subjects the sides of the shell to tensile stress. Note that it is equal to the pressure on only one end of the shell. This pressure tends to cause failure in a transverse section *MN* (Fig. 2.2*a*).

Consider a circular or transverse cross section through the drum as shown in Fig. 2.2*b*. This area of metal shown must resist the total pull exerted on the head. The area of the metal in square inches equals the diameter of the shell in inches times 3.1416 times the thickness of the plate in inches. The total pressure on the head divided by the cross-sectional area of the metal in the transverse section equals the force exerted on each square inch of the metal.

Second, the pressure exerted on the heads of the shell has a tendency

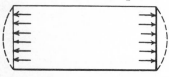

FIG.   2.3   *Bulging ten-dency of flat heads.*

to make them bulge. The outer circumference of the heads is held by the sides of the drum, but the center is not supported; it tends to bulge or form a hemisphere. This same condition occurs in any flat surface of a boiler.

The heads of water-tube-boiler drums are made to conform to the elliptical shape they would take as a result of the pressure to which they are subjected. This is illustrated by the dotted lines in Fig. 2.3.

The heads of fire-tube boilers must be flat so that the tubes can enter at right angles and be expanded to provide a tight joint. The tubes are used to resist the bulging tendency of the heads. In the head areas where there are no tubes the force on the head must be overcome by the use of stays, rods, or braces. Other flat surfaces in the boiler must likewise be supported.

Third, there are stresses in a boiler drum or shell which tend to make it fail along the longitudinal axis *MN* (Fig. 2.4*a*). The pressures on the inside of the drum or shell exert forces on the cylindrical sides that may be considered radial (Fig. 2.4*b*). If we consider a diameter drawn through the points *A* and *B,* the pull on the metal at *A* and *B* will *not* be equal to the entire pressure on the semicircle *AB* because the radial forces near *A* and *B* act in almost opposite directions. To illustrate this point we shall cut the cylinder through the longitudinal

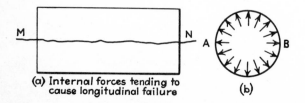

(a) Internal forces tending to cause longitudinal failure

(b)

FIG. 2.4   *Tension in a longitu-dinal section of a drum or shell.*

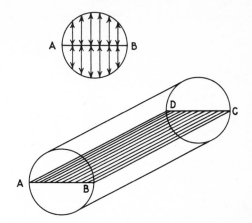

Transverse forces act on projected area A.B.C.D. and produce
tension stresses in sections A.D. and B.C.

FIG. 2.5   *Diagrammatic illustration of transverse forces
in a cylinder.*

axis and replace the lower half with a thick plate; the forces will then
act as shown in Fig. 2.5.   The force exerted on the side of the rectangular surface *ABCD* will be equal to that exerted on the curved surface.[1]
Then the total pressure tending to cause rupture along the longitudinal
axis is equal to the length of the cylinder in inches multiplied by the
diameter of the cylinder in inches multiplied by the unit pressure in
pounds per square inch.   This pressure is held by two strips of metal,
each equal in length to the length of the cylinder and equal in width
to the thickness of the metal in the cylinder.   The force on a unit cross
section of metal is equal to the total pressure on the projected area
divided by the total area of metal in the two strips.

An example will show the application of this reasoning to the calculations of the stresses in an actual cylindrical shell.

Example   A cylindrical shell, as in Figs. 2.2*a* and *b*, has an inside diameter of 6 ft
and is 20 ft long; it is made of ½-in. plate and is subjected to a pressure of 100
psi.   (1) Find the stress in pounds per square inch on the metal in the sides
of the shell due to the longitudinal pressure.   (2) In considering flatheads, what
area must be stayed, and what must be the strength of the stays?   (3) What
is the stress in pounds per square inch in the cylindrical sides due to the transverse
pressure?

Solution   (1) Total pressure on the head is tending to cause failure which must
be held by the metal in the sides.
Total pressure on the head:

---

[1] The rectangular surface is known as the "projected area" of the curved surface.

$$6 \times 6 \times 0.7854 \times 144 = 4,071.51 \text{ sq in.}$$
$$4,071.51 \times 100 = 407,151 \text{ lb total pressure}$$

Area in metal ring (Fig. 2.5b) that must support this load:

$$6 \times 12 \times 3.1416 \times \tfrac{1}{2} = 113.10 \text{ sq in. of metal}$$

Force in each square inch of metal:

$$\frac{407,151}{113.10} = 3,600 \text{ psi}$$

In practice, if tubes and through stays are used to support the heads, this force on the side of the shell is reduced.

(2) Total pressure to be stayed equals the total pressure on the head.

$$6 \times 6 \times 0.7854 \times 144 = 4,071.51$$
$$4,071.51 \times 100 = 407,151$$

In actual practice part of this pressure is stayed by shell and part by the tubes (see Appendix E).

(3) Projected area:

$$6 \times 20 \times 144 = 17,280 \text{ sq in.}$$

Total pressure exerted on this area:

$$17,280 \times 100 = 1,728,000 \text{ lb}$$

But since this pressure is held by both sides of the shell, one side must hold half, or $1,728,000 \div 2 = 864,000$ lb. The area of the strip of metal that supports this pressure is $12 \times 20 \times \tfrac{1}{2} = 120$ sq in. The force on this area equals $864,000 \div 120 = 7,200$ psi.

From these calculations the force on each square inch of metal tending to pull the drum apart in a transverse section is 3,600 psi. The force on the longitudinal section is 7,200 psi. The force on the longitudinal section is 2 times as much as the force on the transverse section. For this reason longitudinal seams are made more efficient and the welding given closer attention than in circumferential or girth seams.

In determining the maximum allowable working pressure of a boiler drum or shell, consideration must be given to the tensile strength of the steel used in the construction, the thickness of the plate, the dimensions of the drum or shell, the permissible factor of safety, and the efficiency of the longitudinal joint or the tube ligament.

The tensile strength of steel and its ability to withstand temperature have been increased by the use of alloys and the improved methods of manufacture. Steel varies in tensile strength from 50,000 to 100,000 psi. When the tensile strength of a boiler plate is not known, it may be assumed as 45,000 psi for wrought iron and 55,000 psi for steel in calculating the maximum allowable working pressure. The metal in a boiler must never be subjected to a pressure greater than, or even approaching the elastic limit. This is the reason there is a definite limit to the pressure used in applying a hydrostatic test.

The required thickness of boiler plates is reduced proportionally by the use of high-tensile-strength steel. Sometimes drums are designed with plates of a greater thickness in that section which is drilled to receive the tubes. This procedure, together with welded construction, results in a decrease in drum weight. The thickness of the plate must always be sufficient to provide the necessary working pressure; in addition, there is also a minimum-thickness requirement based on drum diameter (see Sec. 2.3).

The maximum stress to which a boiler plate may be subjected varies from $\frac{1}{4}$ to $\frac{1}{7}$ of the tensile strength of the material. This comparison of tensile strength to actual working stress is known as the "factor of safety." If the tensile strength of a boiler plate is 55,000 psi, the stress to which it may be subjected varies from $55,000/7 = 7,857$ to $50,000/4 = 12,500$ psi, depending upon its age, type of construction, and condition. The maximum allowable working pressure may likewise be obtained by dividing the bursting pressure by the factor of safety. If a boiler drum has a bursting pressure of 1,000 psi and a factor of safety of 5, the safe working pressure will be $1,000/5 = 200$ psi. A minimum factor of safety of 4 may be used in determining the plate thickness of new boilers. In no case may the maximum allowable working pressure be increased above that allowable for new boilers.[1] Secondhand boilers shall have a factor of safety of at least $5\frac{1}{2}$ unless constructed according to American Society of Mechanical Engineers (ASME) rules, when the factor shall be at least 5.

The *efficiency of a joint or tube ligament* is found by dividing the strength of the section in question by the strength of the solid plate. When calculations or destructive tests show that a joint or tube ligament fails when subjected to $\frac{1}{2}$ as much force as the solid plate, the efficiency would be 50 per cent. The efficiency of a seamless shell is 100 per cent. Welded joints with the reinforcement removed flush with the surface have an efficiency of 100 per cent, but when the reinforcement is not removed, the efficiency is 90 per cent. The arrangement of plates and rivets determines the efficiency of a riveted joint (see Sec. 2.4). When a drum is drilled for tubes in a line parallel to the axis, the efficiency of the ligament may be calculated as follows:

---

[1] When new boilers having a plate thickness in excess of $\frac{1}{2}$ in. are being designed, the following formula is used in determining the maximum allowable working pressure and the minimum required plate thickness:

$$\text{Efficiency of ligament} = \frac{P - d}{P}$$

where P = pitch of tube hole, in.

d = diameter of tube hole, in.

The maximum allowable working pressure of existing boiler installations may be determined by the following formula:

$$P = \frac{\text{T.S.} \times t \times E}{R \times \text{F.S.}}$$

where P = maximum allowable working pressure, psi on inside of drum or shell

T.S. = ultimate strength of plate, psi

t = thickness of plate, in.

R = inside radius of drum, in. (inside diameter divided by 2)

F.S. = factor of safety (ultimate strength divided by allowable working stress or bursting pressure divided by safe working pressure)

E = efficiency of joints or tube ligaments (ultimate strength of joint or ligament divided by ultimate strength of plate). When more than one joint or tube ligament is involved, the lowest efficiency must be used

The bursting pressure is obtained by omitting the factor of safety (F.S.) from the above-mentioned formula.

$$P = \frac{SE(t - C)}{R + (1 - y)(t - C)} \quad \text{or} \quad t = \frac{PR}{SE - (1 - y)P} + C$$

where P = maximum allowable working pressure, psi

S = maximum allowable unit working stress of material, psi. These values are given, for a wide range of steel specifications and operating temperatures, in the ASME Boiler and Pressure Vessel Code Section 1 (Table PG-23.1)

E = efficiency of longitudinal joints or of ligaments between openings

t = minimum thickness of plate in weakest section, in.

R = inside radius of weakest course of shell or drum, in.

y = 0.4 for temperatures of 900°F and below

C = minimum allowance for threading and structural stability, in.

Example   A drum is constructed of SA-285 Grade B steel by welding, approved methods being used and the reinforcement being removed flush with the surface. The plate is 0.75 in. thick, and the inside radius is 24 in.   C = 0.01.   What is the maximum allowable working pressure?

Solution   From Table PG-23.1 (referred to above), we find that for temperatures not exceeding 650°F the maximum allowable unit working stress is 12,500 psi.   The joint efficiency is 100 per cent.

$$P = \frac{SE(t - C)}{R + (1 - y)(t - C)} = \frac{12{,}500 \times 1.00(0.75 - 0.1)}{24 + (1 - 0.4)(0.75 - 0.1)} = 333.13$$

The required thickness of shell or drum may be calculated as follows:

$$t = \frac{P \times R \times \text{F.S.}}{\text{T.S.} \times E}.$$

Example A boiler drum 4 ft in diameter is made of 1-in. plate; the tensile strength of the steel is 55,000 psi; the efficiency of the joint is 85 per cent. What is the bursting pressure?

Solution Bursting pressure in pounds per square inch:

$$\frac{\text{T.S.} \times t \times E}{R} = \frac{55,000 \times 1 \times 0.85}{2 \times 12} = 1,949$$

Working pressure in pounds per square inch:

$$\frac{55,000 \times 1 \times 0.85}{2 \times 12 \times 6} = 325$$

2.3 DRUM AND SHELL CONSTRUCTION. The plates used in the construction of drums and shells are formed to the correct curvature by passing them through rolls. The thickness in most cases is determined by the pressure that is to be used, as explained in Sec. 2.2. Other factors sometimes determine the thickness of the plates. It is necessary that the metal be thick enough so that the joint can be properly calked. If the plates are too thin, they will spring apart between the rivets. The drum must be strong enough to withstand the mechanical load of itself and the water as well as the expansion stresses. The minimum thicknesses of plates for various drum and shell diameters are:

	Minimum Thickness of
Diameter, In.	*Plate, In.*
36 or under..................	¼
Over 36 to 54...............	⁵⁄₁₆
Over 54 to 72...............	⅜
Over 72.....................	½

Whenever possible, seams must be located where they will be protected from the direct radiant heat of the furnace. The flow of heat is restricted at a riveted joint since it must travel through two or more layers of metal. In water-tube boilers the drums are, in most cases, located so that they will be protected from the heat of the furnace. In fire-tube boilers, some seams must be exposed to the hot gases. For example, the longitudinal seam of a return tubular boiler must be located where it will be protected from the direct heat of the furnace, but the girth seams must be partly exposed to the high-temperature zone. Since the stress on the girth seam is only one-half of that on the longitudinal seam, this arrangement is permitted.

The heads in return tubular boilers and some other fire-tube boilers are called "tube sheets"; they are flanged and riveted or welded to the cylindrical section of the shell. Since the tube sheets are flat the entire area must be stayed or supported. Three different methods are used for supporting this area: (1) The shell supports an area around the edge of the sheet. (2) The tubes are beaded over on the ends so that they support an area of the head. (3) The remaining area is supported by braces or stays.

The heads of water-tube-boiler drums are formed from boiler plate by a die which dishes and forms the manhole and flange around the outside, all in a single operation. The flange is then machined so that it fits the inside of the cylindrical section for riveting or equals the diameter of the cylindrical section for welding.

Steam, upon being liberated from the surface of the water in a boiler drum or shell, has a tendency to carry particles of water along with it. This water contains impurities in the form of soluble salts and insoluble particles which are very objectionable.

Some of the design features which decrease the tendency of a boiler to carry over water with the steam are (1) adequate circulation so that the steam generated near the furnace will be quickly carried to the steam drum, (2) sufficient surface area of the water in the steam drum to liberate the steam from the water without excessive agitation, and (3) provisions in the steam drum for retaining the water while allowing the steam to flow into the boiler outlet pipe.

Fire-tube boilers are sometimes fitted with steam domes which consist of a small drum riveted to the highest point of the shell. Steam is taken from the top of this dome, thus providing an opportunity for the water to separate from the steam. A center vertical row of tubes is sometimes omitted from an HRT boiler to improve circulation and thereby reduce carry-over.

Several methods are employed to reduce carry-over from water-tube boilers. One of the simplest but least effective of these is the dry pipe, which is located at the highest point in the drum and connected to the steam-outlet nozzle. The upper side of this pipe contains many holes,

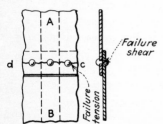

FIG. 2.6 *Single-riveted joints showing the two points of failure.*

through which the steam must flow as it leaves the drum. Another method is to install a separator in the upper portion of the drum to remove the particles of water. In some cases boilers are provided with separate drums in which these separators are installed. Efficient separating devices reduce the required drum size for a given capacity and thereby decrease the cost of the boiler (see Sec. 1.4).

2.4 RIVETED JOINTS. Riveted joints must withstand the force imposed upon them and also hold the boiler plates firmly together to prevent leakage. To accomplish this the rivets must be made of high-grade steel and the workmanship must be the best. The rivet holes must be drilled or reamed to size, not punched. Burrs must be removed from the holes. The space between the plates must be free from all cuttings so that the plates make good contact. Plates must also be properly curved to fit.

A riveted joint may fail in two distinctly different ways. This can best be explained by considering a simple lap joint (Fig. 2.6). The joint may fail by having the rivets shear at section C. The plate may fail under the tension load at section D. A joint should be made so that the rivets and plate will have about equal strength.

Example Figure 2.6 is a single-riveted lap joint, the plate being ⅜ in. thick; the tensile strength, 50,000 psi. Rivets are ½ in. in diameter, spaced on centers 1 in. apart. Shearing strength of the rivets is 48,000 psi. What is the efficiency of the joint?

Solution Consider a strip 1 in. wide, as shown by the dotted lines (Fig. 2.6). This strip contains one rivet and two strips of metal. The joint, regardless of length, is made up of 1-in. strips like the one being considered. Strength of the plate at the joint:

$$(p - d)T = A_t$$

where p = distance from center of one rivet to center of next, in. (pitch)
   d = diameter of rivet, in.
   T = thickness of plate, in.
   A_t = area of plate at rivets (ligament), sq in.
     $(1 - \frac{1}{2})\frac{3}{8} = 0.1875$ sq in. of plate subjected to tension
     $0.1875 \times 50,000 = 9,375$ lb, strength of ligament

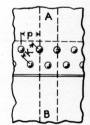

FIG. 2.7 *Double-riveted lap joint.*

Strength of rivet:

$$d \times d \times 0.7854 = A_s$$

where d = diameter of rivet, in.

A_s = area of rivet subjected to shear, sq in.

$$\tfrac{1}{2} \times \tfrac{1}{2} \times 0.7854 = 0.1964 \text{ sq in. of rivet subjected to shear}$$
$$0.1964 \times 48,000 = 9,427 \text{ lb, strength of rivet}$$

Strength of the solid plate:

$$1 \times \tfrac{3}{8} \times 50,000 = 18,750 \text{ lb}$$

With the 1-in. strip the plate would fail first with a pull of 9,375 lb; the rivet would ˙fail if subjected to 9,427 lb. The solid plate would fail if subjected to 18,750 lb. The efficiency of the joint would be 9,375 divided by 18,750 and would be equal to 50 per cent; that is, the joint is only about half as strong as the solid plate.

The low efficiency of the single-riveted lap joint makes a different type of joint necessary. The first step in increasing the efficiency is to use a double-riveted lap joint (Fig. 2.7). Consider the section shown by the dotted lines. Here the cross-section area of only one rivet must be deducted from the area of the plate while. there are two rivets to hold against the shearing stress. This is true because the rows of rivets are placed far enough apart so that the distance k is greater than the distance p. The plate is not reduced in area by the rivets so much as in the case of the single-riveted joint. The efficiency is therefore higher.

Joints may be classified according to the arrangements of plates. When one plate overlaps the other, it is called a "lap joint." When two ends of the plates meet end to end and are joined by a single plate, we have a "single butt-strap joint." If two plates are used to join the plates, one inside and one outside, we have a "double butt-strap joint."

Joints are also classified as single-, double-, triple-, and quadruple-riveted, according to whether one, two, three, or four rows of rivets

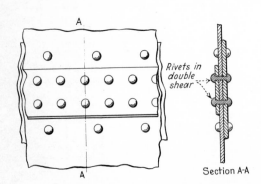

Rivets in double shear

FIG. 2.8 *Double-riveted butt joint.*

Section A-A

are used. In the double butt-strap joint the rivets hold three sheets of boiler plates together. For the rivets to fail they would have to shear in two places. They are said to be "in double shear" (Fig. 2.8) and have about twice the strength of a rivet in single shear. The use of rivets in double shear increases the efficiency of the joint and reduces the required number of rivets, and the plates have fewer holes to weaken them.

The efficiency of any of the joints can be determined by the same method of reasoning as that used in the case of the single-riveted lap joint. The following is a table of efficiencies which are the average for the different types of joints. The values are only approximate, because the efficiency depends upon the ultimate strength of the plate and rivet, thickness of the plates, and diameter of the rivets.

Type of riveting	Lap	Butt
Single.................	57	
Double...............	73	82
Triple................	..	88
Quadruple............	..	94

The double-riveted butt-strap joint consists of a wide strap inside the drum and a narrow strap on the outside (Fig. 2.8). The wide strap is placed inside the drum, where it can easily be made to conform to the curvature of the shell. The joint is made tight by driving the metal along the edges of this strap tightly against the shell, using a chisel having a blunt, rounded tip. This operation is referred to as "calking." The joint is calked inside the drum to prevent the possibility of boiler water's entering the joint, becoming concentrated, and causing the metal to crack. If the water leaks past the calked edge of the inner strap, it will appear at the exterior of the drum.

If it were not for the possibility of the water's becoming concentrated in the joint, the narrow outside strap would be calked. This narrow outside strap has the outside row of rivets on closer center than the inner strap, providing more rigidity to prevent the plates from separating as a result of calking. However, even moderate-pressure boilers are made of sufficiently thick plates to permit satisfactory calking of the inner strap.

Machine riveting is superior to hand riveting because the rivets are drawn down uniformly. The hole is completely filled with the rivet. Welding has completely replaced riveting in the construction of boilers.

2.5   WELDED CONSTRUCTION.   The fusion-welding process consists of applying metal in the molten or vapor state without the use of mechanical pressure or blows. If specified procedures are followed and the necessary precautions taken, it is permissible to employ fusion welding in all pressure parts of a boiler except stays and braces.

Fusion welding has been almost universally adopted in the construction of boiler drums.   The construction of high-pressure boilers by use of riveted joints requires extreme thickness of plate. Welded joints relieve the operators of the inconvenience caused by joint leakage and of the necessity for calking.   Welding also lessens the possibility of caustic embrittlement, which is partially attributed to boiler water's entering the joint, evaporating, and thus producing a solution containing a large amount of impurities.

Rigid specifications cover the application of fusion welding to the pressure parts of boilers.   This work must be done in the manufacturers' shops with special equipment operated by a qualified welder.   The completed work must be subjected to rigid inspection and tests.   It is not permissible to field-weld the pressure parts of a boiler unless provisions are made to stress-relieve the welded portion.   Field welding *must not* be employed in places where its failure would result in a boiler rupture.   A patch may be welded in the firebox of a boiler provided the stay bolts are adequate to hold the pressure.

Terms used to describe the welding process are defined as follows: The "throat" of a weld is the joint where the filler material is of minimum thickness.   The "fillet weld" is approximately triangular in cross section with the throat in a plane which is 45° to the surface joined.   A "double-welded butt joint" is formed by adding welding metal from both sides of the joint and reinforcement on both sides.   A "single-welded butt joint" is formed by adding filler metal and reinforcement on only one side.

In the fabrication of a fusion-welded drum, the carbon content of materials used must not exceed 0.35 per cent.   The plates to be welded may be cut to size or shape by machining, by shearing, or by flame cutting. When flame cutting is used, the edges must be uniform and free from slag.

In the design of boiler drums it is frequently necessary to join plates of unequal thickness.   When welded joints are used, this is permissible within the following limits: the offset of the plates from each other must not exceed one-quarter of the plate thickness, with a maximum permissible offset of $\frac{1}{8}$ in. for the longitudinal joints and $\frac{1}{4}$ in. for the girth joints.   In all cases when plates of unequal thickness are joined together, the thicker plate must be reduced by tapering until the plates are equal in thickness at the joint.   The design of welded joints must be such that bending stresses are not imposed directly upon the welded joint. Backing strips are used when one side of the joint is inaccessible for welding. The filler metal is then added from only one side of the joint. In this case, the reinforcing metal must be at least $\frac{1}{16}$ in., although it may be machined off if so desired.

After fusion-welding a drum or shell (including any connections or attachments) it is always necessary that the assembly be stress-relieved. Stress relieving is accomplished by heating the drum uniformly to a temperature of 1100 to 1200°F. The temperature to which the drum is heated for stress relieving is varied to meet the characteristics of the metal used in the construction. The drum to be stress-relieved must be brought up slowly to the specified temperature and held there for a period of time equal to at least 1 hr per in. of plate thickness. After this period of heating the drum is allowed to cool slowly in a "still" atmosphere. The stress-relieving procedure must be repeated if it is necessary to weld on additional outlets or if repairs are made. Welded attachments may be locally stress-relieved by heating a circumferential band around the entire vessel. The entire band must be brought up to the specified stress-relieving temperature. A similar procedure is followed in stress-relieving welded pipe joints.

The welding of pressure vessels is made possible not only by the improved technique of the process but also by the advancement in the method of inspecting and testing the completed joint. The tests applied to welded pressure vessels may be divided into two classifications: one, known as the "destructive test," in which the metal is stressed until it fails; and the second, which does not injure the metal and may, therefore, be applied to the completed vessel.

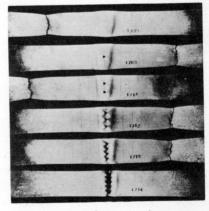

FIG. 2.9 *Tensile-test specimens showing that to cause failure the area of the weld had to be reduced by 30 percent by drilling five holes. (The Babcock & Wilcox Co.)*

Special test specimens are prepared for the application of destructive tests. Plates of the same thickness and material as those used in the actual construction are prepared for welding and attached to one end of the longitudinal seam. The same welding material and technique are used on both the joint and the test plates. Two specimens for tension and one for a bend test are cut from the welded test plates. One of the tension specimens is cut transversely to the welded joint and must show a tensile strength not less than the minimum require-

FIG. 2.10 *Bend-test specimen with the outside fibers elongated 55 per cent.* (*The Babcock & Wilcox Co.*)

ment for the plate material used (Fig. 2.9). The other tension specimen is taken entirely from the deposited weld metal, and the tensile strength must be at least that of the minimum range of the plate welded. When the plate thickness is less than $\frac{5}{8}$ in., an all-weld-metal tension test may be omitted. The bend-test specimen is cut across the welded joint through the full thickness of the plate with a width $1\frac{1}{2}$ times the thickness of the plate. The specimen must bend without cracking until there is an elongation of at least 30 per cent of the outside fibers of the weld (Fig. 2.10).

The x-rays provide a satisfactory nondestructive procedure for examining welds in boiler plate. This method of inspecting boiler plate has been developed to a high degree of perfection. The process consists of passing powerful x-rays through the weld and recording the intensity by means of a photographic film. The variation and intensity of the x-rays as recorded on the film show cracks, slag, or porosity. The exposed and processed films showing the condition of the welded joint are known as "radiographs."

Both longitudinal and circumferential fusion-welded butt joints of boiler drums must be radiographically examined throughout their length. Welded joints are prepared for x-ray examination by grinding, chipping, or machining the welded metal to remove irregularities from the surface. A small specified amount of reinforcement, or crown, on the weld is permissible. Single-welded butt joints can be radiographed without removing the backing strip provided it does not interfere with the interpretation of the radiograph.

Thickness gauges (penetrometers) are placed on the side of the plate to test the ability of x-ray to show defects. These gauges are rectangular strips of about the same metal as that being tested and not more than 2 per cent of the thickness of the plate. They contain identification markings and holes which must show clearly at each end of the radiograph. The entire length of the welded joint is photographed.

Figure 2.11*a* shows a radiograph of a satisfactorily welded joint, while Fig. 2.11*b* shows a joint which is unsatisfactory owing to excessive poros-

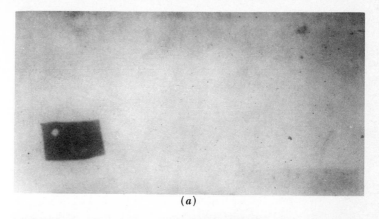

(a)

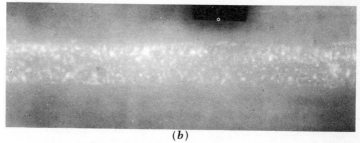

(b)

FIG. 2.11   *Radiographs.   (a) Satisfactory weld.   (b)*
Unsatisfactory weld as a result of slag inclusions.
(The Babcock & Wilcox Co.)

ity.   Welded joints are judged unsatisfactory when the slag inclusion
or porosity as shown on the radiograph exceeds the amount shown by
a standard radiograph reproduction which may be obtained from the
ASME.

The percentage of defective welds in modern boiler shops is very
small.   When the radiograph shows an excessive amount of slag inclusion or other unsatisfactory condition within the joint, the defective
portion must be chipped out.   The section is then rewelded, after which
it must again be x-rayed.

2.6   BRACES AND STAYS.   Since it is not practical to make the metal
in flat boiler surfaces thick enough to withstand the bulging tendency,
these flat surfaces are supported by braces and stays.   The braces and
stays are designed to support the entire force exerted on the flat surface.
They must make the boiler rigid and safe without serious interference
with water circulation and internal inspection.

There are three different kinds of boiler stays: (1) Through stays run from one tube sheet to the other or from one surface to the other. The rods are threaded on both ends. A nut is placed inside and outside the plate. Care must be exercised to see that the tension is correct and that all bolts carry equal loads. Long through stays have a tendency to sag and put excessive strain on the plates. The stay bolt (Fig. 2.12) is really a short through stay. It is used where the surfaces to be stayed are only a few inches apart. The sheets are threaded, and the bolts screwed in. The ends of the bolts are then riveted over to make them tight. A small hole is drilled through the stay bolts so that if they fail, a leak will develop and the failure will be detected at once. (2) Diagonal stays (Fig. 2.13) are used to tie the flat surface to a curved surface. These stays differ widely as to the method used in attaching them to the head and shell. The usual method is to have the stay drawn out on the end so that it can be riveted to the plate. It is poor practice to rivet diagonal stays to the section of a shell that is exposed to the furnace heat. The double thickness of metal is likely to cause the plate and rivets to overheat. It must also be remembered that a diagonal stay must be stronger than a through stay even in the same location. The diagonal stay pulls at an angle, and not all its force is directed perpendicularly to the plate. (3) The gusset stay consists of a piece of boiler plate riveted to the shell and head by means of angle irons (Fig. 2.14). Gusset stays are used in much the same way as diagonal stays. They are, however, more rigid. This to some extent is a disadvantage because it puts expansion stresses on the boiler plate. Gusset stays require careful fitting.

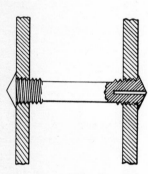

FIG. 2.12   *Stay bolt showing telltale.*

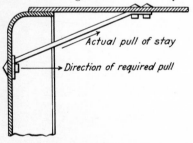

FIG. 2.13   *Diagonal riveted stay (crowfoot stay.)*

The heads of a horizontal-return tubular boiler must be flat so that the tubes can be properly expanded into place. After the tubes have been expanded, they are beaded over on the ends. This makes them act as stays and so prevents the heads from bulging. The beading over of the tube ends keeps them from being burned by exposure to the hot gases.

The area of the heads above and below the tubes must be properly stayed. It is customary to use diagonal stays (Fig. 2.13) to stay the top sections. Through stays are used to stay the bottom sections. The diagonal stays riveted to the shell would not be satisfactory, because the rivets would be exposed to the direct heat of the furnace (Fig. 2.15). (See Appendix E for the explanation of the method used in calculating the area of heads that must be stayed and the number of stays required.)

2.7   MANHOLES, HANDHOLES, AND FITTINGS.   In the construction of boiler drums and shells it is necessary that men work on the inside in making up joints and installing stays; and in the operation of a boiler they must enter in order properly to inspect, clean, and repair the inside. Therefore manholes are made at suitable places in the drum and shell. At other locations it is only necessary to gain access to tubes and for a man to get his hand inside the boiler. In these places handholes are provided; they are placed opposite each tube in the headers of some water-tube boilers and give access to the tubes for replacement, rerolling, and cleaning.

Other connections to the boiler drum and shell include water column, safety valve, blowoff line, feedwater, chemical feed, and steam connections.

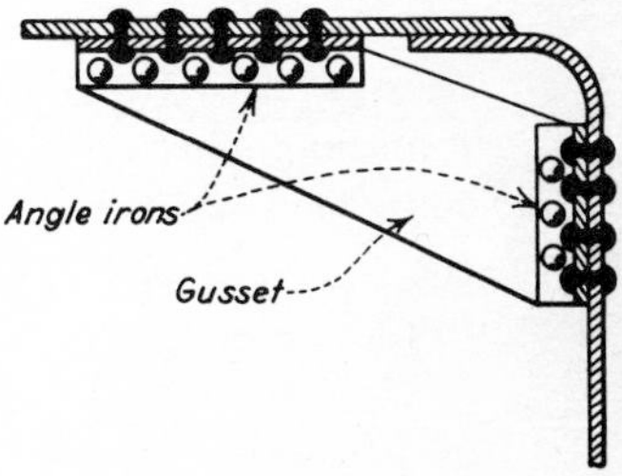

The minimum size of a manhole opening is, for a circular manhole, 15 in. in diameter, and for an elliptical manhole, 11 by 15 or 10 by 16 in. The minimum size of an elliptical handhole opening is 2¾ by 3½ in. However, the holes must be large enough to provide the necessary access for cleaning and for tube replacement.

Manhole openings must be reinforced to compensate for the metal removed. This reinforcement may be obtained either by forming a flange around the hole from the shell or drum metal or by using a frame riveted or welded to the shell or drum. Manhole-reinforcing frames are made of rolled, forged, or cast steel. The minimum seat width for a manhole-opening gasket is $1\frac{1}{16}$ in. The thickness of a manhole gasket when compressed must not exceed ¼ in.

Angle irons

Gusset

FIG. 2.14   *Gusset stay.*

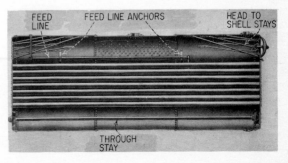

FIG.   2.15 *Section through horizontal-return tubular boiler showing bracing.* (*Combustion Engr. Co., Inc.*)

FEED LINE

FEED LINE ANCHORS

HEAD TO SHELL STAYS

THROUGH STAY

In most cases both manholes and handholes are elliptical in shape. This makes it possible` for the cover plate to be placed on the inside and removed through the hole. Placing the cover plate on the inside puts the pressure against the seat. The yoke holds the cover in place but does not have to carry any pressure. Round handhole cover plates must be placed inside the drum or header through a larger opening than the hole which they are to cover. When outside cover plates are used, the bolt must hold all the pressure. In low- and medium-pressure boilers, gaskets are used to prevent leakage around the handhole cover plates. , In high-pressure boilers the gaskets are eliminated, and the cover plates are seal-welded to prevent leakage.

In addition to manholes, steam drums and shells must have a number of connections, including those for attaching: feedwater inlet, steam outlet, safety valves, water column, chemical feed, and continuous blow-down. These connections are made by welding during fabrication and before the drum or shell is stress-relieved. All welding must be stress-relieved, and radiographic examinations are required for some types of attachments. Connections may also be made by riveting, by use of studs, and by threaded pipe in accordance with accepted standards. Re-inforcing plates and saddles attached to the outside of drums or shells must be provided with a hole, not larger than a $\frac{1}{4}$-in. pipe tap, for testing the tightness of the weld. This hole must be left open during operation of the boiler.

The protruding ends of drum connections may be threaded, flanged, or prepared for either butt or socket welding. The socket welding is preferred for smaller pipe; butt welding, for intermediate sizes; and flanges, for the larger sizes. The steam outlet and safety valve connection on low- and moderate-pressure boiler drums are flanged, but on high-pressure boilers all connections are welded.

All boilers must have a bowoff-line connection to the lowest part of the water-circulating system, which is usually the lower or mud drum. Waterwall and economizer headers which do not drain into the boiler must have separate drain connections.

Superheaters must have connections to both inlet and outlet headers for draining and venting. However, some superheaters have the headers located above the tubes and therefore cannot be drained.

2.8 BOILER ASSEMBLY. It is always advantageous to do as much of the assembly work as possible in the manufacturers' shops. The shop workmen not only are trained on specific jobs but have tools and equipment available for the work. This advantage of shop assembly is applied to cast-iron fire- and water-tube boilers. Cast-iron low-pressure heating boilers are usually delivered fully assembled. However, when restricted conditions make it impossible to install the complete unit, the cast-iron sections may be shipped separately for assembly in the plant. Integral-furnace-type fire-tube boilers are essentially contained within the shell and are readily factory-assembled. Burners and others accessories are also made a part of the factory assembly. After being set on the foundation, and the various services connected, the boilers are ready for operation. Water-tube boilers with integral furnaces for burning gas and fuel oil are factory-assembled and transported to the customer's plant. However, capacities are limited by the physical sizes that can be transported. These transportation limitations are reduced by: designing boilers for increased output for a given size, making them longer but within the restrictions of width and height; adapting transportation facilities to this particular problem, including the use of special railroad cars and barges. As a result of these measures, package water-tube boilers are built with capacities of 200,000 lb of steam per hr.

An application of partial factory assembly is applied to the pressure sections of both fire- and water-tube boilers. The shell of fire-tube boilers and the drum and tube section of water-tube boilers are factory-assembled, pressure-tested, and shipped to the customer's plant. The shell or drum and tubes are installed; then the furnace, boiler enclosure, and combustion equipment are field-erected. The conventional horizontal-return tubular boiler is an example of this type. However, the complete package boiler has replaced the partially field-erected type, for gas and oil firing.

Large power boilers must be field-erected. The components are shipped and assembled in the customer's plant. The boiler drums and large sections are placed in the plant before the building structure is finished. After the supporting steel has been completed, the boiler drums are set and secured in place.

The cost of these large boilers is reduced by modular, or "building-block," design and construction. This consists of designing and fabricating sections which can be employed in the construction of boilers for a range of capacities. The use of modular components permits more fabrication in the boiler shop and reduces the amount of work on the job site, resulting in a decrease in overall construction cost.

The modules are fabricated by means of special tools to shop standards and tested, thus reducing the number of field joints and therefore the possibility of leaks when the assembled boiler is pressure-tested. The period from the time when an order is placed for a boiler to the time when it is put in service is decreased by this modular construction. This decrease in the time required to get the boiler in service is the result of the manufacturer's stocking the modular sections and the fact that the preassembled sections reduce the erection time on the job site.

In top-supported boilers the upper drums are securely anchored in place, and the lower portions of the boiler, including the mud drum, are allowed to expand downward when the boiler is operating. In the case of high boilers, especially those having waterwalls, there is considerable movement of the lower portion of the boiler as a result of temperature change. This movement of mud drums, waterwall tubes, and other pressure parts must be taken into consideration when furnace walls and ashpit hoppers are installed or repaired.

Bottom-supported boilers expand upward, minimizing the problem of sealing the bottom of the furnace to prevent air leaks. This arrangement is used effectively in connection with slag-tapped furnaces and some types of stokers.

The upper portion of the boiler is supported by the tubes. The expansion resulting from increased temperature causes the drums to move upward. This movement must be considered in the design and installation of piping and of ducts which connect to the boiler.

After the drums are in place, the tubes are installed. When headers are used, the tubes and header assembly or module is tested in the factory and shipped as a complete unit. The bent tubes are formed exactly to the required shape in the factory and installed on the job. In the modern boiler, equipped with waterwalls, many different shapes of tubes are required. Regardless of whether the tubes are assembled in the factory or in the customer's plant, the three-roll expander is almost universally used for joining the tubes to the drums and headers. The expander consists of three or more rolls mounted in separate compartments in a cage. The cage contains a hole which allows a tapered mandrel to be forced between the rollers. The appropriate expander is placed in the end of the tube and the tapered mandrel inserted. The mandrel forces the rollers out against the tube. When the mandrel is turned, a similar motion is imparted to the rollers as they are forced against the walls of the tube. Continued turning of the mandrel results in the end of the tube being enlarged and the outer surface pressed against the tube hole.

Expanding has been the accepted method of installing boiler tubes for many years, but there has been some change in the design of the expanders. The first expanders produced a slight taper at the tube ends. This was not serious as the tube sheets were thin. Then the mandrel and rollers were tapered, and this allowed the expanded section to remain parallel with the remainder of the tube wall. Expanders are built with the rollers set in the cage to produce a self-feeding action on the mandrel. When thick-walled high-pressure boiler tubes are expanded, the service on the rollers is most severe.   The mandrel is driven by either an electric or an air-operated motor.   The three-roll expander has been found preferable to those having more than three rolls since the tube holes in the drums and header are seldom exactly round nor is the tube-wall thickness uniform, and only the three-roll expander is capable of conforming to these slight irregularities and pressing the tube tightly against the seat throughout its circumference.   Expanders with more than three rollers tend to roll the tube into a true circle rather than to exert equal pressure against the seat.

At low and moderate pressures it is satisfactory to expand the tube into the smooth seats.   For high-pressure boilers and special application it has been found advisable to cut or expand grooves in the tube seats. This grooving results in considerable improvement in both the resistance to leakage and in the holding power of the tube.   Satisfactory results have been obtained with two grooves, each $\frac{1}{32}$ in. deep and $\frac{1}{16}$ in. wide.   The tube metal is expanded into these grooves.

Several methods have been developed for determining when a tube has been sufficiently expanded, but satisfactory results are obtained by relying upon the judgment of expert workmen.   Mill scale cracking from the surface of the sheet about the tube hole has been found to be a reliable method of determining when the tube has been expanded sufficiently to make a tight joint.

2.9   HEATING SURFACE AND CAPACITY.   The capacity of boilers may be stated in terms of square feet of heating surface, rated boiler horse-power, percentage of rating, and maximum steam-generating capacity in pounds per hour.   The boiler-horsepower and percentage-rating units are restricted to use in expressing the capacity of small, simple types of boilers.   The capacity of medium and large modern boiler units is expressed in maximum pounds of steam per hour at a specified feedwater temperature and outlet-steam pressure and temperature.

The "heating surface" of a boiler consists of those areas which are in contact with heated gases on one side and water on the other.   The boiler code of the ASME specifies that the gas side is to be used when calculating heating surfaces.   It follows that in determining the heating surface for flue and fire-tube boilers the inside diameter of the tubes must be taken; for water-tube boilers, the outside diameter is used.

Unless otherwise stated, the area included as heating surface in a horizontal-return tubular boiler comprises two-thirds of the area of the shell, plus all the tube surface, plus two-thirds of the area of both heads; minus the area of the tube holes.   The boiler heating surface is usually expressed in square feet.   The boiler manufacturer will furnish this in-formation upon request.   If, however, it becomes necessary to calculate the heating surface from the physical dimensions of the boiler, the following procedure may be used:

Example   Find the heating surface of an HRT boiler 78 in. in diameter, 20 ft long, containing eighty 4-in. tubes (inside diameter) bricked in one-third of the distance from the top. It is required to find (1) area of shell, (2) area of tubes, (3) area of heads.

Solution   (1) Multiply the diameter of the shell in inches × 3.1416 × the length of the shell in inches × ⅔ (as only two-thirds is in actual contact with the gases) ÷ 144 (number of square inches in 1 sq ft).

$$\frac{78 \times 3.1416 \times 240 \times 0.66}{144} = 270 \text{ sq ft of heating surface in shell}$$

(2) Multiply the inside diameter of the tube in inches × 3.1416 × the length of the tube in inches × the number of tubes ÷ 144 (number of square inches in 1 sq ft).

$$\frac{4 \times 3.1416 \times 240 \times 80}{144} = 1,675 \text{ sq ft of heating surface in tubes}$$

(3) Multiply the square of the diameter of the boiler in inches × 0.7854 × 2 (as there are two heads) × ⅔ (boiler is bricked one-third from top). From this must be subtracted the area of the tube holes; so to subtract, multiply the square of diameter of the tube in inches × 0.7854 × the number of tubes × 2 (as tubes have openings in each head). Divide this product by 144 (the number of square inches in 1 sq ft).

$$\frac{(78 \times 78 \times 0.7854 \times 2 \times 0.66) - (4 \times 4 \times 0.7854 \times 80 \times 2)}{144}$$

$$= 30 \text{ sq ft of heating surface in heads}$$

Heating Surface of	Sq Ft
Shell...........................	270
Tubes..........................	1,675
Heads..........................	30
Total heating surface...........	1,975

When extreme accuracy is unnecessary, the area of the heads may be omitted from the heating-surface calculations, or it may be assumed as 1.5 per cent of the combined area of the heads and shell.

The "rated horsepower" of a boiler depends upon its type and the number of square feet of heating surface it contains. The square feet of heating surface divided by the factor corresponding to the type of boiler will give the rated horsepower (boiler horsepower). The factors or square feet of heating surface per boiler horsepower are as follows:

Type of Boiler	Sq Ft of Sur- face/Boiler Hp
Vertical fire-tube	14
Horizontal fire-tube	12
Water-tube	10

Example Find the rated horsepower of the HRT boiler in the previous example.

Solution The horizontal-return tubular boiler in the previous example was found to have 1,975 sq ft of heating surface.

$$\frac{1,975}{12} = 164.6 \text{ rated boiler hp}$$

A boiler is operating at 100 per cent of rating when it is producing 34.5 lb[1] of steam per hr from feedwater at 212°F to steam at 212°F (abbreviated "from and at 212°F") for each rated boiler horsepower. The rated steam output of a 164.6-hp boiler is $34.5 \times 164.6 = 5,678.6$ lb of steam per hr from and at 212°F. Modern boilers are capable of generating more than twice this amount of steam per rated boiler horsepower. That is to say, they may be operated in excess of 200 per cent of rating.

The evaporation from and at 212°F is equivalent to adding 970.3 Btu to each pound of water already at 212°F. The steam would be at atmospheric pressure, dry and saturated with heat at this temperature.

The factor of evaporation is the heat added to the water in the actual boiler in Btu per pound divided by 970.3. To determine the heat added by the boiler one must know the feedwater temperature, boiler outlet pressure, and percentage of moisture or degrees of superheat in the outlet steam. When these factors are known, the heat input per pound can be determined as explained in Sec. 2.10.

The equivalent evaporation from and at 212°F is determined by multiplying the actual steam generated in pounds per hour by the factor of evaporation. The equivalent evaporation in pounds of steam per hour divided by 34.5 equals the developed boiler horsepower. The developed boiler horsepower divided by the rated boiler horsepower equals the percentage of rating of the boiler.

The performance of modern high-rating boilers has rendered the boiler-horsepower and percentage-rating method of evaluating boiler size and output obsolete. The feedwater is heated almost to the boiling temperature in the economizers. In fact, some economizers are built

---

[1] When the boiler-horsepower unit (of 34.5 lb of equivalent steam from and at 212°F evaporation per hr) was adopted, it represented the heat input normally required by the average engine to produce a horsepower-hour of energy. This quantity of heat bears no relation to the requirements of modern engines and turbines but is still used to some extent in designating the performance of boilers.

to evaporate a portion of .the water and deliver steam to the boiler drums. The furnaces are constructed with waterwalls which have high rates of heat transfer. The heat is transmitted to these waterwall tubes by radiation, and the convection-heat transfer takes place in the boiler proper, economizer, and air preheater. Under these conditions the boiler horsepower and percentage of rating are inadequate in measuring and stating boiler output. The manufacturers of this equipment have adopted a policy of stating the boiler output in terms of pounds of steam per hour at operating pressure and superheat. When actual conditions vary from design specification, the necessary compensating corrections are made so that the actual operating results can be compared with expected performance.

In evaluating boiler performance and in making comparisons between boilers, it is helpful to consider the relative square feet of heat-absorbing surface and cubic feet of furnace volume involved. The "convection heating surface" of a water-tube boiler includes the exterior surfaces of the tubes, drums, headers, and connection nipples which are exposed to the hot gases. The "waterwall heating surface" includes furnace wall areas and partition walls which are composed of tubes that form a part of the boiler circulating system. The projected area of these tubes and extended metal surfaces exposed to the heat of the furnace are included in the furnace heating surface. The "furnace volume" includes the cubic feet of space available for combustion of the fuel before the products of combustion enter the convection sections of the boiler or superheater. "Superheater and reheater surface" consists of the exterior area of the tubes through which the steam flows. All of the tube area subjected to either convection or radiant heat or a combination of both is included as heating surface. In like manner the square feet of heating surface is an important factor to use when evaluating economizers and air preheaters.

The size of a boiler unit is determined by the amount of heat-transfer surface in the boiler, furnace, superheater, economizer, and air heater and by the heat-release volume in the furnace. However, boiler manufacturers assign varying heat-transfer rates to the heat-transfer surfaces and heat-liberation rates to the furnace volume. The values assigned are influenced by many factors but are essentially based on experience and are increased as the design is improved. These design improvements include better circulation, more effective removal of moisture from steam leaving the drum, and combustion equipment which is more effective in utilizing the furnace volume.

Package water-tube boiler output steam rate has been based on heat-absorption rates as high as 30,000 Btu per hr per sq ft of combined boiler and furnace heating surface and furnace heat liberation of 120,000 Btu per hr per cu ft of volume. More conservative rates are recommended for reliability and low maintenance. The heat-absorption rate by the boiler is based on the heat output while the heat release in the furnace is based on the heat input (see Sec. 2.10).

2.10   BOILER-CAPACITY CALCULATION. The steam tables and charts in Appendix D are useful in determining the heat content of water and steam. These tables are used in calculating the heat content (enthalpy) of steam and water for various pressure and temperature conditions. It must be remembered that the total heat contents of steam (enthalpies) given in the tables are the Btu required to raise the water from 32°F to the boiling point, evaporate into steam, and further increase the temperature when superheaters are employed. The 32°F is an arbitrary point taken for convenience in making the steam tables. Actually a substance contains heat until its temperature is lowered to absolute zero. The steam discharged from a boiler drum may be dry and saturated or wet. When the boiler unit is equipped with a superheater, the steam will be superheated, that is, heated above the saturation temperature corresponding to the pressure.

"Dry and saturated" steam is at the temperature corresponding to the boiler pressure (not superheated) and does not contain moisture. It is saturated with heat, since additional heat will raise the temperature above the boiling point and the removal of heat will result in the formation of moisture. When the absolute pressure (gauge plus 14.7 approximately) is known, the enthalpy may be read directly from the steam tables.

Example   Steam at 150 psia, dry and saturated, is produced by a boiler. What is the enthalpy in Btu per pound?

Solution   From the table in Appendix D, steam at 150 psia is at a temperature (t) of 358.42°F, and the enthalpy (h_g) of saturated vapor is 1194.1 Btu.

When the values given cannot be found in the steam tables, it becomes necessary to interpolate to find the required values.

Example   Steam is at a pressure of 132 psig, dry and saturated. What is the enthalpy in Btu?

Solution

$$132 + 14.7 = 146.7 \text{ psia pressure}$$

Since the enthalpy of steam at 146.7 psi cannot be read directly from the values given in the steam tables, it becomes necessary to calculate the value (interpolate) from the values given.

We note that 146.7 is between 140 and 150 psi, both of which may be found in the tables:

Enthalpy (h_g) at 150 psia...1194.1 Btu
Enthalpy (h_g) at 140 psia...1193.0 Btu
The change in enthalpy for a difference of 10 psi pressure............   1.1 Btu

The change in pressure to be considered is 146.7 − 140 = 6.7.
A change of 6.7 psi pressure causes a change of

$$\frac{6.7}{10} \times 1.1 = 0.74 \text{ Btu}$$

The Btu at 146.7 psi equals the Btu at 140 psi (1193.0) plus the amount resulting from a change of 6.7 psi (0.74) = 1193.74 Btu.

"Wet" steam is in reality a mixture of steam and water. The heat supplied has been insufficient to evaporate all the water. The amount of moisture present in steam can be readily determined by means of a throttling calorimeter.[1] When the pressure and quality are known, the enthalpy can be calculated from the values given in the steam tables. The quality of wet steam refers to the percentage of the mixture which has been evaporated, that is, 100 minus percentage of moisture.

$$h_w = h_{fg}X + h_f$$

where h_w = enthalpy of wet steam, Btu per lb
  h_{fg} = latent heat of vaporization
  X = quality of steam
  h_f = enthalpy of saturated liquid, Btu per lb

Example Wet steam at 150 psia, quality 95 per cent, is produced by a boiler. What is enthalpy in Btu per pound?

Solution From steam tables, h_{fg} = 863.6, h_f = 330.51. X is given as 95 per cent.

$$h_w = 863.6 \times 0.95 + 330.51$$
$$h_w = 1150.93 \text{ Btu/lb}$$

"Superheated steam" has an enthalpy and temperature above that of dry saturated steam at the same pressure. The heat necessary for producing the superheat is applied (by use of a superheater) after the steam has been removed from the presence of water. The enthalpy is the sum of the heat, in the liquid (h_f), used for vaporization (h_{fg}) and that added in the superheater. The amount of superheat is designated either by total temperature or by the temperature above the boiling point. This temperature above the boiling point is referred to as the "degrees of superheat." The superheated-steam tables in Appendix D give the enthalpy when the pressure and total temperature or degrees of superheat are given.

---

[1] See ASME Performance Test Code 19.11.

Example   Superheated steam at 160 psia and a total temperature of 500°F is produced by a boiler.   What are the enthalpy and the degrees of superheat?

Solution   From the steam tables in Appendix D, steam at 160 psia and 500°F total temperature has a total heat of 1273.1 Btu per lb.

The saturation temperature at 160 psia is 363.53°F; hence 500 — 363.53 = 136.47°F of superheat.

After determining the enthalpy of the steam leaving the boiler, we are ready to calculate the heat added by the boiler.   Remember that the starting point of the steam is 32°F while the feedwater temperature is measured from 0°F.   The specific heat of water may be assumed as 1; therefore the heat content of the feedwater may be determined by subtracting 32 from the feedwater temperature.   In this way both the heat content of the steam and that of the water have 32°F as the starting point.

Hence the heat added by the boiler may be determined as follows:

Heat added by boiler (Btu/lb steam)
$$= \text{enthalpy (Btu/lb steam)} - (\text{feedwater temperature} - 32)$$

$$\text{Factor of evaporation} = \frac{\text{heat added by boiler (Btu/lb steam)}}{970.3}$$

$$\text{Developed boiler hp} = \frac{\genfrac{}{}{0pt}{}{\text{actual evaporation}}{\text{(lb steam/hr)}} \times \genfrac{}{}{0pt}{}{\text{factor of}}{\text{evaporation}}}{34.5}$$

$$\text{Percentage boiler rating} = \frac{\text{developed boiler hp}}{\text{rated boiler hp}}$$

Example   A horizontal-return tubular boiler contains 1,200 sq ft of heating surface and generates 5,500 lb of steam per hr at 150 psia.   The water is fed to the boiler at 180°F.   The steam is assumed to be dry and saturated.   (1) What is the rated boiler horsepower?   (2) What is the developed boiler horsepower? (3) At what percentage rating is the boiler operating?

Rated boiler horsepower:

$$\underset{\substack{\text{sq ft} \\ \text{heating surface}}}{1{,}200} \div \underset{\substack{\text{sq ft} \\ \text{boiler hp}}}{12} = \underset{\substack{\text{rated} \\ \text{boiler hp}}}{100}$$

Heat added by boiler:

$$\underset{\substack{\text{heat of} \\ \text{steam, Btu/lb}}}{1194.1} - (\underset{\substack{\text{temp feedwater} \\ °F}}{180} - \underset{\substack{\text{base} \\ \text{steam tables}}}{32)} = \underset{\substack{\text{heat} \\ \text{added, Btu/lb}}}{1046.1}$$

Factor of evaporation:

$$\underset{\substack{\text{heat} \\ \text{added, Btu/lb}}}{1046.1} \div \underset{\substack{\text{Btu/lb from} \\ \text{and at 212°F}}}{970.3} = \underset{\substack{\text{factor} \\ \text{of evap}}}{1.078}$$

Developed horsepower:

$$(\underset{\substack{\text{actual} \\ \text{evap, lb/hr}}}{5{,}500} \times \underset{\substack{\text{factor} \\ \text{of evap}}}{1.078)} \div \underset{\substack{\text{lb equiv} \\ \text{evap/hr/boiler hp}}}{34.5} = \underset{\substack{\text{developed} \\ \text{hp}}}{171.9}$$

Percentage boiler rating:

$$\underset{\substack{\text{developed}\\\text{hp}}}{171.9} \div \underset{\substack{\text{rated}\\\text{boiler hp}}}{100} = \underset{\substack{\text{percentage}\\\text{rating}}}{171.9}$$

Example A water-tube boiler has a capacity of 75,000 lb of steam per hr, and operates at 435 psig and a total steam temperature of 700°F. Water enters the economizer at 235°F and leaves at 329°F. The combined efficiency is 85.00 per cent. The areas of the heating surfaces in sq ft are as follows: boiler, 5,800; furnace waterwall, 750; superheater, 610; and economizer, 3,450. The furnace volume is 1,400 cu ft. Calculate the heat transfer in Btu per hr per sq ft for the (1) combined boiler and furnace waterwall area, (2) superheater, and (3) economizer; (4) also calculate the heat release in the furnace in Btu per hr per cu ft of volume. Neglect blowdown from the boiler and carryover of moisture in steam to superheater.

Solution

1. *Heat-transfer rate in combined boiler and furnace waterwall areas:*

Steam pressure,   450 psia   water inlet to boiler   329°F

$$\underset{\substack{\text{heat of steam,}\\\text{boiler outlet,}\\\text{Btu/lb}}}{1204.6} - (\underset{\substack{\text{temp. econ.}\\\text{outlet water,}\\\text{°F}}}{329} - \underset{\substack{\text{base}\\\text{steam}\\\text{tables}}}{32}) = \underset{\substack{\text{heat}\\\text{added,}\\\text{Btu/lb}}}{907.6}$$

$$\frac{\text{Heat added} \times \text{steam lb/hr}}{\text{Surface, boiler} + \text{furnace wall}} = \frac{907.6 \times 75,000}{5,800 + 750} = 10,392$$

2. *Heat-transfer rate in the superheater:*

$$\underset{\substack{\text{heat of steam,}\\\text{supt. outlet,}\\\text{Btu/lb}}}{1359.9} - \underset{\substack{\text{heat of steam,}\\\text{boiler outlet,}\\\text{Btu/lb}}}{1204.6} = \underset{\substack{\text{heat}\\\text{added,}\\\text{Btu/lb}}}{155.3}$$

$$\frac{\text{Heat added} \times \text{steam lb/hr}}{\text{Surface superheater}} = \frac{155.3 \times 75,000}{610} = 19,094$$

3. *Heat-transfer rate in the economizer:*

$$\underset{\substack{\text{temp.}\\\text{outlet}\\\text{water, °F}}}{329} - \underset{\substack{\text{temp.}\\\text{inlet}\\\text{water, °F}}}{235} = \underset{\substack{\text{temp.}\\\text{rise,}\\\text{°F (approx)}\\\text{Btu/lb added}}}{94}$$

$$\frac{\text{Heat added} \times \text{steam lb/hr}}{\text{Surface economizer}} = \frac{94 \times 75,000}{3,450} = 2,043$$

4. *Heat release per cu ft of furnace volume:*

Heat input is the sum of Btu/lb added by boiler, furnace waterwalls, superheater, and economizer multiplied by the steam generated per hr and divided by the efficiency.

$$\frac{(907.6 + 155.3 + 94)75,000}{.85} = \underset{\substack{\text{furnace heat} \\ \text{release, Btu/hr}}}{102,079,000}$$

$$\frac{\text{Btu/hr released in the furnace}}{\text{Cu ft furnace vol}} = \frac{102,079,000}{1,400} = 72,900$$

QUESTIONS AND PROBLEMS

2.1. What is the cross-sectional area of a rod 1 in. in diameter?

2.2. What is the area of the head on a boiler drum 27 in. in diameter?

2.3. A water-tube boiler operates with a steam pressure of 100 psi and has a head 3 ft in diameter.   What is the total pressure on the head?

2.4. A boiler tube has an outside diameter of 2 in. and a wall thickness of $\frac{3}{16}$ in.   What is the tube's inside diameter?

2.5. What is meant by the tensile strength of a material?   What is shearing strength?

2.6. What is an alloy?

2.7. What are stress and elongation?

2.8. What is meant by factor of safety?

2.9. Of what materials are boilers made?

2.10. Why is a row of tubes sometimes left out of an HRT boiler?

2.11. What is a dry pipe, and what is its purpose?

2.12. Why is it unnecessary to brace and stay the heads of a water-tube boiler?

2.13. What is the advantage of machine over hand riveting?

2.14. What is a butt joint?   A lap joint?   What is the advantage of a butt over a lap joint?

2.15. Which is on the inside, the narrow or the wide strap of a double butt strap?   Why?

2.16. Why is the seam weaker than other parts of a boiler?   What is the strength of a single-riveted lap joint?   A double-riveted lap joint?

2.17. What do you mean by calking, and what are the results of incorrect calking?

2.18. How are these boilers stayed: B. & W. and HRT?

2.19. Is welding used satisfactorily in boiler construction?

2.20. Which is stronger, a rivet in single or a rivet in double shear?

2.21 What is the minimum thickness of boiler plates for boiler drums and shells 36 in. and under?   36 to 54 in.?   54 to 72 in.?

2.22. What is the difference between single-, double-, triple, and quadruple-riveted joints?   How is their efficiency compared?

2.23. A steel rod is ¾ in. in diameter; it reaches its elastic limit when subjected to a force of 12,000 lb and fails under a load of 25,000 lb.   Calculate its elastic limit and ultimate strength in pounds per square inch.

2.24. A 600-lb-pressure boiler has a drum 36 in. in diameter.   What will be the total pressure on the head?

2.25. If the drum in Prob. 2.24 is 30 ft long, calculate the total pressure on the longitudinal projected area as shown in Fig. 2.5.

2.26. A boiler tube is 3 in. inside diameter and ¼ in. thick; the metal has an ultimate tensile strength of 55,000 psi.   Calculate (*a*) the bursting pressure of the tube; (*b*) the safe working pressure, assuming a factor of safety of 6.

2.27. A tube is 6 in. inside diameter and ⅜ in. thick; its ultimate tensile strength is 65,000 psi and its factor of safety 6.   Calculate the safe working pressure.

2.28. A 72-in. boiler shell is ½ in. thick and its ultimate strength 65,000 psi.   The longitudinal joint is a double-riveted butt joint.   Find (*a*) the bursting pressure of the drum in pounds per square inch; (*b*) the safe working pressure if the factor of safety of 6 is used.

2.29. Two boiler drums are constructed of the same grade of steel and of the same thickness of plate.   The joints are the same in each drum.   One is 30 in. in diameter, and the other is 60 in.   Which will stand the greater pressure?   How much greater?

2.30. A boiler shell is to be 50 in. in diameter, with an ultimate strength of 50,000 psi.   Joints are to be double-riveted butt-strap; working pressure, 100 psi.   Calculate the thickness of plate required.

2.31. A boiler shell is to be 36 in. in diameter, and the ultimate tensile strength of the metal 50,000 psi; efficiency of joint is to be 80 per cent, factor of safety 5, and working pressure 200 psi.   What thickness of plate would be required?

2.32. A water tank 30 in. in diameter has a longitudinal single-riveted lap joint.   The rivets are ¾ in. in diameter and are placed on 2-in. centers; the plate is ⅝ in. thick.   What is the pounds-per-square-inch shearing stress on the rivets?   The pounds-per-square inch tension on the plate?   (Assume a pressure of 100 psi on the tank.)

2.33. If the rivets in Prob. 2.32 have an ultimate shearing strength of 45,000 psi and the plate an ultimate tensile strength of 50,000 psi, what would be the efficiency of the joint?

2.34. What is meant by boiler horsepower?

2.35. Calculate the heating surface of a return tubular boiler 6 ft in diameter, 18 ft long, containing seventy 3½-in. tubes.   The boiler is bricked in one-third from the top, and the area of the heads is neglected.

2.36. What would be the horsepower of the boiler in Prob. 2.35?

2.37. What is the heating surface of a boiler 48 in. in diameter and 20 ft long, containing two flues 12 in. in diameter?

2.38. If a boiler 60 in. in diameter and 20 ft long contains fifty 4-in. tubes, what is the horsepower of the boiler?

2.39. An HRT boiler contains 2,256 sq ft of heating surface and generates 9,100 lb of steam per hr at 150 psi gauge pressure. The feedwater temperature is 172°F, and the steam is dry and saturated. Calculate the hosepower developed and the percentage boiler rating.

2.40. A water-tube boiler has a capacity of 150,000 lb of steam per hr, and operates at 385 psig and a total steam temperature of 500°F. Water enters the economizer at 250°F and leaves at 346°F. The efficiency of the unit is 87.00 per cent. The sq ft of heating surfaces are as follows; boiler, 6,500; furnace waterwall, 1,125; superheater, 440; and economizer, 4,500. The furnace volume is 2,000 cu ft. Calculate the heat transfer in Btu per hr per sq ft for (1) combined boiler and furnace waterwall area, (2) superheater, and (3) economizer; (4) also calculate the heat release in the furnace in Btu per hr per cu ft of volume.

CHAPTER 3

COMBUSTION
OF
FUELS

In commercial practice combustion is accomplished by mixing fuel and air at elevated temperatures. The air supplies oxygen, which unites chemically with the carbon, hydrogen, and a few minor elements in the fuel to produce heat.

3.1 THE COMBUSTION PROCESS. The combustion process follows fundamental principles which must be understood by the designers and operators of the equipment to assure satisfactory service and high efficiency.

1. *Control of air supply.* The amount of air required depends upon the fuel, the equipment, and the operating conditions and is determined from manufacturer's recommendations and actual trial. Too much air results in an excessive amount of hot gases being discharged from the stack with a correspondingly high heat loss. A deficiency of air permits some of the fuel, unburned or only partially burned, to pass through the furnace. It is, therefore, important that the best proportion of air to fuel be determined and maintained in order to secure high efficiency.

2. *Mixing of air and fuel.* The air and fuel must be thoroughly mixed since each combustible particle must come into intimate contact with the oxygen contained in the air before combustion can take place. If the air distribution and mixing are poor, there will be an excess of air in some portions of the fuel bed or combustion chamber and a deficiency in others. With this principle in mind consider the combustion equipment with which you are familiar and note the means employed by the designer in an attempt to obtain the mixing of fuel and air.

3. *Temperature required for combustion.* All around us we see combustible material in intimate contact with air, and still it is not burning. Actually a chemical reaction is taking place, but it is so slow that it is referred to not as "combustion" but as "oxidation." The corrosion (rusting) of steel when exposed to the atmosphere is an example of this oxidation.

When the combustible material reaches its ignition temperature, oxidization is accelerated; the process is called "combustion." It is evident, therefore, that it is important to maintain the fuel-and-air mixture at a temperature sufficiently high to promote combustion.

When the flame comes into contact with the relatively cool boiler tubes or shell, the carbon particles are deposited in the form of soot. When boilers are operated at a very low rating, the temperatures are low, resulting in incomplete combustion and excessive smoke.

4. *Time required for combustion.* Air supply, mixing, and temperature determine the rate at which combustion progresses. In all cases an appreciable amount of time is required to complete the process. When the equipment is operated at excessively high ratings, the time may be insufficient to permit complete combustion. As a result, considerable unburned fuel is discharged from the furnace. The rejected material may be in the form of solid fuel or combustible gases. The resulting loss may be appreciable and must, therefore, be checked and controlled.

These principles involving the process of burning (combustion) may be understood by reference to Figs. 3.1 and 3.2. Here the principles of combustion are applied to solid fuels burned on grates and to pulverized fuel, gas, and oil burned in suspension.

Figure 3.1 illustrates a hand-fired stationary grate installed under a water-tube boiler. The coal is supplied by hand through the fire door. Air for combustion enters through both the ashpit and the fire door. For the purpose of illustration the fuel bed may be considered as having four zones. The green coal is added to the top or distillation zone; next are the reduction and oxidation zones and, finally, the layer of ash on the grates. The air which enters the ashpit door flows up through the grates and ash into the oxidation zone, where the oxygen comes into contact with the hot coke and is converted into carbon dioxide. As the gases continue to travel upward through the hot-coke bed, this carbon dioxide is reduced to carbon monoxide. The exposure of the green coal in the upper zone to the high temperature results in distillation of hydrocarbons, which are carried into the furnace by the upward flow of gases. Therefore, the gases entering the furnace through the fuel bed contain combustible materials in the form of carbon monoxide and hydrocarbons. The oxygen in the air which enters the furnace

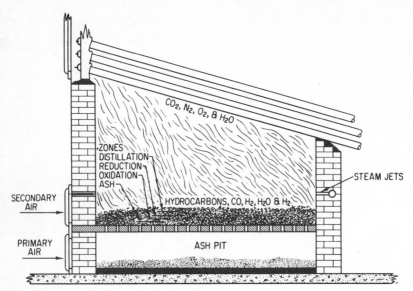

FIG. 3.1   *The combustion process as applied to hand-fired grates.*

through the fire door must combine with these combustibles before they enter the boiler tube bank and become cooled.

From this discussion of the process it is evident that with hand firing a number of variables are involved in obtaining the required rate of combustion and complete utilization of the fuel with a minimum amount of excess air. The fuel must be supplied at the rate required by the steam demand. Not only must the air be supplied in proportion to the fuel, but the amount entering through the furnace doors and the ashpit must be in correct proportions. The air which enters through the ashpit door and passes up through the fuel bed determines the rate of combustion. The secondary air which enters directly into the furnace is used to burn the combustible gases. Thorough mixing of the combustible gases and air in the furnace is necessary because of the short time required for these gases to travel from the fuel bed to the boiler tubes. Steam and high-pressure air jets are used to assist in producing turbulence in the furnace and mixing of the gases and air (see Sec. 4.13). A failure to distribute the coal evenly on the grates, variation in the size of the coal, and a formation of clinkers result in unequal resistance of the fuel bed to the flow of gases. Areas of low resistance in the fuel bed permit high velocity of gases and accelerated rates of combustion, which deplete the fuel and further reduce the resistance. These areas of low resistance are termed "holes in the fire."

As a result of these shortcomings inherent in hand firing, several methods of introducing solid fuel to furnaces and the control of the air supply have been developed.   These are explained in Chap. 4.

Suspension firing of fuel in a water-cooled furnace is illustrated in Fig. 3.2.   This method may be utilized in the combustion of gaseous fuels without special preparations, of fuel oil by providing for atomization, and of solid fuels by pulverization.   The fuel particles and air in the correct proportions are introduced into the furnace, which is at an elevated temperature.   The fine particles of fuel expose a large surface to the oxygen present in the combustion air and to the high furnace temperature.   The air and fuel particles are mixed either in the burner or directly after they enter the furnace.   When coal is burned by this method, the volatile matter, hydrocarbons and carbon monoxide, is distilled off when the fuel enters the furnace.   These combustible gases and the residual carbon particles burn during the short interval of time required for them to pass through the furnace.   The period of time required to complete combustion of fuel particles in suspension depends upon the particle size of the fuel, control of the flow of combustion air, mixing of air and fuel, and furnace temperatures.   Relatively

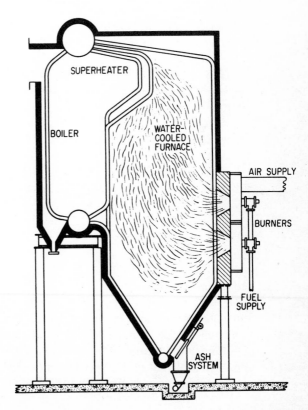

FIG. 3.2 *The combustion process as applied to suspension firing.*

large furnaces are required to assure complete combustion. The equipment used in the suspension burning of solid, liquid, and gaseous fuels is discussed in Chap. 4.

From these illustrations we note that the requirements for good combustion are time of contact between the fuel and air, elevated temperature during this time, and turbulence to provide thorough mixing of fuel and air. This time, temperature, and turbulence are referred to as the three Ts of combustion.

3.2   THE THEORY OF COMBUSTION. Combustion is a chemical process which takes place in accordance with natural laws. By applying these laws the theoretical quantity of air required to burn a given fuel can be determined when the fuel analysis is known. The air quantity used in a furnace, expressed as percentage of excess above the theoretical requirement, can be determined from the flue-gas analysis.

In the study of combustion we encounter matter in all possible forms: solid, liquid, gas, and vapor. Matter in the form of a *solid* has both volume and shape. A *liquid* has a definite volume, in that it is not readily compressible, but its shape conforms to that of the container. A *gas* has neither a definite volume nor shape since both conform to that of the container.

When liquids are heated, a temperature is reached at which *vapor* will form above the surface. This vapor is only slightly above the liquid state. When vapor is removed from the presence of the liquid and heated, a gas will be formed. There is no exact point at which a substance changes from a gas to a vapor or from a vapor to a gas. It is simply a question of degree as to how nearly the vapor approaches a gas. Steam produced by boiling water at atmospheric pressure is vapor because it is just above the liquid state. On the other hand, air may be considered a gas since under normal conditions it is far removed from the liquid state (liquid air). Gases follow definite laws of behavior when subjected to changes in pressure, volume, and temperature. The more nearly a vapor approaches a gas, the more closely it will follow the laws.

When considering gas laws and when making calculations in thermodynamics, we must always express pressures and temperatures in absolute units rather than in "gauge" values (read directly from gauges and thermometers).

Absolute pressures greater than atmospheric are found by adding the atmospheric pressure to the gauge reading. Both pressures must be expressed in the same units. The atmospheric pressure is accurately determined by means of a barometer, but for many calculations the approximate value of 14.7 psi is sufficiently accurate. When, for example, a pressure gauge reads 150 psi, the absolute pressure is 164.7 psia.

The absolute pressures which are below zero gauge are found by subtracting the gauge reading from the atmospheric pressure. When a gauge reads −5 lb, the absolute pressure would be 14.7 − 5.0 = 9.7 psia. Pressures a few pounds above zero gauge and below (in the vacuum range) are frequently measured by a U tube containing mercury and are expressed in inches of mercury. Many of the pressures encountered in combustion work are nearly atmospheric (zero gauge) and can be measured by a U tube containing water.

The zero on the Fahrenheit scale is arbitrarily chosen and has no scientific significance. Experiments have proved that the true or absolute zero is 460° below zero on the Fahrenheit thermometer. The absolute temperature on the Fahrenheit scale is found by adding 460°F to the thermometer reading.

Expressed in absolute units[1] of pressure and temperature, the three principal laws governing the behavior of gases may be stated as follows:

Constant temperature. When the temperature of a given quantity of gas is maintained constant, the volume will vary inversely as the pressure. If the pressure is doubled, the volume will be reduced by one-half:

$$\frac{V_1}{V_2} = \frac{P_2}{P_1}$$

Constant volume. When the volume of a gas is maintained constant, the pressure will vary directly as the temperature. When the temperature is doubled, the pressure will also be doubled:

$$\frac{P_1}{P_2} = \frac{T_1}{T_2}$$

Constant pressure. When a gas is maintained at constant pressure, the volume will vary directly as the temperature. If the temperature of a given quantity of gas is doubled, the volume will also be doubled:

$$\frac{V_1}{V_2} = \frac{T_1}{T_2}$$

In combustion work the gas temperature varies over a wide range. Air enters the furnace, for example, at 70°F, is heated in some instances to 2500°F in the furnace, and is finally discharged from the stack at between 300 and 400°F. During these temperature changes the volume varies because the gases are maintained almost constant at atmospheric pressure. This is most important since fans, ducts, boiler passes, etc., must be designed to accommodate these variations in volume.

[1] When V_1 and V_2 are respectively the initial and final volumes, P_1 and P_2 are respectively the initial and final absolute pressures, and T_1 and T_2 are respectively the initial and final absolute temperatures.

In addition to these physical aspects of matter we must also consider the chemical reactions which occur in the combustion process. Chemistry teaches that all substances are composed of one or more of the 105 existing elements. The smallest particle into which an element may be divided is termed an "atom." Atoms combine in various combinations to form molecules, which are the smallest particles of a compound or substance. The characteristic of a substance is determined by the atoms which make up its molecules. Combustion is a chemical process because it involves a changing about of atoms and hence the molecular structure of the substances concerned.

The trading about and changing of atoms from one substance to another is an exacting procedure. Substances always combine in the same definite proportions. The atoms of each of the 105 elements have a weight number usually referred to as the "atomic weight." These weights are relative and refer to oxygen, which has an atomic weight of 16. Thus, for example, carbon, which is three-quarters as heavy as oxygen, has an atomic weight of 12. Atomic weight, molecular weight, and other data relative to elements and substance are given in Table 3.1.

When oxygen and the combustible elements or compounds are mixed in definite proportions at an elevated temperature under ideal conditions, they will combine completely.[1] This shows that a given combustible element requires a definite amount of oxygen to complete combustion. If additional oxygen is supplied (more than necessary for complete com-

TABLE 3.1   Constants of Quantities Employed in Combustion for Atmospheric Pressure and 32°F

Name	Chemical			Physical					
	Molecular formula	Atomic weight (approx.)	Molecular weight (approx.)	Specific weight, lb/cu ft	Specific volume, cu ft/lb	Specific heat (mean at constant pressure)	Heating value, Btu/lb (higher)	State
Carbon..........	C	12	...	145	0.0069	0.160	14,540	Solid
Hydrogen........	H₂	1	2	0.00565	177.1	3.140	62,000	Gas
Sulfur..........	S₂	32	64	125	0.0080	0.137	4,050	Solid
Carbon monoxide.	CO	...	28	0.0778	12.86	0.2485	4,355	Gas
Oxygen..........	O₂	16	32	0.0891	11.22	0.2175	......	Gas
Nitrogen........	N₂	14	28	0.0778	12.86	0.2485	......	Gas
Carbon dioxide...	CO₂	...	44	0.1230	8.14	0.2025	......	Gas
Sulfur dioxide....	SO₂	...	64	0.18272	5.473	0.1544	......	Gas
Water vapor*....	H₂O	...	18	0.0373	26.82	0.458	......	Vapor
Air.............	....	...	29†	0.0807	12.39	0.240	......	Gas

* Values are for atmospheric pressure and 212°F.
† Air is a mixture of oxygen and nitrogen. This figure is the average accepted molecular weight for air.

[1] The theoretical proportions (no deficiency, no excess) of elements or compounds in a chemical reaction are referred to as the "stoichiometric ratio."

bustion), the excess will not enter into the reaction but will pass through the furnace unchanged. On the other hand, if there is deficiency of oxygen, the combustible material will remain unburned. Briefly the law of combining weights tells us that the elements and compounds combine in definite proportions which are in simple ratio to their atomic or molecular weights.[1]

The following is an explanation of some of the chemical reactions used in combustion:

Carbon Burned to Carbon Dioxide

Substance.................... carbon + oxygen → carbon dioxide
Kind of matter............... solid + gas → gas

Volume and chemical equation.

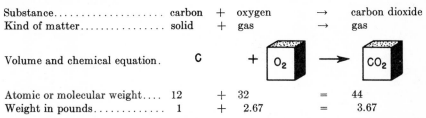

Atomic or molecular weight.... 12 + 32 = 44
Weight in pounds............ 1 + 2.67 = 3.67

The volume of carbon dioxide produced is equal to the volume of oxygen used. The carbon dioxide gas is, however, heavier than the oxygen. The combining weights are 12 lb of carbon and 32 lb of oxygen, uniting to form 44 lb of carbon dioxide, or 1 lb of carbon requires 2.67 lb of oxygen and produces 3.67 lb of carbon dioxide. The combustion of 1 lb of carbon produces 14,540 Btu.

Carbon Burned to Carbon Monoxide

Substance.......... carbon + oxygen → carbon monoxide
Kind of matter..... solid + gas → gas

Volume and chemical equation........

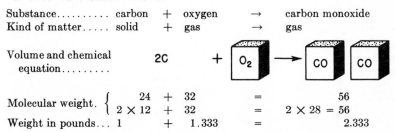

Molecular weight. $\begin{cases} 24 + 32 = 56 \\ 2 \times 12 + 32 = 2 \times 28 = 56 \end{cases}$
Weight in pounds... 1 + 1.333 = 2.333

[1] The atomic and molecular weights of elements and compounds are useful in determining the weights and volume of gases. It has been proved that at the same temperature and pressure a given volume of all perfect gases will contain the same number of molecules. This means, for example, that at a temperature of 32°F and 14.7 psia, 32 lb of oxygen (a weight equal to the molecular weight) will occupy 359 cu ft.

$$32 \text{ lb} \times 11.22 \text{ cu ft/lb} = 359 \text{ cu ft}$$

In like manner, 2 lb of hydrogen gas at the standard temperature of 32°F and 14.7 psia will be found to occupy 359 cu ft. Here again the 2 is the weight in pounds equal to the molecular weight of hydrogen. From this we must conclude that 359 cu ft of hydrogen at standard conditions weighs 2 lb, and an equal volume of oxygen weighs 32 lb. A weight, in pounds, of any substance equal to its molecular weight is known as a "pound mole."

When carbon is burned to carbon monoxide (incomplete combustion), the volume of oxygen used is only one-half of that required for completely burning the carbon to carbon dioxide; the volume of carbon monoxide produced is two times that of the oxygen supplied. The heat released is only 4380 Btu per lb, but it is 14,540 Btu when 1 lb of carbon is completely burned. The net loss is, therefore, 10,160 Btu per lb of carbon and shows the importance of completely burning the combustible gases before they are allowed to escape from the furnace.

Carbon Monoxide Burned to Carbon Dioxide

Substance..........	carbon monoxide	+	oxygen	→	carbon dioxide
Kind of matter	gas	+	gas	→	gas

Volume and chemical equation.........

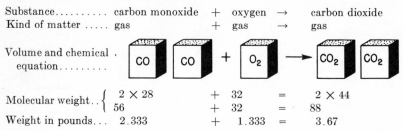

Molecular weight.. {	2×28	+	32	=	2×44
	56	+	32	=	88
Weight in pounds...	2.333	+	1.333	=	3.67

Here the two molecules of carbon monoxide previously produced, by the incomplete combustion of two molecules of carbon, are combined with the necessary one molecule of oxygen to produce two molecules of carbon monoxide. The 1 lb of carbon produced 2.333 lb of carbon monoxide. Finally, however, the 1 lb of carbon produces 3.67 lb of carbon dioxide regardless of whether the reaction is in one or two steps. The total amount of oxygen required, as well as the heat liberated per pound of carbon, is the same for complete combustion in both cases.

Combustion of Hydrogen

Substance..........	hydrogen	+	oxygen	→	water
Kind of matter......	gas	+	gas	→	liquid

Volume and chemical equation.........

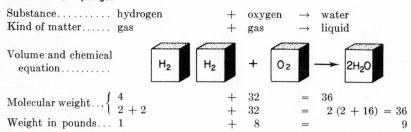

Molecular weight... {	4	+	32	=	36
	$2 + 2$	+	32	=	$2 (2 + 16) = 36$
Weight in pounds...	1	+	8	=	9

Hydrogen is a very light gas with a high heat value. The combustion of 1 lb of hydrogen gas liberates 62,000 Btu. To develop this heat, two molecules of hydrogen combine with one molecule of oxygen to form two molecules of water. One volume of oxygen is required for two volumes of hydrogen. The weight relations are 1 lb of hydrogen and 8 lb of oxygen, producing 9 lb of water. The oxygen used to burn hydrogen appears as water vapor in the flue gases.

The equations for these and some of the other reactions involved in combustion are as follows:

Carbon	plus	oxygen	gives	carbon dioxide
C	+	O_2	→	CO_2
Carbon	plus	oxygen	gives	carbon monoxide
2C	+	O_2	→	2CO
Carbon monoxide	plus	oxygen	gives	carbon dioxide
2CO	+	O_2	→	$2CO_2$
Hydrogen	plus	oxygen	gives	water vapor
$2H_2$	+	O_2	→	$2H_2O$
Sulfur	plus	oxygen	gives	sulfur dioxide
S	+	O_2	→	SO_2
Sulfur	plus	oxygen	gives	sulfur trioxide
2S	+	$3O_2$	→	$2SO_3$
Methane	plus	oxygen	gives	carbon dioxide and water
CH_4	+	$2O_2$	→	$CO_2 + 2H_2O$
Acetylene	plus	oxygen	gives	carbon dioxide and water
$2C_2H_2$	+	$5O_2$	→	$4CO_2 + 2H_2O$
Ethylene	plus	oxygen	gives	carbon dioxide and water
C_2H_4	+	$3O_2$	→	$2CO_2 + 2H_2O$
Ethane	plus	oxygen	gives	carbon dioxide and water
$2C_2H_6$	+	$7O_2$	→	$4CO_2 + 6H_2O$

Sulfur is of little value since its heating content is only 4050 Btu per lb. On the other hand the presence of sulfur causes several serious problems in the utilization of fuel. Therefore low-sulfur fuels sell at premium prices. Sulfur causes corrosion of conveying and storage equipment. Sulfur in coal causes storage problems and adversely affects pulverization. The sulfur oxide produced by the burning of sulfur lowers the dew point of the flue gases, causing corrosion of economizers and air heaters. Most importantly the sulfur oxides are contaminants in the atmosphere, and the amount that can be discharged is limited by law.

In the combustion process, 1 lb of sulfur combines with 1 lb of oxygen to form 2 lbs of sulfur dioxide.[1] The sulfur dioxide in the flue gases can be approximated as follows:

$$SO_2 \text{ lb/hr} = K \times \text{lb fuel burned per hr} \times 2 \times S/100$$

where K = ratio of SO_2 in flue gases to theoretical amount resulting from the combustion of the sulfur in the fuel. (Frequently assumed .95.)

$$S = \text{Per cent sulfur in fuel}$$

It is customary to express sulfur oxide emission in pounds per million Btu of fuel burned.

$$SO_2 \text{ lb/million Btu} = \frac{SO_2 \text{ lb/hr} \times 1,000,000}{\text{lb fuel/hr} \times \text{Btu in fuel}}$$

[1] Actually a portion of the sulfur is converted to sulfur trioxide. The summation of all the sulfur oxides in the flue gases is referred to as SO_x.

Example Coal containing 1.5 per cent sulfur with a Btu content of 11,500 Btu per lb burned at the rate of 3 tons per hr. What is the sulfur emission in lbs per million Btu input to furnace?

$$SO_2 \text{ lb/hr} = .95 \times 3 \times 2000 \times 2 \times 1.5/100 = 171$$

$$SO_2 \text{ lb/million Btu} = \frac{171 \times 1,000,000}{3 \times 2,000 \times 11,500} = 2.48$$

In practice, the oxygen supplied for combustion is obtained from the atmosphere. The atmosphere is a mechanical mixture of gases which for practical purposes may be considered as being composed of the following:

Element	Percentage, volume	Percentage, weight
Oxygen...........	20.91	23.15
Nitrogen.........	79.09	76.85

Only the oxygen enters into chemical combination with the fuel. The nitrogen passes through the combustion chamber without chemical change. It does, however, absorb heat and reduces the maximum temperature attained by the products of combustion.

In order to supply 1 lb of oxygen to a furnace it is necessary to introduce

$$\frac{1}{0.2315} = 4.32 \text{ lb of air}$$

Since 1 lb of carbon requires 2.67 lb of oxygen, we must supply

$$4.32 \times 2.67 = 11.53 \text{ lb of air/lb of carbon}$$

TABLE 3.2 Theoretical Quantities Involved in Combustion of Fuel

All expressed in pounds per pound of fuel

Constituent	Required		Resulting quantities			
	O_2	Air	CO_2	N_2	H_2O	SO_2
Carbon (C)...............	2.67	11.53	3.67	8.86		
Hydrogen (H)...........	8.00	34.56	....	26.56	9.00	
Sulfur (S)...............	1.00	4.32	....	3.32	....	2.00

The 11.53 lb of air is composed of 2.67 lb of oxygen and 8.86 lb of nitrogen.

$$\underset{\text{lb air}}{11.53} - \underset{\text{lb } O_2}{2.67} = \underset{\text{lb } N_2}{8.86}$$

By referring to the equation for the chemical reaction of carbon and oxygen, we find that 1 lb of carbon produces 3.67 lb of carbon dioxide. Therefore, the total products of combustion formed by burning 1 lb of carbon with the theoretical amount of air are 8.86 lb of nitrogen and 3.67 lb of carbon dioxide.

In a similar manner it can be shown that 1 lb of hydrogen requires 34.56 lb of air for complete combustion. The resulting products of combustion are 9 lb of water and 26.56 lb of nitrogen.

When quantities (expressed in percentage by weight) of carbon (C), hydrogen (H), sulfur (S), and oxygen (O) in a fuel are known, the theoretical quantity of air required can be determined by the following formula:

$$\text{lb of air required/lb of fuel} = 11.53\ C + 34.56\ (H - \tfrac{1}{8}O) + 4.32S$$

The quantities of oxygen and air required together with the resulting products of combustion for 1 lb of carbon, hydrogen, and sulfur are given in Table 3.2.

The values given in Table 3.2 will now be used in determining the amount of air required and the resulting products involved in the perfect combustion of 1 lb of coal having the following analysis:

Constituent	*Weight per Lb*
Carbon	0.75
Hydrogen	0.05
Nitrogen	0.02
Oxygen	0.09
Sulfur	0.01
Ash	0.08
Total	1.00

In the case of carbon the values given in the table are found as follows:

$$0.75 \quad \times \quad 2.67 \quad = \quad 2.00$$
lb C/lb coal $\qquad$ O_2 req'd./lb C $\qquad$ O_2 req'd./lb coal

$$2.00 \quad \times \quad 4.32 \quad = \quad 8.64$$
lb O_2 $\qquad$ lb air req'd./lb O_2 $\qquad$ lb air req'd./lb coal

$$0.75 \quad \times \quad 3.67 \quad = \quad 2.75$$
lb C/lb coal $\qquad$ lb CO_2/lb C $\qquad$ lb CO_2/lb coal

$$8.64 \quad - \quad 2.00 \quad = \quad 6.64$$
lb air req'd./lb coal $\qquad$ lb O_2 req'd./lb coal $\qquad$ lb N_2/lb coal

The values for hydrogen and sulfur are found in a similar manner. Note in Table 3.3 that the weight of fuel and air supplied is equal to the weight of the resulting quantities.

$$0.92 \text{ lb} + 10.022 \text{ lb} = 10.942 \text{ lb}$$

fuel less ash   air req'd.   total input

$$2.75 \text{ lb} + 7.722 \text{ lb} + 0.45 \text{ lb} + 0.02 \text{ lb} = 10.942 \text{lb}$$

CO_2 N_2 water vapor SO_2 total output

The condition under which, or degree to which, combustion takes place is expressed as "perfect," "complete," or "incomplete." *Perfect combustion,* which we have been discussing, consists of burning all the fuel and using only the calculated or theoretical amount of air. *Complete combustion* also denotes the complete burning of the fuel but by supplying more than the theoretical amount of air. The additional air does not enter into the chemical reaction. *Incomplete combustion* occurs when a portion of the fuel remains unburned because of insufficient air, improper mixing, or other reasons.

TABLE 3.3   Theoretical Quantities Involved in Combustion of Coal

All expressed in pounds per pound of coal

Constituent	Weight per lb	Required		Resulting quantities			
		O_2	Air	CO_2	N_2	H_2O	SO_2
Carbon (C).............	0.75	2.00	8.64	2.75	6.64	....	....
Hydrogen (H)...........	0.05	0.40	1.728	....	1.328	0.45	
Sulfur (S)..............	0.01	0.01	0.043	....	0.033	....	0.02
Total................	0.81	2.41	10.411	2.75	8.001	0.45	0.02
Correction for							
N_2 in coal............	0.02	......	......	....	+0.02	....	....
O_2 in coal............	0.09	−0.09	−0.389	....	−0.299	....	....
Corrected total........	....	2.32	10.022	2.75	7.722	0.45	0.02

In practice it is necessary and economical to supply more air than the theoretical amount in order to obtain complete combustion. The air supplied to a combustion process in an amount above that theoretically required is known as "excess air." To exercise control over combustion it is necessary to determine the amount of excess air. This may be accomplished by chemical analysis or by measuring the volume of the flue gases.

The flue-gas analysis is effective in determining the amount of air supplied for combustion, as indicated by Fig. 3.3. This graph shows how the amount of excess air used in the combustion process is indicated by either the percentage of carbon dioxide or oxygen in the flue gases. When a single fuel is burned, the carbon dioxide content of the flue gases provides a satisfactory index of the amount of excess air being

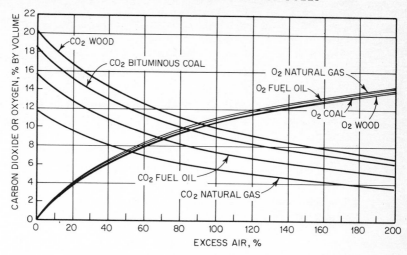

FIG. 3.3 *Carbon dioxide and oxygen in per cent by volume compared with the excess air used when various fuels are burned.*

used. This can be explained by the fact that with the complete combustion of 1 lb of carbon 3.67 lb of carbon dioxide is produced. Therefore the amount of carbon dioxide formed depends upon the amount of carbon burned. When a relatively large amount of air is used, the fixed amount of carbon dioxide gas will be diluted and the percentage correspondingly lowered. Conversely, if only a small amount of excess air is used, there will be less dilution and the percentage of carbon dioxide will be relatively high. For a given percentage of excess air, fuels with higher carbon-hydrogen ratio will have a higher per cent carbon dioxide in the flue gases than fuels with lower carbon-hydrogen ratio. For a given per cent excess air the flue gases from a coal-fired furnace will have a higher percentage of carbon dioxide than when fuel oil is burned. For example, a flue gas containing 12 per cent carbon dioxide will indicate 54 per cent excess air when bituminous coal is being burned and only 27 per cent excess air with fuel oil.

The percentage of oxygen in the flue gases provides an adequate measurement of excess air when either single or multiple fuels are being used. The oxygen in the flue gases represents that portion which entered but did not combine with the combustible elements in the fuel. This oxygen in the flue gases and the nitrogen with which it was mixed are the excess air. The theoretical oxygen and therefore air requirement are approximately proportional to the heat content of the fuel even with variations in the carbon-hydrogen ratio. For a given percentage of oxygen the excess air is approximately the same for either coal or fuel oil. For example, a flue gas containing 6 per cent oxygen will indicate

40 per cent air when bituminous coal is being burned and 38.8 per cent excess air with fuel oil.

The flue-gas analysis is obtained by use of the orsat as explained in Sec. 3.8. The analysis includes carbon dioxide (CO_2), oxygen (O_2), and carbon monoxide (CO). When the sum of these three constituents is subtracted from 100, the remainder is assumed to be nitrogen (N_2).

The excess air in percentage of the theoretical requirements can be calculated by the following formula:

$$\text{Percentage of excess air} = \frac{O_2 - \tfrac{1}{2}CO}{0.263N_2 + \tfrac{1}{2}CO - O_2} \times 100$$

However when there is no carbon monoxide present the formula becomes:

$$\text{Percentage of excess air} = \frac{O_2}{0.263N_2 - O_2} \times 100$$

The use of these formulas will be shown by application to the following flue-gas analyses.

Analysis	CO_2	O_2	CO	N_2
A	13.7	3.5	1.8	81.0
B	13.5	5.5	0.0	81.0

Analysis "A":

$$\frac{\text{Percentage}}{\text{of excess air}} = \frac{3.5 - \tfrac{1}{2} \times 1.8}{0.263 \times 81.0 + \tfrac{1}{2} \times 1.8 - 3.5} \times 100 = 13.9 \text{ per cent.}$$

Analysis "B":

$$\text{Percentage of excess air} = \frac{5.5}{0.263 \times 81.0 - 5.5} \times 100 = 34.8 \text{ per cent.}$$

For analysis A, the excess air is 13.9 per cent; for B, it is 34.8 per cent. Free oxygen is present in both analyses, but the presence of carbon monoxide in A indicates incomplete combustion.

The heating value of fuel depends essentially upon its carbon and hydrogen content. The chemical analyses of coal include, in addition to carbon and hydrogen, ash and other inert materials. From these analyses it is possible to determine the theoretical quantity of air required, the heating value, the products of combustion, and the heat balance.

It is possible to calculate the approximate heating value of fuels from the analyses by using Dulong's formula:

Heating value/lb of fuel $= 14{,}540C + 62{,}000 (H - \frac{1}{8}O) + 4{,}050S$

where C, H, and S represent the percentages by weight of carbon, hydrogen, and sulfur in the fuel.

It is usually preferable to determine the heating value of a fuel by actually developing and measuring the heat. This is accomplished by completely burning a carefully weighed sample of the fuel in a calorimeter. The heat produced causes a temperature rise in a known quantity of water. The temperature rise is indicative of the heating value of the fuel.

Physical characteristics of fuel are, in many cases, more important in practical application than chemical constituents. Means have been devised for determining some of these characteristics by simple control tests which can be made by the plant operators. These procedures are best explained with reference to specific fuels.

3.3 THE AIR SUPPLY. Supplying oxygen, as contained in air, is an important consideration in the combustion process. In a typical case in which 15 lb of air is required per lb of coal, it is necessary to deliver 15 tons of air to the furnace and to pass almost 16 tons of gases through the boiler for each ton of coal burned. In many installations the ability to supply air is the limiting factor in the rate of combustion. The number of pounds of fuel that can be burned per square foot of grate area depends upon the amount of air that can be circulated through the fuel bed. A means must be provided of supplying the required amount of air to the furnace. The products of combustion must be removed from the furnace and circulated over the heat-absorbing surfaces. Finally the excess pollutants must be removed from the resulting flue gases before they are discharged through the stack to the atmosphere.

This circulation of gases is caused by a difference in pressure, referred to in boiler practice as "draft." Draft is the differential in pressure between two points of measurement, usually the atmosphere and the inside of the boiler setting.

A differential in draft is required to cause the gases to flow through a boiler setting. This required differential varies directly as the square of the rate of flow of gases. For example, when the flow is doubled, the difference in draft between two points in the setting will increase four times. For a given amount of fuel burned, the quantity of gases passing through the boiler depends upon the amount of excess air being used. The draft differential across the boiler tube bank and the differential created by an orifice in the steam line are used to actuate a two-pen flowmeter.

Recording flowmeters of this type are calibrated under actual operating conditions by use of the flue-gas analysis so that when the steam-flow-output recording pen and the gas-flow recording pen coincide, the desired amount of air is being supplied.

A draft gauge can be made by first bending a glass tube in the form of a U and then partly filling it with water. A sampling tube inserted in the boiler setting is then connected to one end of the U tube while the other end remains open to the atmosphere. The difference in the height of the two columns of water is a measure of the draft.

The scales on these gauges are calibrated in inches, and the difference in height of the columns is read in inches of water. The draft gauges in common use are mechanical or dry-type gauges. However, the scales are calibrated in inches of water.

A pressure in the furnace slightly lower than that of the atmosphere (draft) causes the air to enter, thus supplying the oxygen required for combustion. A draft at the boiler outlet greater than that in the furnace causes the products of combustion to circulate through the unit.

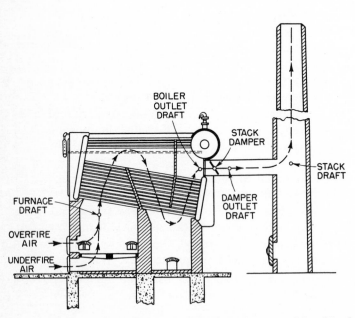

FIG. 3.4 *Flow of air and gases through a typical hand-fired boiler unit.*

The rate of flow or quantity of air supplied can be regulated by varying the draft differential.

The principle of draft and air regulation is explained by reference to the hand-fired boiler shown in Fig. 3.4 and the graph of Fig. 3.5. The stack produces a draft of 1.0 in. of water, which is regulated by the stack damper to give the required furnace draft. As the rating increases, more air is required to burn the additional fuel, and the stack damper must be opened to compensate for the draft loss caused by the increased flow of gases. Draft loss occurs across the fuel bed, the boiler, the damper, and the breeching. Finally, at 152 per cent of rating, the stack damper is wide open and no more air can be supplied, thus limiting the ability of the furnace to burn additional coal efficiently.

The type of fuel determines the amount of draft differential required to produce a given airflow through the fuel bed. This is one of the reasons that more load can be carried with some fuels than with others. Figure 3.6 shows the draft normally required to burn several different types of fuels at varying rates.

Most high-rating combustion equipment employs forced-draft fans for supplying air to the furnace either through the burners or through the grates. The fan supplies the air at a pressure above that of the atmosphere and forces it into the furnace. Figures 3.7 and 3.8 show the application of forced draft to a stoker, the draft, and wind-box pressure at various boiler ratings. This is the same unit as that shown in Fig.

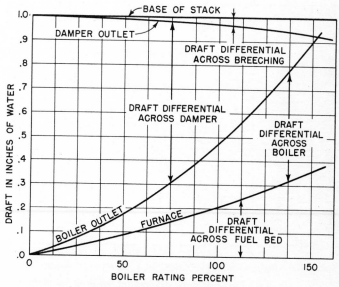

FIG. 3.5   *Draft in a typical hand-fired boiler unit.*

3.4 except that an underfeed stoker has been substituted for the hand-fired grates.   The forcèd-draft fan produces a pressure under the stoker and causes the necessary air to flow up through the fuel bed.   An increase in rating requires an increase in pressure under the stoker to cause the additional air to flow through the fuel bed.   The chimney produces the draft necessary to circulate the gases through the boiler and breeching.   By automatic regulation of the stack damper, the furnace draft is maintained constant at 0.05 in. of water.   Operating a furnace at a constant draft slightly below atmospheric pressure is referred to as "balanced draft."   When the combustion rate increases, more air is added to the furnace and a corresponding increased amount of gases is removed.   This results in a greater flow but maintains a constant pressure in the furnace.   Since the total effect of the chimney is now available in overcoming the resistance through the boiler, a higher rating can be obtained than with the same unit operating without a

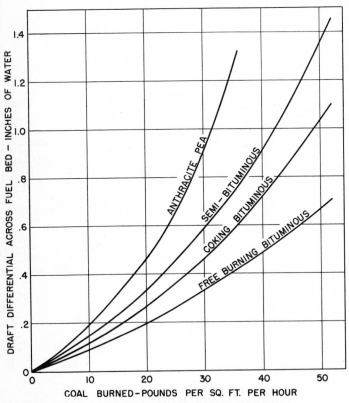

FIG. 3.6   *Fuel-bed draft differential required for various fuels.*

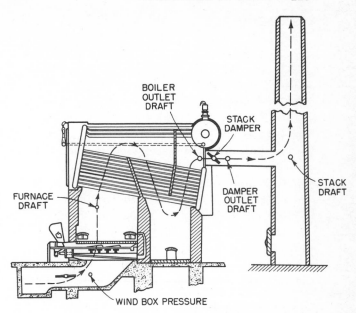

FIG. 3.7 *Flow of air and gases through a typical underfeed stoker-fired boiler unit.*

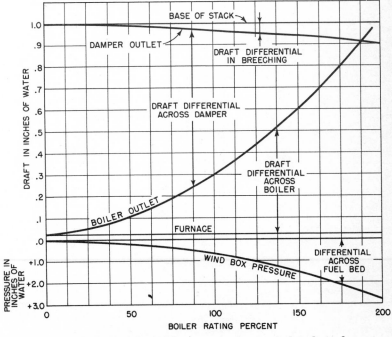

FIG. 3.8 *Draft and wind-box pressure in a typical underfeed stoker-fired boiler unit.*

forced-draft fan and with hand-fired grates. In the examples given (Figs. 3.4, 3.7), the maximum output was increased from 152 to 189 per cent rating.

The chimney provided an adequate means of circulating the air and gases through hand-fired grates or burners and boilers. The application of stokers necessitated the use of forced-draft fans. Induced-draft fans are required when economizers, air heaters, or flue-gas cleaning equipment is applied to balanced-draft boilers.

The pressurized furnaces used in connection with package oil- and gas-fired boilers require only forced-draft fans. Figure 3.9 shows how the pressure developed by the forced-draft fan is used to produce the flow of air and flue gases through the entire unit. Combustion air flows from the forced-draft fan through the supply duct to the wind box, into the furnace, through the boiler, economizer, and interconnecting gas duct to the stack. All the energy required is supplied by the forced-draft fan. At the maximum rating of 200,000 lbs of steam per hr, the static pressure at the forced-draft-fan outlet must be 19.6 in. water

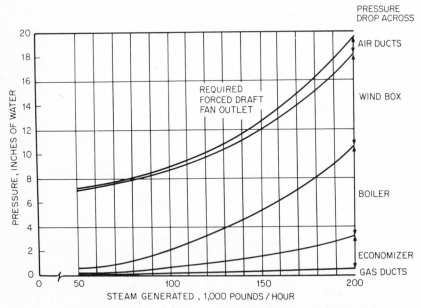

FIG. 3.9 *Air and flue-gas pressures in a 200,000-pph pressurized fuel-oil- and gas-fired package boiler.*

gauge. The pressure drop across the wind box is maintained nearly constant at all ratings by adjustable louvers. This creates a high velocity at the burners and promotes thorough mixing of air and fuel, thereby maintaining good combustion efficiency at partial capacity.

The pressure-furnace principle is also applied to large pulverized-fuel–fired boilers. This makes it possible to develop the necessary air and gas flow with the forced-draft fan, but a number of design and operating problems are introduced.

3.4 COAL. Coal was formed by the decomposition of vegetation which grew in prehistoric forests. At that time the climate was favorable for very rapid growth. Layer upon layer of fallen trees was covered with sediment, and after long periods of aging the chemical and physical properties of the now ancient vegetation deposits were changed, through various intermediate processes, into coal. The process of coal formation can be observed in the various stages on the earth today. However, present-day deposits are insignificant when compared with the magnitude of our great coal deposits.

It is estimated that 100 years is required to deposit 1 ft of vegetation in the form known as peat, and 4 ft of peat is necessary for the formation of 1 ft of coal. Therefore, it requires 400 years to accumulate enough vegetable matter for a 1-ft layer of coal. The conversion from peat to coal requires ages of time. In some areas where other fuel is scarce, the peat is collected, dried, and burned.

The characteristics of coal depend upon the type of vegetation from which it was formed; the impurities which became intermixed with the vegetable matter at the time the peat bog was forming; and the aging, time, temperature, and pressure. It is apparent that the characteristics of coal vary widely.

Preparation plants are capable of upgrading coal quality. In this process foreign materials, including slate and mineral matter, are separated from the coal. The coal may be washed, sized, and blended to meet the most exacting customer demands. However, this processing increases the cost of coal, and an economic evaluation is required to determine whether the cost can be justified. If the raw coal available in the area is unsatisfactory, the minimum required upgrading must be determined. Utility plants obtain the lowest steam cost by selecting their combustion equipment to use the raw coal available in the area. Whether equipment for the coal available in a given region or coal to be used in existing equipment is being selected, all the many factors involved in obtaining the lowest-cost steam production must be taken into consideration.

One satisfactory method of selecting coal for existing equipment is to make plant tests with various coals and in this way determine steam-production costs.   When a plant is in the preliminary design stage, the performance of coal available in the plant area should be checked on equipment similar to that proposed for the new installation.

The proximate analysis of coal is useful in controlling coal quality.   It includes moisture content, volatile matter, fixed carbon, ash, and sometimes sulfur.   This proximate analysis is best understood by an explanation of how the various percentages are determined.

Moisture.   A 1-g sample of coal is placed in an oven where the temperature is maintained at 220°F for 1 hr.   The difference between the weight before and after drying is the amount of moisture removed.

Volatile matter.   The sample is next placed in a furnace in a covered crucible, where the temperature is maintained at 1700°F for a period of 7 min.   The gaseous substance driven off is called "volatile matter."

Fixed carbon.   The lid is now removed from the crucible and the sample returned to the furnace.   The temperature is increased until all the substance has been completely burned.   The difference in weight before and after is fixed carbon.

Ash.   The residue is called "ash."

A proximate analysis is usually all that is required in order to determine the characteristics of a fuel.   This is sufficient to determine the qualities of the fuel for steaming purposes.   When a more detailed and complete analysis is desired, an ultimate analysis is made.   This is necessary when a "heat balance" is to be determined.   (A heat balance is an accurate account of the distribution of the heat lost and absorbed.)   An ultimate analysis requires special equipment and when desired is frequently estimated from the proximate analysis.   The ultimate analysis breaks the fuel up into its constituent elements, which include carbon, hydrogen, nitrogen, sulfur, oxygen, and ash.

The heating value of the fuel is always of utmost importance.   The practical method of determining the Btu equivalent of a pound of coal is by means of a calorimeter.   There are several different types, the bomb calorimeter being most frequently used.

The usual procedure for the determination of the heating value of coal is this: A sample of coal is pulverized and dried.   After drying, a 1-g (a gram is 1/453.6 lb) sample is placed in a tray and the tray placed in position in a steel bomb.   A fuse wire is so arranged that its tip extends into the tray of coal.   The bomb is now closed, and connection made to a tank of oxygen.   The entire bomb is then placed under pressure.

A definite amount of water is next placed in the calorimeter bucket.

The bucket is placed in the calorimeter, and the bomb is set inside the water. A stirring device to keep the water agitated and a thermometer to observe the temperature rise of the water complete the equipment. The coal is ignited by the fuse wire and the temperature rise noted. By means of suitable calculations the heating value of the fuel can then be determined. When a calorimeter cannot be had, the Btu is frequently calculated from the ultimate analysis.

Coal analyses are expressed in three different ways, depending upon the constituents included. The designation "as received" or "as fired" refers to the analysis in which the actual moisture is included. When the expression "moisture-free" or "dry coal" is used, the analysis considers the moisture as having been removed. Since neither moisture nor ash adds to the heating value of coal, analyses are calculated to exclude these constituents and in this way to give a true indication of the nature of the combustible material. When the moisture and ash are not included, the analysis is referred to as "moisture- and ash-free" or "combustible." Sulfur has some heating value but is nevertheless an objectionable constituent of coal for a number of reasons, as explained in Sec. 3.2.

Ash is an inert material, but its characteristics frequently determine the desirability of a coal for a given installation. Owing to the importance of the fusion or melting temperature of the ash, tests have been devised to determine this property. The ash to be tested is molded into a small pyramid, placed in a test furnace, and exposed to a steadily increasing temperature. The atmosphere surrounding the sample is controlled and the temperature measured while the pyramids are observed through a peephole in the side of the furnace (Fig. 3.10). The temperature of the pyramids is noted and recorded at three stages of melting: *initial deformation* (Fig. 3.10a), when the tip of the pyramid first shows a change; *fushion temperature* (Fig. 3.10b), when the pyramid forms in

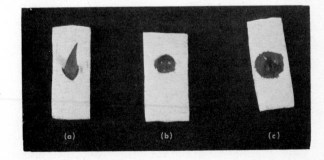

FIG.   3.10   *Ash-fusing-temperature determination.*

(a)   (b)   (c)

[1] See ASTM D720 SEC. 4.1–6.1.

a sphere; and *melting point* (Fig. 3.10c), when the ash becomes fluid and the sphere flattens. These three temperatures are reported in reference to ash fusion.

The coking tendency of coal is expressed by the free-swelling index (FSI). This test is made by grinding a sample of coal to pass a No. 60 sieve and then heating 1 g under specified conditions.[1] The profile obtained by heating the sample is compared with a standard set of profiles to determine the FSI of the sample. The standard profiles are expressed in one-half units from 1 to 9. Coals having an FSI below 5 are referred to as "free-burning," since particles do not tend to stick together and form large lumps of coke when heat is applied but remain separate during the combustion process. Coals having an FSI above 5 are referred to as "caking" or "coking," since the particles swell and tend to stick together when heated.

The caking, or coking, characteristics of coal affect its behavior when it is burned on hand-fired grates or stokers. When a coal cakes, the smaller particles adhere to one another and large masses of fuel are formed on the grates. This action reduces the surface area which is exposed to oxygen and therefore retards the burning. Since these large pieces of coke do not burn, a portion is discharged to the ashpit as undeveloped heat. For efficient combustion, a coking coal requires some agitation of the fuel bed to break up the coke masses in order to maintain uniform air distribution. Free-burning coals, on the other hand, may be burned successfully without fuel-bed agitation. When burned under high-rating high-draft conditions, a friable coal may break up into very small pieces and be carried out of the furnace with the gases. These characteristics (coking and free-burning) need not be considered when coal is burned in the pulverized form.

Coal is classified by rank according to the progressive changes from peat to anthracite as follows:

Peat	Semibituminous
Lignite	Subanthracite
Subbituminous	Anthracite
Bituminous	

Semibituminous, for example, has a higher rank than bituminous; that is to say, it has proceeded further in the progressive aging. This term "rank" must be distinguished from the term "grade," which refers to the relative impurities of ash, sulfur, etc., in the coal. Table 3.4 shows the analysis of the various classifications of coal.

Coal which has a relatively high percentage of volatile matter is termed "soft"; that which has a lower percentage of volatile matter

[1] See ASTM D720 SEC. 4.1–6.1.

is termed "hard." When coal is heated, the volatile matter has a tendency to be distilled off in the form of combustible gases known as "hydrocarbons." These volatile gases liberated from coal must be burned in the combustion space above the fuel bed. A large combustion space must be provided to burn these gases and thereby eliminate fuel loss and smoke. Because hard coal has a relatively lower percentage of volatile matter, it burns with a short flame, and most of the combustion takes place in the fuel bed.

When soft coal is burned in pulverized form, the volatile material is distilled off and burns as a gas. This makes it relatively easy to maintain ignition and complete combustion with a minimum flame travel. Hard coal is also burned in the pulverized form, and in this case each particle is a small portion of carbon which must be burned by contact with oxygen. The combustion of these carbon particles requires an appreciable amount of time, resulting in a long flame travel and a tendency for the fire to puff out at low ratings and when starting up. Other conditions being equal, it is necessary to resort to finer pulverization when hard coal is burned than when soft coal is burned. Thus we see that consideration must be given to the volatile content of the coal, both in designing equipment and in operating.

Coal as it is removed from the mine contains some moisture, and the amount may be increased by exposure to the weather before it reaches the customer. Moisture represents an impurity in that it adds to the weight but not to the heating value of coal. It enters the furnace in the form of water and leaves as steam. Heat generated by the fuel

TABLE 3.4 Representative Analyses of Wood, Peat, and Coal on an "As Received" Basis*

Kind of fuel	Proximate analysis				Ultimate analysis					Calorific value, Btu/lb
	Moisture	Volatile matter	Fixed carbon	Ash	Sulfur	Hydrogen	Carbon	Nitrogen	Oxygen	
Wood.................	.....	.....	.....	.....	.....	6.25	49.50	1.10	43.15	5,800
Peat.................	56.70	26.14	11.17	5.99	0.64	8.33	21.03	1.10	62.91	3,586
Lignite.............	34.55	35.34	22.91	7.20	1.10	6.60	42.40	0.57	42.13	7,090
Subbituminous.......	24.28	27.63	44.84	3.25	0.36	6.14	55.28	1.07	33.90	9,376
Bituminous..........	3.24	27.13	62.52	7.11	0.95	5.24	78.00	1.23	7.47	13,919
Semibituminous......	2.03	14.47	75.31	8.19	2.26	4.14	79.97	1.26	4.18	14,081
Semianthracite.......	3.38	8.47	76.65	11.50	0.63	3.58	78.43	1.00	4.86	13,156
Anthracite..........	2.80	1.16	88.21	7.83	0.89	1.89	84.36	0.63	4.40	13,298

* Adapted from "Coal," by E. S. Moore, John Wiley & Sons, Inc.

must actually be expended to accomplish this conversion. Normally it is to the operator's advantage to burn coal with a low moisture content to prevent the loss of heat that results from converting the water into vapor or steam. However, when coal is burned on grates or stokers, there are conditions which make it advantageous to have a small percentage of moisture present. This moisture tends to accelerate the combustion process, keep the fuel bed even, and promote uniform burning. The advantages gained by the presence of the moisture may then balance the loss resulting from the heat required for its evaporation. Coals having 7 to 12 per cent moisture content are recommended for use on chain- and traveling-grate stokers. The addition of moisture to promote combustion of coal is referred to as "tempering."

Coal with a high moisture content presents some difficult handling problems. During the winter season this moisture freezes while the coal is in transit, making it very difficult to remove the coal from the railroad cars. No simple method has been devised to unload this frozen coal satisfactorily. Heating with steam has been employed, but this increases the moisture content to a point where it causes difficulty in other parts of the system. Coal removed from the storage pile during snowy or rainy weather may also contain a high percentage of moisture. The wet coal adheres to the chutes and refuses to flow to the stoker or pulverizer mill. In many plants this has become a serious operating problem. Improvised methods of wrapping the pipe, using an air lance, etc., have proved ineffective in getting the coal to flow. In the design of the plant, the coal feed pipe from the bunkers to the stoker or mill should be as nearly perpendicular as possible with no bends or offsets. Access openings should be provided so that when stoppages do occur they can be relieved. Electric-driven vibrators have proved beneficial when the pitch of the chute is insufficient to promote flow. In some installations, hot air is passed through the coal pipe with the coal. The larger the coal size, the less water it will retain. Therefore, some relief from freezing in cars and the stoppage of chutes can be obtained by using a coarser grade of coal. This, however, usually means an increase in cost. The practice of "oil-treating" coal has proved quite effective in reducing the moisture pickup and accompanying difficulties.

Particle size is an important consideration in the selection of coal. The size requirement for different equipment varies widely. Coal burned on grates must have a certain size composition to regulate the passage of air through the fuel bed, while for a pulverized-fuel burner the coal must be reduced in size to small particles to promote rapid and complete combustion.

There is a tendency for coal particles to fracture when mechanical force is applied. Dense hard coals resist fracture and retain their size during handling. Soft coals shatter easily, break up into small particles when handled, and are, therefore, said to be "friable."

When coal, containing a range of sizes, is dropped through a chute or other coal-conveying equipment, the fine and coarse particles segregate. If this condition occurs in the coal supply to a stoker, the fine is admitted to one section and the coarse to another. The resistance of the fuel bed to airflow varies, resulting in different rates of airflow through the grates and subsequent variations in the rates of combustion. Coal-conveying systems for supplying stokers should be designed to prevent this segregation.

Fine coal particles have a greater tendency to retain moisture, and wet fine coal will not readily flow through chutes and spouts. The clogging of coal-conveying equipment owing to fine wet coal becomes so troublesome that coal with coarser particles must sometimes be selected. Once in the furnace there is a tendency for the combustion gases to carry the small particles of coal. This results in a loss of heat owing to the unburned carbon and objectionable particulate emission from the stack.

It is unsatisfactory to designate coal size by stating the largest and smallest pieces as "1¼ in. by 0 nut and slack." Some percentage of the various-sized particles should be given, such as 1¼ in. by ¼ in. with 15 per cent minus ¼. A more detailed size distribution is sometimes stated as follows:

Size Range, In.	Percentage
1½–1	10
1–¾	27
¾–½	15
½–⅜	20
⅜–¼	15
¼–0	13

A characteristic known as "grindability" is considered when selecting coal for pulverizer plants. The grindability of coal is tested and the results reported in accordance with the Hardgrove standard. A weighed, screened sample is placed in a laboratory test mill, and a given preestablished amount of energy is applied. The ratio of the fineness produced in the test sample to that produced when the same amount of energy was expended on a sample of standard coal is the Hardgrove value for the coal tested. Many bituminous coals have a Hardgrove rating between 50 and 60. A knowledge of the grindability of coal is used in the selection of pulverizer mills and in the procuring of satisfactory coal for a given pulverizer plant.

When coal for a given installation is being selected, a study should be made of the fusing or melting temperature of the ash which it contains. The exposure of low-fusing-temperature ash to the high-temperature zones of a fuel bed causes hard clinkers to form. These clinkers interfere with the movement of fuel in a stoker and make it difficult to clean a hand-fired grate. This results in an increased amount of carbon being removed with the ash, uneven fuel bed with increased excess air, and burned-out grates or stoker castings. In pulverized-fuel–fired furnaces and in some cases even with stokers, the molten ash is carried by the gases to the walls and into the boiler passes, where it accumulates in large quantities. The molten ash reacts and causes rapid deterioration of the refractory furnace lining. The accumulation of ash on the boiler tubes progresses until it bridges across from one tube to the other. This reduces the effective area of the gas passage, thereby restricting the flow. When the melting temperature of the ash is low, it is sometimes necessary to operate the boiler at reduced ratings or with increased excess air in order to lower the furnace temperature and so to prevent slag formation. The ability of the various commercial combustion equipments to burn coal with high- or low-fusion ash will be discussed in Chap. 4.

The amount of ash in coal should also be given consideration since it is an impurity which is purchased. It produces no heat and must be removed from the furnace and hauled from the plant. Since ash is an inert material, the heating value of coal can be expected to be low when the ash content is high.

A minimum of 4 to 6 per cent ash is required in coal burned on hand-fired grates and on some stokers. The layer of ash forms an insulator which prevents an excessive amount of heat from reaching the grates. When the ash content of the coal is too low, the grates become overheated and maintenance costs are increased. Combustion equipment is designed to use high-ash coal and to operate without difficulty and with a minimum of carbon loss in the ash. Therefore, when coals are selected, the minimum ash requirement to protect the grates must be given first consideration if this is a factor. Then the economics of coals with various ash contents must be considered. Coals having a low ash content can be secured either by selection or by processing. However, can this selection of low-ash coal be justified on an overall economic basis? With the employment of adequate combustion equipment many users find high-ash coal more economical.

When selecting coal and equipment used in the utilization of coal, the sulfur content must be considered. The corrosive effects of sulfur necessitate the use of special materials in the construction of conveyors and bunkers. The dew point of the flue gases is lowered by the pres-

ence of the sulfur oxides. The flue-gas temperature reduction in economizers and air heaters must be limited to prevent the metal temperatures
from being reduced to or below the dew point. This precaution is
necessary to prevent serious corrosion. Furthermore the sulfur oxides
discharged with the flue gases pollute the atmosphere.

Since it is seldom possible to schedule the arrival of coal shipments
to meet plant needs, it is necessary to maintain a storage. This storage
of coal frequently presents a problem in that the coal often overheats
and starts to burn. This self-excited burning is known as "spontaneous
combustion." Actually it is not entirely self-excited because the oxygen
is obtained from the air. At first the oxidation proceeds slowly at atmospheric temperature, but if the heat produced is not carried away, the
temperature increases. This higher temperature accelerates the process,
and the heat is generated more rapidly. When the ignition temperature
is reached, the coal starts to burn. Many tons of coal are wasted as
the result of spontaneous combustion.

Some of the precautions against spontaneous combustion may be listed
as follows:

1. Take the temperature of the coal by means of a thermocouple
inserted in a pointed pipe.

2. In outside areas, use a means of storage which will prevent segregation; tightly pack the coal to exclude air.

3. Prevent the direct radiation of heat from the boiler, etc., against
the side of bunkers by providing an insulating air space between the
inner and outer bunker walls.

4. Do not allow a "dead" storage of coal to remain in excess of one
month in an overhead bunker.

5. Should the temperature excede 150°F, inject carbon dioxide gas
at once, by use of a perforated pipe driven into the coal. The introduction of carbon dioxide should not, however, be considered a final answer,
and the coal must be moved as soon as possible.

The effect of the various characteristics of coal may be divorced
from the combustion process by converting the coal into fuel gas before
it is utilized in the boiler furnace. This procedure has the advantages
of providing a fuel, readily controllable, effecting a reduction in fouling
of heat-transfer surfaces and causing less contamination of the atmosphere. Moreover, gas-fired boilers are less costly to construct and to
operate than solid-fuel-fired boilers.

The economics of solid-fuel firing compared to gasification of solid
fuels involves many factors. The oldest and simplest method of conversion is by means of a gas producer. Air or oxygen is blown through a
preheated and usually agitated bed of crushed fuel. The resulting gas

then passes through a cleaning system to remove tar, particulate, and sulfur oxides. Producer gas contains 100 to 250 Btu per cu ft when air is forced through the fuel bed and 250 to 550 Btu per cu ft when oxygen is used. The heating value of this manufactured gas is low when compared to natural gas. This necessitates larger transmission lines and burners but does not result in a serious decrease in efficiency. The Btu content of this gas can be increased to above 900 Btu per cu ft by a process known as methanation.

3.5 FUEL OIL. The origin of petroleum has not been definitely established, but several theories have been advanced. The combination of carbon and hydrogen as they exist in petroleum could have been generated from either animal or vegetable matter. One theory holds that large amounts of some form of animal or vegetable matter, and perhaps both, have decayed and aged under pressure and in the presence of salt water. Those who advance this theory point to the fact that petroleum deposits are usually found in the presence of salt water. Another theory holds that petroleum was formed by inorganic substances, that is, without the presence of either animal or vegetable life. This theory is based on the fact that material containing hydrogen and carbon has been produced experimentally by the action of certain carbides and water. Petroleum is believed, by the advocates of this theory, to have been formed in the interior of the earth by the action of minerals, water, and gases.

Deposits of petroleum are found floating on subterranean lakes of salt water. A dome-shaped layer of nonporous rock holds the petroleum in place. The dome-shaped rock formation usually entraps a quantity of gas. The deposit consists of natural gas, petroleum, and water, which are separated from the rock down in the order named by virtue of their specific gravity. Oil is obtained by drilling through the layer of rock that covers the oil and gas. This procedure releases the gas, which is stored underground or transmitted through pipelines for use in distant cities. The oil is removed from the ground by a pumping operation which frequently fails to remove all the deposit.

The specific gravity of fuel oil is used as a general index of its classification and quality in much the same manner as the proximate analysis is used to specify the characteristics of coal. In determining the specific gravity of oil or any substance, the weight of a given volume of the substance is divided by the weight of an equal volume of water when both are measured at the same temperature.

Special laboratory determinations of fuel oil are made in terms of specific gravity, but in practical fieldwork the gravity is measured by a hydrometer and read in degrees Baumé or API. The API scale as adopted by the American Petroleum Institute is now generally accepted.

It differs slightly from the Baumé scale. The specific gravity as expressed in relation to water may be converted into degrees API by the use of the following formula:

$$°API = \frac{141.5}{sp\ gr\ at\ 60°/60°F} - 131.5$$

Since water has a specific gravity of 1, we find from this formula that it has an API gravity of 10. The API gravity of commercial fuel oil varies from 10 to 40.

Example A sample of fuel oil has a specific gravity of 0.91 at 60°F. What is its gravity in degrees API?

Solution

$$°API = \frac{141.5}{0.91} - 131.5 = 24.0$$

The viscosity or resistance of an oil to flow is important as it affects the ease with which it can be handled and broken up into a fine mist by the burner. An increase in temperature lowers the viscosity of an oil and causes it to flow more readily. It is therefore necessary to heat heavy oils in order to handle them effectively in pumps and burners.

Light oils contain a larger proportion of hydrogen than heavy oils. These light oils ignite easily and are said to have low flash and fire points. The flash point of an oil is the temperature at which the gases given off will give a flash when ignited. The fire point is the temperature at which the gases given off may be ignited and will continue to burn. An oil which has a low flash point will burn more readily than one with a high flash point.

The heating value of fuel oil is expressed in either Btu per pound or Btu per gallon. The commercial heating value of fuel oil varies from 18,000 to 19,500 Btu per lb. The calorimeter provides the best means of determining the heating value of fuel oil. When the gravity in API degrees is known, the heating value may be estimated by use of the following formula:

Btu/lb of oil = 17,687 + 57.7 × API gravity at 60°F

Example If a sample of fuel oil has an API gravity of 25.2 at 60°F, what is the approximate Btu per pound?

Solution

Btu/lb of oil = 17,687 + 57.7 × 25.2 = 17,687 + 1,454 = 19,141

Some of the impurities in fuel oil which affect its application and should be determined by analysis are ash, sulfur, moisture, and sediment.

There is a wide variation in the composition and characteristics of the fuel oil used for the generation of heat.   Practically any liquid petroleum product may be used if it is feasible economically to provide the necessary equipment.

Fuel oils may be classified according to their source as follows:

1. *Residual oils* are the products which remain after the more volatile hydrocarbons have been extracted.   The removal of these hydrocarbons lowers the flash point and makes the oil safe for handling and burning. This residual oil is usually free from moisture and sediment except for that which is introduced by handling and in transit from refinery to consumer.

2. *Crude petroleum* is the material as it comes from the oil well without subsequent processing.   Sometimes, owing to exposure to the weather or for other reasons, the crude petroleum is too low in quality to justify refining, and it is then used as fuel under power boilers.   Since it may contain some volatile gases, it must be handled with care. Present-day refinery practices make it possible to recover some high-quality product from almost all crude petroleum.   Therefore only a small quantity of this material is now available for use as fuel oil.

3. *Distillate oils* are obtained by fractional distillation and are of a consistency between kerosine and lubrication oils.   Fuel oil produced in this manner does not contain the heavy tar residue found in others. The light grades of fuel oils are produced in this manner.

4. *Blended oils* are mixtures of two or all of the above, in proportions to meet the desired specifications.

For commercial purposes fuel oils are divided into six classes according to their gravity (Table 3.5).   The No. 5 oil is sometimes referred to as "bunker B" and the No. 6 as "bunker C," although bunker C also frequently refers to oil heavier than 14°API.

Fuel oil is used for the generation of heat in preference to other fuels when the price warrants the expense or when there are other advantages which outweigh an unfavorable price difference.   In some instances the cleanliness, ease of handling, small space requirements, low cost of installation, and other advantages derived from the use of fuel oil compensate for a considerable price difference when compared with solid fuels.   The fact that oil can be fired automatically and that there is no ash to be removed makes its use desirable.   The location of the plant with respect to the source of fuel supply and the consequent freight rates are determining factors in the economic choice of fuels. Many plants are being equipped to burn two or more kinds of fuel in order that they may take advantage of the fluctuating costs.

In burning fuel oil, it is essential that it be finely atomized to ensure mixing of the oil particles with the air.   To accomplish this it is neces-

TABLE 3.5 Commercial Classification of Fuel Oil

Fuel-oil No.	Designation	API	Btu/gal
1	Light domestic	38–40	136,000
2	Medium domestic	34–36	138,500
3	Heavy domestic	28–32	141,000
4	Light industrial	24–26	145,000
5	Medium industrial	18–22	146,500
6	Heavy industrial	14–16	148,000

sary to have pumping equipment for supplying the oil and a burner suitable for introducing the fuel into the combustion chamber.

The heavy grades of fuel oil cannot be pumped or atomized properly by the burner until the viscosity has been lowered by heating. The storage tanks which contain heavy oil must be equipped with heating coils and the temperature maintained at 100 to 120°F to facilitate pumping. This oil must then be passed through another heater before it goes to the burners. With intermediate grades, the tank heater is not necessary as the required temperature for atomization is obtained by the heater in the oil lines. The temperature to which the fuel oil should be heated before it enters the burners varies with different equipment, but the generally accepted practice is as follows:

Fuel-oil No.	Temperature at Burner, °F
4	150
5	175
6	275

Fuel oil provides a satisfactory source of heat for many types of services when the correct equipment is used and good operating practices are followed. The ash content is usually low, but even a small amount may react with the refractory furnace lining and cause rapid deterioration. Therefore, the flame should be adjusted to prevent impingement on the furnace walls. Moisture, emulsified oil, abrasive particles, and other foreign matter, referred to collectively as "sludge," settle to the bottom of the storage tank. When this sludge has accumulated in sufficient quantity to be picked up by the oil pump, strainer and burner nozzles become stopped up and abrasive material causes pump and burner wear.

Sulfur is another objectionable constituent of fuel oil. Some of the sulfur compounds formed during the combustion process raise the dew point of the flue gases, mix with the moisture present in the gases, and corrode metal parts. This corrosive action results in plugging of gas passages and rapid deterioration of boiler tubes, casings, and especially economizers and air heaters.   Furthermore, the sulfur compounds discharged with the flue gases contaminate the atmosphere.   The reduction of this source of contamination is the subject of much concern among air-pollution-control authorities.

The corrosive action is reduced by maintaining the flue gases at a temperature higher than the dew point.   The gases must be permitted to discharge at a higher temperature, thus carrying away more heat and reducing the efficiency.

Burner and accessory equipment capable of operating with very low percentages of excess air provide another method of combating the problems encountered in burning fuel oil having a high sulfur content.   For a fuel oil of given sulfur content, the dew point of flue gases can be lowered by operating with a low percentage of excess air.   This method also increases the efficiency, and the contaminates discharged to the atmosphere are less objectionable.

3.6   GAS.   Natural gas usually occurs in the same region as petroleum.   The gas, owing to its low gravity, is found above the petroleum and trapped by a layer of nonporous rock.   Owing to the fact that natural gas and petroleum are usually found together, it is believed that they have a similar origin.   The gas in its natural state is under pressure which causes it to be discharged from the well.

Manufactured substitutes for natural gas may be produced from solid or liquid fuels.   These gases are produced for special industrial application and for domestic use.   Owing to their cost as compared with other fuels, they are seldom used under power boilers.

Producer gas is manufactured from coke or coal.   Steam and air are blown up through the incandescent fuel bed.   Both carbon dioxide and carbon monoxide are generated here and pass up through what is termed a "reduction" zone.   Here the absence of oxygen results in the oxygen's combining with additional carbon to form carbon monoxide.   As the gases pass through this zone, they encounter a heavy layer of green coal.   The volatile gases are distilled from this coal and, together with the carbon monoxide obtained from the region below, pass from the top.   The combustibles are the carbon monoxide, hydrogen, and small amount of volatile gases.   The heating value of this gas is slightly above that of blast-furnace gas.

Water gas and oil gas are manufactured for industrial applications where a gaseous fuel is required and for domestic supply, sometimes being used to augment the natural-gas supply.   These gases are too expensive to use under power boilers unless the unit is small and full automatic control is necessary.

An analysis of gases consists of expressing the chemical compounds of which it is composed either in percentage by volume or in weight. The combustible constituents are composed of various combinations of hydrogen and carbon known as "hydrocarbons." Also there are various inert gases such as carbon dioxide and nitrogen. The chemical analysis of a gas is useful in determining the air required for combustion and in calculating the products of combustion, the heating value, the type of burners required, etc.

The heating value of gas is determined by burning a sample in a calorimeter. This is an instrument in which a given quantity of gas

TABLE 3.6   Analysis of Fuel Gases (by Volume)

Gas	Natural	Producer	By-product coke-oven	Blast-furnace
CH_4...............	83.5	2.5	29.1	....
C_2H_6................	14.4	....	....	....
C_2H_4................	....	0.5	3.1	....
CO_2................	.....	5.5	1.7	12.7
CO................	.....	22.3	4.9	26.7
Nitrogen.............	2.1	59.0	4.4	57.1
Hydrogen............	....	10.2	56.8	3.5
Btu/cu ft at 60°F and atmospheric pressure (higher)............	1080	140	550	98.1

is burned and the resulting heat transferred to a measured quantity of water. Unlike a solid- or liquid-fuel calorimeter, the gas calorimeter utilizes a constant measured flow of gas and water. The heat generated by the combustion of gas is absorbed by the cooling water. The temperature rise of the water is an index of the heat produced and, by use of suitable constants for a given calorimeter, may be converted into Btu per cubic foot of gas.

Gas used for fuel is classified according to its source or method of manufacture. Table 3.6 gives typical analyses of natural and more generally used manufactured gases.

In order for gas to be burned efficiently it must be mixed with the correct proportion of air. Since the fuel is already in the gaseous state, it is unnecessary to break it up, as is the case with solid and liquid fuels. Correct proportioning and mixing are, however, essential. This process is accomplished by the burner and by admission of secondary air. Combustion must be complete before the mixture of gases reaches the boiler heating surfaces. With satisfactory furnace and burner design and with careful operation, it is possible, if mixing is correctly accomplished, to obtain the complete combustion of gas with a very low percentage of excess air.

Gas is a desirable fuel because it is easy to control, requires no handling equipment, and leaves no ash to remove from the furnace.

3.7 BY-PRODUCT FUELS. Industrial processes frequently produce materials which may be burned in industrial furnaces. The utilization of these by-products not only reduces the cost of fuel but frequently solves a disposal problem at the same time. However, the economic advantage to be derived from the use of these fuels must be evaluated in terms of added investment for equipment and of increased operating costs. By-product fuel is seldom available in quantities sufficient to meet the total steam demand. It is necessary, therefore, to provide for a supplementary fuel.

Wood is available as a by-product fuel in sawmills, paper mills, and factories which manufacture articles made of wood. The sawmill waste products consist of slabs, edgings, trimmings, bark, sawdust, and shavings in varying percentages. This material is passed through a hogging machine which reduces it to shavings and chips in preparation for burning. A hogging machine consists of knives or hammers mounted on an element which rotates within a rigid casing. The knives shred the wood waste and force it through a series of spacer bars mounted in a section of the casing. Dry wood waste has a heating value of 8000 to 9000 Btu per lb. This shredded material (hog fuel) contains approximately 50 per cent moisture and, therefore, has a low calorific value. Woodworking plants that manufacture furniture and similar articles use seasoned wood. The moisture content of the refuse material is from 20 to 25 per cent. This waste material consists of sawdust and shavings and only a small amount of trimmings. Precautions are necessary in handling, storing, and stoking the furnace to prevent fires and explosions.

Bark is removed from logs before they are ground into pulp in paper mills. This bark may contain 80 per cent moisture, a part of which must be pressed out before it can be utilized for fuel.

A by-product solution (black liquor) containing wood fibers and residual chemicals is produced in some paper mills. The wood fibers are burned, and the chemicals are recovered as ash in the furnace. In this operation a saving is effected both by the reclamation of heat and by the recovery of the chemicals. The black liquor is concentrated before being introduced into the furnace, where it is atomized and burned in a manner similar to fuel oil.

Bagasse is a by-product material produced when the juice is removed from the cane in a sugar mill. The moisture content is 40 to 45 per cent, and the heating value 4500 to 5200 Btu per lb. The utilization of this by-product material is an important factor in the economic operation of a sugar mill. Special furnaces have been developed to provide a means of utilizing the heat from the furnace to evaporate some of the moisture before attempting to burn the fuel. Bagasse is frequently supplemented with auxiliary fuels.

Municipal solid waste is produced in our cities in very large quantities. This waste material consists of paper, plastics, and miscellaneous combustibles combined with noncombustibles, including metal and glass. The nature of this material varies in different locations and seasonally in a given location. The glass and metal have reclaim value, while the combustible portion has value as fuel. Each pound of waste has the ability to produce from 2 to 3 lb of steam. However there are numerous problems involved in handling and burning this material. The noncombustibles are difficult to separate, and even with the best methods a considerable amount remains to pass through the furnace as refuse. Some of the materials present, plastic for example, are converted into vapors and gases which cause corrosion of heating surfaces.

Coke-oven gas is the volatile material removed when coal is heated to produce coke. These volatile gases are collected and burned as a by-product of the coke production. Coke-oven gas has a heating value of 460 to 650 Btu per cu ft.

Blast-furnace gas is produced as a by-product of steel-mill furnaces. The air is blown up through the hot fuel bed of the blast furnace, where it combines with carbon to form carbon monoxide. The resulting gases contain carbon monoxide and small quantities of hydrogen as combustibles in addition to inert gases consisting of nitrogen and carbon dioxide. The heating value varies from 90 to 110 Btu per cu ft. Owing to the irregular production of this gas and the fact that the steam demand does not coincide with the gas production, gas storage is necessary and supplementary fuel is required.

Refinery gas is a by-product of crude-oil refining operation. The heating value of this gas is high owing to the large amount of heavy hydrocarbons present, and it will vary from 1000 to 1800 Btu per cu ft, depending upon the refining process. The gas produced is frequently adequate to generate the steam required by the refinery. Refineries also produce petroleum coke, which has a high carbon but low hydrogen and ash content. This fuel is used by the refinery or sold for use in neighboring plants.

3.8 CONTROL OF THE COMBUSTION PROCESS. Each unit of fuel contains a given amount of heat in the form of chemical energy. The amount of this energy is readily determined in the laboratory and is expressed in Btu per unit. The burning of fuel and the subsequent operation of the boiler make it necessary to convert the chemical energy into heat, apply it to the water in the boiler, and thereby generate steam.

Fuel supplied to the furnace of a power boiler may be rejected unburned, lost as heat, or absorbed by the boiler. The effectiveness of the process can be determined by noting the losses. When the losses have been reduced to a practical minimum, the highest efficiency is being maintained.

When solid fuel is burned, a certain portion of the combustible carbon becomes mixed with the ash and is removed without being burned. The mixture of carbon and ash discharged from a furnace is correctly referred to as "refuse." The magnitude of this loss is determined by noting the reduction in weight when a sample of refuse is completely burned in a laboratory muffle furnace.

The following formula may be used to estimate the Btu discharged with the refuse:

$$h_1 = 14,540 \times \frac{ab}{100\,(100 - b)}$$

where a = percentage of ash in coal
b = percentage of combustible in dry refuse
h_1 = loss, Btu per lb of coal

One accustomed to operating a furnace can estimate the amount of combustible in the refuse by observation. When large quantities of combustible are being discharged, pieces of coke will be mixed with the ash. The procedure for reducing the "carbon loss" to a minimum is discussed in Chap. 4.

It has been shown that water is formed during the combustion of hydrogen. In the conventional boiler, the gases are discharged to the stack at a temperature in excess of the boiling point of water ($212°F$), and the water therefore leaves in the form of vapor or steam. A part of the heat produced by the combustion of the fuel is utilized in vaporizing the water and is therefore lost with the stack gases. This heat loss is high when fuels containing large amounts of hydrogen are burned. A similar loss occurs in the evaporation of moisture contained in the fuel. The operator cannot do much to reduce the loss except to have the fuel as dry as possible.

By far the largest controllable or preventable loss occurs in the heat rejected from the stack in the dry products of combustion. These losses occur in three different ways as follows:

1. Combustible gases, mainly carbon monoxide, contain undeveloped heat. This loss can be determined by the following formula:

$$h_2 = C \times \frac{10,160\,CO}{CO_2 + CO}$$

where h_2 = heat loss, Btu per lb of fuel burned as result of carbon monoxide in flue gases

  CO = percentage by volume of carbon monoxide in flue gases

  CO_2 = percentage by volume of carbon dioxide in flue gases

  C = weight of carbon consumed per lb of fuel

2. The temperature of gases discharged from the boiler is normally higher than that of the incoming air and fuel. The heat required to produce this increased temperature is lost. The magnitude of this loss can be decreased by maintaining the lowest possible stack temperature.

3. The amount of gases per unit of fuel discharged from the boiler depends upon the percentage of excess air used. The more gas discharged, the greater will be the heat loss. In the interest of economy, the excess air should be kept as low as practicable. The relation of the carbon dioxide content of flue gases to heat loss for several flue-gas temperatures is given in Fig. 3.11.

The heat carried to waste by the dry gases may be calculated by using the following formula:

$$h_2 = w(t_s - t)0.24$$

where h_3 = heat loss, Btu per lb of fuel

  w = weight of dry gases, lb per lb of fuel

  t_s = temperature of stack gases, °F

  t = temperature of air entering furnace, °F

  0.24 = specific heat of flue gases

This heat loss can, therefore, be reduced by lowering the stack temperature or by decreasing the quantity of gases discharged, thus lowering the percentage of excess air used.

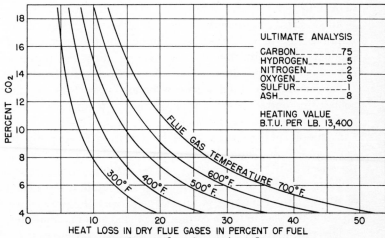

FIG. 3.11 *Dry-flue-gas loss for various flue-gas temperatures and per cent of carbon dioxide.*

FIG. 3.12   *Model 621A portable orsat gas analyzer.* (*Milton Roy Company, Hays Republic Div.*)

Conventional oil and gas burners controlled by standard systems require 10 to 20 percent excess air for complete combustion. However, "low-excess-air" burners and accessory control equipment are available for reducing the air necessary for combustion to nearly the theoretical requirement. Advanced methods of atomizing oil are employed. The temperature of the oil supplied to the burners is carefully controlled. Both the fuel and air supply to the burners are metered. This low-excess-air operation can be obtained with a clear enough stack to meet stringent ordinances and with not more than a trace of carbon monoxide.

The necessary equipment for low-excess-air operation is more costly to install and requires more precise adjustment and care in operation than standard equipment. The reduction in excess air results in an improvement in efficiency, a reduction in the corrosive effect of the products of combustion on the heating surfaces, and a reduction in objectionable contaminants (including nitrogen oxides).

The measure of stack losses is based on analysis of the products of combustion and a determination of the exit temperature of the gases. A commercial gas analyzer is shown in Fig. 3.12.

A sample of gas is obtained by inserting a sampling tube into the boiler pass. A ¼-in. standard pipe is satisfactory if the temperature of the gases in the pass does not exceed 1000°F. Care must be exercised to locate the end of the sampling pipe to obtain a representative sample of gases. It is advisable to take samples with the pipe located at various places in the pass to make certain that the analysis is representative.

The procedure for operating the gas analyzer in Fig. 3.12 is shown diagrammatically by steps in Fig. 3.13.

Step 1. The inlet tube is carefully connected to the sampling tube to prevent air leakage from diluting the sample. The gas is then drawn into the measuring burette by means of the rubber bulb. The excess is expelled through the level bottle. The aspirator bulb is operated long enough to assure a fresh sample of gas in the burette.

Step 2. The handle of the three-way cock is then moved to a vertical position, closing off the sampling line and opening the burette to the atmosphere. The sample is measured by bringing the water level in the measuring chamber exactly to zero. The excess gas is discharged through the three-way valve to the atmosphere, and the amount remaining is measured at atmospheric pressure to ensure accuracy.

Step 3. The handle of the three-way cock is next moved away from the operator to a horizontal position, entrapping the sample of gas in the measuring chamber. The valve above the absorption chamber is now opened and the leveling bottle raised to force the gas sample into contact with the chemical which absorbs the carbon dioxide. Care must be exercised in this operation to prevent the chemical and water from becoming mixed.

It is necessary to raise and lower the leveling bottle and expose the gas to the chemical from two to four times to ensure complete absorption of the carbon dioxide.

Step 4. The remainder of the gas sample is drawn back into the measuring chamber, and when the chemical level in the absorption chamber has been restored to zero, the valve closes. The leveling bottle is held with the water at the same level as that in the measuring chamber. The percentage of carbon dioxide in the sample is then read directly on the etched wall of the measuring chamber.

In order to obtain the percentage of oxygen and carbon monoxide respectively, the same sample of gas is exposed to the chemicals in the second and third absorption chambers of the gas analyzer. See Fig. 3.13 for the step-by-step procedure.

To determine the percentage of excess air, as explained in Sec. 3.2, and for heat-balance calculations, it is necessary to make a complete flue-gas analysis. For routine check on operation, when there is no change in the character of the fuel, carbon dioxide determination is sufficient for comparison with established satisfactory readings. The carbon monoxide absorption chamber has been omitted from the gas analyzer shown in Fig. 3.12. The solutions have a limited gas absorption ability and must be replaced after continued use or long periods of storage. They may be purchased mixed ready for use or compounded using standard chemicals.[1]

[1] ASME PTC 19.10 Part 10, Flue and Exhaust Gas Analysis—Instruments and Apparatus.

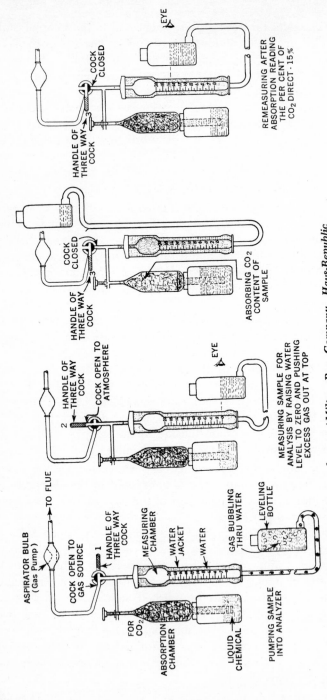

ASPIRATOR BULB
(Gas Pump)

TO FLUE

COCK OPEN TO
GAS SOURCE

HANDLE OF
THREE WAY
COCK

1

FOR CO₂

ABSORPTION CHAMBER

LIQUID CHEMICAL

MEASURING CHAMBER

WATER JACKET

WATER

GAS BUBBLING
THRU WATER

LEVELING BOTTLE

PUMPING SAMPLE
INTO ANALYZER

HANDLE OF
THREE WAY
COCK

COCK OPEN TO
ATMOSPHERE

2

EYE

MEASURING SAMPLE FOR
ANALYSIS BY RAISING WATER
LEVEL TO ZERO AND PUSHING
EXCESS GAS OUT AT TOP

COCK
CLOSED

HANDLE OF
THREE WAY
COCK

3

ABSORBING CO₂
CONTENT OF
SAMPLE

COCK CLOSED

HANDLE OF
THREE WAY
COCK

3

EYE

REMEASURING AFTER
ABSORPTION READING
THE PER CENT OF
CO₂ DIRECT · 15%

FIG. 3.13 *Operation of a gas analyzer.* (*Milton Roy Company, Hays-Republic Div.*)

Gas analyzers are available which greatly simplify the procedure for determining the percentage of either carbon dioxide or oxygen (see Fig. 3.14). The instrument must be supplied with the absorbent solution for either carbon dioxide or oxygen. The oxygen determination is pre-

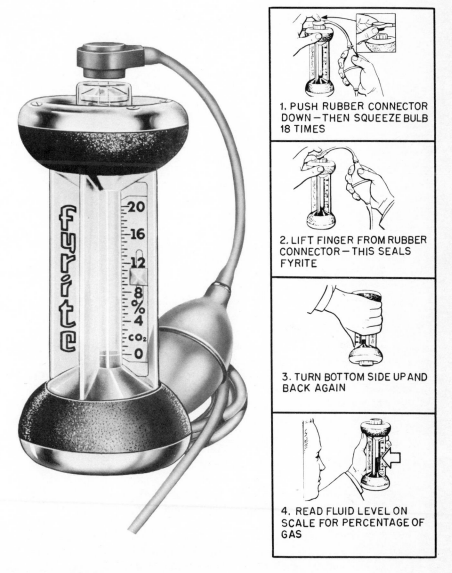

1. PUSH RUBBER CONNECTOR DOWN—THEN SQUEEZE BULB 18 TIMES

2. LIFT FINGER FROM RUBBER CONNECTOR—THIS SEALS FYRITE

3. TURN BOTTOM SIDE UP AND BACK AGAIN

4. READ FLUID LEVEL ON SCALE FOR PERCENTAGE OF GAS

FIG. 3.14 *Gas analyzer and operating procedure.* (*Bacharach Instrument Company Div. AMBAC Industries, Inc.*)

ferred since it is less influenced by a change in fuel. The diagrams in Fig. 3.14 show the four steps required to determine the oxygen content of flue gases with this analyzer.

The use of automatic recording, carbon dioxide, oxygen or steam flow-air flowmeters is recommended. They are a valuable aid to the operator in maintaining the optimum combustion conditions at all times.

Typical Example of Results Obtained by Use of a Gas Analyzer

Absorption chamber to which gas was exposed......................	Right	Center	Left
Reading on measuring chamber indicates.	CO_2	$CO_2 + O_2$	$CO_2 + O_2 + CO$
Example............................	13.7	17.2	19.0

From these readings the analysis would be totaled as follows:

$$CO_2 \qquad\qquad = 13.7 \text{ per cent}$$
$$O_2 = 17.2 - 13.7 = 3.5 \text{ per cent}$$
$$CO = 19.0 - 17.2 = 1.8 \text{ per cent}$$
$$N_2 = 100.0 - 19.0 = 81.0 \text{ per cent}$$

QUESTIONS AND PROBLEMS

3.1. List the conditions necessary for good combustion.

3.2. What gases are normally found in the combustion space directly above the fuel bed of a hand-fired furnace?

3.3. Why is secondary air necessary in a hand-fired furnace?

3.4. Discuss the causes of smoke.

3.5. What are the four states of matter?

3.6. Which of the gas laws is most important in analyzing combustion problems?

3.7. The ultimate analysis of coal is as follows: carbon, 60 per cent; hydrogen, 7 per cent; nitrogen, 3 per cent; oxygen, 12 per cent; sulfur, 5 per cent; and ash, 13 per cent. How many pounds of oxygen would be required to burn this fuel? How many pounds of air (theoretically) would be required?

3.8. Define perfect, complete, and incomplete combustion.

3.9. What is meant by the expression "excess air"?

3.10. If the flue-gas analysis shows CO_2, 15.2 per cent; O_2, 4.7 per cent; and N_2, 80.1 per cent, what is the percentage of excess air?

3.11. Explain two methods of indicating or measuring the percentage of excess air.

3.12. How many pounds of air would be required to burn a pound of coal with the analysis given in Prob. 3.7 if 50 per cent excess air were used?

3.13. Using the analysis in Prob. 3.7 and Dulong's formula, determine the heating value of the fuel.

3.14. What is the purpose of draft in connection with the combustion of fuel?

3.15. Why must the draft be carefully controlled?

3.16. In what units is draft expressed?

3.17. What is balanced draft?

3.18. What are the three ways of expressing coal analysis?

3.19. Discuss the characteristics to be considered when buying coal.

3.20. List and discuss the objections to sulfur in fuels.

3.21. What are the important characteristics to be considered when selecting fuel oil for a given plant?

3.22. Under what condition must fuel oil be heated?

3.23. A fuel oil has a specific gravity of 0.92. What is its gravity on the API scale?

3.24. Discuss the various gaseous fuels sometimes used in firing power boilers.

3.25. Discuss the four most important losses which occur in the operation of a power boiler.

3.26. How can the dry-gas loss be reduced?

3.27. What precautions are necessary when analyzing flue gases?

3.28. What losses might be taking place even when the stack is clear?

3.29. What are the results when a sample of air is analyzed with a flue-gas analyzer?

3.30. Why is too much excess air objectionable?

CHAPTER 4

SETTINGS,
COMBUSTION EQUIPMENT,
AND
HEATING SURFACES

In order to apply and utilize the principle of combustion it is necessary to design, construct, and operate physical equipment. A wide variety of equipment is required to carry out the process involved in satisfactorily burning fuels. The functions that must be performed may be listed as follows: supplying and mixing fuel and air, confining the high temperature during combustion, directing the heat to the absorbing surfaces of the boiler and heat-recovery equipment, and discharging the ash and products of combustion from the unit at a contaminate emission rate within the limits imposed by environmental controls.

4.1 BOILER SETTINGS. The setting includes the structure that forms the furnace and encloses the boiler proper. The boiler may be either supported by the brickwork setting or independently suspended from overhead steel beams. Figure 1.3 shows a typical horizontal-return tubular boiler supported by the brickwork. Note that the boiler rests on rollers so that it can expand independently of the setting. The overhead suspension is considered best because it relieves the brickwork of extra load, keeps the boiler at the proper level, and permits repairs to the brickwork without disturbing the boiler. The drums and upper sections of water-tube boilers are supported from the steelwork in this manner. The tube sections are then free to expand downward without interfering with the setting.

At least 50 per cent of the combustion of soft coal consists of burning gases distilled from the fuel and takes place in the space above the fuel bed. It is therefore necessary that the space above the grates be large enough to ensure complete combustion before the gases reach the heating surface.   The combustible gases rising from the fuel bed are composed of hydrocarbons and carbon monoxide.   Air must be admitted above the fuel bed to burn these gases.   If they reach the heating surface before combustion has been completed, they become cooled. The result is smoke, soot, and poor efficiency.   The setting must provide not only the proper furnace volume but also bridge walls, baffles, arches, etc., for properly directing the flow of gases with respect to the heating surfaces.

The *bridge wall* is located behind the grates.   It keeps the fuel bed in place on the grates and deflects the gases against the heating surface.

Baffles direct the flow of gases and ensure maximum contact with the heating surface.

Arches are in reality roofs over parts of the furnace.   They are made of refractory material and are used to direct the flame and to protect parts of the boiler from the direct heat of the furnace.

In the "flush-front-type" HRT boiler setting the front tube sheet sets back from the front of the boiler casing (Fig. 1.3).   The gases leave the tubes and flow through this space to the breeching or smoke pipe.   A small arch is required at the front of the furnace to seal off this space and prevent the furnace gases from short-circuiting the boiler and going directly from the furnace to the stack.

Other horizontal-return tubular boilers (Fig. 1.4), known as the "extended-front type," have their front tube sheets set in line with the front of the boiler setting.   The bottom half of the shell extends beyond the tube sheet and forms part of the breeching.   This part of the shell that extends past the tube sheet has hot gases on one side but no water on the other and is called the "dry sheet."

In some designs, the combustion space is enclosed within the boiler and no setting is required.   The combustion space is surrounded by two sheets of boiler plate, held together by means of stay bolts.   The space between the plates is filled with water and really is part of the boiler.   Figures 1.1 and 1.2 show a vertical boiler with firebox enclosed within the boiler.   This self-contained feature makes the unit semiportable.

Early boilers were set very close to the grates, and the combustion space was limited.   The result was smoke and poor efficiency.   As more was learned about combustion of fuels, furnaces were made larger by setting the boilers higher above the grates.   A hand-fired furnace with large combustion space will burn a wide variety of coal and operate over a wider range of ratings at higher efficiencies than one with a small combustion space.

Large-capacity stokers operate at high burning rates and heat release. This high-capacity operation requires furnaces having large volume to ensure complete combustion before the gases reach the heating surfaces and to prevent excessive ash deposits from forming on the lower boiler tubes. Furthermore, severe conditions imposed by high rates of combustion make it necessary to use water-cooled furnace walls and arches in preference to refractory construction.

Arches are installed in furnaces to protect parts of the boiler from the direct heat of the furnace and to deflect the flow of gases in the desired path. Ignition arches provide surfaces for deflecting the heat of the furnace to ignite the incoming coal. Advanced furnace design has reduced the requirement for arches. However, some spreader stokers and chain-grate stokers use ignition arches in connection with over-fire air jets to assist in igniting the coal and to reduce the emission of smoke. These arches are usually constructed of boiler tubes with refractory covering on the furnace side. The refractory surface permits the desired high surface temperature required to ignite the coal, while the water-cooled construction results in low maintenance.

In hand-fired furnaces and those equipped with underfeed or chain-grate stokers a considerable portion of the coal is burned on the grates. However, the proportion depends upon the percentage of volatile in the coal. When coal is introduced into a high-temperature furnace, the volatile is separated to form a gas containing hydrocarbons which must be burned in suspension. Coals which contain a low percentage of volatile matter burn on the grates. The furnace volume can, therefore, be less than when low-volatile coals are used, but a larger grate area must be provided. When fuels are burned in suspension, furnace volumes must be larger than when a portion of the fuel is burned on the grates.

Pulverized coal, oil, and gas are examples of fuels burned in suspension. The spreader stoker employs a combination of suspension and on-grate burning. Fine coal particles and high volatile increase the percentage burned in suspension, while large particles and low volatile reduce the amount of fuel burned in suspension and, therefore, increase the amount that must be burned on the grates.

Furnaces may be designed to maintain the ash below or above the ash-fusing temperature. When the ash is below the fusing temperature, it is removed in dry or granular form. When low-fusing-ash coal is burned, it is difficult to maintain a furnace temperature low enough to remove the ash in the granular state. Pulverized-coal and cyclone furnaces are operated at temperatures high enough to maintain the ash in the liquid state until it is discharged from the furnaces. These units are referred to as "intermittent" or "continuous" slag-tapped furnaces, depending upon the procedure employed in removing the fluid ash from the furnace.   Since this process permits high furnace temperatures, the excess air can be reduced and the efficiency increased.

High rates of combustion, suspension burning, large furnace volume, and liquid-state ash discharge are all factors which increase the severity of service on the furnace lining.   Refractory linings are adversely affected by high temperature and the chemical action of the ash.   Therefore, walls and arches are constructed of tubes which form a part of the boiler.   A large percentage of the heat required to generate the steam is absorbed in these tubes, thus reducing the amount of heating surface required in the convection banks of the boiler.

4.2   HAND FIRING.   Although hand firing is seldom used, it is presented to explain a simple application of combustion fundamentals.   The grates (Fig. 4.1) serve the two-fold purpose of supporting the fuel bed and admitting air.   The front end of the grate bars is supported from the dead plate and the rear end by the bridge wall.   The grate bars are not set level but are inclined toward the bridge wall as much as 1½ in. per ft of length.   This is done to aid the fireman in getting the fuel to the

Shaking grate

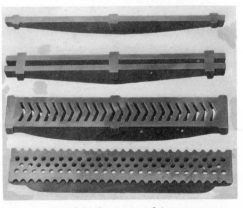

No. 9 grate, 1 in. wide

No. 8 grate, 3 in. wide

Tupper grate for coal, 6 in. wide

Sawdust grate, 6 in. wide

FIG. 4.1   *Hand-fired grates and furnace parts.*
(Combustion Engr. Co. Inc.)

rear of the furnace and to carry a slightly heavier fire near the bridge wall, where the effect of the draft is greatest.

Hand-fired grates are made of cast iron. They vary in design to suit the type of fuel to be burned (Fig. 4.1). The air openings in the grates vary from $\frac{1}{8}$ to $\frac{1}{2}$ in. in width. Larger air openings can be used for burning bituminous than for burning anthracite coal. The bridging action of bituminous coal prevents it from sifting through the grates. Grates designed to burn bituminous coal have air openings equal to 40 per cent of the total grate area, while those intended for anthracite have air openings equal to but 20 per cent.

The width of the boiler limits the width of the furnace. The length of the furnace is limited to 6 or 7 ft by the physical ability of the fireman to handle the fire. The actual grate area required for a given installation depends upon its heating surface, the rating at which it is to be operated, and the kind of fuel to be used.

Shaking grates, as shown in Fig. 4.1, are an improvement over stationary ones. The ash may be discharged into the ashpit without greatly disturbing the operation of the boiler. The agitation of the fuel bed by the grates helps to keep it even and to prevent holes.

To start a fire on hand-fired grates, cover them first with coarse and then with fine fuel to a total depth of 3 to 4 in. Cover this coal with a layer of wood and shavings. Place oily rags or waste on the wood, and light. Partly open the stack damper and ashpit door to permit the air to circulate through the fuel. By this procedure the coal starts to burn from the top, and the volatile gases that are distilled off must pass through the hot fuel bed. They ignite and burn, thus reducing to a minimum the amount of smoke produced. Coal should be added only after the entire fuel bed has been ignited.

The *coking method* of maintaining the fire consists of placing the fresh coal on the dead plate and allowing it to be exposed to the heat of the furnace. The volatile matter is distilled off, mixed with the air entering the furnace through the door, and burned. After coking in this manner the coal is pushed back over the active area of the grates. This method is satisfactory if the coal is not of the caking kind and if sufficient coking time is permitted. If the coal cakes and runs together, the volatile matter in the interior will not be distilled off, and excessive smoking will result when the coal is pushed over the hot fuel bed. The coking process is slow because only one side of the coal pile is exposed to the heat of the furnace.

Some advantages are claimed for the *alternate method* of firing. The green coal is applied evenly to one side of the furnace at a time. The other side is not disturbed; so the heat generated by it helps to burn the green coal. In fact just after the coal has been applied, the volatile gases may be distilled off so rapidly that they cannot be properly burned.   If this is the case, smoke will be produced.

Another method is to *spread* the coal over the fuel bed evenly or in such a manner that holes are eliminated.   The coal comes into direct contact with the hot fuel bed, and the volatile gases are quickly distilled off.   If smoke is to be prevented, the coal must be fired in relatively small quantities.   Care must be exercised to admit sufficient air over the fire to burn these volatile gases.   Any tendency of coal to clinker is likely to be increased by this method of firing because the ash is exposed to the direct heat of the burning fuel.   Caking coals will form a surface crust and cause unequal distribution of air.   This is especially true if too much green coal is added at one time.

Hand firing has been almost completely replaced by mechanical devices for introducing the fuel into the furnace.   These include chain-grate, traveling-grate, underfeed, and spreader stokers, pulverized coal, and cyclone burners.   Consideration will be given to these methods of utilizing solid fuels.

4.3   CHAIN- AND TRAVELING-GRATE STOKERS.   The *chain-grate stoker* employs an endless chain which is constructed to form a support for the fuel bed.   The *traveling-grate stoker* also employs an endless chain but differs in that it carries small grate bars which actually support the fuel bed.   (See Figs. 4.2, 4.3.)   In either case the chain travels over two sprockets, one at the front and one at the rear of the furnace.   These sprockets are equal in length to the width of the furnace.   The front sprocket is connected to a variable-speed driving mechanism.   The air openings in the grates depend upon the kind of coal burned and vary from 20 to 40 per cent of the total area.

Coal is fed by gravity from a hopper located in the front of the

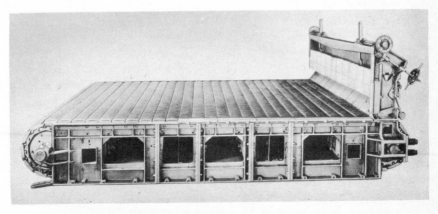

FIG. 4.2   *Traveling-grate stoker showing air compartments and coal grate.*   (*Combustion Engr. Co. Inc.*)

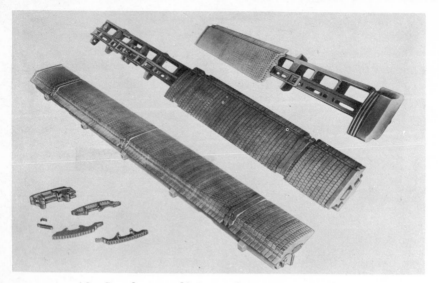

FIG. 4.3 *Grate-bar assembly in traveling grates.* (*Combustion Engr. Co. Inc.*)

stoker. The depth of fuel on the grate is regulated by a hand-adjusted gate (Fig. 4.2). The speed of the grate varies the rate at which the coal is fed to the furnace. The combustion control automatically regulates the speed of the grate to maintain the steam pressure. The burning progresses as the grate travels through the furnace. The refuse, containing a small amount of combustible material, is carried over the rear end of the stoker and deposited in the ashpit.

Air enters the furnace through the openings in the grates and through over-fire jets. These over-fire air jets enter the furnace through the front arch or front wall above the arch. The intense heat at the front of the stoker results in distillation of gases from the fuel bed. The combination of over-fire air and the air passing through the fuel bed provides turbulence required for rapid combustion. In some instances, over-fire air jets are also located at the rear wall to provide a counterflow of gases in the furnace, promoting increased turbulence and further reducing the emission of smoke.

The grates are entirely enclosed by a suitable casing to prevent air infiltration at undesirable points and to promote combustion. Air leakage along the sides of the moving grates is reduced by the use of adjustable ledge plates. Some difficulty is encountered with excessive air leakage at the point where the refuse is discharged into the ashpit. This leakage is reduced or the effect minimized by employing a damper to seal off the ashpit, by using close clearance between the stoker and the setting, and by a rear arch which forces the air to pass over the fuel bed, mix with the combustible gases, and be utilized in the combustion process.

When the clearance between the rear of the furnace and the stoker is reduced to control air leakage, the hot ash comes into contact with the wall and adheres to it, forming clinkers.   This condition is prevented by the installation of water-cooled tubes, or "water backs," in the rear wall where the refuse is discharged from the stoker.   These tubes may be cooled by the flow of low-pressure water supplied for this purpose, or they may be connected to the boiler and form a part of the heating surface.

The chain-grate stoker was originally designed to operate with natural draft, but it is now extensively used in connection with forced draft. When forced draft is employed the entire grate is enclosed.   The air is brought in from the sides, for it is then necessary to force the air only through the upper grate.   The air ducts under the stoker are divided into sections, so that the air supply to different parts of the stoker is regulated to meet the demands of rating, fuel, etc.

The vibrating-grate stoker (Fig. 4.4) operates in a manner similar to that of the chain- or traveling-grate stoker, except that the fuel feed and fuel-bed movement are accomplished by vibration.   The grates consist of cast-iron blocks attached to water-cooled tubes.   These tubes are equally spaced between headers that are connected to the boiler. The connecting tubes between these headers and the boiler circulating

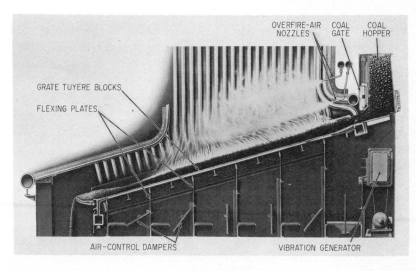

FIG. 4.4   *Vibrating-grate stoker.   (Adapted from "Steam, Its Generation and Use," The Babcock & Wilcox Co.)*

system have long bends to provide the flexibility required to permit vibration of the grates. The movement of the grates is extremely short, and no significant strains are introduced in this system. The space beneath the stoker is divided into compartments by means of flexible plates. Individual supply ducts with dampers permit regulation of the air distribution through the fuel bed. The grates are actuated by a specially designed vibration generator which is driven by a constant-speed motor. This consists essentially of two unbalanced weights which rotate in opposite directions to impart the desired vibration to the grates.

The depth of the coal feed to the grates is regulated by hand adjustment of the gate at the hopper. However, the rate of fuel feed is automatically controlled by varying the off-and-on cycle of the vibrating mechanism. The vibration and inclination of the grates cause the fuel bed to move through the furnace toward the ashpit. The compactness resulting from the vibration conditions the fuel bed and promotes uniform air distribution. This action tends to permit the use of a wider range of fuels. The water cooling decreases the minimum ash requirement for grate protection and permits the use of preheated air for combustion.

Chain-grate stokers are used for burning noncaking free-burning high-volatile high-ash coals. Traveling-grate stokers will successfully burn lignite, very small sizes of anthracite, coke breeze, etc. Vibrating-grate stokers are suitable for medium- and high-volatile bituminous coals and for low-volatile bituminous coals and lignites at reduced burning rates. Coking coal requires agitation and is, therefore, not suited for use on these stokers.

The most satisfactory coals for use on chain- and traveling-grate stokers have a minimum ash-softening temperature of 2100°F, ash content of 6 to 15 per cent, free-swelling index of 3 to 5, and a size range of $1\frac{1}{4}$ or 1 in. by 0 with not more than 30 per cent less than $\frac{1}{4}$ in. For natural-draft stokers the ash content should be 5 to 9 per cent and there should be a minimum amount of coal in the size range less than $\frac{1}{4}$ in. Similar specifications apply to coal for vibrating stokers, except that owing to the water-cooled grates the ash content may be lower.

The rate of combustion for natural-draft chain-grate stokers is 30 to 40 lb of coal per sq ft of grate area per hr. With forced-draft stokers the rate is increased to 40 to 50 lb of coal per sq ft of grate area per hr.

To start a fire on a chain-grate stoker, set the feed and start the stoker. Cover the grates with 3 to 4 in. of coal for two-thirds of its length. Start a wood fire on top of the coal. Do not restart the stoker until the coal has been thoroughly ignited and the ignition arch is hot; then operate it at its lowest speed until the furnace becomes heated.

There are two ways of regulating the rate of coal supplied by a chain-grate stoker: the depth of coal on the grates and the rate of travel of the grates. The depth of coal on the grates is maintained between 4 and 6 in. by means of an adjustable gate (see Fig. 4.2). The average operating speed of the chain or traveling grate is 3 to 5 in. per min. Variations in rating are obtained by changing either or both of these fuel-feed adjustments.

Dampers are provided in both natural- and forced-draft stokers to permit the operator to control the air supply to the various zones. These air-supply adjustments enable the operator to control the rate of burning in the various zones and thereby reduce to a minimum the coke carry-over into the ashpit. If satisfactory operation cannot be accomplished by adjusting these dampers, the next step is to adjust the fuel-bed depth; finally one can experiment by tempering the coal. A combustible content of 15 to 20 per cent is normally anticipated in the ashpit of these stokers, but excessive coke in the refuse results not only in a fuel loss but also in burned stoker-grate links. Moreover, the coke continues to burn in the ashpit, and with low-fusion-ash coal this causes a clinkering mass of refuse difficult to remove.

Improvements in operation can be obtained by tempering (adding water) when the coal supply is dry, contains a high percentage of fine particles, or has a tendency to coke. Once coke has formed on the grates, the flow of air is restricted, making it very difficult to complete combustion and prevent an excessive amount of the coke from being carried over the end of the grate into the ashpit. Best results are obtained when the moisture content of the coal is 8 to 10 per cent. The water should be added 10 to 24 hr before the coal is burned. The addition of water at the stoker hopper results in excessive amounts of surface moisture.

The approximate amount of fuel consumed by a chain- or traveling-grate stoker can be calculated when the grate speed and thickness of coal on the grate are known. The grate speed can be determined by timing with a watch or by a revolution counter on the driving mechanism. When the revolution counter is employed, a factor must be used to convert the reading into grate speed. This factor may be determined by comparing the revolutions of the driving mechanism with the grate speed. This may be done when the unit is out of service and at room temperature. In converting from volume to weight, assume bituminous coal as 48 lb per cu ft, anthracite as 60 lb per cu ft, and coke breeze as 40 lb per cu ft.

Example A chain-grate stoker 8 ft wide is burning bituminous coal. The grate speed is 4 in. per min., and the coal is 6 in. deep. At what rate is coal being consumed in tons per hour?

Solution

$$\underset{\substack{\text{width stoker,}\\ \text{ft}}}{8} \times \underset{\substack{\text{rate of feed,}\\ \text{fpm}}}{\tfrac{4}{12}} \times \underset{\substack{\text{depth of}\\ \text{coal, ft}}}{\tfrac{6}{12}} = \underset{\substack{\text{coal burned/}\\ \text{min, cu ft}}}{\tfrac{4}{3}}$$

$$\underset{\substack{\text{coal burned/}\\ \text{min, cu ft}}}{\tfrac{4}{3}} \times \underset{\substack{\text{min in hr}}}{60} \times \underset{\substack{\text{lb coal/}\\ \text{cu ft}}}{48} \div \underset{\substack{\text{lb/ton}}}{2{,}000} = \underset{\substack{\text{coal burned/}\\ \text{hr, tons}}}{1.92}$$

To bank the fire on a chain- or traveling-grate stoker, close off the air supply under the stoker a little at a time. Allow the fire to burn down before closing off the air supply in order to prevent damage from overheating. Fill the hopper and close off the coal supply. When the furnace is relatively cool, raise the gate and run all the coal from the hopper into the furnace. In order to guard against possible overheating do not leave the coal in contact with the gate. Admit a small amount of air under the stoker at the end of the supply of green coal. Open the stack damper sufficiently to maintain a slight draft in the furnace. As the coal burns and the fire moves from the rear to the front of the furnace, the point of admission of air must also be varied by adjusting the zone dampers. Before the fire reaches the gate, proceed to fill the hopper and introduce another charge of coal. In starting from bank, it is frequently necessary to use a bar to break up the coal that has caked near the gate. The coal feed should not be started until the air has blown through the grates for a period of 3 to 5 min in order to increase the furnace temperature.

The chain should be tight enough to keep it from dipping down appreciably when it comes over the front sprocket. The chains are adjusted by a take-up screw that moves the sprocket. The plates between the side walls and the grates must be adjusted to prevent excess-air leakage. The clearance between the plate and grate is adjusted for ⅛-in. clearance. Excessive burning of the link ends is an indication that combustible material is passing over the refuse end of the stoker. Broken or burned links must be replaced at the earliest opportunity.

4.4 UNDERFEED STOKER. As the name implies, underfeed stokers force the coal up underneath the burning fuel bed. Small and medium-sized boilers are supplied with single- or, in some cases, with twin-retort stokers. Large boilers are equipped with multiple-retort stokers.

The front exterior of a single-retort underfeed stoker is shown in Fig. 4.5. The motor drives a reduction gear which in turn operates the main or feed ram, secondary ram or pusher plates, and grate movement. The two horizontal cylinders located on the sides of the stoker front are actuated by either steam or compressed air to operate the dumping

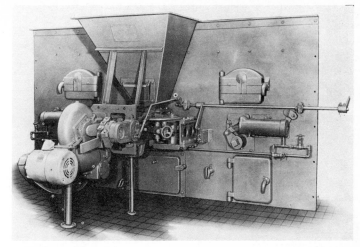

FIG. 4.5 *Front exterior of a single-retort underfeed stoker.* (*Detroit Stoker Co.*)

grates.   The doors above the dump-grate cylinders are for observing the fuel bed.   The doors below these cylinders are for removing the ash after they have been dumped into the ashpits.

Figure 4.6 shows a longitudinal sectional view of this stoker with the various external and internal components.   Figure 4.7 is a cross section of the stoker showing the retort, grates, dumping grates, and ashpits. The illustration also shows how the grates are moved up and down to break up the coke formation and to promote air distribution through the burning fuel bed.

The feed ram forces the coal from the hopper into the retort.   During

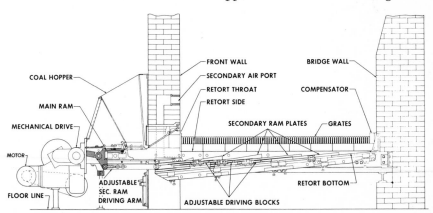

FIG. 4.6 *Longitudinal section of a single-retort underfeed stoker.* (*Detroit Stoker Co.*)

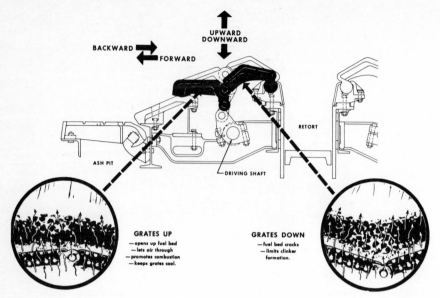

FIG. 4.7   *Cross section of a single-retort underfeed stoker.*   (*Detroit Stoker Co.*)

normal operation the retort contains green coal which is continually pushed out over the air-admitting grates (tuyères). The heat absorbed from the fuel bed above and the action of the air being admitted through the grates cause the volatile gases to be distilled off and burned as they pass up through the fuel bed. The burning fuel slowly moves from the retort toward the sides of the furnace over the grates. As the fuel moves down the grates, the flame becomes shorter since the volatile gases have been consumed and only coke remains. A portion of the coke finds its way to the dump grates, and a damper is provided to admit air under the grates in order to further complete combustion before the ashes are dumped. The combustion control automatically regulates the rate of coal supplied. The secondary ram driving arm is adjustable so the movement can be varied to obtain optimum fuel-bed conditions.

Underfeed stokers are supplied with forced draft for maintaining sufficient airflow through the fuel bed. The air pressure in the wind box under the stoker is varied to meet load and fuel-bed conditions. Means are also provided for varying the air pressure under the different sections of the stoker in order to correct for irregular fuel-bed conditions. When too much coke is being carried to the dumping grates, more air is supplied to the grate section of the stoker. The use of forced draft causes rapid combustion, and when high-volatile coals are employed, it becomes necessary to introduce "over-fire air" to burn the resulting volatile gases and prevent smoke. Figure 4.6 shows how air from the wind box is admitted above the fuel bed. A further improvement in combustion and reduction in smoke can be obtained by use of steam or high-pressure air jets, as explained in Sec. 4.13.

All single-retort stokers feature the principles explained above, but there are numerous variations in details to comply with specific requirements. The feed ram may be directly driven by a steam or hydraulic piston. The use of transmission gears permits the use of all types of rotative power, including steam engines and electric motors. A screw may be used in place of the reciprocating ram to feed the coal into the retort. Stationary grates may be used in place of the moving grates shown in Fig. 4.7. When stationary grates are used, the action of the coal as it is pushed out of the retort is utilized to obtain movement over the grates and for maintenance of the correct fuel bed. The grates at the sides of the stoker may be stationary, making it necessary to remove the ashes from the furnace rather than dumping them into the ash pit.

Single-retort stokers correctly applied and operated provide a satisfactory means of burning coal to produce steam. The capacity is dependent upon the characteristics of the coal. Ash content in the order of 8 to 12 per cent and ash-fusing temperatures above 2600°F are desirable. Low ash coal may result in exposure of the grates to high furnace temperatures and excessive maintenance. Coal containing ash with a low fusion temperature results in the formation of clinkers. It is advisable to "carry" a thick fuel bed on the grates to prevent the formation of holes through which the air can pass without coming in contact with the fuel. The underfeed stoker is quite responsive to variations in steam demand. When there is a decrease in steam pressure, the control increases the airflow through the fuel bed; the resulting increase in heat liberated quickly restores the pressure. Conversely when the pressure increases the decrease in airflow reduces the rate of heat liberated.

Multiple-retort stokers have essentially the same operating principle as that of single- and twin-retort stokers. They are used under large boilers to obtain high rates of combustion. A sufficient number of retort and tuyère sections are arranged side by side to make the required stoker width (Fig. 4.8). Each retort is supplied with coal by means of a ram (Fig. 4.9). These stokers are inclined from the rams toward the ash-discharge end. They are also equipped with secondary rams which, together with the effect of gravity produced by the inclination of the stoker, cause the fuel to move toward the rear or refuse discharge. The rate of fuel movement and hence the shape of the fuel bed can be regulated by adjusting the length of the stroke of the secondary rams. Some stokers produce agitation of the fuel bed by imparting movement to the entire tuyère sections. After the fuel leaves the retorts and tuyères, it passes over a portion of the stoker known as the "exten-

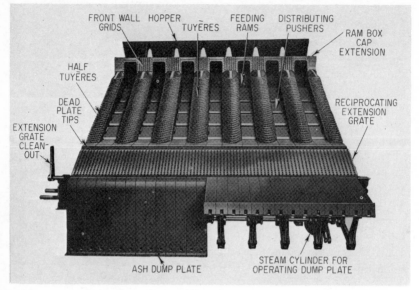

FRONT WALL HOPPER   FEEDING DISTRIBUTING
GRIDS   TUYÈRES   RAMS   PUSHERS
                                                      RAM BOX
                                                      CAP
                                                      EXTENSION

HALF
TUYÈRES

DEAD
PLATE
TIPS   RECIPROCATING
                                                      EXTENSION
EXTENSION   GRATE
GRATE
CLEAN-
OUT

                                              STEAM CYLINDER FOR
            ASH DUMP PLATE   OPERATING DUMP PLATE

FIG. 4.8 *Multiple-retort stoker showing steam-operated ash-dumping plates and coal-and- air-distribution mechanisms.* (*Detroit Stoker Co.*)

sion grates" or "overfeed section" (Fig. 4.8). By this time the volatile gases have been consumed, and only a part of the carbon remains to be burned.

These large stokers have mechanical devices for discharging the refuse from the furnace. Dumping grates receive the refuse as it comes from

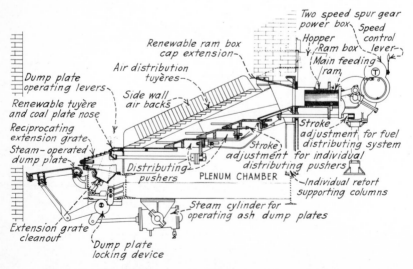

Two speed spur gear
power box
Hopper   Speed
Renewable ram box   Ram box   control
cap extension   lever
Air distribution   Main feeding
tuyères   ram
Dump plate
operating levers
Renewable tuyère
and coal plate nose   Side wall
Reciprocating   air backs
extension grate   Stroke
Steam-operated   adjustment for fuel
dump plate   distributing system
                    Distributing   Stroke
                    pushers   adjustment for individual
                         PLENUM CHAMBER   distributing pushers
                                          Individual retort
                                          supporting columns
Extension grate
cleanout   Steam cylinder for
          Dump plate   operating ash dump plates
          locking device

FIG. 4.9 *Cross section of a multiple-retort stoker with dumping grates.* (*American Engineering Co.*)

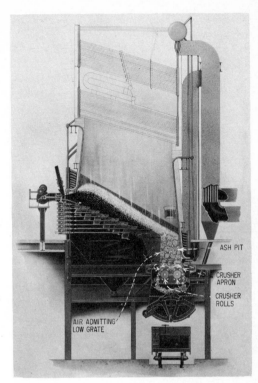

FIG. 4.10   *Multiple-retort stoker with rotary ash discharge. (American Engineering Co.)*

the extension grates.   When a sufficient amount has collected, the grates are lowered and the refuse dumped into the ashpit.   These dumping grates, even when carefully operated, frequently permit large quantities of unburned carbon to be discharged with the ash.   Furnace conditions are disturbed during the dumping period, and efficiency is lowered. Power-operated dumps have been introduced to shorten the dumping period (Fig. 4.9).   The clinker grinder (Fig. 4.10) consists of two rolls that replace the dumping grates.   These rolls can be operated at any desired speed, and the ash is discharged continuously.   The combustible in the refuse from a stoker employing dumping grates will vary from 20 to 40 per cent; with clinker grinders it will vary from 5 to 15 per cent.

Multiple-retort stokers are driven by motors, turbines, or engines. Provisions are made for varying the stoker speed either by changing the speed of the driving unit or by use of a variable-speed device between the driving unit and the stoker.   Automatic controls vary the stoker speed to supply coal as required to meet the steam demand.

Irregularities in the fuel bed are corrected by controlling the distribution of air under the stoker.   The wind box is divided into sections, and the air supply to each section is controlled by a hand-operated damper. This arrangement enables the fireman to correct for thin or heavy places in the fuel bed by reducing or increasing the wind-box pressure at the required spot.

The setting of an underfeed stoker is relatively simple in that no special arches are required. Mixing of fuel and combustible gases with air is accomplished by means of the high-pressure air supply and, in some cases, by the addition of over-fire air. Since there are no arches, the stoker is directly under the boiler and a large quantity of heat is transmitted by radiation, resulting in a decrease in the temperature of stack gases and improved efficiency. Multiple-retort stokers are provided with sidewall tuyères or air blocks, which improve combustion and prevent the formation of clinkers (Fig. 4.9). The application of waterwall tubes to the furnace of a stoker-fired boiler as a means of preventing slagging difficulties and reducing maintenance is shown in Fig. 4.10

With certain limitations as to the type of coal used, the underfeed stokers are well suited for continuous operation at high ratings. They are especially adapted to the burning of high-volatile coals, which can be burned economically by the underfeed principle. In some instances it has been found advisable to use over-fire air jets to obtain smokeless combustion. The fuel bed attains a very high temperature, to which the ash is exposed as it is moved to the surface. Therefore, clinkers are formed when coal having a low-fusing-temperature ash is burned on these stokers. Clinkers shut off the air openings in the tuyères and interfere with the movement of fuel on the stoker. This results in increased operating labor, reduced economy, and high maintenance. In general these stokers will successfully burn coking coal; however, stokers which provide for agitation of the fuel bed are best suited for this type of coal.

Underfeed stokers can be built in any desired size and applied to a wide variety of conditions. They vary in size from the small domestic type to the central-station multiple-retort unit. The use of forced draft and the relatively large quantity of fuel on the stoker make them responsive to rapid changes in load. When the fusing temperature of the ash exceeds 2500°F, combustion rates in excess of 60 lb of coal per sq ft of stoker area may be successfully maintained. These stokers have been developed to a high state of perfection and when correctly applied and operated give very satisfactory results.

Before a stoker is operated, it should be lubricated in accordance with the manufacturer's instructions. Reduction gears run in lubricants, and the gear housing must be filled to the required level with the grade of lubricant specified by the manufacturer. When the stoker is in operation, the oil level should be checked daily and the lubricant changed in accordance with the manufacturer's recommendation. The crankshaft bearings and other slow-moving parts are lubricated by medium grease applied with a high-pressure grease gun. These bearings should receive daily attention.

When preparing to put a stoker in service, operate the mechanism and check to see that it is in satisfactory condition. Feed coal into the furnace and spread it manually to a uniform thickness of 3 in. over the tuyères. If the stoker is steam-driven and the entire plant is out of service, it will be necessary to shovel the coal into the furnace. Place oily waste, shavings, and wood upon the bed of green coal. Then open the stack damper or, if mechanical draft is used, start the induced-draft fan to ventilate the setting and furnace, consequently eliminating the possibility of an explosion which might result from the presence of combustible gases. At this stage the wind-box door may be opened and the fire lighted. The drafts are then regulated to keep the fire burning brightly.

When the coal is burning freely, the wind-box door is closed, the forced-draft fan is started, and the air supply is regulated to maintain the desired rate of combustion. It may be necessary to stop the fans occasionally to prevent the boiler from heating up too rapidly. The rate at which the unit can be brought "on the line" depends upon the design of the boiler, but one should always err on the safe side and extend rather than shorten the prescribed time. During this period enough coal must be fed into the stoker to replace that which is being burned and to establish the fuel bed.

Underfeed stokers operate with furnace pressures at, or only slightly below, atmosphere; the condition is referred to as a "balanced draft." This procedure prevents excessive air leakage into the setting and does not cause excessive heat penetration into the furnace setting. The balanced draft is maintained by utilizing the wind-box pressure, as created by the forced-draft fan, to produce a flow of air through the fuel bed. Similarly the induced-draft fan is used to remove the gases from the furnace by overcoming the resistance of the boiler economizer, etc. Actually the draft in the furnace should be maintained between 0.05 and 0.10 in. of water during both operating and banked periods. The wind-box pressure varies from 1 to 7 in. of water, depending upon the stoker design, fuel, and rating. Control equipment will maintain the desired draft conditions automatically. As a safety precaution, the furnace draft should be increased when blowing soot or dumping ashes. Remember that stokers depend upon the flow of air to keep them cool, and a consideration of this fact will greatly reduce maintenance.

The following characteristics are given as a general guide to the selection of coal. They should be supplemented by plant checks and experience to determine the most satisfactory selections, considering both the ease of operation and the cost. Coals for use on single-retort stokers should have a minimum ash-softening temperature of 2350°F, ash content of 5 to 7 per cent, free-swelling index of 4 to 6, and a size range of 1¼ or 2 in. by 0 with not more than 30 per cent less than ¼ in. Multiple-retort stokers use coals having the same ash-softening temperature with an ash content between 7 and 8 per cent, free-swelling index of 5 to 7, and a top size of 2 in. with the same minimum-size limitation of not more than 30 per cent less than ¼ in. As the free-swelling index increases, the top size should be increased and the amount of fines decreased. Coarse coal tends to segregate in the coal-conveying equipment, causing uneven feeding and burning. Wet coal may clog in the chutes and refuse to feed to the stoker. Stokers are supplied with shear pins which prevent serious damage to the mechanism as a result of foreign material in the coal, but the breaking of a shear pin and the removal of the foreign material cause interruption to service; consequently every effort should be made to have the coal supply correctly sized and free from foreign material. Never allow the coal hopper to run empty or the supply chute to clog while the stoker is in operation, since this will result in the coal's burning in the retort and causing the tuyères to become overheated.

In manipulation of the underfeed stoker one must adjust the airflow in proportion to the coal and regulate the supply to the various sections of the grates in order to maintain the correct contour of the fuel bed (Fig. 4.11). For example, insufficient air to the tuyère section or too great a travel of the secondary rams will cause the fuel to pass over the tuyères before it is burned. This will result in an excessive

FIG. 4.11   *Contour of fuel bed on a multiple-retort underfeed stoker.* (*Detroit Stoker Co.*)

amount of carbon reaching the dumping grates, where the combustion activity must be accelerated to prevent an excessive amount of carbon from being rejected with the ash. The travel of the secondary rams must be adjusted by observing the fuel bed and the flue-gas analysis until satisfactory results have been obtained. The fuel-bed depth, 3 ft from the front wall, should be 16 to 24 in. The correct depth depends upon a number of factors, including the coking tendency of the coal and the amount of coarse and fine particles that it contains. The fireman must take into account the fact that, upon entering the furnace, coking coals tend to swell and produce a deep fuel bed. If compensation is not made for this tendency, the fuel bed will become too heavy and excessive amounts of coke will be discharged with the ash.

Regardless of whether stationary grates, dumping grates, or clinker grinders are used to remove the refuse from the furnace, the fireman must exercise judgment in "burning down" before discharging the refuse. Insufficient burning down will cause high carbon content in the refuse; on the other hand, the indiscriminate admission of air will result in a high percentage of excess air (low carbon dioxide) and consequent high stack loss. The air chambers under the stoker should be inspected daily and the accumulation of siftings removed before they become excessive and interfere with the working parts of the stoker or form a fire hazard.

To bank the fire on an underfeed stoker the fans are stopped, but the stack damper is not completely closed. There is danger of the coal's caking in the retort and blocking the movement of the rams. If a fire is to be banked for several hours, the stoker should be run one revolution at a time at regular intervals during the banked period.

Stoker maintenance may be divided into three classes: (1) coal feeders, speed reducers, and driving mechanisms; (2) internal parts, including tuyères and grates; and (3) retorts and supporting structures.

Satisfactory and long maintenance-free operation of the driving and external coal-feeding mechanism of a stoker depends upon the correct lubrication and the elimination of foreign matter from the coal. If the driving mechanism should fail to start, examine the stoker carefully to determine whether foreign material is lodged in the ram or screw. Do not allow coal siftings and dirt to accumulate to a point where they impede the operation. Repairs to these external parts can be made without waiting for the furnace to cool.

The tuyères and grates have to be replaced occasionally. The frequency of replacement depends upon the fuel used, the type of service, and the care employed in the operation of the stoker. Coal with a low-fusion-temperature ash results in clinkers which adhere to the grates, restrict the airflow, and cause the stoker castings to become overheated.

When the ash content of the coal is low, there is insufficient protection and maintenance will be high. Even when suitable coal is used, the stoker castings may be overheated if the fuel bed is allowed to become so thin that parts of the grates are exposed. These castings are easily replaceable, but it is necessary to have an outage of equipment long enough to allow the furnace to cool sufficiently for one to enter.

Unless the stoker is improperly operated, the retorts and structural parts should not have to be replaced for a long period of time. The replacement of these parts is expensive and requires an extended outage.

4.5   SPREADER STOKER.   The spreader-stoker installation consists of a variable-feeding device, a mechanism for throwing the coal into the furnace, and grates with suitable openings for admitting air. The coal-feeding and -distributing mechanism is located in the front wall above the grates. A portion of the volatile matter and the fine particles in the coal burn in suspension, and the remainder falls to the grate, where combustion continues. As a result of the pressure created by the forced-draft fan, air enters the furnace through the openings in the grates. A portion of this air is used to burn the thin layer of fuel on the grates, and the remainder passes into the furnace, where it is utilized to burn the volatile matter and fine particles. The over-fire air fan supplies additional air to the furnace through jets in the wall. This adds to the air supply for suspension burning and produces turbulence.

The conventional mechanical spreader-stoker feeding-and-spreading device is arranged to supply coal to the furnace in quantities required to meet the demand. The feeder delivers coal to the revolving rotor with protruding blades. These revolving blades direct the coal and throw it into the furnace. The spreader system must be designed to distribute the coal evenly over the entire grate area. This distribution is accomplished by the shape of the blades which throw the coal. Variations in performance of the spreader, owing to changes in coal size, moisture content, etc., are corrected by means of an external hand adjustment of the mechanism.

The suspension burning results in a portion of the unburned carbon particles' (cinders') being carried out of the furnace by the gases. Some of these cinders collect in the hoppers beneath the convection passes of the boiler. However, a portion is retained in the exit boiler gases and must be intercepted by a fly-ash collector to prevent excessive pollution of the atmosphere. Since these particles contain carbon, the efficiency of steam generation can be improved by returning the cinder to the furnace for reburning.

Spreader stokers have a wide application with respect both to the fuel that can be handled and to the boiler sizes to which they can be applied. A wide variety of coal can be burned on a spreader stoker, but the variation must be considered when the unit is specified. It is inadvisable and uneconomical to provide a stoker which will burn almost any coal in a location where a specific coal is available. In addition to coal, spreader stokers are used to burn bark, chips, and many other kinds of by-product fuels.

Spreader stokers are used with high- and low-pressure boilers from the smallest size to those having a steam-generating capacity of 300,000 lb of steam per hr. At lower capacities the installation cost of spreader stokers is less than that for pulverized fuel. However, between 200,000 and 300,000 lb of steam per hr the cost tends to equalize, and at higher capacities pulverized-fuel plants are less costly. Owing to the relatively high percentage of fuel burned in suspension, spreader-stoker–fired boilers will respond quickly to changes in firing rate. This feature qualifies them for use where there is a fluctuating load. The fact that these stokers discharge fly ash with the flue gases at one time subjected them to severe criticism as contributors to air pollution, but with efficient dust-collecting equipment they can be made to meet the most stringent air-pollution code.

In the smaller sizes spreader-stoker installations are comparatively costly owing to the necessity for ashpits, fly-ash reinjection, and fly-ash-collection equipment required for efficient and satisfactory operation. Difficulty is also encountered in designing a unit which will operate satisfactorily when the lowest steam requirement is less than one-fifth of the designed capacity (turned-down ratio less than 5). The "turned-down ratio" is defined as the design capacity divided by the minimum satisfactory operating output.

The spreader-stoker feeder functions to vary the supply of coal to the furnace and to provide even distribution on the grates. In the feeder-and-distribution system shown in Fig. 4.12, there is a reciprocating feed plate in the bottom of the hopper. The length of stroke of this plate determines the rate at which coal is fed into the furnace. The automatic combustion regulator varies the length of the feed-plate stroke. The coal leaving the hopper drops from the end of the spilling plate into the path of the rotor blades which distribute the coal on the grates. The distribution is regulated by hand adjustment of the spilling plate and by the speed of the rotor. The in-and-out adjustment of the spilling plate changes the point at which the coal comes in contact with the rotor blades. Moving the spilling plate back from the furnace allows the coal to fall on the rotor blades sooner. The blades impart more energy to the coal, and it is thrown farther into the furnace. In a like manner increasing the speed of the rotor imparts more force to the coal, throwing it farther into the furnace.

The ability of spreader stokers to utilize coal having a high moisture content is limited to the performance of the feeder-and-distribution mechanism. Figure 4.13 is a drum-type feeder which employs a pocket wiper to prevent the recycling of wet coal. When there is a decided change in either the size or the moisture content of the coal, the feeder mechanism must be adjusted to maintain a uniform fuel bed on the grates. Air enters the furnace both above and below the feeders to keep the feeder cool and provide air for combustion. In addition to being air-cooled, the rotor bearings are water-cooled.

Spreader-stoker feeders are divided into two classes: overthrow, when the coal comes into contact with the upper part of the rotor blade assembly (Figs. 4.12, 4.13), and underthrow, when it comes into contact with the lower part.

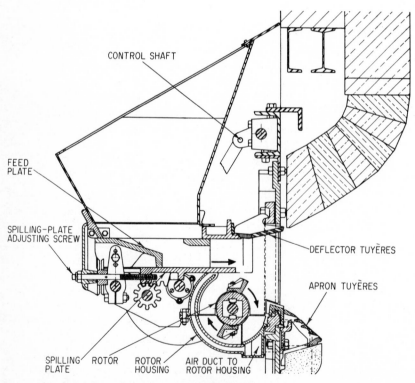

FIG. 4.12   *Spreader-stoker reciprocating-type coal feeder and rotary distribution mechanism.*   (*Detroit Stoker Co.*)

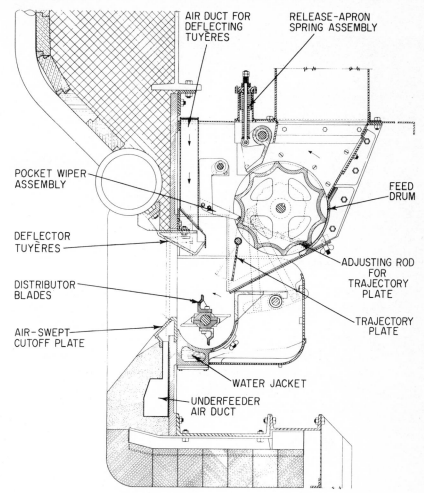

AIR DUCT FOR
DEFLECTING
TUYÈRES

RELEASE-APRON
SPRING ASSEMBLY

POCKET WIPER
ASSEMBLY

FEED
DRUM

DEFLECTOR
TUYÈRES

DISTRIBUTOR
BLADES

AIR-SWEPT
CUTOFF PLATE

ADJUSTING ROD
FOR
TRAJECTORY
PLATE

TRAJECTORY
PLATE

WATER JACKET

UNDERFEEDER
AIR DUCT

FIG. 4.13 *Spreader-stoker drum-type coal feeder and rotary distribution mechanism.* (*Riley Stoker Corp.*)

The simplest types of grates are stationary and similar to those employed for hand firing. The feeder automatically deposits the coal on the grates, and air for combustion enters the furnace through the holes in the grates. At least two feeders are used, and before the ash deposits become deep enough seriously to restrict the airflow, one of the feeders is taken out of service. The fuel on the grate is allowed to "burn down,"

FIG. 4.14 *Four-unit spreader stoker with power-operated dumping grates.* (*Detroit Stoker Co.*)

and the ash is raked through the furnace door. The feeder is then started, and after combustion has been reestablished and stabilized, the remaining grate sections are cleaned in a similar manner.

Dumping grates (Fig. 4.14) provide a means for tipping each grate section and thereby dumping the ash into a pit. The tipping of the grates can be accomplished either by hand operation or by means of steam- or air-powered cylinders. The procedure of taking one feeder section out of service long enough to remove the ash is the same as when stationary grates are used, but the time required for the operation and the amount of labor involved are reduced. The intermittent discharge of ash from the furnace has the following disadvantages: considerable labor is needed; skill is required to prevent a drop in steam pressure; efficiency is reduced owing to the high excess air introduced during the cleaning period; and there is a tendency to discharge considerable combustible with the ash, produce smoke, and increase the discharge of cinder while the grates are being cleaned.

For these reasons several methods have been devised to discharge the ash continually from these furnaces and to permit uninterrupted operation of the feeders. These devices are adjustable to compensate for variations in ratings and ash content of the coal.

A traveling-grate spreader stoker installed in a furnace is shown in Fig. 4.15. The coal falls on the grate, and combustion is completed as it moves slowly through the furnace. The ash remains and falls into the pit when the grates pass over the sprocket. It is customary to locate the ashpit in the front of the stoker (beneath the feeder), but successful performance has been obtained by having the ash dis-

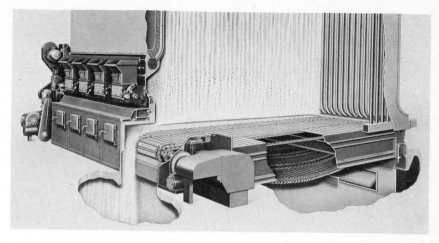

FIG. 4.15 *Traveling-grate continuous-ash-discharge spreader stoker.* (*Detroit Stoker Co.*)

charge at the rear of the furnace. The ashpit is located in the basement to provide adequate storage for the ash produced on an eight-hour shift. The rate of grate movement is varied to produce the required depth of ash at the discharge end.

The principle used in vibrating conveyors is also employed in removing ash from spreader-stoker furnaces. The grates are mounted on a pivoted framework, a motor vibrates the assembly, and the ash is moved along the grates toward the ashpit. The motor which produces the vibration is run at intervals by a timer. The off-and-on cycles are varied to obtain the desired depth of ash at the discharge end of the grate.

Another method of obtaining continuous ash discharge consists of overlapping grates similar to shingles on the roof of a house. The grate bars are mechanically driven and move back and forth, alternately increasing and decreasing in the amount of overlap. This motion causes the ash to shift from one grate to the other and slowly move toward the ashpit. The rate of ash discharge is varied by changing the amount of travel of the grate bars.

These continuous methods of removing the bottom ash from the stoker furnace are capable of discharging the cinders with a combustible content of 4 to 8 per cent. However, the cinders carried from the furnace with the gases contain 40 to 60 per cent combustible. Provisions must be made to recover this combustible material in order to reduce air pollution and to reclaim the heat.

It is customary to collect a large portion of these cinders in hoppers beneath the boiler passes and in dust-collector hoppers and return them to the furnace for reburning. A high-pressure stream of air is used to remove the cinders from the hoppers and blow them into the furnace, preferably through a number of openings in the rear wall. The necessary air is delivered by high-pressure blowers referred to as "over-fire air fans." They are sized to deliver approximately 10 per cent of the total air required for combustion and, in addition to returning cinders, also supply a system of over-fire air jets.

Over-fire air jets are used to produce the turbulence required to accelerate the suspension burning and reduce smoke emission. The air jets are located in the rear of the furnace, resulting in an air movement in counterflow to the gases. Over-fire air jets are also frequently installed in the front wall above the feeders and directed downward to intercept the fine particles of coal which are burning in suspension. Steam jets and air induced by the steam jets are also used to obtain turbulence in boiler furnaces. (See Sec. 4.13.)

The stoker shown in Fig. 4.16 has forced draft, power-operated dumping grates, and a combination over-fire-air and cinder-return system. Cinders deposited in the hoppers beneath the convection passages of the boiler are transferred to the side of the unit by screw conveyors and then picked up by streams of air and returned to the furnace. Cinders from the dust-collector hopper are fed directly into a stream of air and also blown into the furnace. The high-pressure blower supplies air to both the cinder-return system and the over-fire air jets. Ash deposits on the grates are dumped into the shallow pit and transferred by hand to the "pickup" conveyor located in front of the stoker.

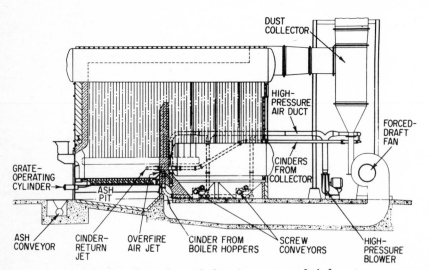

FIG. 4.16   *Spreader stoker with dumping grates and cinder return.*
(*Detroit Stoker Co.*)

The reinjection of cinders in a spreader stoker increases the efficiency of the unit by 3 to 5 per cent, but this practice introduces operating and maintenance problems which limit the extent to which it is advisable to seek the added efficiency.   When the cinders from an efficient dust collector are reinjected, the dust recycles, causing a high concentration in the flue gases.   This dust is very abrasive, resulting in increased maintenance of fans, dust collectors, and reinjection piping.   In some cases this reinjection of cinders interferes with combustion in the furnace, causing slag deposits to form on the furnace walls and clinkers on the fuel bed.   To reduce these difficulties it has sometimes been found advisable to discard the cinders deposited in the collector hoppers.   A compromise is to install two dust collectors or a collector having two compartments, reinjecting the cinders collected in the low-efficiency unit and discarding that from the one having the higher efficiency.   The coarse particles collected in the unit having the lower efficiency contain the highest percentage of carbon.

Another method employed to reduce the recycling is gravity return of the cinders to the traveling grate.   The cinders from the various sources are collected in a hopper extending across the rear of the furnace and permitted to flow by gravity, in a number of streams, onto the traveling grate just after it has passed over the rear sprocket and entered the furnace and before the air from the forced-air plenum has been applied.   This procedure of depositing the ash directly on the grates lessens the adverse effect upon furnace conditions and decreases the tendency to recycle.

Strict limitations on the permissible discharge of particulate into the atmosphere dictates careful consideration to the selection and operation of control equipment.   In some instances, adequate particulate reduction can be accomplished by the use of mechanical collectors if the fly ash collected is not reinjected into the furnace.   The decrease in efficiency owing to loss of heat in the carbon contained in the fly ash is preferable to the installation of more costly collection equipment.   Another possible solution is the installation of both mechanical collectors and electrostatic precipitators.   The flue gases pass through the mechanical collectors and electrostatic precipitators before being discharged to the atmosphere.   Even with this combination, all of the fly ash should be discarded or at most only the coarser portion from the mechanical collector should be reinjected into the furnace.

A simple combustion control for a spreader stoker consists of a pressure-sensing element in the steam header which activates a controller on the fuel-feed rate and changes the flow of air from the forced-draft fan.   The fuel-feed change is accomplished by adjusting the length of stroke of the feed plate (see Fig. 4.12).   The amount of air supplied under the grates is usually regulated by a damper or by inlet veins on the fan, although sometimes the speed of the fan is varied. The induced-draft-fan damper or the fan speed is varied to maintain a constant furnace draft of 0.05 to 0.10 in. of water. A recommended refinement in the control consists of a means by which the forced-draft-fan air supply is readjusted to maintain a definite rate of airflow to fuel regardless of the depth of the fuel bed. Other refinements are the control of the rate of ash discharge from the grates and the over-fire air pressure. Adjustments to vary the distribution of coal on the grates are made by hand. On stokers equipped with stationary dumping grates manipulation of the feeder and air supply, as well as the dumping of the grates, is made by hand during the cleaning period.

Consideration must be given to size when selecting coal for use on a spreader stoker. A distribution of size is helpful in obtaining a uniform depth of coal on the grates. The largest pieces are thrown to the rear, while the fine particles are deposited in front. The size of the largest pieces should not exceed 1½. Most manufacturers recommend a top size of ¾ in. with some limit on the amount smaller than ¼ in. Fine coal can be burned on these stokers, but the cinder carryover with the flue gas and the associated problems are greatly increased. Owing to the effect of size upon distribution in the furnace, it is essential that segregation of the coarse and fine particles of coal be prevented. Coal conveyors and bunkers serving spreader-stoker–fired boilers must be arranged to reduce segregation to a minimum.

The spreader stoker will burn a wide range of fuels varying from lignite to semianthracite. Clinkering difficulties are reduced, even with coals which have clinkering tendencies, by the nature of the spreading action. The fuel is supplied to the top of the fuel bed, and the ashes naturally work their way to the bottom, where they are chilled below the fusing temperature by contact with the incoming air. The coking tendency of a coal is reduced before it reaches the grates by the release of volatile gases which burn in suspension. The utilization of high-ash coal lowers the efficiency and at the same time increases the work of the operator because there is a greater volume of refuse material to be handled. The cleaning periods must be more frequent when the ash content of the coal being burned is high. In general the spreader stoker will burn good coal with an efficiency comparable with that of other stokers and in addition will successfully utilize poor coal but at reduced efficiency.

The spreader-stoker principle can be successfully applied to burn refuse-type materials. Both coal and refuse may be burned in the same furnace. This system performs the two-fold purpose of disposing of

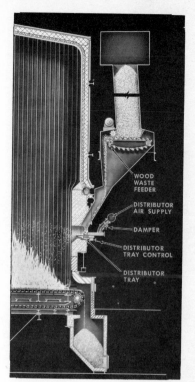

FIG. 4.17   *Spreader stoker for burning wood waste. (Riley Stoker Corp.)*

waste material and at the same time reducing the consumption of costly fossil fuel.   Moreover, the waste fuel frequently has a lower sulfur content, and its use reduces the sulfur dioxide content of the flue gases. Figure 4.17 shows wood waste being blown into a furnace.   The initial burning occurs in suspension, and the final burning takes place on a traveling-grate stoker.   The chain feeder regulates the flow of wood waste from the overhead bin to the burner.   As the fuel falls on the distribution trays it is blown into the furnace by streams of air.   Fuel trajectory is controlled by adjusting the distribution trays.   Figure 4.18 shows refuse feeders installed in a furnace above the conventional coal feeders.   This arrangement provides for the use of refuse material and coal in varying proportions.   It results in ultimate utilization of refuse material and at the same time supplies a variable steam demand.   Almost any kind of combustible refuse can be used provided it is milled to a 4-in. top size.   The partial list of refuse material suitable for use as fuel includes municipal solid waste, bark, bagasse, wood chips, shaving, sawdust, and coffee grounds.

FIG. 4.18 *Conventional coal-burning spreader stoker with airswept refuse distribution spouts.* (*Detroit Stoker Co.*)

Before placing a spreader stoker in operation, check it over carefully to see that it is in good condition. This precaution may prevent a forced outage when the unit is urgently needed. Check the oil level in the reduction-gear case; grease and oil the bearings of all moving parts. Turn on and adjust the flow of cooling water through the bearings, check the dampers and automatic-control mechanism, and operate the fans. Start the stoker and see that it operates without sticking or binding. Set or adjust the over-fire-draft controller to maintain a draft of 0.1 in. of water draft in the furnace.

When a stoker is found to be in satisfactory condition, operate the feeder until an approximately ½-in. layer of coal has been deposited on the grates. Cover the coal with a generous supply of kindling wood and paper. Operate the fans and adjust the dampers to create a draft in the furnace. Now light the fire and regulate the air and draft to produce a bright, smokeless flame. Then start the stoker (if of the moving-grate type) and continue to feed coal slowly to sustain ignition. Bring the unit on the line slowly to avoid rapid temperature changes. Start the over-fire air fan to avoid smoke and to keep the cinder-reinjection piping from overheating. If smoke becomes excessive during this start-up period, reduce the coal feed and adjust the over-fire air. Proceed in accordance with specific start-up instructions for the unit; to speed up the stoker, increase the fuel feed and adjust the airflow and draft.

Even when spreader stokers are provided with controls by which both the fuel feed and the air supply are automatically regulated to meet the steam demand, there are numerous details to which the fireman must attend in order to obtain good operation. The spreader mechanism must be adjusted to distribute the coal evenly over the grates. Irregularities in the fuel bed caused by segregation of coal, wet coal, clinker formation, etc., must be corrected. The fuel bed should be maintained at sufficient thickness to sustain combustion for 2 to 3 min with the fuel feed shut off. The fuel-bed thickness may be changed by adjusting the ratio of fuel to air. An increase in the air pressure under the grates with the same fuel feed will decrease the thickness of the fuel bed; a decrease in air pressure will allow the fuel-bed thickness to increase. The application of over-fire air must be regulated to obtain the lowest amount of excess air (highest carbon dioxide) without permitting temperatures detrimental to the furnace lining, clinkers in the fuel bed, and excessive smoke. Regulate the over-fire air jets to supply additional air for burning the fuel in suspension as well as to increase turbulence and consequently to improve combustion. The furnace draft must be maintained at between 0.05 and 0.10 in. of water to prevent overheating of the furnace lining, the coal-feeding mechanism, and the furnace doors. Frequent inspections must be made of the refractory piers at the spreader outlet to see that there is no accumulation of coke. The bearing-cooling-water flow must be adjusted to maintain an outlet temperature not exceeding 150°F.

Refuse is removed from spreader stokers equipped with a stationary grate by using a hand hoe. The refuse should be removed when it is approximately 3 in. deep on the grates. The cleaning should be performed as quickly as possible in order to minimize disturbance to boiler operation. These stationary-grate stokers should have at least two sections with separate feeders, coal distributors, and divided wind boxes so that they can be cleaned separately. This gives the fireman an opportunity to burn up the carbon before removing the ash and enables him to maintain the steam pressure during the cleaning operation. The cleaning operation is performed by first stopping the coal feed to the unit to be cleaned. When the fuel on the grates has been burned down, close the hand-operated damper to this unit. If the fuel bed is of the correct thickness, the grates will be ready for cleaning in 2 to 3 min. After the fuel has been consumed, quickly hoe the refuse from the grates and remove any clinker which has formed on the side of the furnace or bridge wall. Immediately start the feeder and spread a thin layer of coal on the grates. After the green coal has been ignited, slowly open the hand-operated forced-draft damper. After the fire has become fully established and loss in steam pressure restored, clean the other grate units in a similar manner.

The removal of refuse from hand- or power-operated dumping grates is similar to the procedure with stationary grates except that less labor is involved and the operation can be performed more quickly. After the fire has burned down and the forced-draft damper closed, operate the dumping grates. Some firemen open and close the grates several times to make sure they are clean. Be certain that the grates are back in position and perfectly flat before applying the green coal to reestablish the fire. When the stoker is composed of more than two units, it is advisable to clean the units alternately. After cleaning one unit, permit steam pressure to recover before dumping the next section of grates.

When traveling grates are provided, the refuse is continuously discharged from the furnace into the ashpit, which is maintained under pressure. The coal is thrown to the rear of the furnace, and the speed of the traveling grates is adjusted to give sufficient time for combustion to be completed before the refuse is discharged at the front. When the grates are correctly adjusted, the process is continuous and there is no interference with boiler operation.

The furnace, cyclone, and cinder hoppers must be emptied frequently enough to prevent the accumulation of refuse from interfering with the operation. Excessive refuse in the furnace ashpit will blank off the air and cause the grates to be overheated. When too much material accumulates in the hopper under the cyclone or cinder catcher, its effectiveness will be reduced and an excessive amount of cinder and ash will be discharged to the stack. When means are provided for returning the refuse from the cyclone hopper to the furnace, frequent checks must be made to see that the nozzles do not become clogged. If these transfer nozzles stop operating, the refuse will accumulate in the hopper.

If a spreader-stoker–fired boiler is not required, it is normally advisable to allow the fire to go out and then restart it when there is need for steam. For short periods and in anticipation of sudden increased demands for steam, however, it may be advisable to resort to banking. During banking periods care must be exercised to prevent injury to the grates and to the coal-distribution rotor.

Prepare a spreader stoker, equipped with stationary or dumping grates, for banking by allowing a layer of ashes to accumulate on the grates. Reduce the air supply to the grates, and adjust the feeder and speed of the rotor to deposit a layer of coal on the front end of the grates. Then stop the coal feed and forced-draft fan, but continue to operate the rotor to prevent its overheating. Maintain a slight furnace draft (about 0.1 in. of water) during the banking period by manipulating the dampers. Operate the feeder at intervals to restore the fuel on the grates, and regulate the draft to control the rate of combustion in order to prevent the fire from going out and the steam pressure from decreasing below the level necessary for the anticipated standby requirements. The operators of small spreader stokers sometimes prefer to resort to hand firing when banking. When ready to return the boiler to service, distribute the remaining fuel over the grates and then start the fans to ignite the raw coal quickly as it is fed into the furnace. Hold-fire controls are sometimes provided to apply coal automatically to the grates in sufficient quantities to keep the fire from going out and to maintain the steam pressure required for quickly returning the boiler to service.

To bank a traveling-grate spreader stoker, stop the grate and decrease the coal feed slowly until the steam output is nearly zero and the fuel bed appears nearly black. Close both the forced- and induced-draft-fan dampers and reduce the furnace draft to nearly zero. Allow the fuel bed to darken and get "spotty" with live coals. Then feed raw coal to the front section of the grates. This raw coal will smother the live coals. Waiting until the fuel bed has cooled sufficiently before introducing the raw coal reduces the tendency to smoke and increases the time interval between feedings. Stop the forced-draft fan, and maintain a slight draft in the furnace either by operating the induced-draft fan or by manipulating the damper. When the steam pressure has dropped to 50 per cent of the line pressure, run the fans to restore the pressure and then apply raw coal as before.

The coal-feeder bearing-cooling-water flow should be checked several times each shift. The bearings should be lubricated daily with a solid lubricant which will not thin when heated to 500°F. Dampers and accessory-control and refuse-dumping equipment should be checked whenever the stoker is out of service. The maintenance of grates and furnace linings requires an outage of the boiler, but some repairs can be made to the feeding-and-spreader mechanism when the unit is in operation.

Careful attention to the following details reduces maintenance: Prevent excessive furnace temperature by avoiding overloading the unit and by supplying an adequate amount of air to the furnace; maintain a draft in the furnace at all times; follow established procedures when dumping ashes, blowing down, blowing soot, banking fires, placing the unit in service, and taking it out of service. The cinder-return system must be checked for possible stoppage resulting from slagging over in the furnace or from large pieces of slag becoming lodged in the cinder-inlet nozzles.

4.6   PULVERIZED FUEL. The pulverization of fuel is a means of exposing a large surface area to the action of oxygen and consequently accelerating combustion. The increase in surface area, for a given volume, can be expressed by reference to an inch cube of coal. The cube has six faces; each has an area of 1 sq in., or a total of 6 sq in. of surface. Now, if this 1-in. cube of coal is cut into 2 equal parts, each part will have 2 sides with an area of 1 sq in. and 4 with an area of $\frac{1}{2}$ sq in., making a total of 4 sq in. of surface. The 2 pieces together have a total surface of 8 sq in. as compared with the original 6 sq in. Imagine the increase in surface when the process proceeds until the coal particles will pass through a 200-mesh sieve. The improved combustion and flexibility of the unit must, of course, justify the cost of power required for pulverization.

Electric-generating steam plants were among the first to use pulverized fuel under boilers. They employed the "central system" which consists of a separate plant for drying and pulverizing the fuel. This system provides a high degree of flexibility in the utilization of equipment and in boiler operation. However, it requires additional building space to house the coal preparation equipment and presents fire and explosion hazards. Both the initial installation and operating costs are high. A separate crew of operators is required in the coal preparation plants. For these reasons the central system has been superseded by the "unit system."

With the unit system of pulverization, each boiler is equipped with one or more pulverizing mills through which the coal passes on its way to the burners. The coal is fed to these pulverizing mills by automatic control, to meet the steam demand. No separate drying is necessary since warm air from an air preheater is supplied to the pulverizer mill, where drying takes place. This stream of primary air carries the fine coal from the pulverizer mill through the burners and into the furnace. Combustion starts as the fuel and primary air leave the burner tip. The secondary air is introduced around the burner, where it mixes with the coal and primary air. The velocity of the primary and secondary air creates the necessary turbulence, and combustion takes place with the fuel in suspension.

The satisfactory performance of the pulverized-fuel system depends to a large extent upon the mill performance. The pulverizer mill should deliver the rated tonnage of coal, have a nominal rate of power consumption, produce a pulverized fuel of satisfactory fineness over a wide range of capacities, be quiet in operation, give dependable service with a minimum of outage time, and operate with low maintenance cost.

There is considerable variation in the pulverized-fuel fineness requirement. Coals with low volatile content must be pulverized to a higher degree of fineness than those with higher volatile. For normal conditions bituminous coal will burn satisfactorily when 65 per cent will pass through a 200-mesh sieve and 99 per cent will pass through a 40-mesh sieve. It is wasteful of energy to pulverize coal finer than required to obtain satisfactory combustion.

The three fundamental principles employed in the construction of mechanical pulverizer mills are contact, ball, and impact.

Contact mills. These mills contain stationary and power-driven elements, which are arranged to have a rolling action with respect to each other. Coal passes between the elements again and again, until it has been pulverized to the desired fineness. The grinding elements may consist of balls rolling in a race or rollers running over a surface. A stream of air is circulated through the grinding compartment of the mill (Fig. 4.19a). The rotating classifier allows the fine particles of coal to pass in the airstream but rejects the oversize particles, which are returned for regrinding. The airstream with the pulverized coal flows directly to the burner or burners.

The sectional views (Fig. 4.19) show how the primary air fan supplies a mixture of room and heated air for drying the coal and for transporting it from the mill, through the burners and into the furnace. The proportion of room air to heated air is varied, depending upon the moisture content of the coal, to effect sufficient drying. The fan is on the inlet and therefore is not subject to wear owing to air and pulverized coal mixture. However, the mill is pressurized and casing leakage causes pulverized coal to be blown into the room.

The mill shown in Fig. 4.19 employs steel balls and rings or races as grinding elements. The lower race is power-driven but the upper is stationary. Pressure is exerted upon the upper race by means of springs. The force exerted by these springs can be adjusted by the screws which extend through the top of the mill casing. The coal is pulverized between the balls and the lower race.

The grinding elements of these mills are protected from excessive wear and possible breakage by heavy foreign objects in the coal. These heavy particles resist the upward thrust of the stream of primary air and collect in a compartment in the base, from whence they are removed periodically.

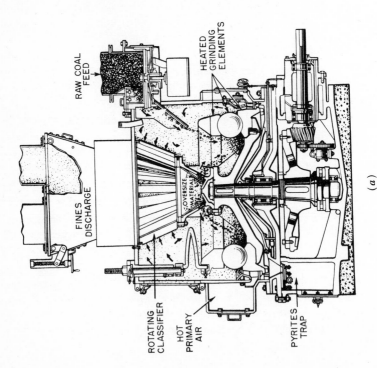

(b)

RAW COAL
FEED

HEATED
GRINDING
ELEMENTS

FINES
DISCHARGE

OVERSIZE
MATERIAL

ROTATING
CLASSIFIER

HOT
PRIMARY
AIR

PYRITES
TRAP

(a)

FIG. 4.19 *Ball-and-race type pulverizer mill.* (a) *Cross section showing path of coal and air.* (b) *Mill interior and primary air fan. The Babcock & Wilcox Co.*)

The coal supply to the burners is automatically regulated by the combustion control. When additional coal is required the flow of primary air is increased. When this control senses an increased demand for steam, more primary air is supplied to carry additional coal to the burners. In this way the residual coal in the mill is utilized to give quick response to a demand for more steam. The mill coal-feed controller senses a drop in differential pressure across the grinding elements, owing to the decrease in quantity of coal, and increases the coal feed to supply the deficiency.

The mill shown in Fig. 4.20 employs rolls and a ring as grinding elements. The ring is power-driven, but the rolls axles are stationary. The revolving ring causes the rolls to rotate and the coal is ground between the two surfaces. The segmented grinding ring and replaceable "tires" on the rolls are provided to reduce and facilitate maintenance.

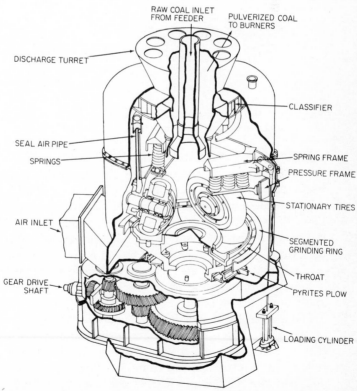

FIG. 4.20   *M.P.S. type pulverizer.* (*The Babcock & Wilcox Co.*)

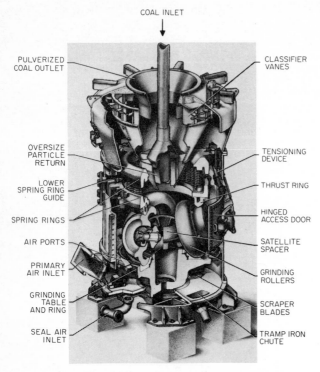

COAL INLET

PULVERIZED
COAL OUTLET

CLASSIFIER
VANES

OVERSIZE
PARTICLE
RETURN

TENSIONING
DEVICE

LOWER
SPRING RING
GUIDE

THRUST RING

SPRING RINGS

HINGED
ACCESS DOOR

AIR PORTS

SATELLITE
SPACER

PRIMARY
AIR INLET

GRINDING
ROLLERS

GRINDING
TABLE
AND RING

SCRAPER
BLADES

SEAL AIR
INLET

TRAMP IRON
CHUTE

FIG.   4.21   *M.B.   Ring-and-roller   pulverizer   mill.*
(*Foster Wheeler Corp.*)

In the ring-and-roller-type mill shown in Fig. 4.21, coal enters through the center tube and is deposited on the power-driven grinding table and ring.   Pulverization is accomplished by this ring and three contacting rollers.   These rollers are held 120° apart by a three-prong spacer.   Pressure between the rollers and the grinding ring is maintained by a stationary thrust ring and adjustable spring assembly.   The motion of the grinding ring is transmitted to the rollers, and they rotate at about one-half the speed of the ring.   The pulverized coal is carried by primary air up through the adjustable classifier and out the angular opening at the top of the mill.

The pulverizer shown in Fig. 4.22 has grinding elements consisting of stationary rollers and a power-driven bowl in which pulverization takes place as the coal passes between the sides of the bowl and the rollers.   A primary air fan draws a stream of heated air through the mill, carrying the fine particles of coal into a two-stage classifier located in the top of the pulverizer.   The vanes of the classifier return the coarse particles of coal through the center cone to the bowl for further

FIG. 4.22   *Bowl-type pulverizer with fan. (Combustion Engr. Co. Inc.)*

grinding, while that which has been pulverized to the desired fineness passes out of the mill, through the fan, and into the burner lines. The automatic control changes the supply of coal to the bowl of the mill by adjusting the feeder speed and the flow of primary air by regulating a damper in the line from the pulverizer to the fan.

These mills have several mechanical features which warrant consideration. The construction details of the roller assembly in Fig. 4.23 show how the bearings are pressure-lubricated to exclude coal dust. This figure also shows how the roller-support bearing and spring may be adjusted to hold the rollers in place relative to the grinding-ring surface of the revolving bowl. The roller does not come into contact with the grinding ring but allows some space for the coal to pass. The classifier located in the top of the mill may be adjusted to change the coal fineness while the mill is operating. The primary air fan is located on the outlet, which results in the mill's operating under a draft or negative pressure; this eliminates the leakage of coal dust from the mill casing. In this location, the fan must handle primary air after it has picked up

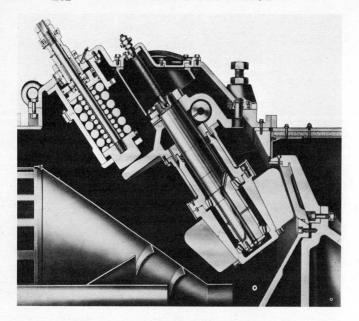

FIG. 4.23 *Roller assembly and spring-pressure mechanism of bowl-type pulverizer. (Combustion Engr. Co. Inc.)*

the pulverized coal; consequently there is some wear of fan blades and casings. The impurities in coal compose the heavy particles, and when these enter the rotating bowl, they are thrown over the side by centrifugal force. These heavy particles then fall into the space below the bowl and are discharged from the mill through a specially provided spout.

Pulverized coal may be readily transported through a pipe by means of a high-velocity stream of air, but the coal will not be equally distributed through the cross section of the pipe. This causes difficulty when two burners are supplied from one pulverizer mill. The distributor (Fig. 4.24) divides the cross section of the main pipe into several narrow strips. The mixture of pulverized coal and air which enters these strips is directed alternately into the two burner pipes in order to secure equal amounts of coal to the burners.

Ball mills. The typical ball mill, shown in Fig. 4.25, consists of a large cylinder or drum partly filled with various-sized steel balls. The cylinder is slowly rotated while coal is fed into the drum. The coal mixes with the balls and is pulverized. Hot air enters the drum, dries the coal during pulverization, and then carries the pulverized fuel out of the drum through the classifiers and to the burners. The classifiers return the oversized particles to the drum for further grinding. In ad-

FIG. 4.24 *Pulverized-coal distributor.* (*Combustion Engr. Co. Inc.*)

dition to the mills, raw-coal feeders (a source of heated air), primary air fans, burners, and controls are required for a complete pulverized-coal system.

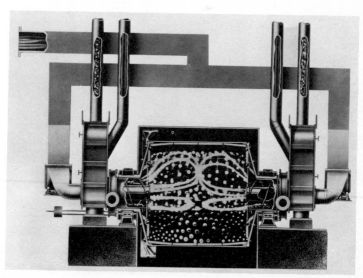

FIG. 4.25 *Double-classifier ball mill.* (*Foster Wheeler Corp.*)

These pulverizers have several features worthy of note. Unlike the grinders in most other pulverizers, the grinding elements in these units are not seriously affected by metal and other foreign matter in the coal. The pulverizers contain enough coal for several minutes' operation. This feature prevents the fire from going out when there is a slight interruption in fuel feed caused by coal's clogging in bunkers or spouts. These mills have been used successfully for a wide range of fuels including anthracite and bituminous coal, which are difficult to pulverize.

Impact mills. In a pulverizer mill which employs the impact principle the coal remains in suspension during the entire pulverizing process (Fig. 4.26). All the grinding elements and the primary air fan are mounted on a single shaft. The primary air fan induces a flow of air through the pulverizer, which carries the coal to the primary stage, where it is reduced to a fine granular state by impact with a series of hammers, and then into the final stage, consisting of pegs carried on a rotating disk and traveling between stationary pegs, where pulverization is completed by attrition. The now finely pulverized coal is carried to the center of the pulverizer, where it encounters the rapidly rotating scoop-shaped rejector arms, which throw the large particles

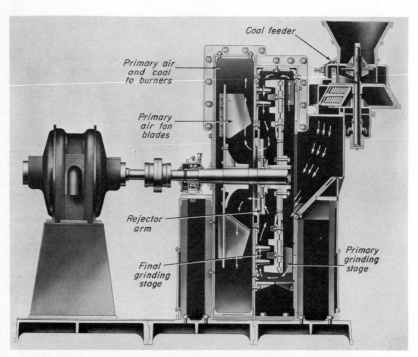

FIG. 4.26　*Cross section of Atrita unit pulverizer.*　(*Riley Stoker Corp.*)

back into the grinding section while those of the desired fineness are passed through the fan and discharged into the burner lines.

The principle of operation of the impact mills results in several interesting characteristics. These pulverizers may be directly connected to the motor drive and operated at high speed. There is only a few seconds' coal supply in the mill, and any variation in the rate of feed to the mill is almost immediately reflected in the output. This means that the power required to drive the pulverizer is nearly proportional to the coal pulverized, over a wide range of rating. Owing to high-speed operation and to the fact that the fan is integral with the pulverizer, a mininum of floor space is required.

Control of pulverizer output is accomplished by varying the coal feed and the flow of primary air, either by hand or by automatic control.

Mill feeders. Care must be exercised to prevent scrap metal from entering pulverizer mills. Figure 4.27 shows a feeder with a magnetic separator used in connection with an impact pulverizer. The rotating drum *A* has eight coal pockets. The spring-loaded apron *B* levels off the coal and prevents flooding past the feed, yet it will release if there is a large lump of coal in one of the pockets. The wiper *C* is geared to the

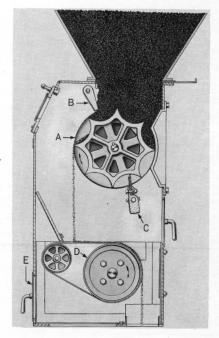

FIG. 4.27 *Pulverizer feeder with magnetic separator.* (*A*) *Rotating drum.* (*B*) *Spring-loaded apron.* (*C*) *Wiper.* (*D*) *Magnet pulley.* (*E*) *Inspection door.* (*Riley Stoker Corp.*)

main feeder and rotates eight times as fast as the main feeder shaft. This wiper completely removes the coal from each pocket, thus helping to maintain uniform feed even when the coal is so wet that it has a tendency to stick in the feeder. From the feeder the coal drops onto a belt which runs over the magnetic pulley D. The coal drops vertically from the belt, while magnetic material is carried to the left and is deposited in the box, from whence it may be removed periodically. The raw coal to all pulverizers should be passed over a magnetic separator.

Burners. The efficient utilization of pulverized coal depends to a large extent upon the ability of the burners to produce mixing of coal and air and turbulence within the furnace. Figure 4.28 shows the construction and operation of a pulverized-coal burner. The coal-and-air mixture, as it comes from the pulverizer, enters the burner chamber at B. The mixture then flows through the adjustable vanes D, which divert the stream and evenly distribute the coal in the nozzle C. At the tip of the pipe the stream of coal and air is deflected by the spreader E and enters the furnace with a whirling motion in the form of a conical flare. This rich mixture ignites readily and burns with a stable flame. The spreader is attached to the end of a tube which extends from the back of the burner. This makes it possible for the operator to adjust the spreader in or out with respect to the tip of the burner pipe in order to vary the shape of the flame.

The secondary air is supplied under pressure by the forced-draft fan. The quantity entering the burner box is controlled by damper M. The amount of whirl imparted to the secondary air as it enters the furnace is regulated by the position of the vanes. Just inside the furnace the burning stream of coal and primary air is met by the whirling secondary air, and combustion proceeds rapidly. The primary and secondary vanes and the primary air pressure must be adjusted to prevent the flame from striking the furnace walls and boiler tubes.

This burner is equipped with an opening Z through which a gas or kerosine torch may be inserted for lighting the burner. Many pulverized-fuel burners have oil burners which extend through the spreader carrying tube (Figs. 4.34, 4.35). These auxiliary burners are effective in igniting the pulverized fuel and add to the safety of operation.

Adequate combustion space must be provided for satisfactory combustion of pulverized coal. Flame temperatures are high, and the conventional types of refractory-lined furnaces are inadequate. For high-rating units it is necessary to resort to some form of water-cooled walls (see Chap. 1). The use of waterwalls permits higher rates of heat release with a consequent reduction in furnace size. The furnace volume must be sufficient to complete combustion before the gases reach the relatively

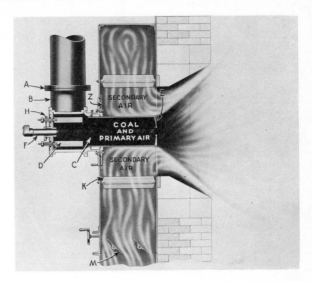

FIG. 4.28 (*a*) *Cross section of flare-type pulverized-coal burner showing flame.* (*b*) *Phantom view of the flare-type pulverized-coal burner showing primary and secondary air-deflecting vanes.* (*A*) *Burner pipe flange.* (*B*) *Inlet chamber.* (*C*) *Nozzle.* (*D*) *Adjustable vanes.* (*E*) *Coal spreader.* (*F*) *Spreader carrying tube.* (*H*) *Spreader-carrying-tube locking screw.* (*K*) *Secondary air vanes.* (*M*) *Secondary-air plenum-chamber damper.* (*Z*) *Opening for inserting lighting torch.* (*Riley Stoker Corp.*)

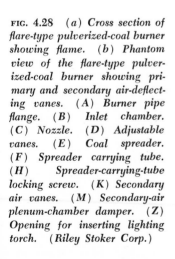

(*a*)

(*b*)

cool heating surface. However, the use of burners which produce turbulence in the furnace makes it possible to complete combustion with flame travel of minimum length.

Operation. The ash in the coal presents a problem which must be given careful consideration in furnace design. Small and medium-sized units are arranged for the removal of ash from the furnace while it is in the dry state. In this case, care must be exercised to prevent the temperature of the ash from exceeding its fusion temperature. If this precaution is not taken, large masses of slag will form. In some large units the ash in the furnace is maintained at a temperature in excess of the melting point. The slag is "tapped" and allowed to flow from the furnace in the liquid state. After leaving the furnace, it is rapidly chilled by the application of water. This quick change in temperature causes slag to shatter into small pieces.

The use of slag-tapped furnaces enables the operator to use a minimum of excess air because the furnace can be maintained at a high temperature without slagging difficulties. Since the usual slagging difficulties are overcome, it is possible to use a wide variety of coals, including those having a low-fusing-temperature ash. Heat-release rates in excess of 50,000 Btu have been successfully maintained in these slag-tapped furnaces.

The principle of pulverized-coal firing has been widely accepted, first in large utility and then in industrial and large heating plants. The choice of pulverized in preference to other firing methods depends upon the size of the boiler unit, the type of coal, the kind of load (constant or fluctuating), the load factor, the cost of fuel, and the training of personnel.

Pulverized-fuel equipment has not been adapted generally to small units. The benefits in efficiency and flexibility do not warrant the complications in equipment and operating technique.

Some of the characteristics of coal which are so important when it is burned on grates need less consideration in the pulverizer plant. Coal in sizes less than 2 in. can be pulverized without difficulty, and segregation does not usually affect operation. Coking and caking characteristics, so important in stoker operation, require less consideration. In general, high-volatile coals are harder to grind (have lower grindability) than low-volatile coals, requiring larger mills for the same capacity, more power per ton, and more maintenance. The disadvantages of high-volatile coal are partly compensated for by the fact that the gases are more easily distilled and burned, making it unnecessary to pulverize the coal as finely as is required for low-volatile coal. Furthermore, it is easier to maintain ignition with high- than with low-volatile coal, owing to the presence of the volatile gases.

Because there is only a small amount of fuel at any one time in a pulverized-fuel furnace, it is possible to change the rate of combustion quickly to meet load demands. This is frequently one of the factors worthy of consideration when combustion equipment is selected. The possible rate of change in output, from the banked condition to full load, is usually limited by the boiler rather than by the pulverizer and combustion equipment. Automatic control applied to pulverized-fuel–fired boilers is effective in maintaining an almost constant steam pressure under wide load variations.

A pulverizing plant costs more than some of the other methods employed in burning coal, and this difference in cost must be repaid in decreased maintenance and increased efficiency. To effect the greatest possible financial benefit from this increased efficiency the equipment must be utilized at high ratings and during a large percentage of the time. A pulverized-coal plant can utilize a wide variety of coal, including the lower grades. The use of low-cost fuel frequently justifies the increased cost of pulverizing equipment. When the fuel cost is high, the installation refinements in combustion and heat-recovery equipment can be justified. The question is: will the cost of equipment required for increased economy be justified by fuel saving?

The pulverized-fuel plant requires trained personnel who are familiar not only with the dangers involved but also with safe operating practices. To control the rapid combustion of the pulverized fuel, as it is burned in suspension, requires an alert operator.

In order to have an explosion with pulverized coal there must be the correct mixture of coal dust and air, as well as a source of ignition. It follows that for safe operation the mixture of coal and air must be either too "lean" (too much air) or too "rich" (too much coal) to explode except in the furnace. This principle is applied both in the boiler room and in the pulverizer. In the boiler room all areas are kept clean so that there is no dust to be agitated by a draft or minor explosion and so to produce sufficient concentration in the air to be exploded by a chance spark or an open flame. Inside the pulverizer a high percentage of coal to air (rich mixture) is maintained so that there will not be enough air to cause an explosion even if a spark is produced in the pulverizer. Once a combustible mixture has been established, the application of a spark or an open flame will cause an explosion. The pulverized coal must be burned as fast as it is introduced into the furnace or an explosive mixture will be produced.

Reasons for and against the Use of a Pulverized-fuel Plant

Advantages

Can change rate of combustion quickly to meet varying load

Requires low percentage of excess air

Reduces or eliminates banking losses

Can be repaired without cooling the unit down, since equipment is located outside the furnace

Can be easily adapted to automatic combustion control

Can utilize highly preheated air successfully

Can burn a wide variety of coal with a given installation

Limits use of fine wet coal only through the ability of the conveying-and-feeder equipment to deliver the coal to the pulverizers

Disadvantages

Is costly to install

Requires skilled personnel because of explosion hazards

Requires electrostatic fly-ash precipitators to reduce stack emission to acceptable limits

Requires multiple mills and burners to obtain a satisfactory operating range

Gives difficulty from slag deposits on the lower rows of boiler tubes

Requires extra power to pulverize the fuel

Before attempting to operate a pulverized-fuel boiler, thoroughly inspect the following: fuel-burning equipment, controls, fans, and safety interlock equipment. The failure of any one might result in an explosion. If, for example, the induced-draft fan should fail and the pulverizer mill and primary air fan continue to operate and supply coal to the furnace, an explosion might occur. In order to reduce such a possibility the boiler auxiliaries are equipped with interlocks, by means of which when one auxiliary fails, the others stop automatically. When the induced-draft fan fails, the interlocks are arranged to stop the mill feeder, the mill, and the forced-draft fan in the order listed. Electric-eye flame detectors are used to indicate flame failure or to disrupt the fuel supply when the flame failures occur. Television receivers are installed in the control panel to permit constant observation of the flame. Study the safety equipment on your boiler, check it at regular intervals, and make sure that it is in good operating condition at all times. Remember it was placed there for your protection, very probably as the result of difficulty encountered with similar equipment.

The exact procedure for lighting a pulverized-fuel furnace depends to some extent upon the types of pulverizer, burners, and other equipment. There are, however, certain general practices which are common to all. First, operate the induced-draft fan and set the dampers to produce a furnace draft sufficient to prevent a pressure from being developed when the fire is lighted. Then insert the lighted torch and make sure it is producing sufficient flame in the path of the pulverized coal. Now start the pulverizer and feeder, and adjust the flow of primary air and the mill-coal level to supply a rich mixture to the burner.   Regulate the coal-and-air supply to the burner and maintain as rich a mixture as possible without smoke until the furnace has been heated.   Do not depend upon hot walls or the fire from an adjacent burner to light a pulverized-fuel burner.   Use a lighting torch for each burner.   If the coal as it is discharged from a burner fails to ignite after even a few seconds, the furnace will be filled with an explosive mixture.   When this occurs, it is necessary to operate the induced-draft fan until the explosive mixture has been removed.

Care and judgment on the part of the operators are necessary during the warming-up period.   There is a low limit of output below which a given pulverizer mill cannot be satisfactorily operated.   When the primary air drops below a minimum velocity, it fails to remove sufficient coal from the mill and a lean mixture results.   This minimum mill output usually results in too high a firing rate during the warming-up period. The only solution is to fire the boiler at the lowest rating, intermittently and frequently enough to bring the unit on line in the specified time. Safety precautions and procedures must be carefully followed during each lighting.

Owing to the possible high rate of change in firing and to the possibility of the flames being extinguished by momentary interruptions of fuel feed, more precautions are necessary with pulverized fuel than when coal is burned on grates.   Most pulverized-fuel boilers are equipped with automatic controls which relieve the operators of tedious adjustments of fuel and air.   These controls do not, however, lessen an operator's responsibility for the equipment.   In case of failure or emergency, the unit must be operated safely by hand control.   The control regulates the rate of fuel feed to produce the required amount of steam in order to maintain the established pressure.   The air is regulated to burn the coal with the amount of excess air considered most satisfactory for the given installation.   The steam-flow-airflow meter and the carbon dioxide and oxygen recorders are used to indicate the amount of excess air.   Burners, louvers, spreaders, and primary air pressure must be adjusted to produce a flame that will not strike the furnace walls yet will result in almost complete combustion before the gases enter the boiler passes.

When the mixture of coal and primary air becomes too lean, the flame will have a tendency to leave the burner tip and then return.   This condition may become so severe that the combustion consists of a series of small explosions.   This pulsating condition can be corrected by adjusting the ratio of air to fuel.

Slag deposits in the furnace are a guide to the operator in adjusting the flame shape. Excessive temperatures in the furnace are caused by a deficiency of air, unequal distribution of fuel, or excessive ratings and result in slag formation and high furnace maintenance. The heated and tempering air to the pulverizer mill must be proportional to produce an air and pulverized-coal outlet-mill temperature of 150 to 180°F. The moisture content of the coal entering the mill determines the amount of preheated air which must be mixed with the room tempering air to produce the desired outlet temperature.

Pulverized-fuel–fired boilers are banked by closing off the coal supply and thus extinguishing the fire. Insert the lighting torch and stop the coal feed to the mill. Allow the pulverizer to operate until all the coal has been removed. When the pulverizer is nearly empty, the coal feed will be irregular and it is necessary to have the lighting torch in place to prevent the flame from being extinguished. If it is necessary to have the boiler ready for service on short notice, light the burners at intervals to maintain the steam pressure. After each firing period close the dampers to prevent the escape of heat and thus to increase the length of time between firing periods.

Practically all the pulverized-fuel equipment may be located outside the furnace and may, therefore, be serviced, as has been said, without waiting for the boiler to cool down. Large units are equipped with two mills (in a few installations with even more), one of which will supply sufficient coal to operate the boiler at three-quarters of its maximum output. This makes it possible to adjust and maintain the pulverizer mills without taking the boiler out of service. Most small units are arranged so that they can be supplemented with fuel oil if the pulverizer equipment fails.

The principal parts of a unit pulverizer system which require adjustment and maintenance are the grinding elements, the mill liners, the primary air-fan rotors and liners, the piping and burners through which mixtures of coal and air flow, the dampers, and the classifiers.

In the contact mill shown in Fig. 4.19 the spring tension must be adjusted to compensate for wear of balls and races. This adjustment is made by means of the bolts which extend through the top of the mill. These bolts are adjusted to maintain a specified spring length. Once arranged to give the correct fineness, the classifier does not require further adjustment. The primary air fan handles the clean air to the mill and, therefore, requires very little maintenance. The satisfactory operation of the mill depends, to a considerable extent, upon the quantity of coal maintained in the grinding compartment as determined by the mill-feeder-controller setting. When there is insufficient coal in the mill, the mixture supplied to the burner is too lean for satisfactory combustion. If there is too much coal in the mill, a considerable amount will be discharged to the impurities-reject compartment or pyrites trap (Fig. 4.19), causing the operator unnecessary work.

The roller-type contact pulverizer requires occasional adjustment of the grinding elements. During operation the bowl is partly filled with coal. The lower end of the rollers and the bottom of the bowl liner ring are subjected to more wear than the other parts of these grinding elements. As this wear increases, the space between the rollers and the ring increases, and the coal level in the bowl must be higher for a given output. It then becomes necessary to adjust grinding surfaces so that they will be parallel and $\frac{3}{8}$ to $\frac{1}{4}$ in. apart. Failure to maintain this adjustment will result in overloading the driving motor and in spillage of coal from the impurities-discharge spout, especially at high ratings. The roller-spring tension is established by the manufacturer to satisfy specific load and coal conditions. This tension is determined by the spring length and should not be altered without thorough investigation. When it is impossible to adjust the rollers to maintain the relation between the grinding surfaces, the rollers and grinding ring must be replaced. Some operators have found it practical to replace the rollers twice to one replacement of the grinding ring.

The primary air fan in this roller-type contact pulverizer handles the mixture of air and coal as it comes from the mill and is, therefore, subject to wear. This fact has been taken into consideration in the design of both the fan blades and the casing in that they have replaceable liners. The stationary classifier by which the fineness of the coal can be regulated may be readily adjusted, while the mill is in operation.

The ball mill shown in Fig. 4.25 is supplied with the necessary assorted sizes of steel balls when first placed in service. Wear reduces the size of the balls, and the grinding capacity is maintained by adding a sufficient number of the large balls. Normal wear reduces the size of the balls and thus provides the necessary variation in size. The liners in the cylinder are wear-resistant and seldom require replacement. The coal-and-air mixture passes through the primary air fan, which is, therefore, subject to some wear, but the parts subject to wear are easily replaceable.

The Atrita pulverizer shown in Fig. 4.26 operates at the same speed as the motor and, therefore, eliminates the use of reducing gears. Wear occurs in the pins which perform the secondary or final pulverization. These parts are designed, however, so that considerable wear can take place before the fineness is appreciably affected. All parts, including the primary air fan, are readily accessible for inspection and replacement by removal of the upper casing in much the same manner one would use to inspect a centrifugal pump.

Whatever the type of mill, every effort should be made to keep the pulverizer system tight since the leakage of pulverized coal is dangerous; many explosions have resulted from the careless practice of allowing pulverized-coal dust to settle on the equipment. When pulverized-coal dust leaks out of the system, it should be cleaned up immediately and the leaks repaired.

The tips of pulverized-fuel burners are exposed to the radiant heat of the furnace and are, therefore, subject to some deterioration. The burner shown in Fig. 4.28 utilizes a spreader to obtain turbulence in the furnace. Spreaders of this type burn and warp and must, therefore, be repaired or replaced if the highest efficiency is to be maintained.

Also vanes and dampers must be kept in good operating condition.

4.7   FLUIDIZED-BED COMBUSTION. Fluidization offers a unique method of mixing fuel and air to obtain combustion. The fluidized furnace consists of an enclosed space with a base having openings for admitting air. The space above this perforated base has a layer of sand or other inert granular material. The plenum beneath the base is supplied with combustion air by a blower. The flow of air through the perforated base is sufficient to lift (fluidize) the bed of granular material and suspend it in the airstream. This creates a violent mixing and agitation of the bed material.

After the bed has been fluidized, oil or gas is introduced and ignited. The bed temperature is raised to above the ignition point of the fuel to be burned (1100 to 1200°F). Almost any fuel, solid or liquid, can now be introduced into the furnace. Solid fuel must be crushed or shredded.

The temperature in the furnace must not be allowed to reach the fusion temperature of either the granular bed material or the ash in the fuel. The optimum operating temperature is between 1500°F and 1600°F.

When municipal refuse, sewage-plant sludge, and other high-moisture fuels are burned in the fluid-bed furnace, the heat required to evaporate the moisture maintains the bed temperature within an acceptable range. When the temperature tends to become excessive it is lowered by adding either moisture or air. When the temperature tends to become less than optimum, it is restored by the addition of supplemental fuel. The heat generated from the combustion of these waste materials can be utilized in waste-heat boilers or dissipated in the water used in the flue-gas scrubbers.

If coal and other high-heating-value fuels were burned in a refractory-lined fluidized-bed furnace, the temperatures would be excessive. To avoid this condition the furnace walls are made of tubes containing water, which absorbs the heat directly from the bed. The agitated bed results in rapid heat transfer to these surfaces; therefore the unit is smaller than a conventional one for a given output. In order to take advantage of this heat transfer and to limit the temperature, the bed is divided into a number of cells each surrounded by waterwalls. The heat in the gases leaving the furnace is removed by convection surfaces and heat-recovery equipment in the conventional manner.

The low temperatures maintained in these fluid-bed furnaces offer some advantages in the reduction of air pollution. The formation of nitrogen oxides is reduced by these low furnace temperatures. Limestone can be introduced into the furnace to reduce the discharge of sulfur oxides to the atmosphere.

The high velocities tend to carry most of the ash from the furnace. When coal is burned, provisions are made to return the carbon for reburning. Provisions must also be made to clean the gases with scrubbers or electrostatic precipitators.

4.8   FUEL OIL.   The satisfactory utilization of fuel oil in a furnace requires consideration in the selection of equipment and grade of oil as well as attention to operating details. The necessary amount of plant storage capacity depends upon the amount of oil consumed, availability of the supply, and transportation facilities. The oil-heating equipment must be adequate to heat the heaviest oil that is to be burned at the maximum rate of consumption in the coldest weather. The pumps should deliver the maximum quantity at the pressure specified for the type of burner used. The pumps should be installed in duplicate, each having sufficient capacity for maximum requirement, and preferably with two sources of power, such as steam and electricity. The burners must be selected to deliver the required quantity of oil to the furnace in a finely divided mist to assure quick and thorough mixing with the air.

Recommendations and requirements for safe storage and transport of fuel oil are explained by the National Fire Protection Association.[1]   In addition the installation must also comply with the local codes and insurance company requirements. Meeting these requirements is the responsibility of the designer but the operator must also be familiar with the requirements to make certain that codes are not violated by maintenance changes, operating practice, or neglect.

Underground steel tanks afford a generally accepted method of storing fuel oil, and the added safety justifies the added cost. However, there are situations where underground storage is too costly and otherwise not feasible. Aboveground tanks are used with restrictions on how close they can be placed to other structures. Moreover, dikes or other type of secondary containment must be provided to prevent the spread of oil if the tank should rupture or leak.

[1] National Fire Codes, 470 Atlantic Avenue, Boston, Ma. 02210.

Fuel-oil tanks require connections for the following: filling, venting, pump suction, oil return, sludge removal, and level measurement. When the heavier grades of oil are to be used, the tank must have steam coils for heating the oil, thereby lowering its viscosity, to facilitate pumping. This is accomplished by use of a tank suction heater. The heating coils, suction lines, and return lines terminate within an enclosure in the tank. A small portion of oil is heated to operating temperature while the remainder is at a lower temperature. The oil to the pumps is maintained at optimum temperature.

When the tank is located above the pump level, special precautions should be taken to prevent the oil from being siphoned from the tank if a leak should develop. One method is to provide an antisiphon valve which will open under the suction action of the pump but resist siphoning. In event of a fire, oil could be siphoned from a tank located above grade. To guard against this possibility a shutoff valve is installed in the discharge line. This valve is held open by a system containing a fusible link. A fire would open the fusible link closing the valve and so prevent the flow of oil.

When the tank is below the suction of the pumps, a check valve must be installed in the suction line above the tank to prevent oil from draining back into the storage tank. The suction line must be free from leaks to assure satisfactory operation of the pumps. The discharge line must be tight to prevent fire hazard and unsightly appearance caused by leakage.

Supply and return lines intended for use with heavy oil must be insulated and traced to prevent the oil from congealing. Tracing consists of applying heat beneath the insulation, either by a tube supplied with steam or by passing electricity through a resistance wire.

Fuel oils frequently contain sludge which settles in the bottom of storage tanks, causing irregular pump and burner operation. Additives are available which hold this sludge in suspension and reduce operating difficulties. A manhole must be provided to permit cleaning and inspection of the interior of the tank. Since the sludge and water collect in the bottom of the tank, it is advantageous to have a connection to permit its removal without draining the tank.

A typical fuel-oil-supply system is shown in Fig. 4.29. Oil flows from the storage tank through the suction strainers, steam- or electric-driven pump, steam or electric heater, discharge strainer, meter, regulating valve, and safety shutoff valve, to the burners or return line to the tank. Atomizing steam is supplied from the main steam header through a regulating valve to the burners. The combustion control system positions the regulating valves in both oil and steam lines to proportion the fuel to boiler output steam requirements.

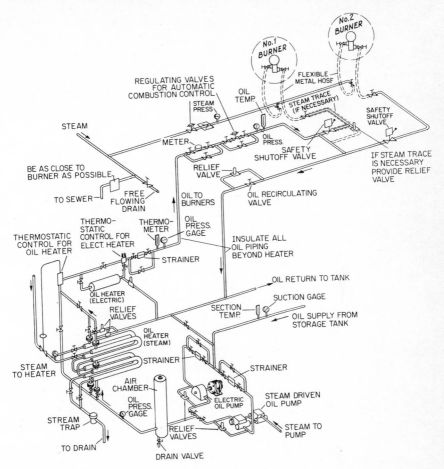

FIG. 4.29 *Typical fuel-oil pumps, heaters, and piping arrangement. (Factory Mutual Engineering Corp. Factory Mutual System.)*

Strainers are provided to prevent foreign material from being delivered to the burners.   It is good practice to use a coarse screen, 16- to 20-mesh, in the suction line, and a finer screen, 40- to 60-mesh, in the discharge line.   The discharge screen is located in the line after the oil heater. These screens are of the twin type, which enables the operator to clean one screen basket while the oil is passing through the other.

There are several types of pumps which may be employed to deliver oil satisfactorily from the storage tank to the burners.   Owing to their low first cost and reliability, steam-driven duplex pumps are extensively used for pumping oil.   These pumps are equipped with governors, which regulate the flow of steam to maintain the desired oil pressure. Electric-driven gear pumps are also used frequently for pumping oil (see Fig. 7.14). These displacement pumps must be selected for the viscosity of the oil to be used. They are operated at constant speed and deliver a quantity of oil in excess of that required by the burners. The excess is discharged through a spring-loaded bypass valve to the pump suction or the storage tank. This bypass valve is set to maintain the desired pressure at the burners, and since the pump has a capacity in excess of the burner requirements, it is possible to supply the demand automatically.

Oil is usually heated with steam, but electric heaters are used in small installations and for standby. Tank heaters raise the oil temperature sufficiently to reduce the viscosity, thus facilitating straining and pumping. The heater located on the high-pressure side of the pump lowers the viscosity of the oil so that it can be atomized effectively by the burners. Oil heaters are supplied with thermostats which regulate the steam supply and maintain a constant outlet-oil temperature. Thermometers in the lines permit the operator to check the temperature of the oil and hence the performance of the regulator. (Recommended temperatures for the various grades of fuel oil are given in Chap. 3.)

The efficient and satisfactory utilization of fuel oil depends to a large extent upon the ability of the burners to atomize the oil and mix it with air in the correct proportions. Burners are classified as rotary, steam, air, or pressure according to the method employed to atomize the oil.

Rotary burners. The rotary burner shown in Fig. 4.30 is a self-contained unit in which the motor drives the oil pump, the primary air fan, and the atomizing cup. The pump supplies oil to the atomizing cup 4, which is rotated at 3,450 rpm. The centrifugal force created by this rotation causes the oil to leave the edge of the cup in a fine spray. A current of primary air produced by the fan 7 is introduced through the passage 8 into the fine spray of oil. This high-velocity air introduces a secondary flow through the passage 10, which serves the dual purpose of cooling the burner and aiding combustion.

The spray-type burner employs air or steam as the atomizing agent. In the inside-mixing-type burner, the atomizing agent and oil mingle before they leave the tip, whereas in the outside-mixing type they come together after leaving the tip.

Steam-atomizing burners. A steam-atomizing inside-mixing-type burner is shown in Fig. 4.31. Atomization is accomplished by projecting steam tangentially across the jets of oil. This results in the formation of a conical spray of finely divided oil after the mixture has left the orifice plate. This atomizer may be used in connection with the multiple-fuel burner shown in Fig. 4.34. The vanes control the flow of secondary air and make it possible to use forced draft. The burning rate

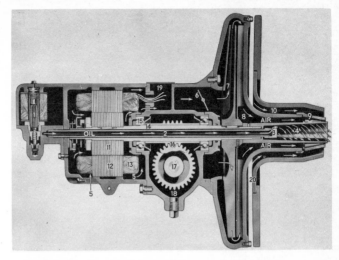

FIG. 4.30 *Rotary oil burner with integral pump.* (*1*) *Oil-metering valve.* (*2*) *Fuel-oil tube.* (*3*) *Oil-distribution head.* (*4*) *Rotary atomizing cup.* (*5*) *Cooling-air passages around motor.* (*6*) *Adjustable primary air shutter.* (*7*) *Primary air-fan rotor.* (*8*) *Primary air passage to burner.* (*9*) *Angular-vaned air nozzle.* (*10*) *Passage for induced air to cool burner front plate and sleeve.* (*11*) *Motor rotor.* (*12*) *Motor stator.* (*13*) *Motor-stator windings.* (*14*) *Motor shaft for driving oil pump and rotating cup.* (*15*) *Ball bearings.* (*16, 17*) *Oil-pump driving gears.* (*18*) *Splash-feed lubricating-oil reservoir.* (*19*) *Motor-leads junction box.* (*20*) *Sleeve to provide clearance for water legs, refractory walls, or dead plates.* (*Petro Oil Burner Co.*)

can be varied from 500 to 3,500 lb of fuel per hr without changing the spray plate and with pressure variation of 20 to 70 psi in the steam supply and 100 to 140 psi in the oil supplied. With careful operation, the steam required for atomization does not exceed 1 per cent of the quantity generated by the boiler.

A proportioning inside-mixing-type oil burner using low-pressure air as an atomizing agent is shown in Fig. 4.32. The oil enters at the rear of the burner and flows through a central tube. Upon reaching the end of this tube, the oil combines with primary and secondary atomizing air. The primary atomizing air passes through tangential openings, which impart a swirling motion to this air as it passes around the stream of oil. The mixture next meets the secondary air and leaves the burner in a converging cone to complete atomization. The control lever at the side of the burner regulates both the oil flow and the airflow.

When once adjusted, the correct ratio of air is maintained over the entire range of operation. The effective operating range of this burner is approximately 5 to 1, and an air pressure of 32 oz is required for maximum capacity.

Pressure-type burners. The pressure-type burner shown in Fig. 4.33 effects atomization without the use of steam or air. The oil flows at high pressure in the center tube and is discharged through tangential slots in the sprayer-plate swirling chamber. The swirling oil then passes with undiminished energy through the sprayer-plate orifice and into the space between the two plates. The centrifugal action forces some of the oil into the openings which lead to the return line. The amount

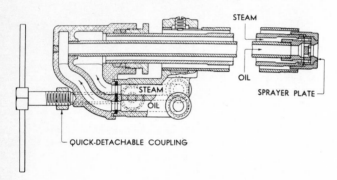

(a)

(b)

FIG. 4.31 (a) *Cross section of a steam-atomizing oil-burner assembly.* (b) *Sprayer plate and orifice assembly of a steam-atomizing oil burner.* (*The Babcock & Wilcox Co.*)

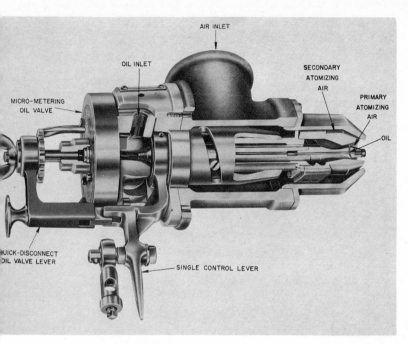

FIG. 4.32 *Proportioning air-atomizing oil burner.   (Hauck Oil Burner Co.)*

of oil thus returned is determined by the position of the return-line control valve.   The oil which is not returned passes through the orifice plate and into the furnace in the form of a hollow conical-shaped spray. From this it will be seen that the same amount of oil flows in the burner at all ratings; therefore, the whirling action within the burner is maintained constant at all rates of oil flow.   The pump supplies an excess of oil, and the amount which passes through the outer orifice to be burned depends upon the opening of the return valve.   This type of atomizer can be used with burners shown in Figs. 4.34 and 4.35.

Fuel oil is used in the generation of steam and in general industrial applications.   Compared with solid fuel, it has various advantages but also some disadvantages.

Advantages and Disadvantages of Fuel Oil as Compared with Solid Fuel

Advantages

Oil can be stored without deterioration or danger of spontaneous combustion
Plant can be operated with less labor (ease of operation and no ashes to
    be removed)
Combustion process can be controlled automatically

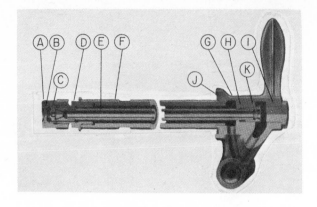

(a)

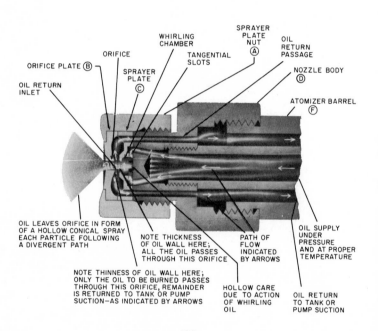

(b)

FIG. 4.33 *Mechanical pressure-atomizing fuel-oil burner.* (a) *Atomizer assembly.* (A) *Sprayer-plate nut.* (B) *Orifice plate.* (C) *Sprayer plate.* (D) *Nozzle body.* (E) *Inlet tube.* (F) *Atomizer barrel.* (G) *Atomizer handle.* (H) *Ferrule.* (I) *Plug.* (J) *Gasket.* (K) *Packing.* (b) *Detail of atomizer tip.* (*Todd Combustion Equipment, Inc.*)

Initial cost of the plant is less because coal- and ash-handling equipment
is unnecessary
Plant can be kept clean
Less air pollution

Disadvantages

Price of oil in some localities is higher than that of solid fuel
Storage tanks are more expensive than storage for solid fuel
Cleaning of oil tanks and disposal of sludge are costly and troublesome
Impurities in oil result in operating difficulties caused by plugged strainers
and burners
Sulfur in the fuel oil causes corrosion of boiler casings and heating surfaces
and pollution of the atmosphere

Oil unsuited for other purposes may be used successfully as fuel under
power boilers. The equipment selected must, however, be suitable for
handling this inferior grade of waste oil, which usually contains impurities in the form of sludge and grit. These clog the strainers, cause
wear of the pumps, and must be periodically removed from the storage
tank.

Oil is used extensively as a standby fuel for emergencies when there
is an outage of the other combustion equipment. It is also employed
in lighting pulverized-fuel burners. The oil burner is lighted with a
torch or electric spark and the pulverized fuel safely ignited by the
oil burner.

Fuel oil may be delivered to the plant in barges, tank cars, or tank
trucks. Except for very high rates of consumption, tank trucks are
used. The haul from supplier to customer is usually comparatively
short, and enough heat is retained in the oil for it to flow into the
storage tank without the use of further application of heat. When the
heavy grades of oil are shipped in railroad tank cars, it is usually necessary to supply steam to the coils in the tank before the oil will flow
or can be pumped from the car.

Before attempting to operate a fuel-oil–burning plant, check the piping, pumps, strainers, heaters, and valve arrangement. The burners
should be inspected and cleaned if necessary. The conventional mechanical steam- and air-atomizing burners can be readily removed and
the tip disassembled for cleaning. The rotating-cup-type unit is
mounted on a hinge and may be swung out of the furnace position
for inspection and service.

When No. 1, 2, or 3 oil is used, no heat is required, but for heavier
oil it is necessary to admit steam to the storage-tank heating coil. Start
the pump and circulate the oil through the system; observe the pressure
and temperature.

There is a difference of opinion regarding the correct temperature to which oil must be heated because of the various types of burners. In general, satisfactory results are obtained with burner-oil temperatures of 150°F for No. 4 oil, 175°F for No. 5, and 275°F for No. 6. Another rule is to heat the oil to within 10°F of the flash point of the oil being used. Mechanical-atomizing-type burners require temperatures near the flash point. Steam-atomizing burners require lower oil-admission temperatures since some heat is available in the steam. Rotary-type burners will utilize unheated No. 4 oil, and Nos. 5 and 6 at a lower temperature than the spray-type burner.

An ignition flame must be placed in front of the burner before the oil is admitted. A gas flame provides a satisfactory means of igniting an oil burner. A torch may be made from a ¼-in. pipe by wrapping asbestos rope around one end for about 10 in. Wire should be used to hold the asbestos rope tightly in place. This torch should be kept in a convenient place in a container of kerosine or light oil. A suitable container can be made by welding a plate on the end of a short section of 3-in. pipe.

With the fuel supply in readiness, operate the draft equipment to free the furnace and setting from all combustible gases. Insert the torch and make certain that the flame is in front of the burner. Admit oil to the burner tip and watch to make certain that it ignites immediately. If the fire should go out, turn off the oil and allow the draft to clear the unit of combustible gases before attempting to relight.

As soon as ignition has taken place, the oil-and-air mixture must be adjusted to produce a stable flame. At this point it is best to have a flame slightly yellow, indicating a deficiency of air. If steam is being used, the supply should be reduced to the minimum required for atomization. Insufficient steam or improper functioning of the burner may result in oil dripping on the furnace floor, creating a dangerous condition that must not be tolerated. The operator must learn how to adjust the fuel-and-air supply to obtain the lemon-colored flame required during the starting-up period. After the correct flame conditions have been obtained, the boiler is brought up to temperature slowly as with other fuels.

It is essential when burning fuel oil that the operator know the equipment and attend to certain details to assure satisfactory operation. The burners must be regulated to prevent the flame from striking the boiler heating surface or furnace walls. If this precaution is not taken, localized heating will cause a burned-out tube or rapid deterioration of the furnace wall. The burner tips should be cleaned frequently enough to ensure good atomization. When retrieving-type burners are used, a spare set should be available at all times. Even with automatically controlled heaters, the oil temperature should be observed frequently. If the oil temperature is too low, atomization will be poor, and if it is allowed to get too high, it will exceed the flash point. Remember that an appreciable amount of steam is required in burners which use this atomizing agent. Economies may be effected by keeping the amount of steam used for atomization to a minimum.

The flow of air and gases in an oil-fired furnace may be produced by: natural draft created by a chimney, forced-draft fan to introduce the air, and induced-draft fan to remove the gases; or by only a forced-draft fan having enough pressure to supply the air and to discharge the products of combustion (see Sec. 3.3). In either case the air supply must be regulated to correspond to the rate of fuel feed. This is accomplished by movable dampers in the air supply duct or wind box. Normally the supply of both fuel and air are automatically controlled.

The operator of a fuel-oil-fired unit is always confronted with the possibility of foreign material interrupting the flow of oil. Sediment may collect in the bottom of the tank and enter the suction line. The strainers either before or after the pumps may become clogged and restrict or stop the flow of oil. Then, too, the sediment may reach the burner tip and cause trouble. An interrupted and self-restored flow of oil from a burner is very dangerous. When the oil flow stops, the fire goes out, and then if the flow is restored, it may not reignite until the furnace has been filled with oil vapor. If this happens to be a combustible mixture, a disastrous explosion will result. Safety devices are available which employ a photoelectric cell (electric eye) so that when the fire goes out, the absence of light on the sensitive element of this device trips a relay which closes off the oil supply. This prevents the restoration of the oil flow until the operator has applied the torch and manually tripped out the safety device.

A summary of the difficulties which may be encountered in operating an oil-fired plant is as follows:

1. Failure of the oil to flow to the pumps owing to ineffective venting of the tank
2. Clogged strainers caused by sludge deposits in the tank
3. Failure of pumps to operate owing to air leaks in the suction line
4. Vapor formation in the pump suction line as a result of too high a temperature in the tank
5. Oil too heavy to flow to the pumps because of insufficient heating
6. Faulty atomization, caused by
   a. Too low an oil temperature
   b. Carbon formation on burner tips
   c. Worn burner tips

    d. Too low an oil pressure

    e. Insufficient steam

7. Smoke, poor combustion, and flame, striking furnace walls, caused by

    a. Faulty atomization of oil

    b. Incorrect adjustment of air supply

    c. Automatic control's not functioning properly

8. Fire hazard and unsightly appearance, caused by leaks in the pump discharge lines

Oil-fired boilers are banked by closing off the oil-and-air supply. The fans are shut down and the stack damper is closed to prevent natural circulation of air and thus retain the heat in the unit. If it is necessary to have the boiler available for service on short notice, the fire must be relighted often enough to maintain the steam pressure at a predetermined minimum.

When sufficient sediment collects in the fuel-oil-supply tank, it will frequently clog the strainers and perhaps even pass on to the burners, causing irregular firing. The real solution to this situation is to drain and thoroughly clean the tank.

Some oil contains abrasive materials which cause wear to the pumping equipment and result in low oil pressure. It is not economical to run with low oil pressure, especially when mechanical atomization is employed. When wear in the pumps causes a reduction in pressure sufficient to affect operation adversely, the worn parts of the pumps should be replaced. This condition should be anticipated by selecting pumps which have wearing parts that can be replaced cheaply.

The importance of maintaining the oil system free from leaks cannot be overemphasized. Socket welding of oil-piping joints is an accepted method of reducing leakage. Many an oil-system failure has been traced to a small leak in the suction pipe. The flexible connection at the burners should be replaced or repaired before serious leaks develop.

In most plants the cleaning of burner tips is considered an operating function. In time, with cleaning and wear from impurities in the oil, it becomes necessary to replace the tips.

Air registers and automatic-control equipment must be checked frequently to assure satisfactory operation.

4.9 GAS. The gaseous fuels available are natural, manufactured, water, oil, by-product coke-oven, blast-furnace, and refinery gas. Natural gas is widely distributed by means of high-pressure pipeline systems. Nevertheless, cost varies widely and depends upon the distance the gas must be transmitted. In some instances gas at a reduced cost is available for steam generation during periods when primary customer demand is low. By-product gas is usually used in the same plant in which it is produced or in a neighboring plant.

Natural gas from high-pressure pipelines is reduced in pressure and distributed to local customers through a low-pressure network of piping. When the pressure in the supply mains is above that required by the burners, pressure-regulator valves are used to maintain constant pressure at the burners. However, when the supply-main pressure is less than that required at the burners, it becomes necessary to install booster compressors. These boosters are either multistage centrifugal blowers or constant-volume units similar in principle to a gear pump. The outlet pressure of constant-volume units must be controlled by spring-loaded bypass valves. In some instances it is necessary to install coolers in these bypass lines to remove the heat developed in the compressor. Gas burners are designed and regulated for a given pressure, and for efficient and satisfactory operation it is imperative that this pressure be maintained. In like manner, provision must be made to supply by-product gas to the burners at a controlled pressure.

Gas is a high-grade fuel and is extensively used for domestic heating and in industrial plants for heating and limited process applications. Industries which require large quantities of steam usually find other fuels more economical. Power boilers which burn gas generally have combustion equipment suitable for two or more kinds of fuel. The use of multiple fuels make it possible to take advantage of market conditions, provide for shortage in supply, and guard against possible transportation difficulties. It is difficult to operate a pulverized-fuel boiler safely at widely variable ratings. When low ratings are necessary, it is advisable to use oil or gas as a substitute for pulverized coal. The burner shown in Fig. 4.34 is designed to use pulverized fuel, oil, or gas. The gas is introduced through many small holes in a ring which is located in front of the burner proper. The gas mixes with a stream of air passing through the ring.

Industries that have a supply of by-product gas utilize it to advantage as a fuel in their power boilers. They use the by-product fuel first, and when the supply is insufficient to meet the demand, they use a secondary fuel which they must buy. This method of operation is most effective in reducing the cost when there is a source of waste or by-product fuel. Combustion controls have been developed which will admit the by-product and secondary fuels to the furnace to meet the requirements. A burner arrangement for either oil or blast-furnace gas is shown in Fig. 4.35. Owing to the low heating value of this gas, a correspondingly large volume must be handled. The gas is discharged into the burner through a circular ring formed by the burner tube and throat. The air enters through the burner tube with a whirling motion and mixes with the gas. The burner is designed to use forced draft, although gas may be burned with natural draft.

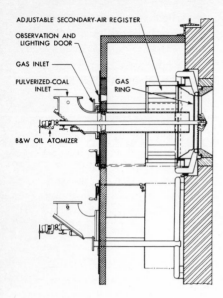

ADJUSTABLE SECONDARY-AIR REGISTER

OBSERVATION AND
LIGHTING DOOR

GAS INLET

PULVERIZED-COAL
INLET

GAS
RING

B&W OIL ATOMIZER

FIG. 4.34 *Multifuel burner.*
(*The Babcock & Wilcox Co.*)

Sufficient furnace volume is essential to the proper operation of gas-fired furnaces (see Sec. 2.9). If the fuel has not been entirely burned before the gases reach the heating surface, serious smoking and loss of efficiency will result. Thorough mixing at the burners generally reduces the length of the flame. The design should be such that the entire furnace volume is utilized for combustion. Dead spaces tend to cause pulsation of the flame.

Many of the procedures and precautions that are required for oil firing also apply to gas. Before attempting to light the burners make certain that the gas-supply valves are closed, and then operate the draft equipment to remove all possible accumulations of gas. Adjust the shutters to admit a small amount of air. Then insert the lighted torch and turn on the burner valve. If the fire should go out, never relight until you are sure the draft has removed all combustible gases from the unit.

The appearance of the flame is an unsatisfactory method of adjusting the air going to the burner and the furnace draft. The best procedure is to analyze the flue gases with an orsat and regulate the air accordingly. With conventional burner design and careful adjustment, gas-

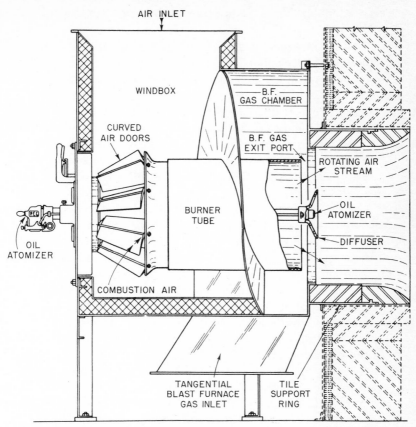

FIG. 4.35   *Combined blast-furnace-gas and oil burner.   (Peabody Engineering Corp.)*

fired boilers can be successfully operated with from 10 to 20 per cent excess air.[1]   Improper mixing of air and gas results in long flame travel. If the flames strike the heating surface, soot will be deposited and smoke will be produced.   Gas flames frequently have a tendency to pulsate.   This may be caused by poor distribution of flame in the furnace.   If the burner and furnace designs are correct, these pulsations can be eliminated by increasing the furnace draft.   When the furnace is started up, pilot burners and the main burners should be adjusted to produce a yellow flame which will not be easily extinguished.

The accurate control of the flow of gases to the burners depends upon functioning of the pressure regulator.   In some installations the pressure of the gas entering the burners is changed to vary the rate of firing, while in others the number of burners in service is varied. Automatic control is effectively employed not only to regulate the fuel

[1] For a discussion of "low excess air" see Sec. 3.8.

and maintain the steam pressure but also to maintain the correct ratio of air to fuel.

The maintenance of gas burners and accessories is less than that for other types of combustion equipment, but there are a few points that must receive attention. The air-register operating mechanisms and gas regulators must be kept in good operating condition, the burners must be kept clean so that the flow of gas will not be obstructed, and the boiler setting must be checked to reduce air leakage to a minimum. The use of gas as a fuel eliminates stokers or pulverizers and coal- and ash-handling equipment, thus decreasing plant maintenance.

4.10 AUTOMATIC OPERATION OF BOILERS. Combustion control regulates the fuel to meet the demands for steam or hot water and maintains the ratio of air to fuel required for the best economy. The operator performs the remaining functions, which include purging the furnace, lighting the burners, adjusting dampers, opening and closing fuel valves, starting and stopping pumps, and turning burners on and off.

The entire combustion process is made automatic by having the unit come in service and shut down, as required, without hand manipulation. The degree of automatic control is made possible by the use of safety monitoring which checks every step of the operation. This is accomplished by means of sensing devices which detect abnormal conditions, including flameout, high pressure, low water level, high temperature, and low fuel pressure. The control either corrects the situation or warns that trouble exists and shuts the unit down safely.

Automatic controls of this type may be applied to units of all sizes. However, extensive economic justification has been found in the small and medium-sized packaged steam boilers and high-temperature water generators. The package boiler (Fig. 4.36) is equipped with automatic controls capable of starting up and maintaining the firing rate as required for varying demands of steam. The control panel includes pressure and draft gauges, buttons for initiating the start-up, lights for indicating the sequence of the cycle, and an annunciator to indicate specific causes of trouble. When two or more boilers are installed, the controls can be arranged to start additional units if a unit in service fails or the load increases above the capabilities of the units on the line.

The schematic drawing in Fig. 4.37 shows the control arrangement for an oil- and gas-fired boiler. This control is also applicable to high-temperature water generators. The steam pressure from the boiler or temperature of the water from the generator actuates the submaster 1. This impulse moves a shaft having linkages for controlling the forced-draft-fan damper through the airflow regulator 2 and the power unit and relays 3, burner-register louvers, oil supply to the burners by adjusting the orifice-control valve 5, and gas supplied to the burners by

FIG. 4.36 *Package boiler equipped with burners and controls for automatic oil and gas firing. (Coen Company, Inc.)*

adjusting the orifice gas valve 8.   The atomizing steam pressure is controlled from the oil-supply pressure applied to the burner, by the differential steam-control valve 6.   The gas supply to the pilot is turned on and off by the solenoid valve 16.

Automatic burners of this type are equipped with safety controls which sequence the start-up and shutdown and override the operating controls to close the fuel valves if any of the components malfunction. These safety controls will shut the unit down when there is trouble, including low water, flame failure, low gas or oil pressure, overpressure, and high water temperature.   They must be selected to meet the insurance underwriters' specifications.

These controls provide an option of recycle or nonrecycle for restarting the unit once it has tripped out owing to a failure.   A recycling control automatically attempts to restart the burner, while a nonrecycle control allows the burner to remain out of service until the start switch has been closed by hand.   There are some insurance restrictions on the use of the recycle control.

The automatic start-up sequence and timing depend upon the type of burners, specific conditions, and the use of oil, gas, or a combination.

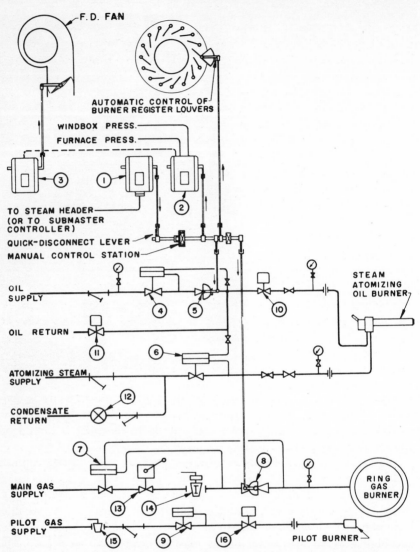

FIG. 4.37 *Schematic control diagram—automatic gas- and oil-fired unit.*
(1) Master actuator. (2) Airflow regulator. (3) Power unit and
relays. (4) Differential oil-control valve. (5) Adjustable orifice oil-
control valve. (6) Differential steam-control valve. (7) Differential
gas-pressure-control valve. (8) Adjustable-orifice gas-control valve.
(9) Pilot-gas-pressure regulator. (10) Oil solenoid valve. (11) Oil-
return solenoid. (12) Steam trap. (13) Main-gas solenoid valve.
(14) Supervisory cock. (15) Pilot shutoff cock. (16) Pilot-gas sole-
noid valve. (Coen Company, Inc.)

In general, however, the forced-draft fan will start immediately when the start switch is turned on. The air pressure produced by the forced-draft fan will close the air interlock. The flame safety control will be engaged for a preset purge period with the fuel-control valve at high fire. Then the firing-rate controller will be adjusted to minimum fire in preparation for pilot ignition. The ignition transformer will be energized to light the pilot. After the pilot-proving period has elapsed, the main fuel valve will open and the main burner will be placed in operation. After the main burner has been "proved," the ignition transformer will be de-energized. The firing-rate controller will then be released from the minimum position to respond to the demand placed upon the unit. In the event that the pilot or main valve has not been proved, the fuel valve or valves will close and an alarm sound.

The safe operation of automatic burners depends upon the reliability of the flame-detecting device. Four principles are employed in sensing elements used in connection with flame-detection devices:

1. The photoelectric cell is not suited to use with gas burners and is subject to interference when two or more burners are installed in the same furnace. If it is not carefully adjusted, it will sense the hot furnace wall and not indicate a flame failure.

2. Flame rectification employs a flame rod upon which alternating current is imposed. When this rod is exposed to gas flame, the alternating current is converted into direct current. Sensing only direct current, the control device detects the absence of flame on the rod. This detection device cannot be used in connection with oil firing.

3. Flame-frequency sensing is accomplished by use of a lead sulfide cell. This sensing device requires a flame having a frequency of at least 10 cps (cycles per second), thus reducing the effect of hot walls. It is widely used to detect the presence of both oil and gas flames.

4. An element sensitive to only the ultraviolet portion of the light spectrum has been developed for sensing burner flame. This device will indicate the presence of either an oil or a gas flame and is not affected by the red rays emitted from hot refractory. The fact that these devices will sense sunlight is rarely an adverse factor.

Packaged automatic boilers and high-temperature water generators require the same careful consideration to design limitations as apply to field-erected units. Satisfactory operation and low maintenance requires limitations on fuel-burning rates, heat-absorption rates, and gas velocities. However, some decrease in the size of units for a given output can be obtained as a result of improvement in combustion equipment, the extended use of water cooling in the furnace, and better circulation of water in the tubes. The operation of automatic boiler and high-temperature water-generator units depends upon carefully designed and located sensing devices, electric circuits, and electronic controls.

Electronic controls once considered unreliable by plant operators have been improved by the use of "solid-state" (transistor) components. Replacements are made possible by preassembled "plug-in" components which can be stocked and quickly installed by plant personnel. This automatic equipment should be checked for performance of the safeguards on a regular schedule. In addition to these checks by plant personnel it is advisable to have semiannual or annual inspections by qualified control servicemen.[1]

4.11   CHIMNEYS.   Chimneys serve a twofold purpose: (1) to produce the draft necessary to cause the air to flow into the furnace and discharge the products of combustion to the atmosphere and (2) to deliver the products of combustion and fly ash to a high altitude. The gases within the chimney are at a higher temperature than that of the surrounding air. The weight of a column of hot gases in the chimney is less than that of a column of air at outside temperature. The intensity of draft produced by the chimney depends upon the height and the difference between the outside air and inside gas temperatures.

The following formula may be used for calculating the static, or theoretical, draft produced by a chimney:

$$D = 0.52H \times P\left(\frac{1}{T_a} - \frac{1}{T_s}\right)$$

where D = draft at base of chimney, in. of water
H = height, ft., of chimney above point where draft measurement is taken
P = atmospheric pressure, psia (can be taken as 14.7 up to elevation of 1,500 ft above sea level)
T_a = outside temperature, absolute °F
T_s = gas temperature in chimney, absolute °F

Example   A chimney is 175 ft high, and the exit-gas temperature is 610°F. The outside temperature is 80°F, and the plant is located at a sea-level elevation of 550 ft. What static, or theoretical, draft in inches of water will it produce?

$$D = 0.52\,H \times P\left(\frac{1}{T_a} - \frac{1}{T_s}\right)$$

$$H = 175\text{ ft} \qquad P = 14.7 \qquad T_a = 80 + 460 = 540$$
$$T_s = 610 + 460 = 1,070$$

Substituting values in the equation:

$$D = 0.52 \times 175 \times 14.7\left(\frac{1}{540} - \frac{1}{1,070}\right)$$

$$D = 1.23\text{ in. water}$$

[1] For further discussion of automatic controls see Sec. 5.11.

This represents the static, or theoretical, draft; when gases flow, it will be reduced, owing to friction.   The amount of reduction will depend upon the size of the chimney, upon the quantity of gases, and, to some extent, upon the material used in the construction of the chimney.   The actual available draft may be considered as 80 per cent of the calculated static draft.

Whereas the height of a chimney determines the amount of draft it will produce, the cross-sectional area limits the volume of gases it can successfully discharge.   The quantity of gases that a chimney must expel depends upon the amount of fuel burned and the amount of air used for combustion.   Air leaks in the boiler setting, breeching, etc., dilute the chimney gases, increase the volume, lower the temperature, and thus lower the effective draft.

It frequently becomes necessary to study the performance of a given chimney installation.   Questions arise as to whether the chimney has sufficient capacity for the existing boiler.   Will it be adequate if the existing boilers are rebuilt to operate at a higher rating?   Can more boilers be added without building another stack?   In these instances draft- and gas-temperature readings should be taken at the base of the stack while the boilers are operating at full rating.   These results should be compared with the expected draft as calculated from the height of the stack, the temperature of gases, and the temperature of the outside air.   If the results show the draft as read to be less than 80 per cent of the calculated value, the chimney is overloaded.

The adjustment of the stack damper is a good indication of the available draft.   Must the damper be wide open when the boilers are operating at full load?   A chimney may become so badly overloaded that it is impossible to maintain a draft in the furnace when operating at maximum output.

Sometimes difficulties develop, and a plant that has formerly had sufficient draft finds that the furnaces develop a pressure before the boilers reach maximum rating.   This condition, although a draft problem, might be caused, not by a defective chimney, but by any one or combination of the following conditions:

1. Stoppage of boiler passes, breeching, etc., with soot, slag, or fly ash, resulting in abnormal draft loss

2. Baffles defective or shifted so that they restrict the flow of gases

3. The use of more excess air than formerly

4. Operating at higher ratings than indicated owing to the flowmeter's being out of calibration

5. The damper out of adjustment so that it will not open wide even though the outside arm indicates that it is open

6. Air infiltration through leaks in the setting

There are limitations to the application of the natural-draft chimney. The trend in design of modern units is toward high rates of heat transfer, which result in increased draft loss. High fuel costs justify the installation of heat-recovery equipment to reduce the heat loss by lowering exit-gas temperature. The limits imposed upon the emission of fly ash by local ordinances and by public relations necessitate the installation of high-efficiency mechanical collectors in many cases. These collectors have a high draft loss. This increase in draft loss and the lowering of gas temperature necessitate a higher chimney. These conditions soon demand a chimney higher than is physically or economically practical, and mechanical-draft equipment must be used.

However, chimneys are used in connection with both induced-draft and pressurized-furnace units. Fans are used to produce the necessary draft, higher velocities, and correspondingly higher pressure drops that are developed in the breechings and chimneys. The products of combustion are discharged at high elevations in order to reduce the ground-level concentrations of pollutants.

4.12 MECHANICAL DRAFT. There are essentially three methods of applying fans to boiler units:

Balanced draft, in which a forced-draft fan delivers air to the furnace and an induced-draft fan or a chimney produces the draft to remove the gases from the unit. The furnace is maintained at 0.05 to 0.10 in. of water gauge below atmospheric pressure.

Induced draft, in which a fan or chimney is employed to produce sufficient draft to cause the air to flow into the furnace and the products of combustion to be discharged to the atmosphere. The furnace is maintained at a pressure sufficiently below that of the atmosphere to induce the flow of combustion air.

Pressurized furnaces, in which a fan is used to deliver the air to the furnace and cause the products of combustion to flow through the unit and out the stack. The furnace is maintained at a pressure sufficiently above that of the atmosphere to discharge the products of combustion.

In addition to these applications, fans are used to supply over-fire air to the furnace and in the case of pulverized-fuel units to deliver the fuel to the furnace.

Centrifugal fans are widely used to handle combustion, air, and flue gases. Pressure is developed by the rotation of the rotor blades within the casing. There are three principal types of fan blades: radial, forward-curved, and backward-curved. These principal types are modified to meet specific conditions.

A straight-radial-blade fan wheel is shown in Fig. 4.38, and the characteristic curves for a fan with this type of blading are shown in Fig. 4.39. This fan was selected to deliver 10,000 cu ft per min of air

FIG. 4.38 *Straight-radial-blade fan wheel. (Green Fuel Economizer Co. Inc.)*

at 13 in. of water pressure. The efficiency is 69 per cent and the shaft horsepower 29.60. The requirements are 10,000 cu ft per min at a pressure of 13 in. of water, but consideration must be given to the performance of this fan under varying conditions. If the damper were completely closed (zero flow) the pressure would increase to 17.5 in. of water. If, on the other hand, the resistance in the system were reduced (a bypass damper opened) to permit a flow of 13,000 cu ft per min at a pressure of 6.7 in. of water, the fan would require 35 horsepower. Furthermore the horsepower of a fan in a given installation is influenced by the temperature and pressure of the gases (see Sec. 3.2). Under normal operating temperatures, an induced-draft fan may not overload its driving motor with the damper in the full open position. However if this same fan is used to handle air at room temperature to provide ventilation during maintenance, the motor may be overloaded. The horsepower requirements of this radial-blade fan increase in an almost straight-line ratio with the volume of air delivered.[1]

The backward-curved-blade fan may be designed to decrease the rate of increase in horsepower, after the design gas-flow and pressure conditions have been reached. This feature reduces the possibility of the motor or other driving units being overloaded owing to high rates of flow. These fans operate at relatively high speed, thus eliminating the necessity for speed-reducing mechanisms when high-speed turbine drives are employed.

[1] This applies to constant-speed operation. When the speed of the fan changes, capacity varies directly as the speed, pressure varies as the square of the speed, and horsepower varies as the cube of the speed.

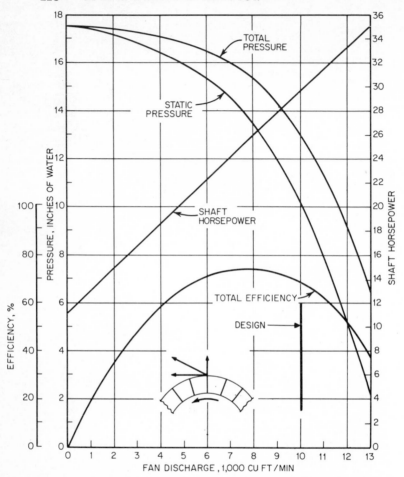

FIG. 4.39 *Characteristic curve for fan with radial blades.*

Power is required to operate fans, and the cost of this power is charged to plant operation. An effort must be made in both the design and operation of a plant to reduce the cost of power to operate the fans. Fan shaft horsepower is a function of the quantity of air or gases, the total differential pressure (static plus velocity pressure), and the efficiency of the fan.

$$\text{Fan shaft horsepower} = \frac{5.193 \times Q \times H}{33,000 \times E}$$

where Q = quantity of air in actual cu ft per min (ACFM)[1]
H = total pressure, in. of water
E = efficiency of the fan

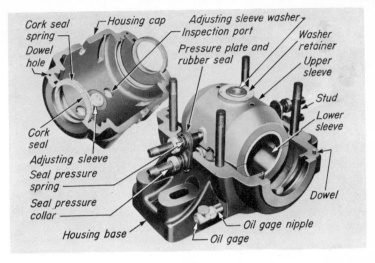

(a)

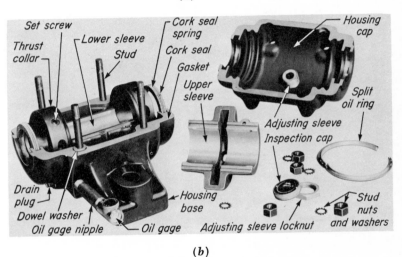

(b)

FIG. 4.40   (a) Self-aligning water-cooled sleeve bearing.   (b) Ring lubricated sleeve bearing with thrust collar.   (American-Standard Industrial Div.)

[1] ACFM—actual cu ft per min of gases at the temperature, pressure, and relative humidity existing.   This is in contrast to SCFM, standard cu ft per min of gases at 29.92 in. of mercury pressure and 68°F, 50 per cent saturated with moisture and weighing 0.075 lbs per cu ft.

The design of fans must provide not only sufficient capacity to obtain maximum rating but also regulation when the unit is operated at partial capacity. There are three principal ways of controlling the output of a fan: by the use of dampers, by speed variation, and by the use of inlet vanes.

Damper control consists of operating the fan at constant speed and dissipating the excess pressure by means of a variable obstruction (damper) in the flow of gases or air. This provides a low-cost installation but is wasteful of operating power. To reduce this waste two-speed motors are sometimes used in connection with dampers.

Variable-inlet vanes provide a method of controlling a fan which is less wasteful of power than a damper and less costly than variable-speed drives. This method of control has been widely applied to forced-draft fans, but fly ash introduces difficulties when it is used in connection with induced-draft fans.

The output of a fan can be varied by changing the speed to meet the draft requirements, but variable-speed drives increase the installation cost. Variable speed can be obtained with constant-speed motor drives

FIG. 4.41 *Fan wheel constructed to resist fly-ash erosion.* (*Westinghouse Elec. Corp., Sturtevant Div.*)

by use of magnetic or hydraulic couplings. The motor runs at constant speed, and the coupling is controlled to vary the fan speed. The speed of a steam-turbine–driven fan is varied by regulating the steam flow to the turbine. It is customary to employ dampers in addition to these variable-speed devices.

Fan outputs are invariably regulated by the combustion controls. The combustion controls must be compatible with the type of fan control selected. For example, a variable-speed fan and damper arrangement requires a more complex combustion control than when only a damper is used.

In building a fan it is necessary to consider not only the correct characteristics but also the mechanical features. Fans are subjected to adverse conditions, such as heat, dust, and abrasive material carried by the air or gases.

The bearings are a most important part of a fan from a standpoint both of selection and of operation. Fans are supplied with a variety of bearings, depending upon the use to which they are to be put. Sleeve-ring oiled bearings and antifriction oil- and grease-lubricated bearings are used.

The sealed grease-packed antifriction bearings require least attention and are recommended for out-of-the-way places where periodic lubrication would not be feasible. For high-speed operation, oil lubrication is desirable. Antifriction bearings are subject to quick and complete failure when trouble develops but can be easily and quickly replaced.

Sleeve-ring oiled bearings shown in Fig. 4.40 are extensively used in fans. Figure 4.40a shows a sleeve bearing with water cooling. This type of bearing would be used in an induced-draft fan. The heat in the flue gases is conducted to the bearing by the shaft and must be carried away by the cooling water to prevent overheating. Figure 4.40b shows a fixed sleeve bearing provided with a thrust collar and setscrews for attaching to the shaft and thus restricting the axial movement of the shaft. Oil should be drained from the bearings at regular intervals and the sediment flushed out. After they have been cleaned, the bearings must be filled with a light grade of oil. When in operation the oil level and operation of the oil ring must be checked regularly.

As with all rotating machinery, the alignment and balance must be correct. The fan must be aligned with the motor or turbine. Consideration must be given to expansion. Do not depend upon flexible couplings to compensate for misalignment. Even a slight out-of-balance condition will produce vibration detrimental to the bearings and to the entire unit. Wear will sometime cause a fan to get out of balance.

The erosion of fan blades and casing by fly ash and other abrasive particles in the air and gases can cause serious and costly maintenance.

There are no materials available which will successfully withstand the erosive action of fly ash. These problems are solved by installing the straight-paddle-wheel type of fan, which permits quick and cheap replacement of the blades, and by special construction as shown in Fig. 4.41. Here the center plate has been cut away to present a continuous blade to the action of the ash. The ribbed liners can be replaced, thus decreasing maintenance cost. In a similar manner, removable wearing plates are bolted to the casing in locations where the wear is excessive. These plates are replaced as they wear out, thereby prolonging the life of the fan casing indefinitely. Periodic inspections of the removable wearing parts are essential so that replacements can be made before wear has progressed beyond the renewable sections.

4.13   STEAM AND AIR JETS. The energy derived from a jet of expanding air or steam may be utilized in producing a draft. Devices to accomplish this consist of a high-pressure jet of steam or air discharging into a larger pipe. The aspirating effect produced causes an additional amount of gases to be delivered.

Jets are seldom used as a primary means of producing draft in stationary practice. However, they have been used for many years to introduce over-fire air, create turbulence, and thereby promote economy and reduce smoke.

The secret of smokeless combustion is in burning the hydrocarbons before they form soot particles. To this end jets must supply the required over-fire air, distribute it advantageously, and at the same time create the required turbulence. As a result of research,[1] it is now possible to design and successfully apply jets to specific furnace installations.

While over-fire air jets are not a cure-all for faulty design and operation, they are capable of performing a very important function in boiler operation. Early installations were often makeshift arrangements, and the results were disappointing. Consideration must be given both to the quantity of air to be supplied and to the distance it is to be carried into the furnace. The jets must be so located and spaced that their effects will overlap to prevent lanes of unburned gases from escaping. The air must be introduced close to the fuel bed to prevent the formation of smoke because once smoke has formed, it is very difficult to burn. However, the jets should not be close enough to the fuel bed to cause interference with the solid mass of burning fuel. Care must also be exercised to see that the jets do not impinge upon the brickwork or waterwall tubes. The most advantageous location of jets can sometimes

[1] Aid to Industry Bulletin 500-300, Bituminous Coal Research, Inc., Monroeville, Pa.

be determined by exploring the firebox with a pipe supplied with steam through a hose. This pipe should be inserted through the available openings and the most effective location of the pipe determined by observation.

The jet action must be modified to meet specific conditions. In some cases the actual amount of air present in the furnace may be sufficient, and only turbulence need be provided. This calls for the installation of steam jets without the air tubes. In this case the steam jets must be placed near the inner furnace wall to obtain maximum penetration. Air may be supplied by a steam-actuated jet or by a high-pressure air blower. The cost of installation of the blower is higher than that of the steam jet. The over-fire system as applied to spreader stokers usually employs a high-pressure blower.

In many cases the jets are required only a small part of the time, but it is difficult to get the operators to use them when required. The fireman should be provided wth a means of determining when he is making smoke. The requirements vary from a case in which the fireman can easily observe the chimney to plants where a smoke indicator or recorder can be justified. It is possible to install a control which, in conjunction with a smoke-measuring device, will automatically put the jets into service when the smoke exceeds a certain density. Equipping the jets with silencers to eliminate the objectionable noise will assist in getting the operators to use them when they are needed.

4.14   AIR POLLUTION CONTROL.   The most troublesome air pollution resulting from the combustion of conventional fuels are visible smoke, particulate matter, sulfur oxides, and nitrogen oxides. Pollution problems are solved by preventing the formation of the pollutants or by removing them from the flue gases. The prevention or reduction in the amount of pollutants is accomplished by improved combustion equipment and techniques of operation and by a change in the type of fuel or modifications to the fuel being used. A number of processes and devices are available for reducing the amount of pollutants in flue gases.

Government agencies regulate the permissible discharge of pollutants to the atmosphere. It is the plant management's responsibility to assure that pollutant emissions are in compliance. This often necessitates testing the stack gases. Samples are taken from the stack in an approved location and manner and analyzed to determine the nature and amount of pollutants. These tests are made in accordance with, and to conform to, the ordinances which apply.

There are several methods of reporting the concentrations of pollutants in stack gases but the *pounds per million Btu input to the furnace* is preferable.[1]

Example A boiler plant generates 225,000 lb of steam and burns 12.5 tons of coal per hr. The coal has a heating value of 12,200 Btu per lb. A test of the stack gases shows that 145 lb of particulate is being discharged per hr. (1) What is the particulate discharged per million Btu to the furnace? (2) What must the efficiency of a collector system be to reduce this emission to 0.10 lb per million Btu?

Solution Furnace heat release

1. $12.5 \times 2,000 \times 12,200 = 305,000,000$ Btu per hr

$$\frac{145}{305} = 0.475 \text{ lb particulate per million Btu}$$

2. $\dfrac{\text{Removal by collector}}{\text{Total emission}} = \text{efficiency}$

$$\frac{0.475 - 0.10}{0.475} \times 100 = 78.95 \text{ per cent efficiency}$$

Smoke. The most objectionable of pollutants is smoke mainly because it is easily observed by the "neighbors." It is composed of unburned carbon particles from either carbon or hydrocarbons. Smoke can be formed from the combustion of gaseous or liquid fuels because they contain hydrocarbons.

Density of smoke is measured by comparing the effluent from the stack with Ringelman smoke charts. There are five of these charts, referred to as Ringelman 1, 2, 3, 4, and 5, or smoke density 20, 40, 60, 80, and 100 per cent. They are composed of different weights of horizontal and vertical black lines. When these charts are viewed at a distance of 50 ft they appear as varying shades of gray. Smoke density is determined by visually comparing the effluent from the stack with these charts. Similar measurements are made with instruments which employ comparable densities of gray (smoked) lens. Smoke density is also measured by passing a beam of light through the breeching (carrying the flue gases) and measuring the intensity by means of a photoelectric cell. As smoke density increases, the intensity of light reaching the photoelectric cell is reduced and an alarm device warns the operator of the unsatisfactory condition.

Particulate emission. Solid particles carried by stack gases consist essentially of a portion of the ash that was in the fuel with some unburned carbon particles. The amount and nature of this material depend upon the fuel, combustion equipment, rating, and operating skill. Restrictive ordinances make it necessary to provide some type of collection equipment with solid fuel and in some instances with liquid fuels.

---

[1] Pollutants in lb per million Btu input times 0.85 equals the approximate lb per 1,000 lb of flue gases adjusted to 50 per cent excess air.

The small particles, aerosol in nature, may be carried a considerable distance, or during an atmospheric inversion may remain in suspension and become a factor in creating the smog sometimes prevalent in cities.

The quantity and particle size must both be considered when evaluating and controlling particulate emission. Particle size is designated in microns.[1] A size distribution analysis of a particulate sample states the percentage of the material that is less than specified microns. For example, an analysis showed that 30 per cent of the particulate from a pulverized-fuel boiler was less than 10 microns. These data are required in the selection of collection equipment.

There are four principal types of equipment used to remove this fly ash from the flue gases before they are discharged from the stack: (1) mechanical collectors, (2) electrostatic collectors, (3) washers, (4) cloth collectors.

Mechanical collectors consist of multitube units which remove the fly ash by centrifugal action (Fig. 4.42). The gases flow downward

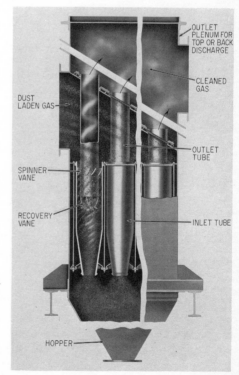

FIG. 4.42 *Mechanical dust collectors.* (*American-Standard Industrial Div.*)

---

[1] A micron equals one-millionth of a meter, or 1 in. = 25,400 μ.

through the spinner vanes in the annular space between the outer and inner tubes. Then the gases reverse direction and flow up through the inner tube and out of the collector. The combination of centrifugal force created by the spinning of the gases and the action of gravity when the flow is reversed separates the dust particles, and they fall through the bottom end of the outer tube into the hopper beneath. These collectors can be designed to remove 92 per cent of the fly ash from spreader-stoker flue gases. However, their efficiency decreases rapidly on that portion of the fly ash with a particle size less than 10μ. A draft loss of 3.0 in. of water gauge is required to obtain high efficiency. The induced-draft fan must be selected for this added resistance. Operating costs are increased by the added power required to operate the fan.

As the load on the boiler decreases, the velocity of gases through the collector also decreases, lowering its efficiency. This loss is overcome by providing two or more sections in the collector. All but one of these sections have dampers which shut off the flow of gases. As the load decreases on the boiler, the collector dampers are closed progressively to maintain the velocity and hence the efficiency at the reduced capacity.

The ash hoppers must be emptied on a regular schedule and frequently enough to prevent the ash from building up and sealing off the bottoms of the tubes. Should this occur, the tubes may remain stopped up even after the ash has been removed from the hopper. Failure to remove the ash regularly may also result in a smoldering fire in the hopper.

Electrostatic precipitators are effective in removing small dust particles from the gases (Fig. 4.43). These precipitators consist of a series of plates or tubes which traverse the gas passage. Alternate elements are charged with high-voltage direct current and grounded. This high-voltage direct current causes the dust particles to become ionized and attracted to the grounded element. At intervals these grounded elements are automatically vibrated to shake the dust into a hopper.

Maximum efficiency is obtained by automatic control of the high voltage. The voltage is maintained at the maximum value without excessive sparking between the wires and plates. When correctly sized, applied, and operated, these collectors will remove more than 98 per cent of the fly ash from the flue gases. However, the efficiency depends upon a characteristic of ash known as the resistivity. This determines the susceptibility of the fly-ash particles to the influence of the electrostatic field. Unfortunately this ability is increased when there are sulfur oxides present. When a plant changes to low-sulfur coal, to reduce

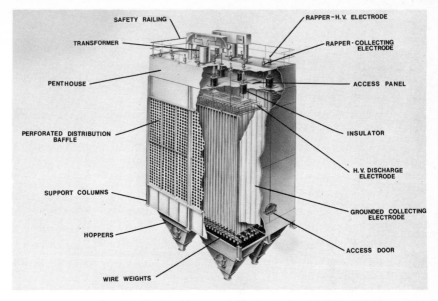

FIG. 4.43 *Electrostatic precipitator.* (*Air Correction Division UOP.*)

sulfur oxides emission, the collection efficiency of the electrostatic precipitator is reduced. In new plants this problem can be solved by locating the electrostatic precipitator ahead of the heat-recovery equipment, where the gas temperature is in the order of 600°F. In existing plants the efficiency has been at least partially restored by introducing additives to the gases. Electrostatic collectors are sometimes installed after mechanical collectors to obtain high overall collection efficiency.

The high-voltage rectifying equipment and insulated elements are costly to install. The charged wires break, necessitating shutdown for replacement. Unlike the mechanical collectors, the draft loss is nominal, on the order of 0.5 in. of water gauge. Safety systems are provided to prevent personnel exposure to the high voltage, but a thorough knowledge of the equipment and safety precautions is essential.

Washers using water have been employed in many ways to remove particulate from flue gases. One of the simplest of these methods is water sprays in the stack which have been successful in reducing dust emission during soot-blowing periods.[1] These stack sprays can be used with either hand or stoker plants when the breeching enters the side of the stack and there is a collection chamber in the base of the stack. It must be recognized that with certain combinations of temperature, flue

---

[1] "Stack Sprays to Reduce Dust Emission during Soot Blowing," Bituminous Coal Research, Inc., Monroeville, Pa.

gases, and stack construction the introduction of water could result in deterioration of the stack lining.

The passing of gases through a spray of water is relatively ineffective in removing particulate. The more intimate contact of dust-carrying gases and water is obtained in scrubbers. Many types are available, but the venturi and packed tower are prevalent.

In the venturi scrubber, the water and gases are forced through the funnel-shaped restriction into a diverging section. The energy expanded results in extreme turbulence and thorough mixing. The dust particles are carried away in the water. The cleaned gases pass through eliminators which remove some of the moisture.

The packed tower consists of a vessel containing inert granular material. Contaminated gases enter the bottom and pass up through the packed section. Water enters the top and flows downward washing the dust from the gases and discharging it at the bottom of the tower. The clean gases flow through a "demister" and out the top of the tower.

Scrubbers are used to clean air and gases discharged from industrial processes but are used far less in boiler plants. A considerable pressure drop in the gases with accompanying increase in fan power consumption is required. The fly ash is in the form of a wet slurry, making disposal difficult. Flue gases entering the scrubber are above the boiling point of water; therefore a considerable amount of cooling water is evaporated, and makeup must be added. The water evaporated leaves with the gases. This moisture-laden flue gas produces a visible "plume," especially noticeable in cold weather. Since this plume is readily visible it is objectionable. Sometimes the gases are superheated so the vapors will be less noticeable. The combination of water and flue gases is very corrosive. Special corrosion-resistant materials must be used. These may have very restrictive temperature limitations, necessitating the automatic bypassing of the gases should the cooling water fail.

Scrubbers may be used to satisfactorily clean gases heavily laden with particulate matter. By addition of a lime slurry to the wash water it is possible to remove a portion of the sulfur oxides from the flue gases.

The *baghouse* provides another method of removing particulate matter from air and gases. The filter elements are made of thick felted fabric which retains the dust particles but allows the air or gases to pass through. The baghouse is composed of many filter elements to provide adequate surface. The amount of surface must be large to reduce the pressure drop. Filter elements are made in the form of tubes, or the fabric is supported on frames to provide flat, rectangular pocket-type elements for maximum surface area in a given space.

Several methods are used for removing the dust from the surface of the filter cloth. One consists of flexing the elements and the other of blowing clean air through the elements counterflow to the dirty gases. In either case the operation is automatic and performed without taking the unit out of service. Hoppers are provided in the base of the bag-house to receive the dust.

By use of suitable bag material very fine particles can be removed by these units. The higher gas temperatures encountered in power-boiler applications necessitate the use of special high-temperature fabric. It is very important to select fabrics suitable for the application as replacement of these elements necessitates equipment outages and maintenance expense.

Sulfur oxides. When fuels containing sulfur are burned, the sulfur is converted into sulfur dioxide and sulfur trioxide; these gases are discharged to the atmosphere with the other products of combustion (see Sec. 3.2). The obvious simple method of reducing the discharge of sulfur oxides is to burn a fuel with a lower sulfur content. This may be an unsatisfactory answer since the low-sulfur fuel is invariably more costly and in some cases unavailable. Another possibility is to process the fuel to reduce the sulfur content. Then there is the possibility of removing enough of the sulfur oxides from the flue gases to bring the concentration down to acceptable limits. Sulfur oxide concentration in flue gases may be reduced by the use of lime or limestone. One method of application is to feed the pulverized limestone into the furnace. The quantity of limestone feed is varied in proportion to the rate of fuel usage, sulfur content of the fuel, and necessary reduction in sulfur oxide content of the flue gases. This obviously increases the amount of particulate which must be removed from the gases. Another method is to feed lime slurry into a wet scrubber. The sulfur oxides are partially removed and appear as solids (mostly calcium sulfate) in the scrubber water. After settling they are removed as a slurry. The disposal of this slurry presents a difficult operating problem. Other systems provide for the recovery of the sulfur oxides in the form of sulfuric acid.

Nitrogen oxides. The nitrogen oxides in flue gases are from two sources: the nitrogen in the fuel, which is oxidized by the combustion air; and the oxidization of nitrogen in the combustion air. Reduction in the amount of nitrogen oxides formed has been accomplished by having the combustion take place at a reduced temperature and by the use of a low percentage of excess air.

This principle is applied to large pulverized-fuel furnaces by two-stage combustion. The pulverized coal is introduced through burners in the lower portion of the furnace with a deficiency of air. The additional air required for complete combustion is introduced into the furnace at a higher elevation. The admission of air must be carefully controlled to prevent the formation of soot in the first stage: Soot is difficult, if not impossible, to burn when the additional air is admitted to the furnace.

QUESTIONS AND PROBLEMS

4.1. What is the first duty of a combustion chamber?

4.2. When an HRT boiler is set on lugs, how are expansion and contraction taken care of?

4.3. Which is the best way to support a boiler: setting on lugs or the suspended method?

4.4. What is the advantage of setting an HRT boiler high?

4.5. How are grate bars set: level or inclined?

4.6 What is meant by the dry sheet? What is the purpose of a bridge wall?

4.7. How can the kind of grate bars to be used be determined?

4.8. What conditions are necessary when burning hard coal? When burning soft coal?

4.9. Name several advantages of mechanical stokers over hand firing.

4.10. What is meant by furnace volume? By grate surface?

4.11. What is meant by the rate of combustion?

4.12. Explain the procedure for regulating the coal feed to a chain- or traveling-grate stoker.

4.13. Specify the characteristics of coal for use on a chain-grate stoker.

4.14. Explain the procedure for banking the fire on a chain-grate stoker.

4.15. What are the causes of a high percentage of combustible in the refuse discharged from an underfeed stoker?

4.16. What are some of the characteristics of coal to be avoided when it is to be burned on an underfeed stoker?

4.17. How would you start a fire on an underfeed stoker?

4.18. How would you operate an underfeed stoker to pick up a load quickly?

4.19. What determines the necessary length of travel of the underfeed secondary rams?

4.20. How would you operate a spreader stoker, with dumping grates, when removing ashes?

4.21. Why is it necessary to use over-fire air with a spreader stoker?

4.22. Explain how you would adjust a spreader stoker to obtain satisfactory fuel distribution on the grates. Under what conditions does a change in these adjustments become necessary?

4.23. What is the advantage gained by reinjection of cinder into a spreader-stoker–fired furnace?

4.24. List the difficulties which may be encountered as a result of cinder reinjection.

4.25. Specify a coal suitable for use on a spreader stoker.

4.26. What is the correct banking procedure to use when operating a spreader stoker?

4.27. Write a set of safety instructions or precautions for a fireman of a pulverized-coal–fired plant.

4.28. State the possible causes of pulsations (variations in pressure) in a pulverized-coal–fired furnace.

4.29. What determines the amount of primary air to use in a pulverized-coal burner?

4.30. List the difficulties likely to be encountered in a pulverized-fuel–fired plant together with the possible means of overcoming them.

4.31. Discuss the operation and possible adjustments of a pulverized-coal burner.

4.32. Under what conditions is it necessary to heat fuel oil?

4.33. Make a sketch showing the tank, pumps, piping, etc., for a simple fuel-oil installation.

4.34. Explain the principle of operation and the advantages of rotary, steam-atomizing, and mechanical atomizing oil burners.

4.35. What are the essential points to remember in operating a fuel-oil- –burning plant?

4.36. What precautions should you take before attempting to light a pulverized-fuel, oil, or gas burner?

4.37. Under what conditions is gas used as a fuel for power boilers?

4.38. If you were encountering difficulty in carrying the load on a boiler, explain how you would proceed to determine if insufficient draft were the cause?

4.39. A chimney is 150 ft high, and the exit-gas temperature is 550°F. The outside termperature is 45°F. What draft can be expected at the base of the stack when the plant is operating at the designed capacity?

4.40. What are some of the conditions that make it necessary to employ mechanical draft?

4.41. What are some of the difficulties that may be encountered when operating fans?

4.42. Explain the three methods of controlling the output of fans as applied to boiler operation.

4.43. Discuss the two possible ways in which over-fire jets may improve combustion.

4.44. State and explain two methods of determining smoke density.

4.45. Write instructions for operation of a mechanical dust collector.

4.46. What precautions are required when operating scrubbers?

CHAPTER 5

BOILER
ACCESSORIES

5.1 WATER COLUMNS. Maintaining the correct water level in the boiler at all times is the duty of the boiler operator. Gauge glasses are provided to assist him and are installed to indicate the level of water in the boiler or boiler drum.

For the small boiler (Figs. 5.1, 5.2) the gauge glass is attached directly to the drum or shell by screwed connections, or a water column may be used. The water column is a vessel to which the gauge glass or other water-level–indicating devices are attached (Figs. 5.3, 5.4). The water column permits the gauge glass to be located where it can easily be 'seen and makes the installation accessible for inspection and repairs. The location of the gauge glass and water column varies for different types of boilers, but wherever they are located, the water in the glass must be maintained at the level required to avoid overheating of boiler surfaces.

The try cock is employed to check the level of the water in the boiler when the gauge glass is broken or out of service. So, in addition to the gauge glass, each boiler must have three or more gauge cocks (Figs. 5.2, 5.3) located within the range of the visible length of the water glass when the maximum allowable pressure exceeds 15 psi, except when such a boiler has two water glasses with independent connections to the boiler, located on the same horizontal line and not less than 2 ft apart. Boilers 36 in. in diameter and under, in which the heating surface does not exceed 100 sq ft, need have but two gauge cocks; the lower cock is located at the lowest permissible water level, and the upper cock at the highest desired water level. If the cocks are located properly, water should always come from the lower cock and steam from the upper cock.   Depending on the water level in the boiler, either water or steam will come from the center cock.

The water-gauge-glass connections are fitted with valves at the top and bottom, so that if the glass breaks, the steam and water may be shut off for repairs. The hand valves are frequently chain-operated, so that the operator may remain out of danger. Many gauges have both hand and automatic shutoff valves. The automatic shutoff valves consist of check valves located in the upper and lower gauge-glass fittings. Should the glass break, the rush of steam and water would cause these valves to close, the pressure holding them against their seats.

The ball check valves are of the nonferrous type. They are designed to open by gravity. These automatic valves (balls) must be at least ½ in. in diameter. It must be possible to remove the lower valve for inspection with the boiler in service. For added safety, the gauge glass is sometimes enclosed by wire-insert plate glass to protect the operator in the event that the glass breaks.

For higher pressures, up to 3,000 psi, water columns are made of flat glass (Fig. 5.5) having sheet mica to protect the water side of the glass from etching action of the steam.

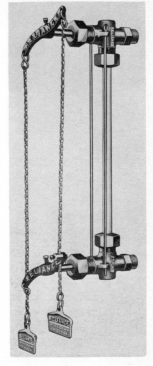

FIG. 5.1 *Water-gauge valves.* (*Clark-Reliance Corp.*)

The water column (Fig. 5.4) contains a high-low alarm which provides a signal (whistle) when the water rises or falls below the safe water level. Two floats are located in the water column, one above the other. To each is attached a rod opening a valve and actuating the same whistle. Should the water rise high enough to raise the upper float, the rod would open the valve, admitting steam to create a whistling noise. The same thing would happen if the water level fell; here the lower float would drop and repeat the warning. There are many different designs of high-low alarms that operate on the same principle. Whistles may be so designed that the tone denotes either a high- or a low-water alarm.

The Simpliport bicolor gauge (Fig. 5.6) is employed for high-pressure service. It has the advantage of being sectionally constructed; replacement of port assemblies is quickly accomplished without removing the gauge from the boiler. Normally but one port need be serviced at a time. The gauge is illuminated to secure maximum color contrast.

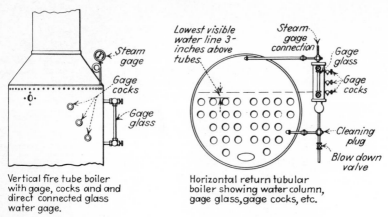

FIG. 5.2   *Glass water-gauge and water-column connections.*
(*a*) *Vertical boiler.*   (*b*) *HRT boiler.*   (*Clark-Reliance Corp.*)

Water is green, and steam is red; the water level is always where the colors meet.

The operation of the Simpliport gauge is based upon the simple optical principle that the bending of a ray of light differs as the ray passes obliquely through different mediums; hence when light passes through a column of steam, the amount of bending to which it is subjected is not the same as that when it passes through a similar column of water. If steam were to occupy the space between the windows ahead of the glass, the green light (water) would be bent out of the field of vision and the red light (steam) would appear in the glass; if water were to occupy the space between the windows, the red light would be bent out of the field of vision and the green light would appear in the glass.

At times the water column is far removed from the operating-floor level. This problem is overcome by using a periscope to bring the image of the gauge glass down to the operating level. The colors in the glass are directed into a hooded mirror, which in turn is reflected in a mirror at operating level. Distance is not a factor, and assurance is had of a positive water-level indication at all times.

Other devices to indicate the water level are actuated by the height of the water in the drum. This differential pressure can be used to operate a pointer on a gauge (similar to a pressure gauge), or the instrument can be made to record the level. Devices of this type do not replace the gauge glass but provide an additional aid for the operator.

The gauge glass and water column are piped to the boiler (Figs. 5.2, 5.3) so that the water level is the same in the glass as it is in the boiler. For the small vertical boiler, the gauge glass is attached directly to the shell; it must be so located that when water shows halfway in the glass, the boiler shell will be three-quarter full. For the HRT boiler (Fig. 5.2b), the lowest visible level in the water glass is 3 in. above the top of the tubes. Water-column piping is brass, extra-heavy iron, or steel; it must be at least 1 in. in diameter. Crosses are used instead of ells (Fig. 5.3) so that the line may be easily cleaned. When

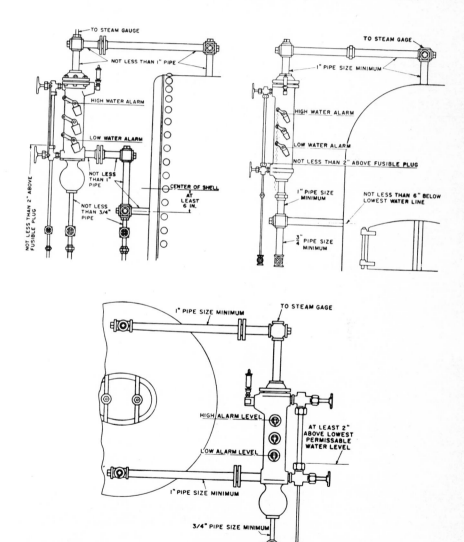

FIG. 5.3   *Installation of water columns.*   (*Clark-Reliance Corp.*)

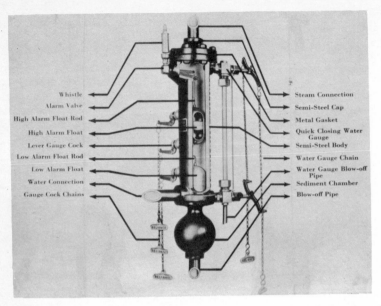

Whistle
Alarm Valve
High Alarm Float Rod
High Alarm Float
Lever Gauge Cock
Low Alarm Float Rod
Low Alarm Float
Water Connection
Gauge Cock Chains

Steam Connection
Semi-Steel Cap
Metal Gasket
Quick Closing Water Gauge
Semi-Steel Body
Water Gauge Chain
Water Gauge Blow-off Pipe
Sediment Chamber
Blow-off Pipe

FIG. 5.4   *Safety water column.   (Clark-Reliance Corp.)*

the correct level is maintained in the boiler, the water column should show the water approximately in the center of the glass.

In the horizontal fire-tube boiler (Figs. 5.2*b*, 5.3), the water-column connection must enter the front top of the shell or as high as possible in the head. The lower water-column connection must be at least 6 in. below the center line of the boiler. Now, with the water column properly located, the lowest visible part of the water-column glass must be at least 3 in. above the top row of tubes (Fig. 5.2*b*) and at least 2 in. above the fusible plug (Fig. 5.3). The gauge glass must be at least ½ in. in diameter, with the blowoff connecting pipe not less than ¾ in. in diameter. The correct locations of the gauge glass and the water column are checked by filling the boiler with water to the normal operating level and measuring the height of water above the tubes. This reading can then be compared with the level as it appears in the glass.

For the firebox-type boiler (Fig. 5.3), the lower water-column connection must be taken at least 6 in. below the lowest permissible water level but in no case closer than 18 in. to the mud ring. Boilers of the horizontal fire-tube type shall be so set that when the water is at the low level in the water-gauge glass, there shall be at least 3 in. of water over the highest point of the tubes, flues, or crown sheet.

If shutoff valves are used between the boiler and the water column, they must be either outside screw-and-yoke-type gate valves or stopcocks. In either case the valves must be locked or sealed open. All water columns must be provided with a blowdown line of ¾-in. minimum diameter. This line must be run to the ash sluice or other suitable drain.

No connection must be made to the water column except for the pressure gauge, feedwater regulators, or damper regulators. Pipelines that are to supply steam or water must never be connected to the water column, since the flow of steam or water would cause the column to record a false level.

When replacing a broken glass, make sure no broken pieces remain in the gauge fittings. Prior to inserting a new glass, blow out the piping connections. Make certain that the glass is of the proper length. A glass that is too long will break because of its inability to expand; a short glass will contine to leak around the packing glands. Give leaks around the glands immediate attention, taking care first to close the valves before using the wrench. Dirty and discolored glasses should be replaced at the first opportunity. At the time of the annual inspection, check all component parts of the water column carefully, paying particular attention to floats and linkage, alarms, whistles, etc.; inspect all connecting pipes, removing scale and dirt which may have collected on them. For the high-pressure port and flat-type gauge glasses, replace leaky gaskets.

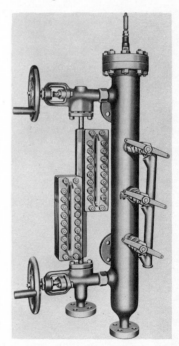

5.2   FUSIBLE PLUGS. Fusible plugs (Fig. 5.7) are used to provide protection against low water and consequent damage to the boiler. They are constructed of brass or bronze with a tapered hole drilled through the plug. In the ordinary direct-contact fire-actuated plug, this hole is filled with tin, which has a melting point of 445 to 450°F. "Fire-side" plugs are those inserted from the fire side of the plate, flue, or tube to which they are attached. On the water side they are to project at least ¾ in. on the other (gas) side, as little as possible but not more than 1 in. "Water-side" fusible plugs are those inserted from the water side of the plate, flue, or tube to which they are attached. Fusible plugs are installed with the large end of the core

FIG. **5.5** *Water gauge with water column. (Diamond Power Specialty Co.)*

FIG. 5.6   *Simpliport bicolor gauge.* (*Clark-Reliance Corp.*)

exposed to the water and made to blow through the narrow bore of the plug. For fire-actuated plugs, the least diameter of tapered hole is not less than ½ in. unless the maximum allowable pressure is over 175 psi or unless such a plug must be placed in a tube, in which case the smallest diameter of fusible metal is not less than ⅜ in. If a fire-actuated fuse plug is inserted in a tube, the tube wall shall not be less than 0.22 in. thick, or sufficiently thick to give four threads. For pressures over 225 psi, the fire-actuated plug is not recommended.

One side of the fuse plug is exposed to the hot furnace gases; only water cooling on the other side of the plug permits the heat to be carried away fast enough to keep the plug from melting. Should the water drop below the fuse plug, the heat melts the tin and is blown out of the core. The pressure, upon being released, sounds an alarm and warns the operator of low water level. The fuse plug can be replaced only by taking the boiler out of service, cooling, and draining; for the "fuse-alarm"-type plug (Fig. 5.7c), we can shut the valve and insert another fuse plug. Before inserting the plug and continuing with the operation, we should determine the extent of damage due to low water.

The fusible plug must be located at the lowest permissible water level and extend to make contact with the hot combustion gases. In the HRT boiler, the plug is located in the rear head 1 in. above the top row of tubes, as measured from the upper surface of the tubes to the center of the plug. In a vertical fire-tube boiler with nonsubmerged tubes, it is located in an outside tube not less than one-third of the length of the tube above the lower tube sheet. In a vertical fire-tube boiler of a submerged-tube type, it is located in the upper tube sheet. In the horizontal-drum water-tube boiler, the plug is located not less than 6 in. above the bottom of the drum, projecting through the sheet not less than 1 in., and above the first pass of the products of combustion. For other boilers and fusible-plug locations, consult the boiler code.

If fusible plugs are used, they should be inspected frequently, since scale and dirt on the water side and soot on the fireside may foul the plug. A plug in this condition will not function properly and will not

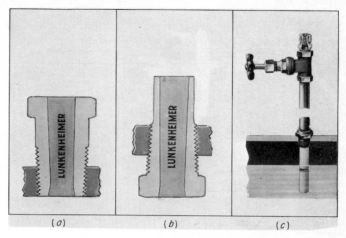

FIG. 5.7 *Fusible plugs and fuse alarms. (a) Inside type. (b) Outside type. (c) Fuse alarm. (The Lunkenheimer Co.)*

provide the necessary protection.

With each boiler outage, the plug should be examined, cleaned, and scraped to a bright surface of the fusible metal. If it is not sound or if its soundness is in question, replace it; do not fill it. All fusible plugs should be replaced at least once a year.

5.3 STEAM GAUGES. All boilers must have at least one pressure gauge. It should have a range and dial graduation of at least 1½ times the maximum allowable working pressure. The connection may be made to the steam space or attached to the upper part of the water column. The gauge itself must be so located that it can be seen easily by the fireman. Piping should be as direct as possible. If a valve is used in the gauge line connected to the boiler, it should be locked or sealed open. If a cock is used in place of a valve, it should be of the type that indicates by the position of its handle whether it is open or closed (open with the handle in line with the pipe). The piping to the gauge should be so arranged that it will always be full of water. A siphon coil is employed to make certain the gauge stays full of water; never connect the steam line so that steam can enter the tube, since the high steam temperature will damage the gauge. A ¼-in. branch connection and valve are provided so that a "test" gauge can be installed without removing the permanent gauge. If the temperature exceeds 406°F, brass or copper pipe or tubing should not be used. The connection to the boiler should be made with not less than a ¼-in. standard pipe size, but when steel pipe or tubing is used, with not less than a ½-in.-inside-diameter size.

Steam gauges used in modern practice are usually of the bourdon

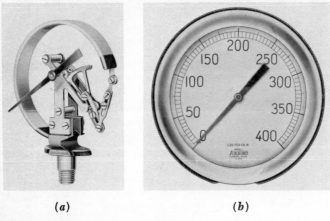

(a)   (b)

FIG. 5.8   *Foxboro pressure gauge.*   (a) *Bourdon tube
and linkage.*   (b) *Exterior view.*

or spring-tube type (Fig. 5.8).   The bourdon tube consists of a curved
tube with an oval cross section.   One end of the tube is attached to
the frame and pressure connection; the other end is connected to a
pointer by means of links and gears.   Movement of the pointer is directly proportional to the distortion of the tube.   An increase in pressure
tends to straighten out the tube; as the tube position changes, its motion
is transmitted to rack and pinion through connecting linkage to position
the pointer.   Motion from the tube-connecting linkage is then transmitted to the pinion to which the pointer is attached, moving it over
the range of the gauge.

The steam gauge for a small boiler is usually mounted directly on
top of the water column, and hence the gauge reads the correct pressure
in the boiler.   However, on many installations, particularly large boilers,
the gauge is brought down to the operating level.   At this level the
gauge reads the steam pressure plus the hydraulic head of water in
the line.   Hence the gauge is inaccurate unless this head is compensated
for.   In this case we must measure the vertical distance between the
point at which the connection is made (assume that it is the top of
the water column) and the center of the gauge and correct for this
water column.   For each foot of vertical distance between the connection and the gauge, the gauge reading must be corrected by 0.433 psi
per ft and this correction subtracted from the gauge reading.

Example   A gauge is located 45 ft below the point at which it is connected
to the steam line or water column; the gauge reads 175 psi.   What is the true
gauge reading?

Solution   Pressure due to head of water:

$$45 \times 0.433 = 19.59 \text{ psi}$$
$$175 - 19.59 = 155.41, \text{ actual pressure at point measured}$$

In this case the pointer on the steam gauge is reset by approximately 20 lb to indicate the true header pressure, or 155 (155.41 actual).

Pressure gauges are frequently mounted above the point of pressure measurement and piping connections made with ¼-in. pipe or tubing. In many instances lines may be run horizontally before they proceed vertically to the gauge board and pressure gauge. These lines remain full of water, and the gauge will read inaccurately unless a correction is made for the water column as in the preceding example. For such installations measuring steam pressure, we may want to put a siphon at the take-off point to make certain the line is full of water. Here again, the pressure in the connecting line (vertical distance) is measured as 0.433 lb per ft, and the pressure added to the gauge reading. The pointer on the gauge must be reset.

5.4 FEEDWATER REGULATORS. A boiler feedwater regulator automatically controls the water supply so that the level in the boiler is maintained constant. This automatic regulator adds to the safety and economy of operation and minimizes the danger of low or high water. Uniform feeding of water prevents the boiler from being subjected to the expansion strains which would result from temperature changes produced by irregular water feed. The danger in the use of a feedwater regulator lies in the fact that the operator may place entire dependence upon it. It is well to remember that the regulator, like any other mechanism, can fail; continued vigilance is necessary.

The first feedwater regulator (Fig. 5.9) was very simple, consisting of a float-operated valve riding the water to regulate the level. If the level dropped, the feed valve opened; if the level was too high, the valve closed; at intermediate positions of water level, the valve was throttled. A modern float-type regulator (Fig. 5.10a) is designed with the float box attached directly to the drum.

A thermohydraulic, or generator-diaphragm, type of boiler feedwater regulator is shown in Fig. 5.10b. Connected to the radiator is a small tube running to a diaphragm chamber; the diaphragm in turn operates

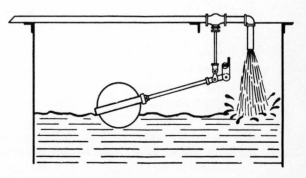

FIG. 5.9 *The origin of the first commercial feedwater regulator.* (*Copes-Vulcan Div. Blaw Knox.*)

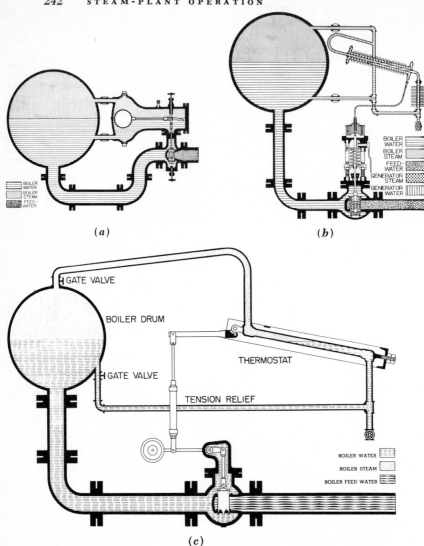

FIG. 5.10.   *Three generic types of boiler feedwater regulators for simple level control.   (Copes-Vulcan Div. Blaw Knox.)*

a balanced valve in the feedwater line.   The inner tube is connected directly to the water column and contains steam and water; the outside compartment, connecting tube and valve diaphragm, is filled with water. This water does not circulate; heat is radiated from it by means of fins attached to the radiator.   Water in the inner tube of the regulator remains at the same level as that in the boiler.   When the water in the boiler is lowered, more of the regulator tube is filled with steam and less with water. Since heat is transferred faster from steam to water than from water to water, extra heat is added to the confined water in the outer compartment. The radiating-fin surface is not sufficient to remove the heat as rapidly as it is generated, so that the temperature and pressure of the confined water are raised. This pressure is transmitted to the balanced-valve diaphragm to open the valve, admitting water to the boiler. When the water level in the boiler is high, this operation is reversed.

The thermostatic expansion-tube-type feedwater regulator is shown in Fig. 5.10c. Because of expansion and contraction, the length of the thermostatic tube changes to position the regulating valve with each change in the proportioned amount of steam and water. A two-element steam-flow-type feedwater regulator (Fig. 5.11) combines a thermostatic expansion tube operated from the change in water level in the drum as one element with the differential pressure across the superheater as the second element; the two combined operate the regulating valve. An air-operated three-element feedwater control (Fig. 5.12) combines three elements to control the water level. Water flow is proportioned to steam, with level as the compensating element, the control being so

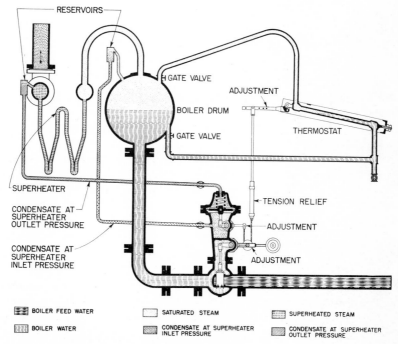

FIG. **5.11** *Two-element steam-flow-type feedwater regulator.* (*Copes-Vulcan Div. Blaw Knox.*)

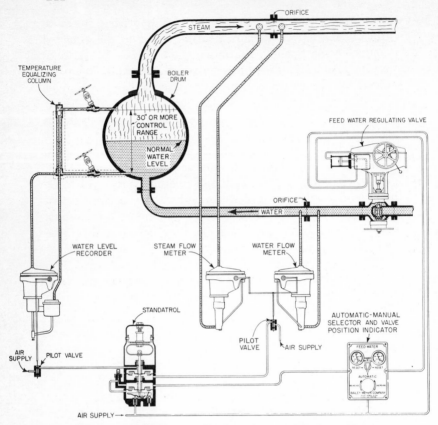

FIG. 5.12 *Diagrammatic layout of three-element feedwater control—air-operated type.* (*Bailey Meter Co.*)

set as to be insensitive to level. In operation, a change in position of the metering element positions a pilot valve to vary the air-loading pressure to a standatrol (self-standardizing relay). The resultant position assumed by the standatrol provides pressure to operate a pilot valve attached to the feedwater regulator. The impulse from the standatrol passes through a hand-automatic selector valve, permitting either manual or automatic operation. The handwheel jack permits manual adjustment of the feedwater valve if remote control is undesirable.

In some installations, a differential pressure valve is placed in series with the regulating valve, operating to maintain a constant pressure drop through the feed valve. When several boilers are operating in battery, this eliminates the possibility of a sudden load change's swinging the load from one boiler to the other until the changed water level has adjusted the valve opening to the new condition.

The simple float-operated regulator is satisfactory for small boilers with large water-reserve capacity. A more modern float-type regulator for the same purpose is shown in Fig. 5.10a. More accurate and dependable control is secured with thermohydraulic, generator-diaphragm, or thermostatic expansion-tube-type regulators, and these are applied to water-tube boilers of moderate size and steam capacity. Such boilers have adequate water storage, and level fluctuations are not critical. The single-element control is affected only by the water level and is capable of varying the water level in accordance with the steaming rate.

Large boilers equipped with waterwalls, having small water-storage space, and subjected to fluctuating loads employ the two-element control, since feed characteristics are dependent on the rate of change rather than on the change in level. This change takes care of "swell" as well as "shrinkage" in boiler water level, and unless operating conditions are very severe, stability of water level can be maintained where load swings are wide and sudden, which is too difficult a condition for a single-element regulator to control. In the two-element unit, steam flow predominates and adjustment is provided from water level.

Three-element feedwater regulators (Fig. 5.12) are employed on large boilers subjected to wide and sudden load fluctuations and in installations where considerable variations in pressure drop across the feed valve are experienced. This type of control is recommended particularly for boilers equipped with "steaming" economizers and boilers that have small water-storage capacity. Three-element control is desirable with an increased rate of steaming because a much greater percentage of the volume below the surface of the water is occupied by steam. Only a small percentage of the total volume of water in a modern boiler is in the drum; hence the drum water level will be seriously affected by changes in the steaming rate.

Maintenance is an important item in connection with control. All control lines should be checked for leakage at frequent intervals. Regulators equipped with a remote manual-automatic selector (Fig. 5.12) can be checked for leakage by positioning the control knob on "reset." In this position, the control is blanked off from either manual or automatic control and should remain in a fixed position. If this position varies (pointer on the gauge), there is leakage between the selector and the control valve. Leakage along the control lines can be detected by the noise of escaping air (in the case of a large leak), or a soap solution can be applied at points suspected of leaking.

Semiweekly. Blow down the water column on the generator; fill condenser (Fig. 5.10b). Take care of water leaks around valves and fittings promptly.

Monthly. Lubricate control parts. Check meters and connections for leaks; check standatrol and automatic selector valves carefully; check flowmeters to zero to determine their accuracy, sensitivity, and response. Check automatic-control system for leakage.

Yearly. Disconnect the meter and all control lines; blow them out. Dismantle the meter; clean, inspect, overhaul, and calibrate by running a water-column test. Carefully inspect all control valves in the system, such as selector and standatrol valves. Also dismantle and inspect the feedwater-regulating valves. If possible, dismantle and overhaul regulators semiannually or at least annually. At such times go over the entire control mechanism to eliminate wear in moving parts. Check valves for wear and replace parts where necessary. Give particular attention to all packing glands.

5.5 SAFETY VALVES. Boilers are designed for a certain maximum operating pressure; if this pressure is exceeded, there is danger of an explosion. This danger is so great that it necessitates equipping all boilers with safety valves to maintain the boiler pressure within design limits. Rules governing safety valves, design, and installation are as follows:

Each boiler shall have at least one safety valve, and if it has more than 500 sq ft of water-heating surface, it shall have two or more safety valves. The safety-valve capacity for each boiler shall be such that the safety valve or valves will discharge all the steam that can be generated by the boiler without allowing the pressure to rise by more than 6 per cent above the highest pressure at which any valve is set and in no case by more than 6 per cent above the maximum allowable working pressure. The safety-valve capacity of new units shall be in compliance with ASME Code P-274 but shall not be less than the maximum designed steaming capacity as determined by the manufacturer.

One or more safety valves on the boiler proper shall be set at or below the maximum allowable working pressure. If additional valves are used, the highest pressure setting shall not exceed the maximum allowable working pressure by more than 3 per cent. The complete range of pressure setting of all the saturated-steam safety valves on a boiler shall not exceed 10 per cent of the highest pressure to which any valve is set.

All safety valves shall be so constructed that failure of any part cannot obstruct the free and full discharge of steam from the valves. Safety valves shall be of the direct spring-loaded pop type. The maximum rated capacity of a safety valve shall be determined by actual steam flow at a pressure of 3 per cent in excess of that at which the valve is set to blow.

If the safety-valve capacity cannot be computed (see ASME code) or if it is desirable to prove the computations, the capacity may be checked in any one of the three following ways. If it is found insufficient, additional capacity shall be provided.

1. By making an accumulation test, shutting off all other steam-discharge outlets from the boiler and forcing the fires to the maximum. The safety-valve equipment shall be sufficient to prevent a pressure in excess of 6 per cent above the maximum allowable working pressure. This method should not be used on a boiler with a superheater or reheater.

2. By measuring the maximum amount of fuel that can be burned and computing the corresponding evaporative capacity upon the basis of the heating valve of the fuel.

The weight of steam generated per hour is found by the formula

$$W = \frac{C \times H \times 0.75}{1,100}$$

where W = weight of steam generated, lb per hr

C = total weight or volume of fuel burned at time of maximum forcing, lb or cu ft per hr

H = heat of combustion of fuel, Btu per lb or per cu ft

3. By determining the maximum evaporative capacity by measuring the feedwater. The sum of the safety-valve capacities marked on the valves shall be equal to or greater than the maximum evaporative capacity of the boiler.

When two or more safety valves are used on a boiler, they may be mounted separately or as twin valves made by placing individual valves on Y bases or duplex valves having two valves in the same body casing. Twin valves made by placing individual valves on Y bases or duplex valves having two valves in the same body shall be of equal size. When not more than two valves of different sizes are mounted singly, the relieving capacity of the smaller valve shall be not less than 50 per cent of that of the larger valve.

The safety valve or valves shall be connected to the boiler independently of any other steam connection and attached as closely as possible to the boiler, without any unnecessary intervening pipe or fitting. Every safety valve shall be connected so as to stand in an upright position. The opening or connection between the boiler and the safety valve shall have at least the area of the valve inlet. The vents from the safety valves must be securely fastened to the building structure and not rigidly connected to the valves, in order that the safety valves and piping will not be subjected to mechanical strains resulting from expansion and contraction and the force due to the velocity of the steam.

No valve of any description shall be placed between the required safety valve or valves and the boiler or on the discharge pipe between the safety valve and the atmosphere. When a discharge pipe is used, the cross-sectional area shall be not less than the full area of the valve outlet or the total of the areas of the valve outlets discharging thereinto.

The pipe shall be as short and straight as possible and so arranged as to avoid undue stresses on the valve or valves.

All safety-valve discharges shall be so located or piped as to be carried clear of running boards or platforms. Ample provision for gravity drain shall be made in the discharge pipe at or near each safety valve and at locations where water of condensation may collect. Each valve shall have an open gravity drain through the casing below the level of the valve seat (Fig. 5.13).

Safety valves shall operate without chattering and shall be set and adjusted to close after blowing down not more than 4 per cent of the set pressure but not less than 2 lb in any case. For spring-loaded pop safety valves at pressures between 100 and 300 psi, the blowdown shall be not less than 2 per cent of the set pressure. Safety valves used on forced-circulation boilers of the once-through type may be set and adjusted to close after blowing down not more than 10 per cent of the set pressure.

Each safety valve shall have a substantial lifting device by which the valve disk may be lifted from its seat when there is at least 75 per cent of full working pressure on the boiler.

The spring in a safety valve in service for pressures up to and including 250 psi shall not be reset for any pressure more than 10 per cent above or below that for which the valve is marked. For higher pressure, the spring shall not be reset for any pressure more than 5 per cent above or below that for which the safety valve is marked.

Screwed openings can be used to attach the valve to the boiler when the proper number of threads is available. A safety valve over 3 in. in size used for pressures greater than 15 psi gauge shall have a flanged inlet connection or a welded-end inlet connection. Safety valves may be attached to drums or headers by fusion welding, the welding being done in accordance with the ASME code.

Every attached superheater shall have one or more safety valves near the outlet. If the superheater-outlet header has a full and free steam passage from end to end and is so constructed that steam is supplied to it at practically equal intervals throughout its length, resulting in a uniform flow of steam through the superheater tubes and header, the safety valve or valves may be located anywhere in the length of the header.

The discharge capacity of the safety valve or valves attached to the superheater may be included in determining the number and size of the safety valves for the boiler, provided there are no intervening valves between the superheater safety valve and the boiler and provided the discharge capacity of the safety valve or valves on the boiler, as distinct from the superheater, is at least 75 per cent of the aggregate valve capacity required.

Every independently fired superheater which may be shut off from the boiler and permitted to become a fired pressure vessel shall have one or more safety valves with a discharge capacity equal to 6 lb of steam per sq ft of superheater surface measured on the side exposed to the hot gases.   The total number of safety valves installed shall be such as to meet capacity requirements.

Every reheater shall have one or more safety valves such that the total relieving capacity is at least equal to the maximum steam flow for which the reheater is designed.   At least one valve shall be located on the reheater outlet.   The relieving capacity of the valve on the reheater outlet shall be not less than 15 per cent of the required total.   The capacity of the reheater safety valves shall not be included in the required relieving capacity for the boiler and superheater.

Every safety valve used on a superheater or reheater discharging superheated steam at a temperature over 450°F shall have a casing, including base, body, and spindle, of steel, steel alloy, or equivalent heat-resisting material.   The valve shall have a flanged inlet connection or a welded-end inlet connection.   The seat and disk shall be of suitable heat-erosion- and corrosion-resisting material, and the spring shall be fully exposed outside the valve casing so that it is protected from contact with the escaping steam.

When two or more boilers which are allowed different pressures are connected to a common steam main and all safety valves are not set at a pressure not exceeding the lowest pressure allowed, the boiler or boilers allowed the lower pressure shall each be protected by a safety valve or valves placed on the connecting pipe to the steam main.   The area or the combined area of the safety valve or valves placed on the connecting pipe to the steam main shall be not less than the area of the connecting pipe, except when the steam main is smaller than the connecting pipe; then the area or combined area of the safety valve or valves placed on the connecting pipe shall be not less than the area of the steam main.   Each safety valve in the connecting pipe shall be set at a pressure not exceeding the pressure allowed on the boiler it protects.

The safety valve (Fig. 5.13) has the disk held on its seat by a spring. The tension on the spring can be adjusted to give some variation in popping pressure; this is accomplished by the compression screw, which forces the valve against its seat.   The valve is correctly positioned by the valve extension fitting into a seat; an adjusting ring to regulate the blowback pressure, which in turn can be adjusted and fixed by the ring pin, is provided to control the relieving pressure.   A hand lever is furnished to permit popping the valve by hand.

When the safety valve opens, it opens wide, causing the valve to "pop." This quick opening of the safety valve is accomplished by having the

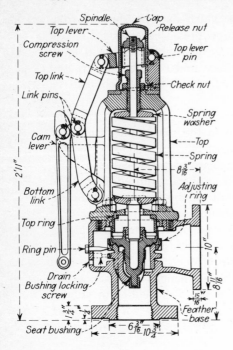

FIG. 5.13 *Construction details of consolidated pop safety valve. (Dresser Industries.)*

steam diverted against a larger area as soon as the valve starts to open. NOTE: when the valve is closed, the steam pressure is directed against the area of the feather which is inside the seat (Fig. 5.13). This area is equal to that of a circle with a diameter equal to the diameter of the valve seat. After the valve feather has lifted, this area is increased by an amount which the feather extends over the seat. The boiler pressure is then exerted on a larger area, giving greater lifting power.

The effect of this steam on the surface outside the valve seat is determined by the position of the adjusting or blowback ring (Fig. 5.13). This ring can be adjusted so that once the valve has opened, the pressure must be lowered to a predetermined pressure before it will close again. The difference between the pressure at which it closes and that at which it opens is known as the "blowdown" or "blowback." Safety valves are set to blow back from 2 to 8 lb before closing but never more than 4 per cent of the set pressure and not less than 2 lb in any case. The popping point tolerance, plus or minus, shall not exceed the following: 2 lb for pressures up to and including 70 lb: 3 per cent for pressures from 71 to 300 lb; 10 lb for pressures over 301 to 1,000 lb; and 1 per cent for pressures over 1,000 lb.

In attempting to determine the capacity of a safety valve, with any degree of accuracy, we suggest reference be made to the valve manu-

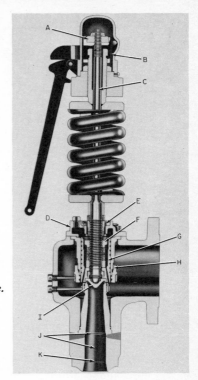

FIG. 5.14 *Consolidated maxiflow safety valve.* (A) *Spring compression.* (B) *Lifting gear.* (C) *Spindle.* (D) *Back pressure.* (E) *Blowdown control.* (F) *Lift stop adjustment.* (G) *Groove disk holder.* (H) *Upper adjusting ring.* (I) *Thermdisk seat.* (J) *Inlet neck.* (K) *Inlet connection.* (*Dresser Industries*)

facturer's "selection chart" for the pressure-temperature range in which the valve is to operate. Valve design varies to a very considerable degree; so this is the most practical approach for determining capacities.

Problems with safety-valve leakage became increasingly severe as steam pressures increased; and so a high-capacity flat-seated reaction-type safety valve was developed (Fig. 5.14) to meet greater discharge capacity, shorter blowdown, etc., as required for high-pressure-temperature steam-generating equipment. Construction details and operation are as shown.

With reference to Fig. 5.14*b*; in Fig. 1 a 100 per cent lift is attained by proper location of the upper adjusting ring G. When full lift is attained (Fig. 2.), lift stop M rests against cover plate P to eliminate hunting, adding stability to the valve. When the valve discharges in an open position, steam is bled into the chamber H through two bleed holes J in the roof of the disk holder.

Similarly, the spindle overlap collar K rises to a fixed position above the floating washer L. The area between the floating washer and the spindle is thereby increased by the difference in the two diameters on the overlap collar.

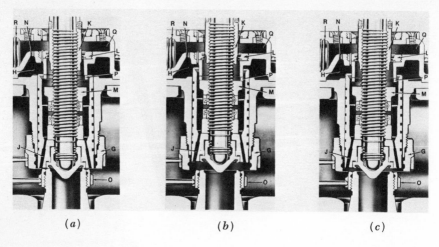

(a)   (b)   (c)

FIG. 5.14 (cont.)   *Operation of consolidated safety valve.*   (*Dresser Industries*)

Under this condition, steam in chamber H enters into chamber Q through the secondary area formed by the floating washer L and the overlap collar K on the spindle, through orifice N, and escapes to the atmosphere through the pipe discharge connection R. When closing (Fig. 3), the spindle overlap collar K is adjusted so that it moves down into the floating washer L, thereby reducing the escape of steam from chamber H effectively.

The resulting momentary pressure buildup in chamber H, at a rate controlled by orifice N, produces a downward thrust in the direction of spring loading. The combined thrust of the pressure and spring loading results in positive and precise closing. Cushioning of the closing is controlled by the lower adjusting ring O.

The valve includes several features such as: (a) back pressure closing—lift and blowdown are separate valve functions and accurate control of each is possible; (b) thermodic seat—provides tight closure and compensates for temperature variations with thermal stresses minimized; (c) spherical-tip spindle with a small flat on the extreme end—provides a better point for pivoting than does a ball; (d) welded—neck is forged; the nozzle is stainless steel; bypass leaks around a seal weld cannot occur with the three-piece construction.

The superjet safety valve (Fig. 5.15) is designed for steam pressures up to 3,000 psi and temperatures to 1000°F. The spring here is pro-

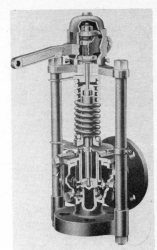

FIG. 5.15 *Superjet safety valve.* (*Foster Engineering Co.*)

tected from the heat by a shield and cradled by a special spring-saddle design to prevent distortion under compression. There are of course many other safety valves, all offering special features and advantages, which are to be determined with use.

When boilers are equipped with superheaters and with safety valves on both the superheater outlet and the steam drum, the safety valve on the superheater outlet should open first. This produces a flow of steam through the superheater and prevents the superheater tubes from being damaged by high temperatures which would result if all the steam were discharged directly from the steam drum.

Safety-valve springs are designed for a given pressure but can be adjusted. However, if the change is greater than 10 per cent above or below the design pressure, it becomes necessary to provide a new spring and in some cases a new valve. If a valve is adjusted for insufficient blowback, it is likely to leak and simmer after popping. Leakage after popping is also caused by dirt which gets under the seat and prevents proper closing of the valve. Safety valves should have the seat and disk "ground in" to prevent leakage. When valves are disassembled for grinding, the springs should be compressed by suitable clamps and held in place so that the adjustment of the spring will not be altered.

When a safety valve leaks at a pressure less than that at which it is set to close, try to free the valve by operating the lifting lever; if this does not stop the valve from leaking, repair or replace the safety valve as soon as possible.

After changing the valve setting, adjusting the spring or the blowback ring, test the safety valve. This can be accomplished by slowly raising the steam pressure and noting the pressure-gauge reading when the popping pressure is reached. At the instant the valve pops, read the pressure gauge, after which the rate of steaming (or firing) should be reduced. Again read the pressure gauge when the valve closes, to note the blowback. Continued and repeated adjustments (if necessary) should be made to adjust the spring and blowback ring to secure the desired popping pressure and blowback. To do this will require that the pressure be raised or lowered until the correct setting has been secured.

When testing and setting the safety valve, maintain a water level below the top gauge cock. Care should be exercised to prevent an accumulation of scale, dirt, or other foreign matter from collecting between the coils of the valve spring. Drains should be kept open. If a hydrostatic test is made on the boiler, the safety valves should be removed and the openings blanked; or clamps should be applied to hold the valve closed.

5.6 BLOWDOWN APPARATUS. Water fed to the boiler contains impurities in the form of suspended and dissolved solids. A large portion of these impurities is left behind when the steam leaves the boiler. Some of these suspended impurities are of such a nature that they settle in the lowest part of the boiler; others are light and float on the surface of the water. This condition frequently calls for the installation of both surface and bottom blowoff lines on some boilers.

There are two types of surface blowoff arrangements. One consists of a pipe entering the drum approximately at the normal water level; the pipe is fixed in location, or stationary. The other arrangement is to have a swivel joint on the end of a short piece of pipe, the free end being held at the surface by a float. This floating-type surface blowoff is frequently referred to as a "skimmer." Surface blowoff is advantageous in skimming or removing oil from the boiler water. A surface blowoff line shall not exceed $2\frac{1}{2}$ in. in diameter.

The surface blowdown is usually made on a continuous basis and after the feedwater has been tested. A Flocontrol valve (Fig. 5.16) is an orifice-type valve equipped with an indicator (for greater accuracy) and regulated by hand to control the quantity of water we wish to discharge, based on the water analysis. Blowing down at a slower rate and over a longer period of time reduces the concentration more effectively than is possible by opening wide the main blowoff valve; hence closer control and more accurate regulation of the blowdown are achieved. Likewise erosion and wear of valve parts are held to a minimum.

The continuous blowdown suggests the use of a flash tank where the high-pressure water can be flashed into low-pressure steam and used for process or feedwater heating, or the continuous blowdown can be passed through a heat exchanger to preheat the makeup water. If required, a small portion of the higher-alkaline blowdown water may be introduced into the boiler feedwater line to raise the pH value of the water and eliminate feed-line or economizer corrosion.

To be effective, however, the continuous-blowdown take-off must be placed at a point in the boiler where the water has the highest concentration of dissolved solids. This point is usually located where the greater part of the steam separates from the boiler water, ordinarily in the steam drum; hence the surface blowdown.

All boilers must be equipped with a bottom blowoff pipe fitted with a valve or cock directly connected to the lowest point in the water space so that the boiler can be completely drained. The blowoff lines must be extra-heavy, not less than 1 in. and not more than 2½ in. in diameter, except that for boilers with 100 sq ft of heating surface or less the minimum size may be ¾ in. Too small a pipe might stop

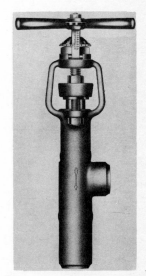

FIG. 5.16 *Continuous-boiler-flow-control blowdown valve. (Dresser Industries.)*

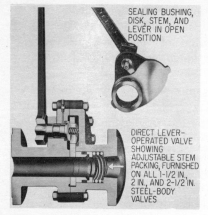

SEALING BUSHING, DISK, STEM, AND LEVER IN OPEN POSITION

DIRECT LEVER-OPERATED VALVE SHOWING ADJUSTABLE STEM PACKING, FURNISHED ON ALL 1-1/2 IN., 2 IN., AND 2-1/2 IN. STEEL-BODY VALVES

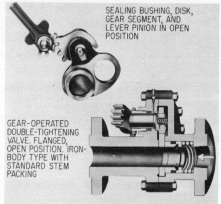

SEALING BUSHING, DISK, GEAR SEGMENT, AND LEVER PINION IN OPEN POSITION

GEAR-OPERATED DOUBLE-TIGHTENING VALVE. FLANGED, OPEN POSITION. IRON-BODY TYPE WITH STANDARD STEM PACKING

FIG. 5.17   *Quick-opening blowoff valve.*   (*Yarnall-Waring Co.*)

up; one too large would discharge the water too rapidly. Since there is no circulation of water in the pipe, scale and sludge frequently accumulate here. Unless this pipe is protected from the hot gases by brickwork or other suitable insulating materials, it may burn out.

On boilers operating at 100 psi or over, two blowoff valves are required; they may be two slow-opening valves or one slow-opening and one cock or quick-opening valve. Over 125 psi both valves and piping must be extra-heavy. A slow-opening valve is one which requires at least five 360° turns of the operating device to change from fully closed to open or vice versa. A quick-opening double-tightening valve is shown in Fig. 5.17; this valve is frequently used in tandem with a seatless valve. The quick-opening valve is installed next to the boiler; it is opened *first* and closed *last.* In the tandem combination the quick-opening valve becomes the sealing valve rather than the blowing valve. The valve shown is designed to operate at 320 psi maximum pressure. NOTE: For all other valves arranged in tandem, the sequence of operation is the reverse. Here the second valve from the boiler is opened first and closed last; blowing down takes place through the valve next to the boiler.

For pressures to 450 psi, a seatless valve (Fig. 5.18) may be employed; for pressures to 600 psi, a tandem arrangement (Fig. 5.19*a*) of a hard-seat blowing valve and a seatless sealing valve is used. With higher pressures (1,500 to 2,500 psi) the hard-head sealing valve finds application. Here the blowing valve (nearest to the boiler) will have flow entering below the seat. The blowing valve (next to the boiler) should be opened *last* and closed *first;* the sealing valve (outside) should be opened first and closed last. For the hard-seat valve, the position of the handwheel above the yoke indicates the location of the disk in the valve, whereas for the seatless valve the position of the plunger indicates whether the valve is open or closed. While these valves are to be operated rapidly, they cannot be opened or closed quickly; water-hammer in the discharge line is thus avoided.

The tandem blowoff valve is a one-piece block serving as a common body for both the sealing and the blowing valves. The seatless valve is equipped with a sliding plunger, operated by the handwheel and nonrising stem; leakage is prevented by packing rings above and below the inlet port. As the seatless valve opens, blowdown is discharged through double ports. In the seatless-valve design, the annular space in the body permits pressure to surround the plunger, making the valve fully balanced and hence easy to operate at all pressures.

In operating, open both valves rapidly and fully to prevent erosion of seat and disk faces and to increase the life of the packing and working parts. Never blow down through a partly opened valve. If a hard-seat valve and a seatless blowoff valve are arranged in tandem, the hard-seat

 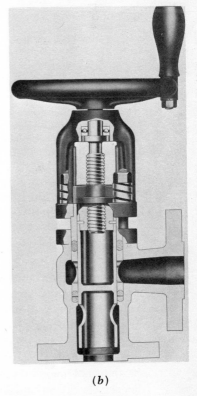

(a)   (b)

FIG. 5.18 *Blowoff valves.* (a) *Angle valve flanged—open position for full and free discharge.* (b) *Angle valve flanged—closed position for drop-tight shutoff.* (Yarnall-Waring Co.)

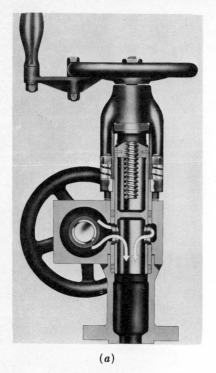 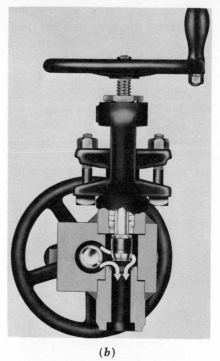

(a)   (b)

FIG. 5.19   *High-pressure-unit tandem blowoff valves.*   (a) *Open position; contains hard-seat blowing valve and seatless sealing valve; 600 psi. (b) Open position; contains hard-seat blowing and sealing valves; welded ends for 1,500 psi.*   (Yarnall-Waring Co.)

valve will be nearest to the boiler. When installing blowoff valves, take care that the piping is not restricted by the boiler setting but left free so that it can expand and contract.

In closing valves, do not force them to close, although firm pressure can be applied. If there appears to be some obstruction, open the valves again before closing them finally. Make certain the blowdown cocks and valves are then shut tight and remain that way when they are not blowing down. Repair leaky valves as soon as possible. The valves should be dismantled at least once a year and worn parts, such as scored plungers, packing rings, valve seats, etc., replaced if necessary. Prior to (and after) taking the boiler out of service for overhaul, it might be well to check for blowoff-valve leakage. Such leakage can be detected (with valves closed) by placing a hand on the discharge line to check the temperature, care being exercised not to get burned. If the line stays hot, leakage is evident. A rod held against the discharge line and used as a listening device will also detect leakage.

Blowoff connections cannot be run directly to a sewer or to the atmosphere. Steam and hot water might damage the sewer; flashing steam

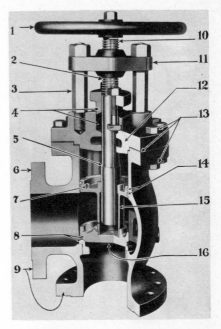

FIG. 5.20 *Edwards nonreturn valve.* (1) *Handwheel.* (2) *Edwards acme threads.* (3) *Hexagonal posts.* (4) *Stuffing box.* (5) *For packing under pressure.* (6) *Cast-iron flanges.* (7) *Threaded, for removal and regrinding.* (8) *Seat and disk.* (9) *Center to face, to permit use for removal.* (10) *Steel stem.* (11) *Steel cross yoke.* (12) *Body and bonnet.* (13) *Bolted bonnet joint.* (14) *Piston rings.* (15) *Narrow-neck piston to reduce pressure drop.* (16) *Full area—disk and piston guided and cushioned.*

might prove harmful to persons in the vicinity. Blowoff lines are run into a blowoff tank, entering at a point above the waterline maintained in the tank. The blowdown water and flashing steam are then discharged above the water level; there is a vent in the top of the tank to avoid back pressure. A discharge line with an opening below the water level and near the tank bottom results in the cooler water in the tank's being discharged first. The discharge line leaves the tank at a point opposite the inlet; the line outside the tank is provided with a vent so that the water cannot be siphoned from the tank. The tank is provided with a drain outlet and valve at the bottom. The blowoff tank also acts as a seal to prevent sewer gas from backing up into a boiler which is out of service. A blowoff tank should never be relied upon as a seal for boiler service. All valves must be closed tightly.

5.7 NONRETURN VALVES. A nonreturn valve is an automatic stop-check valve to prevent a backflow of steam when the boiler pressure falls below the header pressure. Such a drop in pressure may occur as a result of a tube failure or of a decrease in the combustion rate.

A nonreturn valve makes it possible to bank a boiler and return it to service without operating the steam-shutoff valve. As the boiler is placed on the line, the valve opens automatically when the pressure in the boiler exceeds (slightly) the pressure in the steam header on the discharge side of the nonreturn valve. The valve closes when the boiler pressure drops below the header pressure. The use of the nonreturn valve provides additional safety in operation.

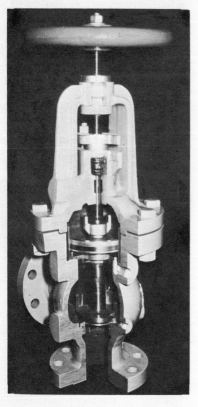

FIG. 5.21 (a) Triple-acting nonreturn valve. (Golden Anderson Valve Spec. Co.)

The automatic nonreturn valve (Fig. 5.20) is in reality a cushioned check valve. The cushioning is provided to keep the valve from chattering when it is being opened or closed. In some valves this cushioning is accomplished by using springs, but the usual method is to employ a dashpot. The nonreturn valve has a stem not connected to the disk. When the stem is screwed down, it forces the disk shut and prevents it from opening. When the stem is screwed out, the disk can open or close automatically, depending upon whether the boiler or the header pressure is the higher.

The triple-acting nonreturn valve (Fig. 5-21a) has an additional feature in that it closes automatically in the event that the header pressure decreases to a predetermined intensity below the boiler pressure; the cause of the decrease might be a break in the header or stop valve located in the steam header. The triple-acting nonreturn valve will (1) automatically open to cut a boiler into the header when the pressures are approximately equal, (2) automatically close when the boiler pressure drops below the header pressure, and (3) automatically close to isolate the boiler when the header pressure drops by approximately 8 lb below the boiler pressure.

With reference to Fig. 5.21b, the main valve is installed with the boiler pressure entering under the valve disk; the autopilot is piped so that pressure in the annular space between dashpots A and B of the main valve will enter under the automatic-pilot piston H. The outlet of the autopilot is connected to the steam header some 10 to 15 ft on the valve discharge side. Test line 4 is run to the operating-floor level for manual testing of the valve by the operator; it remains closed except during testing.

In operation, when the boiler pressure overcomes the header pressure, the main valve D opens to admit steam. Steam from the boiler also passes through the bypass in the center of the disk, through the ball check valve N, and thence to the top of the piston A. Small orifice holes C permit the passage of steam to the annular space between dashpots A and B, as well as to the auto pilot H through valve 3.

In the event of a ruptured boiler tube, the pressure in the steam

line forces the valve *D* to close.   The steam located between the stationary disk *B* and the moving dashpot *A* tends to cushion the movement,
preventing hammer or shock.   If there is a break in the header, the
pressure on the header side of the valve (and to the top of the pilot
valve *H*) begins to drop.   When it drops to about 8 psi below the
boiler pressure, the boiler pressure raises the pilot valve *H*, permitting
steam to escape from between *A* and *B;* the main valve immediately
moves to the closed position.

To test the main valve, the operator at the floor level opens valve
4 to accomplish the same thing as the foregoing (a break in the header);
opening valve 4 adjusts the valve to a closed position.   Opening the
test valve 4 unbalances the main valve to close it, and the boiler-pressure
gauge will immediately record an increase in pressure to prove that
the main valve is closed.   When the test valve is closed again, the main
valve resumes its normal automatic position.

Several typical installations of nonreturn and stop valves are shown
in Fig. 5.22.   When boilers are set in battery and are carrying more
than 135 psi, they must be equipped with two steam valves, one of

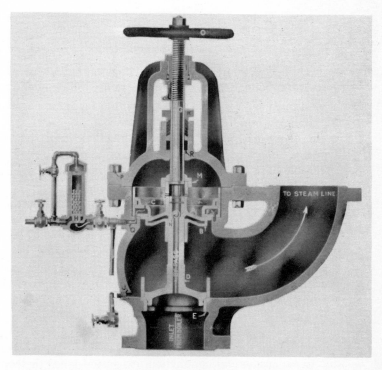

FIG. 5.21 (*b*)   *Triple-acting nonreturn valve; interior view,*
elbow pattern.   (*Golden Anderson Valve Spec. Co.*)

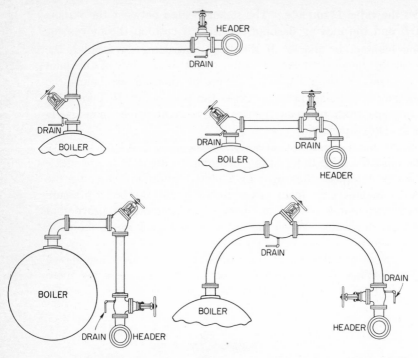

FIG. 5.22   *Typical installations of nonreturn valves.*   (*Crane Co.*)

which can be a nonreturn valve.   The nonreturn valve should be placed nearest the boiler and as close to the boiler outlet as possible.   It should be equipped with a drain or bleeder line for removing the water (condensation) before the valve is opened.

The nonreturn valve should be dismantled, inspected, and overhauled annually.   Replace packing, clean corroded parts, and coat with graphite after painting.   Check valves and seats for leakage, grind in the valve, and replace defective parts.

5.8   STEAM HEADERS.   Engines, turbines, auxiliaries, and process equipment which use steam are usually located at some distance from the boilers, and so it is necessary to transport the steam through pipes or headers.   The main steam header picks up the branch lines from each boiler; take-offs are provided where necessary.   This system of header piping makes it possible to use one boiler or combination of boilers to supply steam for any equipment requiring steam.

Headers must be constructed of pipe sufficiently heavy to withstand the internal pressure and shock due to the velocity of steam passing through them; they must be supported adequately to take care of strain due to expansion and contraction; expansion joints or loops (Chap. 13) must be properly located and installed.   The header is insulated and provided with drains and traps to remove the water of condensation and so prevent the water from entering the pumps, engines, or turbine.

Headers are designed for steam velocities of approximately 6,000 fpm when supplying steam to reciprocating engines and pumps; much higher velocities (10,000 fpm) may be employed for steam lines to turbines. The volume of steam necessary must be conveyed without excessive pressure drop; good design limits the pressure drop, usually to a maximum of 5 per cent. Pressure drop is expressed in terms of loss per 100 ft of lineal pipe; usually 2 to 3 lb is considered a permissible drop based on economic considerations.

5.9 SOOT BLOWERS. Boiler tubes and heating surfaces get dirty because of an accumulation of soot and fly ash. These substances are excellent insulators and so must be removed. Removal can be accomplished by using a hand lance or soot blower. Steam and compressed air are usually employed for blowing, although water and shot are sometimes used to remove certain types of deposits which become baked hard and are difficult to remove with the conventional soot blower.

Soot blowers are made of pipe and special alloys to withstand high temperatures; they are of fixed or movable types. The element itself merely serves as a conduit or mechanical support for the nozzles, through which steam or air is transmitted at high velocity. Soot blowers are designed for many different applications, employing various nozzle contours to meet specific needs under varying temperature conditions. The size, design, and location of soot-blower nozzles are varied to meet the cleaning needs encountered in the generating tube bank and other heat-exchange equipment, including superheaters, air preheaters, and economizers. On bent-tube boilers, nozzles are set at a right angle to the element; on boilers with a staggered tube arrangement, the nozzles are set at an angle to clear the lanes between the tubes.

In Fig. 5.23 is shown a soot-blower installation for a HRT boiler. It consists of a revolving blow arm equipped with special steam-turbine-type nozzles so spaced as to blow directly into each boiler tube as the arm rotates. The arm can be operated from the front of the boiler by means of a chain and operating wheel while the boiler is in service. The spindle carrying the blow arm rotates and is supported by an adjustable bracket fastened to the inner door. Access to the boiler is gained by closing the steam valve and breaking the pipe union.

Automatic valve-head soot blowers control the admission of the blowing medium while rotating the element. The rotary element is attached to the sidewall or boiler casing; it is supported on the tubes by bearings designed for that purpose. This type of soot blower permits blowing only while at the correct angle. Hangers are made of the intimate-con-

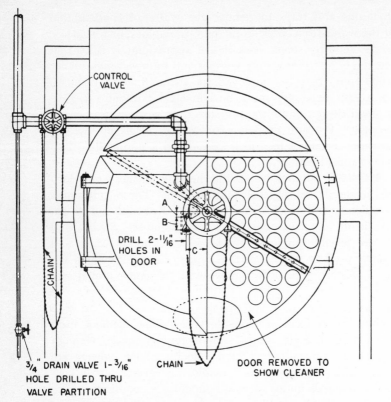

CONTROL VALVE

A
B

DRILL 2-¹¹/₁₆" HOLES IN DOOR

C

CHAIN

³/₄" DRAIN VALVE I-³/₁₆"
HOLE DRILLED THRU
VALVE PARTITION

CHAIN

DOOR REMOVED TO
SHOW CLEANER

FIG. 5.23 *Soot blower for horizontal-return tubular boiler.*
(*Copes-Vulcan Div. Blaw Knox.*)

tact type (Fig. 5.24), in which the bearing is provided with a smooth surface of large area for contact with a (comparatively) cool boiler tube, thus preventing excessive temperature on blower and bearing. Details of bearings and element installations are shown for both the welded and the intimate-contact bearing. Bearings are machined to fit the tube snugly so as to give a heat-conducting bond or contact. With a protective bearing on each boiler tube and a jet between, only a small portion of the element is left exposed to the high temperature of the gases. A clamp-type bearing of the compression type is used where a welded-bearing installation is inconvenient.

For the conventional soot blower only a small section of the element may not be in intimate contact with the tube or bearing (Fig. 5.24), yet such an element frequently overheats and becomes warped and thus inoperative. With higher exit-gas temperatures, hangers and nozzles are damaged by exposure to such temperatures. And so because it was costly to maintain the conventional multijet blower, a fixed-type rotary element (Fig. 5.25) has been converted to a retractable soot blower. The retractable blower element is located outside the furnace; it is designed for lance travels exceeding 40 ft. Paths requiring cleaning are traversed by blowing the jet as the element is being extended into or withdrawn from the furnace. Elements of this type are located in areas in which it was previously difficult to maintain a soot blower in continuous service. The retractable blower element, since it is located outside the furnace, requires no tube clamps, bearings, etc., which formerly necessitated continual maintenance.

Air, saturated or superheated steam, water, or any combination of

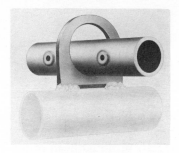

(a) **Welded bearing**

(b) **Intimate-contact bearing**

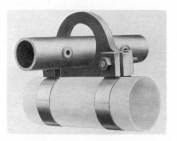

(c) **Crown bearing with straps**

(d) **Protective bearing**

(e) **Compression bearing**

Fig. **5.24**   *Soot-blower bearings.*   (*Copes-Vulcan Div. Blaw Knox.*)

FIG. 5.25   *Long retractable soot blower with dual motor drives.*   (*Copes-Vulcan Div. Blaw Knox.*)

these fluids may be used as the blowing medium without change in equipment. Outside adjustment of nozzle pressure (50 to 425 psi) is made possible by the mechanically operated head of the long retractable blower. Many varieties of nozzle arrangement may be used according to tube spacing. Nozzle inserts are of stainless steel, their shape and diameter depending on design and application. Either the rotating or the traversing speed of the element may be independently adjusted without affecting the other, merely by changing sprockets.

The gun-type deslagger (Fig. 5.26) is also of the retractable type; it is employed for furnace walls and other inaccessible spots previously cleaned by hand lancing. In operation one motor extends and retracts the swivel tube; another rotates the element. To save time, entrance and withdrawal are made quickly, with 2.7 sec for each. The normal blowing cycle is 70 sec; the length of time can be altered to suit requirements. Single or dual nozzles can be employed, depending upon the wall arrangement or slagging conditions. Wall deslagging is one sure way to minimize average slag thickness for maintaining constant high heat-transfer capacity and uniform superheat- and reheat-temperature control.

The Pneu-Blast is a flexible automatic soot-blowing system using air as the blowing medium. The system was developed for application (1) where steam is at too low a pressure to be used satisfactorily (as in low-pressure boilers or hot-water heating) and (2) where compressor capacity is less than required to clean the boiler effectively with other methods of air blowing.

Systems of this type can use a smaller air compressor than the older systems; hence the initial cost is lower. Because blasts are intermittent, fewer particles of soot or other matter are dislodged and suspended in the air at any given moment. Dust-collector systems are not over-loaded during such cycles. Stack emissions with each blast are negligible; hence air pollution is reduced. The small puffs provide more stable boiler-operating conditions, reducing the fluctuations in furnace draft experienced with multijet blowers.

These units can be operated by push button or by automatic sequential control. The automatic sequential control can be applied to a series of blower elements on a single boiler or to multiboiler operation. Blowers are made to operate automatically; they cut in, cut out, and cut in the next head in a series operation. Individual blowers can be cut in or out of service at will. Each blower can be blown individually from a remote push button; indicators (lights) notify the operator which element is blowing or not blowing. The sequence of operation for the single blower or multiblowers can be varied, the number of possible arrangements and combinations being almost unlimited.

Principle of operation. The diagram in Fig. 5.27 shows the system ready to start, at the start of the blast, and at rest between blasts. At the start of the blast, full header pressure is applied through the soot-blower head to the element. The element continues to blow, rotating as it blows, until the receiver pressure has dropped to some lower pre-determined pressure. Here the blast ends, and the system is at rest until the receiver pressure again builds up to the higher predetermined pressure.

Automatic soot-blowing systems are now installed in many power plants; soot blowing is done by remote control. Once the master button has been pushed, the entire system is in correct sequential operation. The operator can, at a glance, tell exactly which blower is being operated; steam or air pressure is recorded or indicated. No longer is the

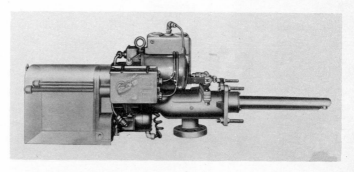

FIG. **5.26** *Retractable gun-type blower, or wall deslagger.* (*Copes-Vulcan Div. Blaw Knox.*)

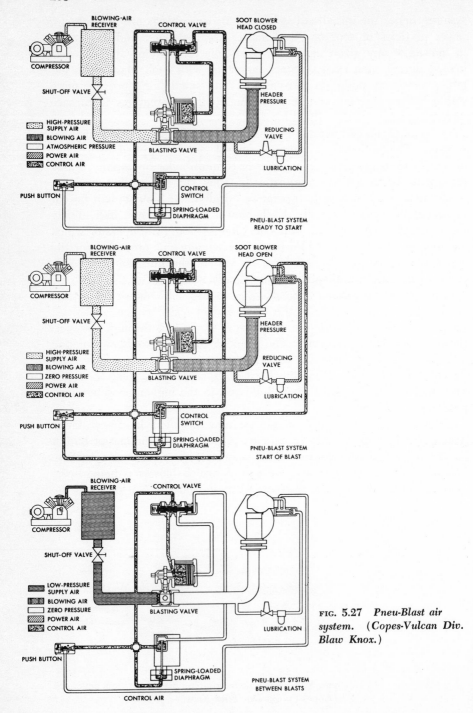

FIG. 5.27   *Pneu-Blast air system.   (Copes-Vulcan Div. Blaw Knox.)*

hard-to-get-at soot blower or the one in the hot location neglected as in the past. Although such systems are expensive to install, they are justified because of high labor costs and because of the fact that without automatic equipment this important operation might be neglected with resulting loss in efficiency, boiler outage, and reduction in capacity.

Dirty boiler heating surfaces result in less heat absorption and higher exit-gas temperatures; a schedule for soot blowing is therefore necessary. Frequency of blowing varies from plant to plant and depends on many factors. To determine the frequency of blowing necessary for a specific installation, reference is most frequently made to draft loss or exit-stack temperature. Practice in general has been to blow boiler heating surfaces once a shift and air preheaters and economizers perhaps once a day.

When operating rotating elements by hand, rotate slowly so as to cover the heating surface thoroughly. "Stops" are frequently provided on many elements; here the intent is to sweep through a definite angle or blowing arc. When blowing soot, maintain a furnace draft of 0.10 to 0.15 in. Start with the blower nearest the furnace and continue to blow progressively through the last pass of the boiler; then blow the economizer and air preheater. For an automatic system, the cycle is prearranged so that all one need do is place the system in operation to carry it through its entire cycle.

Elements which are permanently located in the boiler must be properly aligned to prevent overheating and warping, as well as damage to the tubes by cutting due to misalignment. If the element warps, it becomes inoperative; hence we lose the service of the element. Excessive maintenance is avoided by proper element selection and installation; bearings and elements must withstand high temperatures for prolonged periods of time. Steam piping to the element must be equipped with proper drains and traps to avoid slugs of water which damage the blower and erode the tubes. For those elements which are not intended to blow through the entire 360° range, the arc is established by positioning the stops located in the head of the blower.

5.10 VALVES. Valves are used to control the flow of water to the boiler and steam from the boiler to the header. They are attached in several ways: the body of the valve may be equipped with pipe threads so that the valve can be screwed into position; or the valve may be equipped with flanges and bolted in position; or, on high-pressure boilers, valves are frequently welded into position.

The *globe* valve (Fig. 5.28) consists of a plug or disk which is forced into a tapered hole called a "seat." The angle used on the taper of the seat and disk varies with the valve size and the kind of service to which the valve is applied. Globe valves are used when the flow

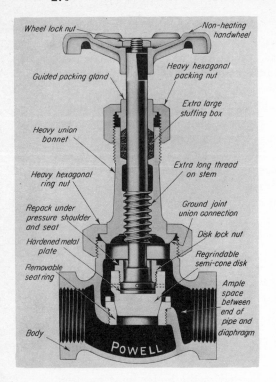

Wheel lock nut

Non-heating handwheel

Heavy hexagonal packing nut

Guided packing gland

Extra large stuffing box

Heavy union bonnet

Heavy hexagonal ring nut

Extra long thread on stem

Ground joint union connection

Repack under pressure shoulder and seat

Disk lock nut

Hardened metal plate

Regrindable semi-cone disk

Removable seat ring

Ample space between end of pipe and diaphragm

Body

POWELL

FIG. 5.28 *Globe valve.* (*Powell Valve Co.*)

is to be restricted or throttled. Whenever a globe valve is used on feed piping, the inlet shall be under the valve disk. The seats and disks on the globe valve are not cut by the throttling action as readily as with gate valves; valve parts are easy to repair and replace. The disadvantages of the globe valve are (1) increased resistance to flow; (2) the fact that, with increased pressure under the disk, more force is required to close the valve; and (3) the possibility that foreign matter may cause plugging of the valve.

The *gate* valve (Fig. 5.29), as the name implies, consists of a gate that can be raised or lowered into a passageway. The gate is at right angles to the flow and moves up and down in slots which hold it in the correct vertical position. It is usually wedge-shaped so that it will tighten against the sides of the slots when completely shut off. A gate valve is used chiefly where the valve is to be operated either wide open or closed. When wide open, it offers very little resistance to flow, and consequently pressure loss, or drop, through the valve is held to a minimum. The pressure acts on one side of the gate so that the gate is forced against the guides and requires considerable force to operate, at least for the larger valves. The gate valve is a difficult valve to repair once the seats have been damaged.

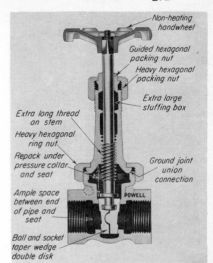

Non-heating handwheel

Guided hexagonal packing nut

Heavy hexagonal packing nut

Extra large stuffing box

Extra long thread on stem

Heavy hexagonal ring nut

Repack under pressure collar and seat

Ground joint union connection

Ample space between end of pipe and seat

POWELL

Ball and socket taper wedge double disk

FIG **5.29** *Gate valve—inside-screw type.* (*Powell Valve Co.*)

For both the globe and gate valves (Figs. 5.28, 5.29), the body contains the valve seat; the valve bonnet is attached to the valve body by a threaded nut or bolted flange (Fig. 5.31), depending upon the size of the valve. The flanged-fitting steel valve shown is employed on a steam header; it is equipped with a bypass valve. The bypass valve is opened prior to opening the main valve, permitting the line to heat up and equalize the pressure on both sides of the valve, thus making the valve easier to open.

The valve stem extends through the bonnet and is threaded and fitted with a hand wheel. The bonnet is provided with a packing gland which prevents leakage around the valve stem. The movable part of the valve element is carried on the end of the valve stem. Turning the valve wheel moves the stem in or out, opening or closing the valve.

The Breech Lock cast-steel-wedge gate valve (Fig. 5.30) with seal-welded bonnet was developed primarily to meet the need for a body-bonnet joint that would be pressure-tight at high temperature and without requiring constant maintenance. The Breech Lock design transmits bonnet thrust to the body through interlocking breech lugs with final pressure-tight closure of the body-bonnet joint made by a small seal weld. This results in pressure tightness and maintenance-free service of a full-strength bonnet, providing ready disassembly and reassembly. For boiler feed conditions the valve is available in pressures to 5,800 psi at 525°F and for main steam conditions up to 2,150 psi and 1050°F. This design is also available for globe and check valves. The unit shown is hand-operated but is also available with gearing or motor unit, for ease of operation or operation by remote control.

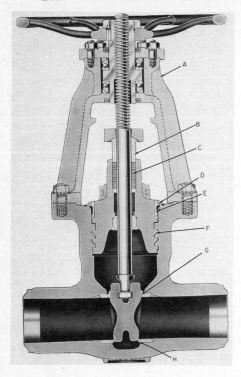

FIG. 5.30 *Breech Lock cast-steel-wedge gate valve; 1500 psi.* (A) *Yoke.* (B) *Packing gland.* (C) *Stuffing box.* (D) *Seal weld.* (E) *Flexible steel ring.* (F) *Breech lugs.* (G) *Stellite faced.* (H) *Seal rings.* (*Lunkenheimer Co.*)

Some of the valve features are: (*a*) The yoke is mounted directly on the body and independent of the bonnet thus relieving the bonnet seal weld of the weight of yoke and operating mechanism; the effect of vibration and stem thrust. (*b*) Packing gland is secured to the bonnet by two swing bolts; packing readily accessible. The stuffing box is outside the high-temperature zone. (*c*) Seal weld provides pressure tightness; the weld can be readily disassembled and reassembled. (*d*) The flexible steel ring, permanently welded to the bonnet, prevents stress due to the seal welding. (*e*) Breech lugs lock the bonnet in the body carrying full internal pressure. (*f*) Both disk and seal rings have stellite-faced seating

FIG. 5.31 *Bypass on steel valve.* (*Crane Co.*)

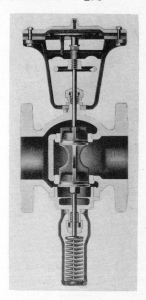

FIG. 5.32 *Balanced valve used in connection with the S.C. feedwater regulator.* (*Swartout Co.*)

surfaces. (*g*) The flexible disk prevents "sticking" or "freezing" with less torque required to open the valve.

Automatic-control valves, such as a valve operated by a feedwater regulator, must be so designed that little force will be required to operate them. This is accomplished by the use of a balanced valve (Fig. 5.32), which is similar to a two-seated globe valve. The balanced valve has two seats and two disks. One of these disks opens against, and one with, the pressure; that is, a balanced valve has the pressure on the top of one disk and on the bottom of the other. Thus the pressures are balanced, and it is possible to open or close the valve with a minimum of effort. Two-seated valves, such as the balanced valve (Fig. 5.32), are not tight-shut or remain that way in operation. They should be checked and inspected

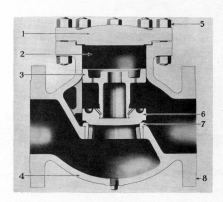

FIG. 5.33 *Cast-steel check valve.* (*1*) *Valve cover.* (*2*) *Body bore.* (*3*) *Piston.* (*4*) *Valve body.* (*5*) *Studs.* (*6*) *Valve disk.* (*7*) *Valve seat.* (*8*) *Flange.* (*The Edwards Valve & Mfg. Co., Inc.*)

frequently. We can determine whether the valve leaks by using a listening rod.

The check valve (Fig. 5.33) is a modified globe valve without a stem. It is usually arranged to close by gravity. The flow is directed under the valve to raise it from its seat; if the flow reverses, gravity plus the pressure above the valve closes it. The check valve is used where a flow in only one direction is desired; one of the main uses of this valve is in the feedwater line to the boiler.

Many forms of reducing valves are employed to operate auxiliary equipment not requiring boiler or header pressures; for example, low-pressure heating system, for heat exchangers, etc. The reducing valve consists of a balanced valve actuated by the pressure on the low-pressure side, the low-pressure acting on a diaphragm to open or close the balanced valve.

The type B steam-pressure reducing and regulating valve (Fig. 5.34) is a single-seated, spring-loaded direct-acting diaphragm valve. This valve automatically reduces a high initial pressure to a lower delivery pressure, maintaining that lower pressure within reasonably close limits regardless of fluctuations in the high-pressure side of the line.

The inner valve assembly is easy to clean or replace by loosening the hex head bottom plug; major repairs are made without removing the valve from the line. Pressure adjustments are readily made simply by turning the top adjusting screw. The valve is protected by a self-supporting inbuilt monel strainer; or a strainer can be installed in the line ahead of the regulator. Preferred installation for this reducing-regulating valve is to provide a shutoff valve on each side of the regulator and then a bypass line around it with pressure gauges at the entrance and behind the regulator so one can observe the operation of the regulator at all times.

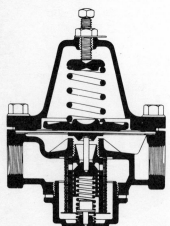

FIG. 5.34 *Steam-pressure type B regulator. (A.W. Cash Valve Mfg. Corp.)*

Stopcocks, or plug valves, are used frequently as blowoff valves, in gas and oil lines, for water softeners, etc. The valve consists of a circular, tapered plug that is ground-fit in a hole in the valve body. There is a hole through this plug at right angles to its axis. When the plug is in one position, this hole lines up with the hole in the body of the valve; the valve is then open. When the plug is turned, the holes are thrown out of line and the valve is shut. The stopcock can be opened or closed very quickly. The chief advantage of this valve lies in the fact that it is not easily affected by dirt in the substance handled. By various combinations of holes in the valve body and plug, several lines can be controlled with a single valve.

5.11 INSTRUMENTS AND AUTOMATIC COMBUSTION CONTROL. In addition to the equipment and appliances necessary for the safe operation of a boiler, there are many others which add to the safety, reliability, efficiency, and convenience of operation. Instruments are available which indicate or give a complete graphic record of draft, pressure, temperature, flue-gas analysis, stoker and fan speeds, steam and water flow, etc.

A draft gauge is necessary or desirable on all boilers for indicating the furnace draft. Indication of the pressure under the stoker fuel bed (forced-draft-fan pressure) as well as draft at the boiler uptake is also desirable. This permits the operator to secure proper furnace conditions, regulate the air supply to the fuel bed, and vary the rating by means of the uptake damper. When multiple-retort stokers are employed, pressures under each section of the grate should be indicated since proper air distribution can be determined accurately only in this manner.

Every boiler must have a pressure gauge to indicate the boiler steam pressure. Pressure gauges are also employed to indicate or record the steam-header pressure, pressure of water in the feed line; they can also be used in the feed or steam line to indicate flow, if reference is made to the pressure drop in the line. Pressure gauges are used for many other purposes and are perhaps the most common instrument available.

Temperature indicators or recorders are employed for steam, water, flue-gas, air, and fuel temperatures and for many other purposes. Also there are carbon dioxide recorders, flowmeters for steam and water, indicators and recorders for boiler water level, smoke-density recorders, etc. The advantage of using a recorder is that we can make reference thereto for changes in operation; in emergencies. The record is then available for analysis and correction of the problem.

Records secured from instruments make it possible for the operator or management to determine if best operating practice is being violated or maintained. Instruments and controls are usually expensive to install and maintain, and it therefore becomes an individual problem for each plant to determine just what is required or necessary to perform the task for which each is intended.

Instruments alone do not improve economy; rather it is the close control of operation and maintenance made possible by a knowledge of the conditions shown by the instruments that results in increased economy. Installation of proper instruments is only the first step toward the efficient operation of a boiler plant. Every instrument must be kept operating and properly calibrated if its cost is to be justified. Records obtained therefrom should have a definite use in the control operation. Recorder charts filed away without being studied never result in increased economy; they represent so much money wasted. The correct instruments, properly operated, and the information derived from them, put to use, pay for themselves in a few months.

Instruments must be accurate, rugged, sensitive, and extremely dependable and must function as precision mechanisms.

Automatic control, developed from these simple instruments, goes a step further; here the instrument actually controls or operates the equipment. By means of automatic control, we can maintain constant steam pressures and uniform furnace draft and can change air or fuel to meet changes in steam demand. We can also alter, adjust, and modify the supply of air and fuel to bring these into proper relationship; at the same time we can vary the load from minimum to maximum rating to secure the most efficient combustion results. It is obvious that the control is no better than the instruments which do the metering and measuring. But by the installation of reliable instruments and with the proper operation of the controls, maintenance can be reduced, the operator can be relieved of repetitive duties, and fuel dollars can be saved.

The problem of boiler control is one of coordinating the following factors: (1) steam pressure, (2) fuel quantity, (3) air for combustion, (4) removal of the products of combustion, (5) feedwater supply, etc. Feedwater control is indeed important, but is not included in this discussion of combustion control.

There are three types of control systems: (1) off-on, (2) positioning, and (3) metering. All three systems are designed to respond to steam-pressure demands, to control fuel and air for combustion so as to secure the highest combustion efficiency.

"Off-on" control is applied to small boilers. A change in pressure actuates a pressurestat or mercury switch to start the stoker, the oil or gas burner, and the forced-draft fan. The control functions to feed fuel and air in a predetermined ratio to secure good combustion. These results can be varied by manually changing the fuel or air setting. The off-on cycles do not produce the best combustion efficiency for reasons that are immediately obvious.

In a plant containing a number of boilers, each equipped with an off-on control operating from a pressurestat, we frequently encounter the problem of one boiler's appropriating most of the load. This is due to the fact that control pressures are difficult to adjust within close limits. Hence it is better to install a single master pressurestat to bring the units off and on at the same time. Or we can employ an off-on sequence control which brings the units on and takes them off the line as required. A selector switch can be installed to vary the sequence in which boilers are placed in operation. Such systems can also be applied to the modulating control.

"Positioning" control is applied to all types of boilers. It consists of a master pressure controller which responds to changes in steam pressure and, by means of power units, actuates the forced-draft damper to control airflow and the lever on the stoker to adjust the fuel-feed rate. Such units usually have constant-speed forced-draft fans equipped with dampers or inlet vanes which are positioned to control the air for combustion. Furnace-draft controllers are employed to maintain the furnace draft within desired limits; this control operates independently of the positioning-control system.

For the normal positioning-type control, the only time the airflow and fuel feed are in agreement is at a fixed point, usually where control calibration was made. This is so because the airflow is not proportional to damper movement. Variables which affect this relationship are the type of damper, variations in fuel-bed depth, variations in fuel quality, and lost motion in control linkage. The necessity of frequent manual adjustment to synchronize the previous control is readily apparent.

For the positioning-type control the variables are in part corrected through the proper alignment of levers and connecting linkage between the power unit and the damper and fuel-feed levers which they operate, by installing cams and rods calibrated to alter the arc angularity of travel from the power-unit levers. These compensate for the movement characteristics of fuel-feed and air-damper control. In addition, this system can be provided with a convenient means for manual control, operation from a central point. This remote manual-control system can be used for changing the distribution of the load between boilers or for making adjustments in the fuel-feed rate to compensate for changes in fuel quality. The positioning-type control has an advantage over the off-on control in that the fuel and air can be provided in small increments to maintain continuous operation, hence eliminating off-on cycling.

"Metering" control is used when the fuel rate and Btu input vary widely because of variations in fuel supply and heat content and when combination fuels are burned. Here we actually meter the fuel and air, maintaining the correct air-fuel ratio for best combustion results, based on design and testing. The steam (or water) flow can be a measure of fuel feed; that is accomplished by measuring the pressure drop across an orifice, flow nozzle, or venturi. Air for combustion can also be metered by passing the air through an orifice, but most frequently airflow is measured by the draft loss across the boiler or air preheater (gas side) or across the air preheater (air side). The air side is frequently chosen as the point of measurement in order to avoid dust and dirt's clogging the lines and fouling the control system.

The metering-type control is more accurate than the positioning system, since compensation for variables is secured through metering without regard for levers, linkage, lost motion, damper position, fuel variables, etc. Also, with the metering control, the fuel-air ratio can be readjusted from the air-steam flow relationship. The metering control usually incorporates a remote manual station wherein the control system can be modified and where hand or automatic operation is possible. There are many varieties of combustion-control systems; they may operate pneumatically, hydraulically, electrically, electronically, and sometimes in combination. Many systems are indeed complex.

For the pneumatic system, all instruments in the control loop are air-activated measurements, to and from central points. Steam pressure is controlled by parallel control of air and fuel; high-low signal selectors function to maintain an air-fuel mixture.

Airflow to the furnace is controlled by automatic positioning of the forced-draft inlet vanes. Furnace draft is maintained at the desired value by controlling or positioning of the induced-draft-fan damper. Feedwater flow to the boiler drum is controlled separately.

Adjustment is provided at the panel board for steam-pressure set point, fuel-air ratio, furnace draft, and drum-level set points. The controls would include pneumatic switches for bumpless transfer from automatic to manual control; and manual control can be accomplished from the control panel.

Figure 5.35 is a schematic diagram for an all-electronic instrumentation system, burning pulverized fuel, employing Babcock & Wilcox ball pulverizers. Electronic transmission permits locating measurement transmitters and final control elements at long distances from the control panel.

As shown in the schematic, steam-header pressure is maintained at the correct value by controlled positioning of air dampers controlling airflow to the pulverizers. The fuel controller automatically corrects for variation in the number of pulverizers in service.

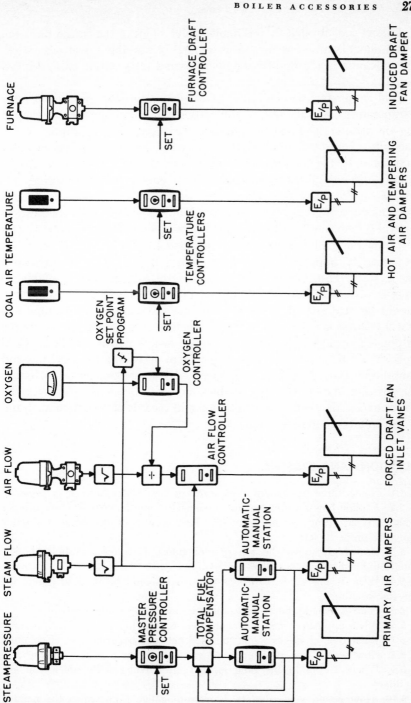

FIG. 5.35   *Schematic diagram of instrumentation for pulverized-coal firing.   (The Foxboro Co.)*

Airflow is controlled by positioning inlet vanes of the forced-draft fan. Steam flow, as an inferential measurement of fuel input, controls airflow. Fuel-air ratio is adjusted by the operator at a ratio station on the control panel.

Coal-air mixture is controlled at a temperature set by the operator. Temperature controller output automatically positions hot-air and tempering-air damper operators to produce the desired temperature. A furnace draft controller positions the induced-draft-fan damper to maintain draft as set by the operator.

Long-distance all-electronic instruments enable fully centralized control, a significant improvement over decentralized arrangements with instruments scattered throughout the plant. The automatic combustion-control system functions to provide efficient operation, with safety.

Automatic combustion control is justified by the benefits it provides: added safety, improved operation, reduction in manpower requirements, lower steam costs, etc. Combustion-control systems are available to meet the needs of both the small and the large power plant. Selection should be made on the basis of justifying installation and maintenance cost by lower overall steam costs. Each system should be as simple as possible to accomplish the purpose for which it was installed. Controls should be so located as to make them accessible for servicing and calibration; they should be kept clean and in working order. Neither the instrumentation nor the combustion control, per se, improves the performance; they must be maintained and records of operation analyzed if best results are to be secured.

QUESTIONS AND PROBLEMS

5.1. Name three appliances that warn of low water level in a boiler.

5.2. Explain how a water column is connected to a horizontal-return tubular boiler. What fittings should be used? How would you determine if a water gauge is properly set?

5.3. How high above the top row of tubes, in a return tubular boiler, should the first gauge cock be placed?

5.4. What is the smallest size of pipe that should be used in piping up a water column?

5.5. Is an automatic shutoff valve allowed on a water column?

5.6. What appliances can be connected to the top of a water column?

5.7. What type of valves should be used between boiler and water column?

5.8. How would you pipe up a pressure gauge to a water column?

5.9. Where is the fusible plug located on a boiler with which you are familiar?

5.10. How would you calibrate a pressure gauge from which the pointer had fallen off?

5.11. A steam gauge is located 25 ft above the steam line. The gauge reads 105 lb. What is the pressure in the line?

5.12. A gauge is installed so that it is 15 ft below the place where it is connected to the boiler water column. In checking this gauge, how would you set the pointer so that it would indicate the correct boiler pressure when installed in this position?

5.13. Explain the operation of a feedwater regulator with which you are familiar.

5.14. Explain the difference between a two- and a three-element feedwater regulator.

5.15. What is a safety valve? Explain how it operates. What makes it pop?

5.16. What are some of the essential requirements and rules governing safety valves and their installation; operation?

5.17. How is a pipe connected to the discharge of a safety valve?

5.18. Why is more than one safety valve sometimes installed on boilers?

5.19. Which safety valve should be set to pop first, the one on the boiler or the one on the superheater, and why?

5.20. How do you determine the size of a safety valve required for a boiler?

5.21. Can the safety valve be connected to some internal pipe in a boiler?

5.22. Which requires the larger safety valve, a high- or a low-pressure boiler, if both boilers are assumed to be the same capacity?

5.23. What is meant by blowback in reference to safety valves? How can this setting be changed?

5.24. The safety valve is 4½ in. in diameter, and the boiler pressure is 250 psi. What is the total pressure on the valve?

5.25. What kind and what size of valves and piping should be used on blowoff lines?

5.26. What is a plug-type valve? Why is it used as a blowoff valve? Why is it seldom used in other locations on the boiler?

5.27. When tandem blowoff valves are employed, which valve is to be opened first; last? Closed first; last?

5.28. What is a surface blowoff or blowdown? When is it employed?

5.29. What is a continuous-blowdown system, and where is it installed?

5.30. What is a nonreturn valve? Where is it placed? What is its purpose?

5.31. How are the steam valves arranged on boilers that are set in battery and carrying 135 lb pressure, or over?

5.32. What is a bleeder? Where is it installed?

5.33. When installing a steam header, what factors must be taken into consideration?

5.34. What are the advantages and disadvantages of steam versus compressed-air soot blowers?

5.35. What is an air-puff soot blower? What advantage does it have over the conventional blower?

5.36. Explain the difference between a globe and a gate valve? Where would you employ a globe valve in preference to a gate valve?

5.37. What is a check valve?   Where is it normally used?   How does it operate?

5.38. How is a globe valve installed in a feedwater line?

5.39. Give five attachments that you think are necessary for the safe operation of a boiler.

5.40. What do you consider to be the most important auxiliaries in steam lines from the boiler to the engine or turbine?

5.41. Explain how a pressure-reducing valve works.

5.42. What is meant by "on-off," "positioning," and "metering" control? What are the advantages of each?

CHAPTER 6

OPERATION
AND
MAINTENANCE
OF
STEAM BOILERS

The procedure to be followed in the operation and maintenance of a boiler plant depends to a large extent upon its size, type of combustion equipment, operating pressures, steam requirements, and other factors pertinent to the specific plant. There are, however, standard practices which the operator should follow to assure safe, continuous service and efficient operation. This chapter is intended to assist the operator to manipulate the equipment correctly, to recognize unsatisfactory conditions, and to take the necessary corrective measures before dangerous, costly emergencies develop.

6.1 PUTTING BOILERS IN OPERATION. The necessary steps to be taken before placing a boiler in service depend upon whether it is new, has been out of service a long time for repairs, or has been "down" for only a few days.

New boilers and those on which extensive replacements of pressure parts have been made must be given a hydrostatic test before being placed in service. Such a test consists of filling the boiler completely with water (being careful to vent all the air) and developing a pressure 1½ times the working pressure. The water temperature must be at least as warm as the temperature of the air in the boiler room and in no case less than 70°F. The test pressure required (1½ times the working pressure) must not be exceeded by more than 6 per cent. A

150-psi-pressure boiler is hydrostatically tested at $150 \times 1.5 = 225$ psi, and the test pressure must not be allowed to exceed $225 \times 1.06 = 238.5$ psi. During the test either the safety valves must be removed and the openings blanked, or the levers clamped so that the valves will not open.

New boilers and those which have accumulated a deposit of oil or grease must be cleaned by boiling out with an alkaline detergent solution. The cleaning must include the boiler, waterwalls, economizer, and superheater, but the cleaning solution must not be allowed to enter nondrainable-type superheaters. Since the caustic solution will injure the gauge glass, a temporary glass must be provided for use during cleaning or the permanent glass replaced after the cleaning operation has been completed. A suggested cleaning solution is 2 lb each of soda ash, trisodium phosphate, and caustic soda per 1,000 lb of water contained in the unit to be cleaned. First, dissolve the chemicals in the boiler drum; then close up the unit and start a light fire. Continue this cleaning operation for 2 or 3 days. Maintain the pressure in the boiler at approximately one-third of the working pressure but not over 300 psig. During this time, periodically blow down the boiler mud drum and waterwalls. After each blowdown period, refill the boiler to the operating level with clean hot water. At the end of this cleaning period allow the boiler to cool, open, and flush out with a hose. Inspect all interior surfaces to see if the cleaning was effective. In addition to receiving the caustic boil-out, boilers designed for 800 psig and over are acid-cleaned before being placed in continuous service.

When the boiler unit is turned over by the maintenance or construction department to the operating personnel, the latter should proceed as follows: See that manhole and handhole covers have been replaced and that the pressure section of the boiler is ready for operation. Inspect the interior of the gas passages to see that all scaffolding, ladders, tools, etc., have been removed. Check the operation of the fans and dampers; if the fans and combustion equipment are supplied with a safety interlock system, check it at this time. After first making sure that no one is inside, close all doors and access openings to the boiler setting.

After these precautionary measures have been taken, check over the various valves and arrange them for starting up as follows: close the blowoff, the water-column gauge-glass drain, the gauge cock, and the feedwater valves; open the drum vent and the cock to the steam gauge. When valves are used on the lines from the boiler drum to the water column, make sure they are open and locked or sealed. The drains should now be opened, except in cases in which superheaters are filled with water during the starting-up period.

Fill the boiler with water, using the auxiliary feed connections if they are installed. Do not fill the drum to the normal operating level since the water will expand when heat is applied and so cause the level to rise. Operate the gauge cock and blowdown valve on the water column and gauge glass to check for possible stoppage.

You are now ready to start the fire in the furnace. For detailed procedure in starting a fire using the various types of combustion equipment, see Chap. 4. When the fire has been started, allow the steam to blow from the steam drum for a few minutes and then close the vent valve. Close all superheater drains except the outlet. This will allow a small quantity of steam to circulate through the superheater tubes and thus prevent excessive temperatures.

Regulate the rate of combustion to allow 45 min for small and medium-sized 150-psi boilers to attain line pressure. For large high-pressure boilers, allow from 45 min to $2\frac{1}{2}$ hr, depending upon the size and superheater arrangement. When pulverized coal or oil is burned, it may not be possible to regulate the rate of combustion low enough to provide the necessary safe time for bringing the boiler up to pressure. When this condition is encountered, allow the burner to operate for a few minutes and then take it off to allow the heat to distribute through the unit, thus preventing excessive temperature differentials and the resulting unequal expansion.

In installations which have pendant nondraining-type superheaters, it is necessary that the warm-up rate be very carefully controlled, since all water must be evaporated from the superheater tubes before the boiler is placed on the line. Water remaining in the superheater would restrict the flow of steam and cause the tubes to overheat. With drainable-type superheaters the drains should be opened for a short blow before the boiler is "put on the line."

It is essential that all water be drained from the boiler steam-line header and that it be filled with steam before the valve is opened and the flow established. This may be accomplished by opening the drain and back-feeding steam from the main through the bypass around the main-header valve. When the pressure in the boiler header approaches that in the main header, open the main-header valve and throttle, but do not close the drain. Before putting the boiler on the line, open the superheater drains for a short blow to make sure all water has been removed. Check the water level, and if it is not sufficiently below standard (2 to 6 in.) to compensate for expansion when the boiler starts to steam, open the blowdown and remove the excess water. When the boiler-drum pressure is from 10 to 25 lb below the header pressure, unscrew the stem of the nonreturn valve so that it can open when the boiler pressure exceeds the steam-line pressure. If there is no nonreturn valve, wait until the pressure in the boiler is approximately equal to the main-header pressure before opening the valve at the boiler outlet. Should waterhammer or vibration occur while the valve is being opened, close the valve at once, allow the pressure on the boiler to drop, and repeat the entire operation. When a boiler is not connected to a common header, it is advisable to open all drains and raise the steam pressure on the entire system at the same time.

Before putting a boiler in service for the first time, each of the safety valves must be checked for the correct setting by allowing the boiler outlet valve to remain closed and raising the boiler pressure until each valve opens. The pressure of the opening and closing of each valve must be recorded and the valve adjusted until specified results have been attained (see Sec. 5.5). Some plants make a practice of opening the safety valves by hand each time the boiler is put in service. Suitable cables attached to the valve handle and extending to the operating floor facilitate this operation. The safety valve should not be opened until sufficient steam pressure is available to prevent dirt from sticking under the seat.

With the boiler in service the drain may be closed and the fuel feed regulated to maintain a low rate of steam flow. Consult the draft gauges and adjust the fans and dampers to establish the required flow of air to, and gases from, the boiler unit. Check the water level and feed-water-supply pressure; then put the feedwater regulator in operation. A skill acquired only by practice is required to perform and time the operation of putting a boiler on the line without a fluctuation in super-heat-steam temperature.

6.2 NORMAL OPERATION. A boiler in service producing steam constitutes a continuous process. Fuel, air, and water are supplied while steam and the waste products of ash and flue gases are discharged. It is the operator's duty to keep these materials flowing in the correct proportion and as required to maintain the steam pressure.

Emphasis is placed upon the responsibility of a fireman, owing to the importance of adequate steam supply, the increased cost of fuel, and the complexity of modern combustion and boiler equipment. Details of manipulation of the various types of combustion equipment during standby, normal operation, and banking are explained in Chap. 4. The following paragraphs explain the correct coordination of the various tasks that must be performed for satisfactory boiler-plant operation.

A drop in steam pressure as indicated by the steam gauge shows the fireman that the fuel supply must be increased; an increase in pressure shows that too much fuel is being supplied. On boilers not equipped with automatic controls the fireman must continually watch the gauge and adjust the fuel as it dictates. If automatic control is used, the change in steam pressure adjusts the fuel feed, relieving the fireman of this repetitive task. However, there is more to firing than simply admitting the correct amount of fuel; consideration must be given to fuel-bed thickness and contour with solid fuel and to flame shape and travel when pulverized fuel, liquid, or gaseous fuels are used.

The draft differential, flue-gas analysis (carbon dioxide), appearance of the fire, emission of smoke, and furnace-slag accumulation are all means by which the fireman can determine the relative quantity of air being supplied to the furnace. Each time the rate of fuel supply is changed, the air supply to the furnace should be adjusted in proportion. The automatic control makes these repetitive adjustments to compensate for load changes and to maintain the correct fuel-air ratio under normal conditions. The fireman must make adjustments for changes in coal size, moisture content, heating value, etc. In any case the combustion control is a device to aid, rather than a substitute for, experienced operation. The operator should be capable of manipulating the boilers "on hand" at any time. It is good practice to use hand control when putting a boiler in service and to change to automatic only after normal operating conditions have been established. The fireman should learn to use all the instruments provided, as they are intended to assist him. The draft gauge is a simple instrument which is seldom used in the most effective manner. Note and memorize the draft-gauge reading (wind box, furnace, boiler outlet, etc.) when the combustion is satisfactory. Use these values as a means of quickly establishing normal conditions, detecting trouble, and changing rating. The draft gauge indicates the airflow to, and the product of combustion from, the furnace. (See Chap. 3 for an explanation of draft as applied to combustion.)

The removal of ashes from the furnace is the responsibility of the fireman. Ashes are removed either as a molten liquid or in the solid state. The design of the equipment determines the method to be employed. It is an operating problem to produce ashes in the solid or liquid state in accordance with the design of the equipment. Early methods involved hand operation of dragging the ashes from the furnace and the use of hand dumping grates. A later development, still in use on small and medium-sized units, consists of power-operated dumping grates. These employ a steam or air cylinder which operates the gates, dumping the ashes into the pit. This system creates a periodic disturbance in the operation when ash dumping is in progress and until stable conditions have been restored. Continuous-ash-discharge systems are employed to reduce the labor of firing and to improve economy.

When operating a furnace equipped with dumping grates, remove the ash as often as practice shows it to be necessary to prevent the formation of clinker, loss of rating, and too heavy a fuel bed. Definite time schedules cannot be established owing to changes in rating and the percentage of ash in the coal. It requires judgment on the part of the fireman to determine when the material on the grate has been burned down sufficiently to permit it to be dumped. A long burning-down period will result in loss of ratings on the boiler (failure to maintain steam pressure) and lowering of efficiency owing to large amounts of excess air. Insufficient burning down allows large amounts of coke to be discharged with the ash. Dumping should be performed as quickly as possible to prevent prolonged interruption of stable operation.

The continuous-dumping devices are adjusted to remove ashes at the same rate as that at which they are being produced. Skill is required to ensure as complete burning as possible without introducing large amounts of excess air. Continuous ash discharge is applied to the various types of stokers as well as to slag-tapped pulverized-fuel furnaces. The molten slag is allowed to run continuously from the furnace into a tank of water which chills the slag, causing it to shatter into small pieces. The tank of water forms a seal which prevents air from entering the furnace.

The water supply is a most important consideration in boiler-plant operation. Familiarize yourself with every detail of this system. Where is the supply obtained, and what are the possibilities of failure? What type of pumps are employed; how are the pumps driven? What type of feedwater regulator is used, and where are the bypass valves located? Is there an auxiliary feedwater-piping system?

The operator's job is to supply hot water to the boilers as required to replace that which is being evaporated. Usually the boilers are equipped with regulators which control the flow to maintain the right level, thus relieving the operator of the tedious task of making repetitive adjustments. These automatic devices do not, however, relieve the fireman or water tender of the responsibility of maintaining the water level. Frequent checks must be made—of the level of water in the heaters, of softeners or other sources of supply, of the temperature of the water, of the performance of the boiler feed pump, of feedwater-pump pressure, and of the level in the boiler-gauge glass. Improperly set water columns or foaming, leaking, or stopped-up connections will cause water columns to show false levels. The water column must be blown down at least once each 8-hr shift and the high- and low-water alarm checked. The feedwater regulator should be blown down and checked according to the manufacturer's instructions. The gauge glass and try cocks are the only water-level–indication devices recognized by law. The recorders and other devices are convenient but must be checked frequently with the water level in the gauge glass. It is desirable to use hot feedwater, and the temperature should be maintained as high as feasible with the equipment available. The rate of flow of water to the boiler should be as nearly uniform as the steam requirements will permit.

The task of maintaining the water level is difficult because during operation the boiler is filled wth a mixture of water and steam bubbles. When the water level in the drum drops, there is an obvious tendency to add water. This tendency occurs both with hand and automatic control. The addition of water, at a temperature lower than the water in the boiler, causes the steam in the bubbles to condense. This action decreases the volume of the steam and water mixture in the boiler and results in a further drop in level and a tendency to add more water. Then when the normal ratio of water to steam bubbles is restored the level in the drum will be too high. The result is a cyclic condition in which the water level in the drum is alternately high and low. This condition is avoided by regulating the water flow to the boiler in proportion to the steam discharge. The three-element feedwater controller accomplishes this by sensing the drum level, feedwater flow, and steam flow and regulating the feedwater supply accordingly. (See Sec. 5.4.)

During normal operation of a boiler plant the water must be conditioned by heating, by the addition of chemicals, and by blowdown to prevent operating difficulties. Inadequate water conditioning results in scale formation, corrosion, carry-over, foaming, priming, and caustic embrittlement.

Scale consists of a deposit of solids on the heating surfaces. The formation of scale is caused by a group of impurities initially dissolved in the boiler feedwater. As the water becomes concentrated and exposed to boiler pressures and temperatures, the impurities become insoluble and deposit on heating surfaces. Water which contains scale-forming impurities is termed "hard water" and will consume a quantity of soap before a lather is produced. These impurities are found in varying amounts in almost all water supplies. The amount of hardness in a given water can be determined by adding a standard soap solution to a measured sample of the water and noting the amount required to produce a lather. Hardness may be classified as temporary, or carbonate, and sulfate, or noncarbonate. Carbonates produce a soft, chalk-like scale, and this temporary hardness may be removed by heating. Sulfate hardness produces a hard, dense scale and requires chemical treatment.

Scale deposits on the tube and shell retard the flow of heat to the water and cause the metal to become overheated; the metal, thus weakened, yields to the pressure, producing a protrusion known as a "bag." Such a bag provides a pocket for the accumulation of sludge and scale, which eventually causes failure. A combination of imperfections in the boiler plate and overheating causes layers of metal to separate and form a "blister."

Corrosion is the result of low-alkaline boiler water, the presence of free oxygen, or both. The boiler metal is converted into red or black powder (iron oxide), which is readily washed away by the water. This action is accelerated at points of greatest stress, and as the corrosion proceeds, the metal thickness is reduced and the stress is further increased.

Carry-over is the continual discharge of impurities with steam. These impurities may be in the form of moisture which contains dissolved solids or of solids from which the moisture has been removed. The amount of carry-over of impurities is determined by condensing a sample of steam and noting the conductivity of the condensate.

Foaming is the existence of a layer of foam on the surface of the water in the boiler drum. This condition is a result of impurities in the water that cause a film to form on the surface. Oil and other impurities which may enter the condensate in an industrial plant produce conditions in the boiler which cause foaming.

Priming is a condition in which slugs of water are suddenly discharged from the boiler with the steam. The condition is caused by boiler design, impurities in the water, the rating at which the boiler is operated, or a combination of these factors. Priming is a serious condition both for the safety of the boiler and for the steam-utilizing equipment. Immediate steps must be taken to correct the condition. Foaming may result in priming, but it is possible to have priming without foam on the surface of the water in the boiler drum.

Caustic embrittlement is the weakening of boiler steel as the result of inner crystalline cracks. This condition is caused by long exposure to a combination of stress and highly alkaline water.

The treatment of boiler-plant water has advanced from the hit-or-miss period of applying "boiler compound" to a scientific basis. Water-control tests have been simplified to color comparisons and titrations which can be quickly and easily performed, and the necessary adjustments in water treatment can be made by the plant operators. The results of these control tests are expressed in parts per million, grains per gallon, or some constant which tells the operator when the treatment is incorrect and the amount of change which must be made. (See Sec. 13.1.)

Water conditioning may be accomplished either by treating the water in a separate unit before it enters a boiler or by adding chemicals directly to the boiler. The combination of these two methods produces the most satisfactory results. The nature of the impurities, quantity of water to be treated, plant operation, pressure, etc., are all factors to be considered in selecting water-treating equipment, procedures, and control. This is a job for a specialist in this field. (See Secs. 13.1, 13.2, 13.3, 13.4, and 13.5.)

External treatment can be accomplished by passing the water through a zeolite softener or demineralizer, by adding lime and soda ash, or by distilling it in an evaporator. The zeolite softener consists of a bed of special granular material through which the water passes. The zeolite bed does not actually remove the scale-forming impurities (hardness) but converts them into non-scale-forming impurities, thus softening the water. When the zeolite bed has been exhausted, the softener unit is taken out of service and regenerated by introducing a salt (sodium chloride) solution. After regeneration the bed is washed free of excess salts and returned to service.

A demineralizer is similar in external appearance to a zeolite softener, except that the water passes through two tanks in series, referred to respectively as "cation" and "anion" units. These systems can be designed to provide very pure water with a low silica content. This demineralized water is suitable for makeup in high-pressure plants, but normally water of this quality is not required for low- and medium-pressure applications. The cation unit is regenerated with acid and the anion unit with caustic.

The hot-process lime–soda-ash softener produces water satisfactory for use as boiler feed. Heating of the water before the addition of the chemical removes some of the temporary hardness, reduces the quantity of lime required, and speeds the chemical action. Softeners of this type provide a means of heating the water with exhaust steam, introducing chemicals in proportion to the flow, thoroughly mixing the water and chemicals, and removing the solid material formed in the softening process. After leaving the softener, the water passes through a filter and into the plant makeup system.

An evaporator consists of a steam coil in a tank. The water is fed into the tank, and the heat supplied by steam in the coil causes the water to boil and leave the evaporator in the form of vapor. The vapor enters a heat exchanger, where the relatively cool condensate passing through the tubes causes the vapor to condense, producing distilled water. The impurities in the original water remain in the evaporator and must be removed by blowing down. Evaporators are connected to the plant condensate-and-steam system to produce distilled makeup water with a minimum of heat loss.

Internal treatment is accomplished by adding chemicals to the boiler water either to precipitate the impurities so that they can be removed in the form of sludge or to convert them into salts which will stay in solution and do no harm. Phosphate salts are used extensively to react with the hardness in the water and thus prevent scale deposits. However, the phosphate reacts with the hardness in the water to form a sludge, which in some cases may result in objectionable deposits. Organic compounds are then employed to keep the sludge in circulation until it can be removed through the blowdown. The amount of sludge formed depends upon the amount of hardness introduced to the boiler in the feedwater. Therefore, it is desirable to have the hardness of the feedwater as low as possible. The phosphate treatment should be supplied directly to the boiler drum with a chemical pump. If introduced into the suction of the boiler feed pump, most phosphates will react with the impurities in the water and cause deposits in the pumps, piping, feedwater regulators, and valves. Therefore the phosphate should be dissolved in the condensate to prevent deposit in the chemical feed pump and lines. The amount of treatment is controlled by analysis of the boiler water for excess phosphate. This analysis consists of a color comparison of a treated sample with standards. An excess of 30 to 50 ppm is satisfactory to ensure removal of the hardness. Since this phosphate treatment removes the hardness, it is not necessary to run a soap hardness test on boiler water if the specified excess is maintained.

Care must be exercised in introducing phosphate into a boiler which contains scale. The old scale may be loosened from the tubes by the action of the phosphate and collected in a mass on the heating surface, causing bags, overheating, and ultimate failure. When phosphate is supplied to a boiler which already contains scale, the boiler should be inspected frequently and the amount of excess maintained at about 20 ppm. In time the phosphate will remove the old scale, but it is better to start the treatment with a clean boiler.

Another approach to internal boiler water conditioning is the use of chemicals which prevent the precipitation of scale-forming materials. These chemicals have chelating power in that calcium, magnesium, and other common metals are tied up in the water. This action prevents the formation of scale and sludge in boilers, heat exchangers, and piping and is effective over the normal range of alkalinity encountered in boiler-plant operation. A compound referred to as EDTA (ethylenediaminetetraacetic acid) is the basic material in a number of products used for this treatment. These products are sold under a variety of trade names.

The treatment is introduced into the boiler feed line by means of a standard chemical feed pump. The pump and piping should be of corrosion-resistant stainless steel for high pressure and of either stainless steel or plastic for low pressure. A continuous feed is desirable, and a slight excess should be maintained in the boiler water at all times.

However, if there is a deficiency for a short time, the deposits will be removed when the excess is restored.

The choice between the use of these chelating materials and phosphate depends upon the condition in the specific plant, the quantity of makeup water, and the amount and type of impurities which it contains. The relative cost of the two methods and anticipated results should be determined by a water-conditioning consultant.

Corrosion may be caused by either a combination of low alkalinity or the presence of oxygen in the boiler water. The alkalinity may be regulated by control of the lime-soda softener, by varying the type of phosphate used, by introduction of alkaline salts (sodium hydroxide), and by treating the makeup water with acid. The alkalinity is indicated by testing (titrating) the boiler water with a standard acid, using phenolphthalein and methyl orange as indicators. Some alkalinity in boiler water is essential, but if allowed to get too high, alkalinity will result in priming and foaming. High concentrations of alkaline water also cause caustic embrittlement. The correct value must be specified for individual plants with due consideration for the many factors involved. A limit of 250 to 500 ppm of total, or methyl orange alkalinity, expressed as calcium carbonate, is required for boilers up to 200 psi pressure.

Free oxygen is most effectively and economically removed from boiler feedwater by heating both the return condensate and makeup water in a vented open-type heater. However, it is good practice to use chemicals to remove the last trace of oxygen from feedwater. Sodium sulfite is added to the deaerated water in sufficient quantities to maintain an excess of 20 ppm in the boiler water. It is not advisable to use sodium sulfite to react with large quantities of oxygen, because of the cost and the fact that it is converted into sodium sulfate and therefore increases the concentration of the boiler water. Hydrazine is another chemical used to neutralize the corrosive effects of free oxygen. It combines with oxygen to form water and nitrogen and therefore does not increase the concentration of solids in the boiler water. Some of the ammonia formed from the hydrazine is carried from the boiler with the steam, where it neutralizes the carbon dioxide, reducing corrosion in the condensate return lines. Hydrazine, being alkaline and toxic, is best handled as a dilute solution.

Caustic embrittlement of boiler metal is the result of a series of conditions which include: boiler water having embrittlement characteristics, leakage and concentration of the solids at the point of leakage, having these solids deposited in contact with stressed metal. It follows that embrittlement can be avoided by preventing the presence of at least one of these factors.

Leakage and the resulting concentration of solids in joints have been eliminated by use of welds in place of riveted joints. All boilers should be inspected for leaks, but special consideration should be given to leaks in riveted joints and where tubes are expanded into drums and headers. Boiler water is rendered nonembrittling by lowering the alkalinity and adding sodium sulfate or sodium nitrate in controlled proportion to the alkalinity and boiler operating pressure.

We have seen how impurities enter the boiler with the feedwater and how, when the steam is discharged, these impurities are left in the boiler water. This action makes it necessary to remove the solids by allowing some of the concentrated water to flow from the boiler. There are two general procedures for removing this water from the boiler. One is to open the main blowoff valve periodically, and the other is to allow a small quantity of water to be discharged continuously. The continuous blowoff is generally favored because it provides a uniform concentration of water in the boiler, may be closely controlled, and permits reclamation of a portion of the heat. Heat is reclaimed from the blowdown water by discharging it first into a flash tank and then through a heat exchanger. The flash steam is used in the feedwater heater or as a general supply to the low-pressure steam system. The heat exchanger is used to heat the incoming feedwater. (See Sec. 13.5.) The continuous-blowdown connection is so located as to remove the water having the highest concentration of solids. The location of this connection depends upon the type of boiler; in many cases it is located in the steam drum.

The operator must exercise careful control over the blowdown, as too high a concentration may result in priming and foaming, in caustic embrittlement, and in scale and sludge deposits, while excessive blowdown wastes heat, chemicals, and water. Blowdown is effectively controlled by checking the chloride concentration of the feed- and boiler water. Chloride salts are soluble and are not affected by the heat of the boiler. When the boiler water contains 120, and the feedwater 12, ppm of chloride, the concentration ratio is 10. That is to say, the boiler water is 10 times as concentrated as the feedwater. This is equivalent to taking 10 lb of feedwater and evaporating it until only 1 lb remains. All the salts that were in the original 10 lb are now concentrated in the 1 lb. For low- and moderate-pressure boilers under average conditions, 10 concentrations is deemed satisfactory, but all factors in an individual case must be considered. Other methods of determining the concentration of boiler water and thus providing a means of controlling blowdowns are (1) a hydrometer which measures the specific gravity in the same manner as that employed in checking storage batteries and (2) a device for measuring the electric resistance of the sample of water as the resistance decreases when the amount of impurities increases. The hydrometer method is satisfactory when high concentrations are encountered. The electric-resistance measurement is advisable when the chlorides in the feedwater are low or subject to wide variation.

In plants where chemical control is not employed, blow down at least one full opening and closing of the blowoff valve every 24 hr. When continuous blowdown is used, the main blowoff should be opened once every 24 hr or oftener if necessary to remove sediment from the mud drum. It has been found that several short blows are more effective in removing sludge than one continuous blow. This is also a good precaution to take when blowing down a boiler that is operating at high rating. In general, boilers should be blown down when the steam demand is least. (See Chap. 5 for a discussion of blowoff valves.)

When the boiler is in service, the skilled operator will devote his time to careful observations and inspections and thus detect possible irregularities before serious situations develop. This ability to discover minor difficulties before real troubles are encountered is a measure of the competence of an operator. It is this skill which prevents boilers from being burned or from exploding when the feedwater supply fails.

Observe the water level in the boiler frequently to check on the operation of the feedwater regulator. When in doubt and in all cases at least once per shift, blow down the water column and gauge glass, open the try cocks, and check the high- and low-water alarms. Check the feedwater pressure frequently, since this may enable you to detect trouble before the water level gets out of hand. Failure of the feed pumps or water supply will be indicated by a drop in pressure before the water gets low in the boilers. The temperature of the feedwater should be checked and adjustments made to supply the necessary steam to the heater to maintain standard conditions.

With hand operation, almost constant observation of the steam pressure is necessary. When automatic controls are used, the steam pressure should be observed frequently as a check on the control equipment. Steam-pressure recorders are helpful in checking the operation of an automatic combustion control. Difficulty in maintaining the steam pressure may indicate trouble developing in the furnace, such as clogged coal-feed mechanism, uneven fuel bed, or clinker formation. Many a disagreeable job of removing clinkers from the furnace could be prevented by closer observation of the fuel bed, ash removal, etc.

The delivery of the required amount of fuel to the furnace does not ensure satisfactory operation. The fireman must ask himself these questions: Is the fuel bed of the correct thickness? Are there holes through which the air is bypassing the fuel? Are clinkers forming which blank off the airflow? Is the flame travel too short or too long? Is smoke, in an objectionable amount, being discharged from the chimney?

The competent operator is always on the alert to detect leaks in the pressure system. An unusual "hiss" may be caused by a leaking tube. A sudden demand for feedwater, without an accompanying increase in load, may be the first indication of a leak. When these symptoms occur, be on the alert, report to your superior, investigate further.

Operators should be constantly on the alert for leaks in the pressure systems and settings of boilers. If the main or water-column blowdown valves and lines remain hot, it is an indication of leakage. Low concentrations of salts in the boiler water without a change in operation or procedure indicate leaking blowdown valves. High exit-gas temperatures or a decrease in draft differential across the boiler may indicate leaking baffles. Report loose brick found in the ashpit when removing ashes or on the tubes when hand-lancing, since these may indicate the beginning of serious failure and their origin should be determined.

Instruments and controls are for the benefit of the plant operators and should be checked, adjusted, calibrated, and kept in good operating condition. Draft and pressure gauges, steam flowmeters, flue-gas temperature recorders, and water-level indicators are useful in making routine adjustments for changes in load as well as in detecting difficulties. Learn to use the instruments available. Keep the automatic control adjusted and operating.

Boiler efficiency represents the ratio between the heat units supplied and those absorbed. Boiler performance may be expressed as the number of pounds of steam produced per pound of fuel. The total pounds of steam generated divided by the pounds of coal burned in a given time equals the pounds of steam per pound of coal. This formula may be valuable in comparing daily performance in a given plant, but it does not take into consideration variations in feedwater temperature, the changes in steam pressure and superheater temperature, or the heating value of the fuel. To calculate boiler efficiency one must know the amount of steam generated and fuel burned, the steam pressure, the quality or superheat at the boiler outlet, and the feedwater temperature.[1]

$$\begin{array}{l}\text{Boiler} \\ \text{efficiency,} \\ \text{percentage}\end{array} = \dfrac{\dfrac{\text{lb steam}}{\text{lb fuel}}\left[\begin{array}{l}\text{heat content,} \\ \text{steam, Btu/lb}\end{array} - \left(\begin{array}{l}\text{feedwater tem-} \\ \text{perature} - 32\end{array}\right)\right]}{\text{heating value, fuel, Btu/lb}} \times 100$$

[1] See Sec. 2.10 for procedure in calculating the heat content of wet or superheated steam.

Example   A boiler generates 50,000 lb of steam and burns 4,880 lb of 13,700-Btu coal per hr. The steam is 150-psig, dry and saturated, and the feedwater is 220°F. What is the boiler efficiency?

Solution

$$\frac{50,000}{4,880} = 10.25 \text{ lb of steam/lb of coal}$$

Heat content of 150-psig, dry and saturated steam = 1,195.6 (see steam tables, Appendix D).

$$\text{Boiler efficiency} = \frac{10.25[1,195.6 - (220 - 32)]}{13,700} \times 100 = 75.39 \text{ per cent}$$

The efficiency of a boiler changes as the output varies. At low ratings large amounts of air must be used, and the unit is inefficient. As the output increases, improved combustion will decrease the losses and maximum efficiency will be obtained. With continual increase in output, the efficiency will decrease owing mainly to an increase in flue-gas temperature. Therefore, there is one output rate at which the efficiency is maximum; however, the use of heat-recovery equipment tends to decrease this variation in efficiency. In addition to design, the type of fuel burned, cleanliness, and general conditions affect the maximum efficiency of the boiler. Besides the operating efficiency, the overall maintenance, boiler outages, and other vital factors must be considered.

When the plant consists of a battery of boilers, a sufficient number should be operated to keep the output of individual units near their points of maximum efficiency. In these instances it is frequently advisable to regulate one boiler to meet the change in demand for steam while the others are operated at their most efficient rating. As the variation in steam demand exceeds the amount that can be satisfactorily obtained with the regulating boiler, others are either banked or brought into service to meet the change. In public utility plants these wide variations occur in daily cycles. Keeping the boilers operating near their most efficient load results in a fuel saving.

The addition of economizers and air preheaters and improvement in combustion equipment have resulted in boiler units which have a wide range of output at nearly maximum efficiency. This is referred to as having a "flat efficiency curve" and is essential to large plants when one boiler supplies all the steam for a turbogenerator. For peak demands boilers can be forced beyond their most efficient output. Consideration of boiler loading improves plant economy and decreases outages and maintenance.

The operators should reduce the air leakage into the boiler setting to a minimum. The air which enters through leaks must be handled by the chimney or induced-draft fan and frequently limits the maximum output of the boiler. The heat required to raise the air to the exit-gas temperature is lost. A check for air leaks should be made while the boiler is operating by placing an open flame near points where leaks are suspected. The flame will be drawn into the setting by the draft. The torch made with pipe fitting, wicking, and burning kerosine is useful in testing for leaks. Large openings are closed by calking with rope asbestos and then covering with a tar-base asbestos compound made for this purpose. The asbestos compound alone is satisfactory for small openings and for covering entire areas of brick settings. Fuel savings of 5 to 10 per cent can often be made by stopping air leaks in the setting. This is a job that requires continual checking, as expansion and contraction cause air leaks to develop.

The use of the correct amount of excess air for the best overall results is an important consideration. Too much air wastes heat to the stack, while too little causes overheating of the furnace, high combustible in the ash, smoke, and unburned gases. (See Chaps. 3 and 4 for a discussion of excess air.) Study the conditions in your plant and regulate the excess air by observing furnace temperature, smoke, clinker formation, and combustible in the ash to give the best overall results. After the correct amount of air has been determined by experience, use draft gauges, carbon dioxide or oxygen recorders, and steam-flow–airflow meters as guides in regulating the supply. In general, the excess air should be lowered until trouble begins to develop. This may be in the form of smoke, clinker, high furnace temperature or stoker maintenance, slagging of the boiler tubes, or high combustible in the ash. The combustible material dumped from the furnace with the ash represents heat in the original fuel which the furnace failed to develop. It represents a direct loss and must be considered by the operator. The designers of combustion equipment are continually striving to devise units that will reduce this loss. In the case of hand firing and with stokers the amount of combustible in the ash depends, to a large extent, upon the skill of the operator. Dump the ash or clean fires as often as required for best overall results. Maintain an even distribution of fuel on the grates so that the combustible matter will be burned nearly completely before dumping. Learn to operate so that the formation of clinkers will be at a minimum. Avoid excessively high rates of combustion. It is difficult to determine the amount of combustible in the ash by observation; the only sure way is to collect a representative sample and send it to a laboratory for analysis.

A deficiency of air or failure to secure mixing of air and combustible gases will result in burning the carbon to carbon monoxide rather than to carbon dioxide. When this occurs, 10,160 Btu of heat per lb of carbon passes out of the stack unburned. A flue-gas analysis provides the only sure means of detecting and evaluating the amount of carbon monoxide. High percentages of carbon monoxide gas are usually accompanied by unburned hydrocarbons, which produce smoke.

After the heat has been developed in the furnace, the largest possible amount must be absorbed by the water in the boiler. Scale formation on the water side of the heating surface and soot and ash on the gas side act as insulators and cause high flue-gas temperature. We have seen how attention to water treatment can eliminate scale from the water side of the heating surface. Failure on the part of the water-treating system to function, owing either to an inadequate system or to unskillful application, will result in the formation of scale. This scale must be removed either by mechanical methods or by acid cleaning. Soot and ash are removed by means of soot blowers. (See Chap. 5.)

To operate mechanical soot blowers, open the drain of the soot-blower header to remove the accumulation of water. Then open the steam-supply valve. Slowly turn each soot blower through its entire arc of travel by means of the chain or handwheel. It is advisable to start with the element nearest the furnace and progress toward the boiler outlet. Sometimes it is found desirable to increase the draft during soot blowing in order to prevent a pressure from developing in the furnace and to carry away the soot and ash removed from the tubes. Care should be exercised in draining the steam lines to the soot blowers, as water will cause warping and breakage of the elements.

On boilers equipped with fully automatic soot blowers the operators first turn on the steam supply or start the air compressor and then initiate the soot-blowing cycle by a switch which is usually mounted on the boiler control panel. The soot-blower elements are inserted, rotated, and retrieved by power drives. The operation proceeds from one element to another in a programmed sequence, and indicating lights on the panel inform the operator of the progress being made and warn if trouble should develop in the system. Many plants find it advantageous to operate the soot blowers once every 8-hr shift. If the boiler is equipped with a gas-outlet thermometer, the effect of blowing soot can be readily determined by the resulting drop in flue-gas temperature. Soot blowers require an appreciable amount of steam and should not, therefore, be operated oftener than necessary as indicated by the outlet-gas temperature.

6.3 OPERATING IN AN EMERGENCY. When emergencies arise, analyze the situation, decide upon the necessary action, and proceed quickly. It is essential that the operator know his plant thoroughly to handle emergencies. A knowledge of the plant will not only make it possible for him to act wisely but also give him the necessary confidence. Learn what constitutes safe practices and how they are applied in your plant. Do not shut down unless it is absolutely necessary, but do not hesitate to shut down rather than resort to unsafe practices.

Thought should be given to the procedure to be followed if difficulty is experienced in maintaining the water level in the boiler. Small oil- and gas-fired boilers equipped for automatic operation have low-water fuel-cutoff devices. In the event of low water the fuel valve automatically closes and shuts down the unit until the difficulty has been corrected and the water level restored. However, industrial- and utility-plant shutdowns must be avoided if at all possible. Alarms are provided, but in event of high or low water the operator must decide upon the action to be taken. When emergencies of this nature arise, quick and decisive action is important and a thorough knowledge of the plant and a preconsidered line of action are essential. Study the various means available for supplying water to the boiler, including pumps, injectors, auxiliary piping systems, regulators, bypasses, and sources of water.

When the water is above the top gauge cock, close the steam-outlet valve from the boiler and check the firing rate. If enough water has gone over to cause waterhammer, shut the stop valve at the engine and open the drain valve ahead of the throttle valve. This will permit all the condensate to be drained before starting up. Restore the correct water level by operating the main blowdown valve, after which the boiler may be returned to service.

When the water is found to be below the gauge glass, stop the fuel and air supplies. The procedure to be followed depends upon the type of combustion equipment employed: for hand-fired units, cover the fire with green coal and wet ashes; for stokers, stop the fuel feed, shut off the air, and open the furnace doors; with pulverized coal, oil, and gas, simply shut off the fuel and air supplies. Do not change the feed-water supply, open the safety valve, or make any adjustments that might result in changes in the stress to which the boiler is subjected. Take the boiler out of service for a thorough internal check by an authorized inspector to determine the extent of damage and repairs necessary as a result of the low water.

Priming and foaming are caused by boiler water high in solids, by high ratings, high water level, or a combination of these. Foaming is serious, as it makes it difficult to determine the water level. Priming can result in a mixture of steam and water in the steam lines, causing waterhammer which may result in rupture of the piping system. When this trouble develops, reduce the rating and blow the boiler down through the surface blowoff, if one is provided; otherwise use the main blowoff. If the priming becomes so serious that the water level can no longer be detected, close the steam valve long enough to determine the correct level. Blow down the gauge glass and water column to make sure the water level indicated is correct. If conditions do not return to normal, take the boiler out of service. Check for possible contamination of the feedwater by material leaking into the condensate-return system from industrial processes. Oil from reciprocating engines and pumps is a frequent cause of foaming. If the condition persists, it is best to have a consultant study the problem.

In the case of tube failure, stop the fuel feed and proceed to retard combustion. Maintain the water level unless this robs the feedwater supply so that you cannot maintain the water level in the other boilers. This is a matter that requires instant decision on the part of the operator, and the correct action depends upon the size of the leak and available reserve supply of water. The pressure should be reduced as quickly as possible to prevent the blow of water and steam from cutting other tubes. Except for the fact that tube failures cause boiler outages, they are not considered serious. An investigation should be made to determine the cause of the tube failure and necessary steps taken to prevent recurrence.

Induced-draft fans are used to create the necessary draft for boiler operation, and when they fail suddenly, a pressure develops in the furnace and setting that causes smoke and sometimes fire to be discharged into the boiler room. This creates a dangerous condition, especially with pulverized fuel, gas, and oil. The immediate remedy is to stop the supply of fuel and air to the furnace until the induced-draft fan is again operating. Many units are equipped with a safety interlock system which shuts off the forced-draft fan and fuel feed in case of induced-fan failure. This interlock system also makes it necessary to start the induced-draft fan before starting the fuel feed and forced-draft fan. When starting up a boiler equipped with an interlock system, always trip the induced-draft fan out and make sure that the forced-draft fan and fuel feed stop. If the induced-draft fan trips out during operation and the automatic fails to operate, trip the forced-draft fan by hand.

Failure of the fuel supply, low rating, too much primary air, and disturbance from blowing soot may result in loss of ignition when fuel is burned in suspension. If the equipment is supplied with a flame detector (electric eye), the fuel feed will be automatically shut off. When this safety device is not provided, the fuel supply must be shut off by the operator. After the flame is out, regulate the airflow to about 10 per cent of the maximum and allow 5 min for the pulverized coal, oil, or gas to be removed from the furnace. Then insert the torch and proceed to relight as in normal starting. Be careful to obtain a rich mixture of fuel for quick ignition.

Power failures or other emergencies make it impossible to remove all the coal from the pulverizing mill before shutting down. When this occurs and a mill is to be returned to service within 3 hr, the partly pulverized coal may be allowed to remain. However, if the outage is to be longer than 3 hr, the coal must be cleaned out to avoid the possibility of fire. If a fire should develop while the pulverizer is out of service, close all the outlets and, one at a time, open the cleanout or access doors and drench the interior with water, chemical fire extinguisher, or steam. Do not stand in front of the doors or inhale these gases, as they may be poisonous. Guard against possible explosion of the gases formed in the mill. After the fire is out, the material in the mill has been reduced to a temperature below the ignition point, and the pulverizer room has been ventilated, remove all coal, coke, and ash from the mill. Never use an air hose or a vacuum system in cleaning the pulverizer mill.

Fires in pulverizers during operation are very infrequent, but they sometimes occur as a result of fire which originates in the raw-coal bunker, of failure to clean up after welding, or of using inlet air to the pulverizer that is of too high a temperature. The general procedure in this emergency is to reduce the supply of air so that the mixture in the mill will be too rich to support combustion. Fire in a pulverizer may be detected by the rapid rise in outlet temperature of the coal-and-air mixture. When this occurs, increase the fuel feed to maximum and supply the pulverizer with cold air. Do not increase the air supply or reduce the fuel feed. If the temperature does not begin to return to normal within 15 min, shut down the pulverizer and close off the air supply; then follow the procedure explained above for a fire in a pulverizer out of service.

When as a safety valve will not close but continues to leak at a pressure lower than that for which it is set, try to free it by operating the lift levers. If this fails to stop the leak, the valve must be repaired or replaced. Never attempt to stop a leaking safety valve by blocking or tightening the spring.

When a gauge glass breaks, shut off the flow of steam and water with the chain-operated valves. While the gauge glass is out of service, check the water level by using the try cock. Before inserting the new gauge glass, blow out the connections to make sure that no broken pieces remain in the fitting, and check the glass and the gasket for the correct length and size. Pull up the packing slowly and not too tightly to avoid breaking the glass. When the new glass is in position, open the top or steam valve first and allow the drain valve to remain open. This permits steam to circulate through the glass, gradually heating the entire surface to a uniform temperature. Then partly open the lower valve. After the water has reached the normal level, open both valves all the way, and the glass is in service. When the lower valve is first opened, the glass is subjected to a variation in temperature which may result in breakage. Always keep a supply of gauge glasses and gaskets available in the boiler room. Replace gauge glasses that become discolored and difficult to read.

6.4 IDLE BOILERS. Since there are no practical means available for storing steam for an extended period of time, it must be generated as it is required. When there are wide variations in steam demand, boilers must be put on and taken off the line to meet the load conditions. (See Chap. 4 for detailed instructions on banking fires with the various types of combustion equipment.) Boilers must also be taken out of service for inspection and repairs.

Before taking a boiler out of service for an extended period of time, allow the coal in the bunkers to be used up. In some cases coal may overheat and start to burn in 2 weeks when allowed to remain in the bunkers. The characteristics and size of the coal and the temperature to which the bunkers are subjected affect the time necessary to cause overheating and burning.

To take a boiler out of service, reduce the fuel feed and slowly decrease the output. With solid fuel this consists in allowing the fuel on the grates to burn out. When oil or gas is used, the fuel supply must be shut off after the rating has been reduced. Pulverized coal requires careful handling. Shut off the coal feed to the pulverizer to allow it to run empty. Observe the flame closely and shut down the pulverizer as soon as the flame goes out. Continued operation of the pulverizer might result in more coal being supplied after the fire was out and in an explosive mixture being formed in the furnace.

When the combustion rate is sufficiently reduced, the steam-drum pressure will drop below the header pressure, the boiler will cease to deliver steam, and the nonreturn valve will close. Screw the stem down on the nonreturn valve or, if there is no nonreturn valve, close the header stop valve nearest to the boiler. Now close the valve to the steam line and open the drain connections to this line between the two valves. Finally close the feedwater-supply valve.

Allow the boiler to cool slowly to prevent injury from rapid contraction. Forced cooling by blowing down and refilling with cold water or using the induced-fan or chimney draft to circulate cold air through the unit should be resorted to only after the pressure has been reduced by normal cooling. When there is no pressure in the boiler, open the drum vent to prevent the formation of a vacuum. Allow the boiler to cool as much as possible before draining. If the boiler is drained when too hot, sludge may be baked on the surfaces or the unequal contraction may cause tubes or seams to leak. When boilers are set in batteries of two or more, check carefully to see if you are opening and closing the proper blowoff valves.

In the case of a water-tube boiler, proceed to wash the drums and tubes. Use the highest water pressure available in order to remove the sludge and soft scale. On a bent-tube boiler, the tubes may be washed by working from inside the drum. To wash straight-tube boilers effectively, the handhole covers must be removed. This is seldom advisable because of the labor required to remove and replace the handhole covers. If, however, these covers are taken off, the deposits should be removed by passing a mechanical cleaner through the tubes. Wash the tubes of the return tubular boiler from both the top and the bottom openings. Allow the wash water to drain through the blowoff lines to the sewer. Exercise care to prevent the water from getting on the brickwork. If this is unavoidable, dry out the brickwork slowly before putting the boiler back into service.

There are two ways to proceed in laying up a boiler: one is to keep the interior dry, and the other is to fill the interior with water. The choice of methods depends upon how long a boiler is to remain out of service. If for a long period of time, the dry method is recommended. If for a short or rather indefinite period, it is better to fill the drums with water.

To take a boiler out of service for an extended period of time, remove the ashes from the interior, especially where they are in contact with metal parts. Ashes contain sulfur, tend to collect moisture, and form acid, which is corrosive. Make sure that all connections, including the steam outlet, feedwater, and blowoff valves, are closed and holding tightly. Repair or blank off leaking valves. After washing, allow the interior to dry. Then place unslaked (quick) lime in the drum in suitable open boxes or containers. About 25 lb of lime per 100 boiler hp has been found satisfactory. Reusable-type desiccants such as silica gel and activated alumina are also suitable. The lime or desiccant will absorb moisture from the air confined in the boiler. Notify the insurance company that the boiler is not in service, and ask them to adjust the insurance coverage accordingly. The insurance company must again be notified when the boiler is returned to service. The interior should be inspected every few months, and the lime should be replaced when it has been reduced to a fine powder (become slaked) by the absorption of moisture. If a desiccant is used it may be dried by application of heat and reused.

When a boiler is out of service for a short or indefinite period, add caustic soda to produce a concentration of 450 ppm and sodium sulfite to produce a concentration of 250 ppm. Fill the boiler to the top of the drum with deaerated feedwater. Do not permit leakage to dilute the water in the boiler, as this will reduce the alkalinity, keep the boiler warm, and, in general, increase the possibility of corrosion. Analyze the boiler water frequently, and if the caustic alkalinity is below 250 ppm and the sodium sulfite below 100 ppm, drop the level in the boiler and add chemicals to increase the concentration to the original values. Heat the boiler to circulate the water and to distribute the chemicals; then again completely fill the drum.

An alternative method of storing a boiler in readiness for almost immediate use is to condition the water as explained above, allowing the water to remain at normal level, and provide a "blanket" of nitrogen gas above the water. The nitrogen is supplied from pressure tanks through a regulating valve which maintains pressure in the boiler slightly above the atmospheric level. This prevents air from leaking into the boiler and contaminating the water with oxygen. This method is suitable for use with a nondrainable-type superheater, which should not be flooded. Boilers prepared for storage by either of these two methods can be readily returned to service by restoring the water level, starting the fire, and venting in the usual manner.

6.5 MAINTENANCE—CAUSES AND REDUCTION. It has been stated that when a boiler is clean and tight, it is properly maintained. At first this may appear simple, but "clean and tight" applies to the entire field of boiler maintenance. "Clean" applies to both the interior and the exterior of the tubes, shell, and drums as well as to the walls, baffles, and combustion chamber. "Tight" refers to the entire pressure section, setting, baffles, etc. When the heating surface of a boiler is free from scale on the water side and from soot and ash deposits on the gas side, it will readily absorb heat. When the boiler is free from steam and water leaks, from air leaks, from air leakage into the setting, and from the short circulating of gases through leaky baffles, it is in excellent condition. However, in the overall operation, the combustion equipment, accessories, and auxiliaries must also be considered.

Boiler units are constructed of several different materials to withstand the conditions encountered in service. The refractory lining in the furnace has but little tensile strength, but it can withstand high temperatures and resist the penetrating action of the ash. On the other hand, the steel in the pressure parts of the boiler must not be allowed to reach furnace temperatures but has great tensile strength. The designer must select materials for the various parts of the boiler which are suited to the specific requirements. If selection of materials is correct and the unit is maintained and operated in accordance with recognized good practice, the service will be satisfactory. When faulty operation, excessive temperatures, or other abnormal conditions cause the safe limits of the material to be exceeded, failures will occur rapidly and the equivalent of years of normal deterioration may take place in a short time.

It is impossible to predict the effect of low water upon maintenance. If only the upper ends of the tubes are exposed, leakage of the expanded joints may be the only ill effect. This can be corrected by reexpanding, provided the inspector decides that the metal in the tube and tube sheet has not been damaged. On the other hand, low water might necessitate a complete retubing job or even cause a boiler explosion, killing the personnel and wrecking the plant. Attention to water level is a most effective means of keeping the boiler in operation. Rapid changes in temperature cause unequal expansion, which may result in steam and water leaks in the pressure section of a boiler and air and gas leaks in the setting and baffle. It has been found that the frequency and rapidity of taking boilers out of service and putting them into service increase maintenance more than actual service hours and quantity of steam generated. In order to reduce outages and maintenance always allow time for the temperature to change slowly and uniformly.

Unequal distribution of combustion in the furnace, sometimes referred to as "flame impingement," can result in several difficulties. Although the average combustion rate may not exceed the manufacturer's specification, localized portions of the furnace can be overloaded. This will cause high temperatures, slagging, and rapid failure of the furnace lining. Severe cases of flame impingement, when accompanied by poor water circulation in the boiler or scale deposits, will result in burned-out water tubes. In high-rating water-cooled furnaces, the exterior surfaces of the tubes have been found to corrode until they become thin enough to burst. This type of failure has occurred when the furnaces were operated at high temperature and with low or zero per cent excess air. Flame impingement is prevented by correct operation and maintenance of the combustion equipment. (See Chap. 4.)

When scale deposits on the tubes and shell of boilers, the water cannot remove the heat and the metal may reach a temperature high enough to reduce its tensile strength. This may cause a bag, tube failure, or boiler explosion, depending upon conditions. Overheating makes it necessary to retube boilers and to patch water legs, shells, and drum. The best method of avoiding these difficulties is to condition the boiler water to prevent the formation of scale. (See Secs. 13.1 to 13.4.)

When a lack of equipment or careless operation has permitted scale to form, it must be removed or loss of efficiency and boiler outages

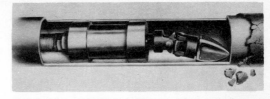

(*a*)

FIG. 6.1 *Liberty tube cleaners.*
(*a*) *Vibrating cleaner for use
in cleaning fire tubes.*
(*b*) *Cutter-head-type cleaner
for water tubes.*

(*b*)

will result. Boilers should be scheduled out of service for cleaning before they are forced out by tube failure. There are two methods of removing scale from the heating surface of a boiler: mechanical, consisting of passing a power-driven cutter or a knocker through the tubes; and chemical, which employs materials that will partly or totally dissolve the scale, thus removing it from the surfaces.

Mechanical cleaners consist of small motors driven by steam, air, or water. The motor is small enough to pass through the tubes that it is to clean. A hose attached to one end of the motor serves the dual purpose of supplying the steam, air, or water and of providing a means by which the operator "feeds" the unit through the tube and withdraws it when the cleaning is complete. Figure 6.1*a* shows a tube cleaner fitted with a "knocker head" for use in fire-tube boilers. The rotation of the cleaner motor causes the head to vibrate in the tube with sufficient force to chip hard scale from its exterior. Such a knocker head is sometimes effective in removing very hard scale from water-tube boilers. The "cutter head" shown in Fig. 6-1*b* is more frequently used in cleaning the scale from water-tube boilers. Heads of this type have a number of cutters made of very hard tool steel. The head is rotated at high speed, and the resulting force presses the cutter against the inner tube surface, crushing and cutting the scale.

When mechanical tube cleaners are being used to remove scale from tubes, several precautions must be taken to secure best results. The tube-cleaner motor must be of the correct size to fit the tube. Special arrangements of motors and universal joints are available for cleaning curved tubes. Although steam and water are sometimes used, compressed air is most satisfactory for operating these motors. A lubricator is attached to the inlet hose to supply oil for lubricating the motor. A stream of water should be introduced into the tube during the cleaning operation to cool the motor and cutter head and to wash away the scale as it is removed from the tubes. The operator feeds the cleaner into the tubes as the scale is removed. The rate of feed depends upon the speed of scale removal and must be determined by trial. Experienced operators can tell by the sound of the motor when the cutter has removed the scale and is in contact with the tubes. Inspect each tube by shining a light through it and observing the interior. Never allow the cleaner to operate in one place for even a short time, or the tube will be injured. In bent-tube boilers the operator remains in the upper drum while cleaning the tubes. In a straight-tube boiler the handhole covers must be removed to gain access to the tubes. The mud drum, lower header, and blowoff line must be thoroughly cleaned to remove the scale before the boiler is returned to service.

Chemical cleaning procedures have been successfully applied to the interior of boilers and heat exchangers. This method of cleaning is particularly adapted to units with small or bent tubes. Mechanical cleaning is tedious and time-consuming, but even large units can be chemically cleaned in a matter of hours. This decrease in outage time is an important factor.

Chemical cleaning is done satisfactorily and safely by specialists who furnish a complete service. Samples of scales are obtained from the tubes by use of a mechanical cleaner or other suitable means. These samples are analyzed, usually by the Hull x-ray method. After the type of scale has been identified, the correct cleaning solution is specified. These solutions consist of acids with added materials known as "accelerators" to ensure attack upon scale and other materials known as "inhibitors" to lessen the attack upon the boiler metal. That is to say, the solution is made selective so that it will dissolve the particular scale without corroding the metal surface. The selection of the cleaning solution involves a knowledge not only of the scale but of the material from which the boiler or heat exchanger is constructed. Classifications for chemical cleaning include the type and strength of solvent solution to be used, the amount and type of accelerators and inhibitors to be used, the temperature at which the solvent is to be applied, the time the solution is to remain in contact with the material, and the method of application. Safety precautions involved are prevention of the solvent and neutralizing solution from coming into contact with the personnel and removal of possible danger due to the formation of poisonous or explosive gases.

Special equipment in the form of tank trucks, pumps, and heat exchangers is required for effective application. The unit to be cleaned is filled with the solvent in the least possible time; then the solvent is circulated through the unit and heat exchanger to maintain the necessary temperature. The solution is analyzed during the operation to check upon the progress in cleaning. When the analysis shows that the unit is clean, the solvent is removed and a neutralizing solution introduced. After the unit has been drained and flushed with water, it is ready for service.

Good results have been reported from acid cleaning. Not only is the usual type of scale readily removed but even difficult forms of silica and plated copper have been successfully handled. Deposit of silica compounds in high-pressure boilers has been very difficult to remove with mechanical cleaners. Copper dissolved from the tubes of feed-water heaters has been deposited in boilers, causing scale and corrosion. Unlike mechanical cleaning, this acid cleaning removes the scale from the headers, water legs, and drums. Chemical cleaning has gone a long way to reduce the time and drudgery involved in scale removal.

Boilers are designed and built for a maximum output; when this is exceeded, maintenance is increased. High rates of steam generation may disrupt the circulation in the boiler. When this occurs, the heat will not be carried away and the metal may become overheated. The high rate of steam flow will result in moisture being carried with the steam that leaves the boiler drum. High gas temperatures may burn or distort baffles and breachings. Slag may form in the furnace or on the tubes, limiting combustion and obstructing the flow of gases. Though a high rating may be obtained for a short time, it may very possibly be necessary to take the boiler out of service because of the formation of slag or the failure of vital parts of the combustion equipment. Guard against excessive ratings.

The correct amount and distribution of combustion air are important considerations in reducing maintenance (see Chaps. 3 and 4). Too much combustion air results in serious loss of heat in the stack gases, while insufficient air causes excessive furnace temperatures. This may cause slagging of the first-pass tubes and early failure of the furnace walls, arches, and grates.

Although soot blowers are valuable in removing soot, ashes, and slag from the gas side of the tubes, they can also be the cause of boiler outages and maintenance. When soot blowers are incorrectly adjusted, steam from the nozzles will cut the boiler tubes and cause failure. To prevent possible damage the soot-blower elements should be so adjusted that the nozzles do not blow directly against the tubes and baffles.

Slag on the boiler tubes restricts the flow of gases and reduces the amount of heat that can be absorbed by radiation. This tends to reduce the maximum output of the boiler, increase the superheater-outlet temperature, and cause high furnace temperatures. The formation of slag may be caused by incorrect operation of the combustion equipment or by fuel unsuited to the furnace. The conventional soot blowers are not effective in removing slag from the first-pass tubes, and specially designed elements are installed in large boilers to remove this slag. A pipe attached to a compressed-air hose and operated through the clean-out doors is usually effective in removing the slag from the first-pass tubes. This hand-slagging operation should be carried out frequently to prevent the slag from bridging over between the tubes. Water is sometimes sprayed on the tubes to assist in removing the slag, but this may injure the baffles and brickwork and is not recommended. When the slag is deposited through the entire first pass and bridged between the tubes, it is advisable to take the boiler out of service for a thorough cleaning.

Boilers depend upon auxiliary equipment to supply the necessary water, fuel, and air. These auxiliaries include pumps, stokers, pulverizers, fans, etc., which have bearings that must be replaced occasionally under normal conditions and frequently if they are not lubricated. Induced-draft-fan rotors and casings wear owing to the abrasive action of fly ash. Boiler feed pumps wear and must be overhauled at regular intervals to assure continuous operation. Piping, valves, controls, and instruments require adjustments and replacement of parts.

Error, on the part of the designer and manufacturer of boilers, in the selection and application of materials will result in premature failure of parts and in high maintenance. When there are repeated failures that cannot be attributed to poor operation, the design or material, and in some cases both, should be changed. If an arch of a furnace wall must be replaced frequently, it would be more economical to use expensive material or to install waterwalls.

Boiler-plant maintenance work should be performed on an established schedule to prevent equipment from being forced out of service owing to failures. Scheduled maintenance work makes it possible to have other boilers available to carry the load or to perform the work when the boilers are not required. Furthermore, the necessary spare parts, tools, and maintenance personnel can be made available to accomplish the work efficiently. In the case of forced outages, on the other hand, production is lost, the necessary parts are not available, and there is probably insufficient maintenance personnel on the job. Large boiler units are now employed in both utility and industrial plants, and since their failure causes loss of a large percentage, and in some cases all, of the plant capacity, it is essential that forced outages be prevented.

The question arises as to how forced outages can be prevented or reduced to a minimum. Details that appear unimportant often cause or contribute to major difficulties. Stoppage of the chemical feed line to the boiler drum is not important in itself, but if the boiler is operated for a period without internal treatment, the resulting scale deposits may cause tube failures. A defective low-water alarm can result in a failure to maintain water level and a boiler explosion. Failure of cooling water or oil on the induced-draft-fan bearings may result in bearing failure and an outage of the complete boiler unit. Recognized standard rules for the operation and maintenance of auxiliary and accessory equipment are explained under the respective headings.

There is nothing that adds more to the impression that a plant is well maintained than cleanliness. Good housekeeping is a part of preventive maintenance. Maintaining a clean plant requires teamwork on the part of management, maintenance, operation, and janitors. The maintenance men must realize that the repair job is not complete until the old and surplus material has been removed. It is desirable to assign specific equipment to individual operators and make them responsible for keeping it clean. In this way the individual can take pride in doing the work and expect to be reprimanded if neglectful. These assignments must be given careful consideration to see that the work is equally distributed among those responsible. The work must be coordinated to prevent one cleaning job from depositing dirt upon equipment already cleaned. In some places the dust is blown out of motors with an air hose; obviously it would be a waste of time to wipe off machinery in the vicinity prior to cleaning motors by this method.

Management should see that all help possible is given to keeping the plant clean. It is difficult for operators to keep up interest in cleaning when improperly designed or maintained equipment continually allows ash, coal, or oil to be discharged. Often the design does not take into account accessibility for cleaning. Adequate equipment, tools, and supplies should be made available for this work. Vacuum cleaners are useful and in a pulverized-fuel plant are classed as a safety necessity. Cabinets for tools, supplies, and lubricants improve the appearance of a plant and facilitate the work.

Establish a lubrication schedule and preventive-maintenance program. Although some operators depend upon memory to determine when an inspection should be made, bearings lubricated, etc., and although in some cases this is satisfactory, in large, rapidly expanding plants, memory is generally inadequate. Sooner or later something is neglected, and an outage results. Base your maintenance and lubrication follow-up system upon manufacturers' recommendations and plant-operating experience. Do not depend upon memory.

6.6 BOILER INTERNAL INSPECTION. Boilers must have at least one internal inspection per year by a qualified representative of the company carrying the insurance. During the year the company inspector should examine the boilers one or more times, depending upon conditions. At the time of inspection, someone representing the owner should be present to confer with the official inspector. During these inspections specific notes should be made of scale deposits, corrosions, erosions, cracks, leaks, and other irregularities. Care of a boiler cannot begin and end with a yearly inspection. The operating personnel must continually inspect the boiler and be on the alert for possible conditions that might lead to trouble. External examinations are made primarily to check upon care and maintenance. An internal inspection should include examination of previous repairs, of the suitability of the unit for the operating pressure, and of the strength of the joints, tube ligaments, and other vital parts; a check for possible deterioration; a listing of necessary repairs; and a decision as to the advisability of continuing the boiler in service at the rated pressure.

When questions arise regarding the possible safe and useful life of boilers, nondestructive tests can be used to advantage. The x-ray examinations (Sec. 2.5) are useful in locating defects, but the equipment and procedures are not readily adaptable to "in-plant" use. The magnaflux method is useful in detecting the presence of cracks and flaws in iron and steel. This test consists of first covering the area to be inspected with fine magnetic powder and then subjecting it to magnetism. A crack or flaw is easily observed by the pattern of the magnetic powder. The penetration method is applicable to the locating of cracks in either metallic or nonmetallic material. This test consists of polishing the area and applying a fluorescent liquid. The excess liquid is wiped off. Cracks are then easily observed by viewing the area while illuminated with an ultraviolet light. Where there is indication of a wasting away of metal it is desirable to determine the remaining thickness. The thickness can be measured by an ultrasonic-sound-wave generator. Access to only one side of the surface is required. These instruments can be calibrated to read the thickness of the metal. Another useful instrument is the scleroscope, which is used to determine the hardness of metal. This metal hardness test determines to what extent the boiler metal has been hardened by the application of repeated stress.

When a boiler is taken out of service for inspection, it should be allowed to cool down slowly. The soot blowers should be operated to clean the tubes and baffles before the temperature has dropped low enough to allow the formation of moisture on the tubes. When the boiler is cool enough, open the blowoff and drain. Remove the manhole covers and some of the handhole caps. Wash the inside of the drum with high-pressure water. Before entering a boiler, close the blowoff, the main-header valve, and feedwater valves, and either lock them in the closed position or tag them to ensure against possible opening. Check extension cords to see that they are in good condition. The inspector should have available a list of the dimensions and other important data, as well as records of the history and previous inspections and repairs to the boiler.

Remove the accumulation of ash, slag, and soot from the furnace and passes of the boiler. Use care in removing hard slag from furnace walls to prevent injury to the refractory lining. The slag adheres tightly to the lining, and it is better to allow part of the slag to remain than to remove a portion of the refractory. Remove the soot and ashes from the horizontal baffle, hoppers, etc. Do not create dust by careless handling or by blowing ashes, fine coal, and soot with compressed air. The careful use of chimney or fan draft while cleaning will reduce the amount of ashes discharged into the boiler room. Do not step into piles of ashes while cleaning, as they may be hot enough to cause serious burns.

An inspection of the water side of a return tubular boiler includes an examination both above and below the tubes. Check for broken stays by testing with a hammer. Note the feedwater inlet and steam connections, including the dry pipe, for possible stoppage. Examine the shell and tube surfaces for scale deposits, bags, blisters, oil, and corrosion. Check for possible pitting or wasting away of the metal at the joints. When a fusible plug is used, note its condition and replace if corroded. In any case, the fusible plug must be replaced once a year. Arrange a light so that it is possible to look down between the tubes and check for possible obstructions that might interfere with the water circulation. Interior inspections of vertical and other small fire-tube boilers must be made by observing conditions through the inspection holes.

Inspect the exterior of a return-tubular-boiler shell from inside the furnace. Cracks in the shell plates are dangerous and should be investigated and repaired. An exception is made in the case of fire cracks that run from the edge of the plate into the rivet holes of the girth seams; a limited number of these cracks is permitted. While in the furnace, check the setting for cracks and note the condition of the grates and bridge wall. If the boiler is the flush-front type, inspect the front baffle for possible leaks. See that the blowoff pipe is adequately protected from hot gases. Open the doors at both ends of the boiler shell and inspect the tube ends for burning, joint leaks, and corrosion and the tube sheets for bulging tendency. Note the exterior appearance of the setting, and check for air leaks where the setting joins the shell.

To inspect a water-tube boiler it is necessary to enter each of the drums. An inspection of the steam drum should include an examination of feedwater inlet, chemical feed, continuous blowdown, and water-column connections for possible stoppage; drum and tubes for scale and corrosion; tube ligaments and joints for thinning or weakening; the dry pipe or separator for corrosion or scale; and drum plates, tubes, and especially rivets for signs of caustic embrittlement. In the mud drum, note the nature of the deposits and the condition of the tubes; check the blowoff connections for scale deposits and possible stoppage. In a straight-tube boiler, it is necessary to remove the handhole caps to inspect the interior of the tubes. Sometimes a mechanical cleaner is run through the tubes to determine the extent and nature of the scale.

Enter the furnace and various passes of the boiler to inspect the setting, baffle, and exterior of the tubes. After the slag and ash have been removed from the furnace walls and first-pass tubes, check for the formation of bags. Examine the tubes for possible erosion from flame impingement and the action of the soot blowers. Have the soot blowers rotated, and observe the angle of travel to see that they do not blow against a wall or baffle and that the nozzles do not blow directly upon the tubes. Have the necessary adjustments made to the blowing angle and alignment with respect to tubes. Examine all baffles to see that the gases are not short-circuiting and thereby failing to come into contact with all the heating surface. Open the stack- or fan-inlet damper to see that it opens wide and closes tightly.

Observe the condition of the exterior of the boiler, piping, and accessories. Check the arrangement of valves, and inspect the following piping systems: main steam header, nonreturn valve and drains, feedwater piping, steam-gauge connections, soot-blower supply lines, main and continuous blowoff, safety valves, discharge and drains, and chemical supply to the boiler drum. Inspect the induced-draft fan for excessive erosion of fly ash. Operate the combustion-control mechanism which moves the damper and coal-feed device. Check the stoker for burned grates and wear or damage to the fuel-feed mechanism.

After the necessary work has been completed, the boiler is ready to be closed and filled with water. Always put a hydrostatic test on a boiler after it has been opened for inspection or repairs. Unless extensive repairs have been made to the drum or shell, it is not necessary to test at $1\frac{1}{2}$ times the working pressure. The pressure developed by the boiler feed pumps is sufficient to check for leaking tube joints and handhole and manhole gaskets. If there are no leaks in the pressure section of the boiler, it is ready for service and the procedure is as explained in Sec. 6.1.

6.7 MAKING REPAIRS. *Tubes.* When a boiler tube fails, it is best to replace rather than weld it or otherwise attempt repairs. There are two methods of procedure in removing tubes: either burn off the faulty tubes with an acetylene torch 1 in. or more inside the sheet or drum or pull the entire tube length through the tube hole. The tube, the type of boiler, and the tube arrangement determine the method that should be used. It is usually advisable to burn out the old tube if this is possible. All bent tubes must be burned out, as they will not go through the tube holes. In these boilers it is sometimes necessary

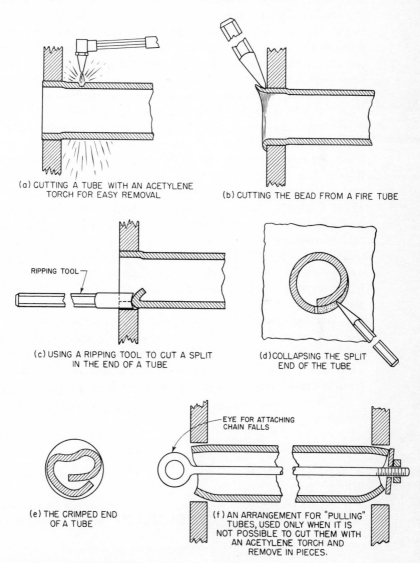

(a) CUTTING A TUBE WITH AN ACETYLENE TORCH FOR EASY REMOVAL

(b) CUTTING THE BEAD FROM A FIRE TUBE

RIPPING TOOL

(c) USING A RIPPING TOOL TO CUT A SPLIT IN THE END OF A TUBE

(d) COLLAPSING THE SPLIT END OF THE TUBE

EYE FOR ATTACHING CHAIN FALLS

(e) THE CRIMPED END OF A TUBE

(f) AN ARRANGEMENT FOR "PULLING" TUBES, USED ONLY WHEN IT IS NOT POSSIBLE TO CUT THEM WITH AN ACETYLENE TORCH AND REMOVE IN PIECES.

FIG. 6.2 *Removing boiler tubes.*

to remove several good tubes in order to reach the faulty ones. In a straight-tube boiler, the tubes in the outside of the pass may be burned off at the end, but those in the inner rows must be pulled through the tube holes, as they cannot be reached without removing the outer tubes.

Decide upon the procedure to be followed in removing the defective tube or tubes. When it is necessary to "pull" rather than burn out the tubes, pass a mechanical cleaner with a knocker head through the inside. This is especially important in the case of a fire-tube boiler, as the cleaner will remove the scale and make it easier to get the tube through the hole. On the other hand, if it is possible to get to the outside of the tube, burn it off with an acteylene torch about 1 in. from each end (Fig. 6.2a). It may be necessary to cut the tube in several pieces to facilitate removal. Exercise caution in removing tubes from water-tube boilers to prevent injury to the baffle.

In either case you are now ready to loosen the tube where it is expanded in the tube sheet, drum, header, or shell. Use a chisel to remove the bead from the tube, being careful not to injure the seat (Fig. 6.2b). Use a ripper (Fig. 6.2c) to cut a slot through the tube at the seat. The tube end may now be collapsed to loosen it in the seat (Figs. 6.2d, e). If the tube has been burned off, the nipple may now be easily removed; if not, the tube must be pulled or driven out. Some skill is required at this point to use the available equipment to pull and drive the tube through the tube hole (Fig. 6.2f).

After the old tube has been removed, clean the hole (seat) and inspect it for possible cuts or injuries. Also clean and polish both ends of the new tube to assure a metal-to-metal contact. Insert the new tube into position and even it up so that the same amount extends out of the seat on each end. Dent the end of the tube with a hammer to hold it in place. Expand both ends of the tube to obtain a tight joint (Fig. 6.3). The detailed procedure to be followed in expanding tubes is explained in Sec. 2.8. Bead over the ends of fire tubes to prevent the

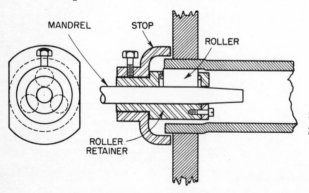

MANDREL   STOP   ROLLER   ROLLER RETAINER

FIG. 6.3 *A three-roller expander in place in a tube.*

ends from being burned off and to allow the
tubes to act as stays for the tube sheets (Fig.
6.4). After replacing tubes in a boiler, always
apply a hydrostatic test to check for leaks.

Drums. Corrosion, caustic embrittlement,
and overheating makes it necessary at times
either to repair or to replace boiler drums,
shells, and water legs. These repairs and replacements should always be made by qualified
boiler mechanics. Bags on the shells may be
repaired by heating and driving back into place
provided the metal is not too badly damaged

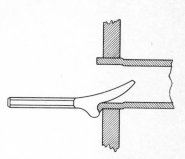

FIG. 6.4 *Forming a bead
on the end of a fire tube.*

by overheating, as determined by the boiler inspector. In any case,
if the metal has lost its tensile strength or the bag extends over a
large area, a patch must be applied. When a blister develops, the
defective metal must be cut away to determine the extent of the
damage. If the plate has been weakened, it will be necessary to apply
a patch. Patches must be applied inside the drum or shell so that
the pressure will tend to hold them in place and to prevent the formation
of a pocket for the accumulation of scale and sludge. There are three
methods of attaching patches to boiler drums and shells: by bolting,
referred to as a "soft patch"; by riveting, known as a "hard patch";
and by welding, when the patch is held in place by stay bolts.

Soft patches are placed in position and fastened with bolts. The
bolts pass through the patch and enter tapped holes in the shell. This
eliminates the necessity of bucking or holding, as is required in the
case of rivets. The patch is applied inside the shell over the defective
section. Packing is used between the patch and the shell to prevent
leakage. These patches are not suited for locations where the heat
transfer is high or on high-pressure boilers. They should be considered
temporary in nature and replaced with a riveted or welded patch as
soon as possible.

Hard or riveted patches make permanent repairs to boilers, drums,
and shells. The faulty section is cut away and the patch applied, using
plate at least as thick as the plate being patched but not over ⅛ in.
thicker. The size of the rivets and spacings must be so arranged that
the seams in an unsupported area will be at least equal in strength
to the weakest part of the shell.

All calking of joints, patches, and rivets must be done inside the
boiler whenever feasible as a precaution against caustic embrittlement
(see Fig. 6.5). If continued difficulty is encountered with leakage at
a joint or around the rivets, have the boiler inspected. Remove a rivet
and have the metal tested in a laboratory to determine its tensile strength
and other physical properties. If a crack is found in a longitudinal

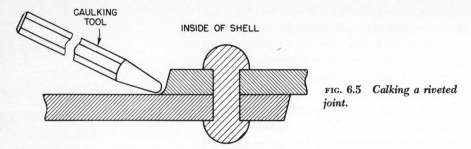

FIG. 6.5   *Calking a riveted joint.*

joint of a boiler, the boiler should be removed from service. This location is difficult to patch satisfactorily owing to the stress to which it is subjected, and in many states a boiler in this condition is permanently discontinued from service.

Failure may occur in the plate and stay bolts of water legs in both fire- and water-tube boilers. The plate cannot be repaired by riveting because only one side is accessible. Field welding is permitted provided sufficient stay bolts are included in the patch to hold the pressure without considering the strength of the weld. The burned metal must be removed and the patch fitted in place and welded according to established standards (Fig. 6.6). With the patch in place, drill and tap the necessary new stay-bolt holes and retap the old ones. Insert the new stay bolts and rivet over the ends to make them tight. Finally drill the telltale holes.

Field welding cannot be generally applied to boiler repairing unless the welded section is stress-relieved according to code requirements and x-rayed. Welding without stress relieving and x-raying is restricted to use as a seal where it is not required to strengthen the pressure parts of the boiler.

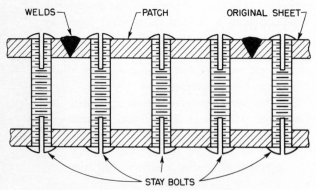

FIG. 6.6   *Section through the water leg of a fire-tube boiler, showing welded patch and stay bolts.*

Baffles. Some types of boilers require baffles to effectively direct the gases for maximum heat transfer to the tubes (see Figs. 1.9, 1.10, 1.17, 1.25, and 1.32). Boiler baffles are made of refractory, alloy, or standard steel, depending upon the temperature of gases to which they are to be exposed. Refractory baffles are used near the furnace when high temperatures are encountered. They are constructed of special refractory shapes which are made to fit the exact location. Some baffles lie on a row of tubes, and others run at an angle across the tubes. To make repairs it is necessary to secure the refractory shapes for the specific application. It requires skill and patience to insert these tile shapes in baffles which are at an angle with the tubes (when the tubes pass through the baffle).

One-piece (monolithic) baffles are shown in a straight-tube boiler in Fig. 6.7. These baffles are made from moldable refractory material. Forms are constructed to retain the refractory and produce baffles which will direct the flow of gases.

When the tubes pass through the baffles, forms can be made by passing wooden strips down both sides of the space where the baffle is to be installed. These strips retain the refractory material as it is worked down between the tubes. The material air-sets, and the wooden strips burn out when the boiler is placed in service. In some cases it is advisable to wrap heavy paper around the tubes before applying the baffle material. This burns out, leaving a space around each tube to allow for expansion and tube replacement. Some baffle material

FIG. 6.7   *Monolithic baffle installed in a longitudinal-drum water-tube boiler.* (*Plibrico Co.*)

shrinks enough to provide the necessary clearance around the tubes. Baffles are also made of heat-resistant metals. One method is to drill a single metal plate to receive the tubes. These single-plate baffles are difficult to repair and cause interference when it is necessary to replace tubes. It is therefore preferable to use baffles composed of metal strips shaped to fit the tubes. These strips are inserted between the tubes and bolted together with tie bars to form a complete baffle. These baffles can be maintained by replacing individual strips. Sections can be taken out to facilitate tube replacement and then reassembled.

Furnace walls. Some boilers and furnaces require refractory walls (see Figs. 1.3, 1.4, 1.9, and 1.10). These refractory walls account for an appreciable portion of boiler maintenance, and it is therefore important that an effort be made to reduce this expenditure.

Solid refractory walls are lined with first-quality firebrick. In many instances this lining can be patched or replaced without rebuilding the entire wall if the repairs are made before the outer section has become too badly damaged. The lining must be "keyed" to the outer wall, and expansion joints must be provided.

Air-cooled refractory walls are constructed to provide an air space between the outer wall and the refractory lining. This lining may be a separate wall constructed of first-quality firebrick, moldable refractory material, or tile supported by cast-iron hangers. The circulation of air through the space between the two walls carries away some of the heat, resulting in a reduction in wall temperature and a possible decrease in maintenance. The lining must be adequately tied to the outer wall to prevent expansion and contraction from causing it to bulge and fall into the furnace. The cast-iron hangers provide adequate support for the tile lining. However, the initial cost of these walls is high; many tile forms must be stocked for replacements; and once the lining has failed, the cast-iron hangers are quickly destroyed by the heat of the furnace.

Careful consideration should be given to the quality of the refractory material. Operating temperatures, ash characteristics, slagging, and flame impingement are all factors to be considered when selecting refractory lining for furnaces. If the fusion temperature of lining is not high enough for the operating temperatures, the inner surface will become soft and in extreme cases even melt and flow, resulting in rapid deterioration. The fusion of ash on the surface of the wall causes the inner face to spall, that is, to crack and fall off. The removal of clinkers either with the unit operating or with it out of service for maintenance will invariably crack off some of the furnace lining. Flame impingement results in a localized high temperature and subsequent failure of the lining.

Furnace walls (either solid or air-cooled) and arches may be constructed or rebuilt by the use of moldable refractory material, which is furnished in a moist condition for ramming into place. Keys made of heat-resistant metal, refractory tile, or a combination of both materials are used to hold the lining in place. The refractory material is rammed into place by an air-operated hammer. Expansion joints must be provided at intervals to prevent excessive forces from developing.

Figure 6.8 shows the steel angles in place for supporting a moldable refractory arch. The hangers are attached to these angles and embedded in the refractory. Wooden forms are used to support the arch during construction. The refractory material is rammed into place around the inserts. Adequate expansion joints must be provided.

Where service is especially severe, causing outage and costly repairs, it is advantageous to use special high-quality, high-cost brick or tile. These materials are frequently used in the walls just above underfeed stokers where there is a combination of high temperatures and contact with ash and clinkers. They are also used in spreader-stoker front-wall arches and in the piers between the feeders.

Waterwalls are installed to reduce maintenance and, unless misapplied or abused, seldom require repairs. The maintenance procedure depends upon the type of waterwall construction. If the tubes themselves fail, the repair is similar to that for replacing any other boiler tube. The cast-iron blocks used on some waterwalls may have to be replaced after long-continued service. In this case replacements are made by first cleaning the tube surface, then applying a heat-conducting cement or bond, and finally bolting the new block securely in place.

Make repairs and replacements to boiler equipment often enough to prevent small jobs from developing into major projects. Repair the

FIG. 6.8 *Supporting steel for a plastic refractory arch in a boiler furnace.* (*Plibrico Co.*)

furnace wall before the steelwork is damaged. Do not remove and replace serviceable parts. Some thinning of the furnace wall and burning of stoker castings are expected and permissible.

QUESTIONS AND PROBLEMS

6.1. How can a boiler with oil and grease in it be cleaned?

6.2. In applying a hydrostatic pressure on a boiler, what are the maximum and minimum pressures permitted?

6.3 How would you start a boiler with a superheater in it? Explain.

6.4. If your gauge glass were full of water and you wanted to bring up the boiler, what would you do?

6.5. What precaution would you take to blow down a boiler with a hot fire under it?

6.6. What causes water columns to show false levels?

6.7. If a boiler shows signs of pitting and corrosion, what should be done?

6.8. Describe two nondestructive tests that may be used to determine the presence of defects in boilers.

6.9. What is meant by priming and foaming? What is the difference between them?

6.10. If distilled water from a heating system were used for boiler feed, would this be satisfactory?

6.11. What is a bag? A blister?

6.12. Name several things that cause high flue-gas temperature.

6.13. What is meant by the maximum capacity of a boiler?

6.14. What is meant by a boiler's being 70 per cent efficient?

6.15. What is the advantage of analyzing flue gas?

6.16. If you were to take charge of a strange boiler room, state the first 10 things that you would check.

6.17. How often do you blow down a boiler? How do you proceed, and what precautions are taken?

6.18. If your water gauge were out of order, what would you be guided by?

6.19. State the main sources of waste in operating a boiler plant.

6.20. A pound of coal will evaporate 10.1 lb of water. If coal is burned at the rate of 30 tons per day, what is the rate of evaporation in pounds of steam per hour?

6.21. What would you do if the water disappeared from sight in the glass?

6.22. How should a boiler and engine be handled when violent foaming occurs?

6.23. How should a new water glass be placed in service?

6.24. What is waterhammer? Is it dangerous?

6.25. If a safety valve leaked, would you take up on the spring? Explain.

6.26. What is the best procedure to use to find the water level if a boiler is foaming?

6.27. If the means of supplying your boiler with water should fail, the water were already low, and you were asked to run the boiler a while longer, what would you do?

6.28. What precautions should you take in laying up a boiler?

6.29. How would you take one of a battery of boilers out of service?

6.30. Which is more injurious to a boiler, heat or pressure?

6.31. How often must boilers be cleaned?   Explain the procedure in cleaning boilers.

6.32. How are seams calked, and what tool is used?

6.33. With a boiler which has been in service for some time, what might cause an explosion?

6.34. How should you get a boiler ready for inspection?

6.35. How should an inspection of a boiler be made?   What are the most important things to look for?

6.36. What is a fire crack?   Is it considered dangerous?

6.37. State the valves you would close and the safety precaution you would take before entering a boiler for inspection.

6.38. Explain how a serious accident might occur in a plant where the blowdown lines from several boilers discharged into a common header.

6.39. How would you inspect a soot-blower installation to make sure it was in good operating condition?

6.40. How would you patch the furnace sheet on a water leg of a vertical boiler?

6.41. How is a blistered boiler plate repaired?

6.42. What is the difference between a hard and a soft patch?

6.43. Where should the patch be placed, on the inside or the outside of the boiler plate?

6.44. What should be done when a crack is found in the longitudinal joint?

6.45. How would you repair a bag on a boiler shell?

CHAPTER 7

PUMPS

Pumps are used for many purposes and a variety of services: for general utility service, cooling water, boiler feed, and lubrication; with condensing water and sumps; as booster pumps, etc. There is a pump best suited to each purpose and individual service.

Turbines and boilers are ever increasing in size, requiring larger boiler feed pumps. At present steam pressures of 3,650 psig are employed, but discharge-header pressures have risen to the supercritical region of 6,500 psig. With increased pump reliability, many generating stations use fewer pumps, perhaps a single pump for each boiler-turbine unit. Steam-turbine drive is receiving favorable consideration. Its advantages are that it (1) decreases power consumption, (2) provides ideal speed operation with the elimination of hydraulic couplings, and (3) furnishes exhaust steam which can be used to improve the station heat balance.

Pumps are classified as reciprocating (piston, plunger, diaphragm), rotary (gear, cam, vane, screw), and centrifugal (radial-flow, mixed-flow, axial-flow); they can be single or multistage, open- or closed-impeller types, or special pumps such as jet pumps, hydraulic-ram pumps, etc.

7.1 PUMPS. There are a great variety of pumps from which to make a selection; each pump has its specific advantages which need be analyzed for a specific application. The simplest pump is the injector, used on small boilers and portable units; its low first cost and its simplicity frequently recommend it. Reciprocating pumps find ready acceptance, particularly in the smaller plant, where first cost is a factor; they are simple in construction, easy to repair, and reliable in operation. Rotary pumps find application in handling oil and lubricants.

Centrifugal pumps are available for a variety of services and purposes having an almost unlimited range of industrial applications.

Pumps may be required to lift or raise water on the suction side; the extent to which this can be done is determined by the type of pump and the atmospheric pressure. Actually the pressure of the atmosphere pushes the water up into the pump suction; the pump creates a vacuum by movement of the tight-fitting piston or action of the impeller into which the water rushes. The height to which water can be lifted is then influenced by the atmospheric pressure and water temperature; consideration must be given these variables when pumping installations are being designed.

The total head developed by the pump, usually called the "total dynamic head," is made up of the following:

1. Total static head of discharge above the level of the suction water
2. Pressure at the point of discharge
3. Pipe friction, including fittings

In addition, the total dynamic head contains velocity head, which is usually unimportant except in large-capacity pumps. The total head which a pump can develop is a matter of pump design and is influenced by its mechanical condition and the factors enumerated.

7.2 INJECTORS. The injector (Fig. 7.1) is perhaps the simplest pump; it is a device to lift and force water into a boiler which is operating under pressure. It operates on the principle of steam's expanding through a nozzle, imparting its velocity energy to a mass of water.

The essential parts of the injector are the steam jet, the suction jet, the combining and delivery tube and ring, and the overflow and discharge tube. In operation, steam issues from the nozzle 8; it drops in pressure as it passes through the jet but gains in velocity. As the steam passes between the steam jet 8 and the suction jet 7, a vacuum is created in the "suction" chamber. As a result, water is drawn into the suction chamber from the supply main, overhead storage tank, or underground reservoir. The high-speed steam jet picks up the water as it crosses the space between 8 and 7, forcing the water along with the steam into the combining tube and finally into the delivery tube 6. Here the steam is condensed; the delivery tube receiving the water (and condensate), owing to its design, changes a considerable amount of the velocity energy of the jet into pressure. The heat energy in the expanding steam not only provides sufficient energy to force the feedwater into the boiler but in addition heats the water, thus providing both a pump and a heater in one operation.

The injector should be installed as shown in Fig. 7.1a. The main lever is used in regulating the inspirator; a check valve is located in the delivery line ahead of the stop valve to prevent the return of water

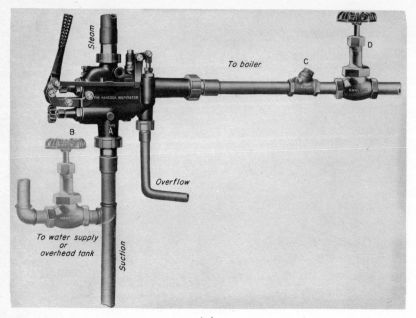

(a)

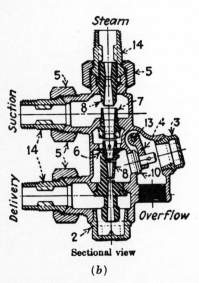

Sectional view

(b)

FIG. 7.1 (a) and (b) Metropolitan automatic injector. (Dresser Industries.)

from the boiler. The steam line should be of the same size as the injector connection; it should be connected to the highest point on the boiler to ensure dry steam. A stop valve is provided so that the injector may be removed. The overflow should be as large as or larger than the injector connection and as straight as possible, it must be open to the atmosphere and not piped below the surface of the water. The suction line must be tight. It should be as large as the take-off of the injector unless the suction lift is more than 10 ft or unless the horizontal length is more than 30 ft, in which case the line should be two sizes larger than the injector take-off; the piping is then reduced in size to match the injector size. If water enters the injector under pressure from an overhead storage tank, equip the line with a globe valve like valve B.

The delivery line should be at least as large as the injector connection. It should be equipped with a check valve C and a globe valve D; the latter is always positioned wide open except when repairs to check valve or injector become necessary.

To operate, pull the main lever back until the resistance of the main steam valve is felt. This will lift the water. When water appears at the overflow, pull the lever back slowly and steadily as far as it will go. To stop, push the lever all the way forward.

The maximum water temperature that an injector is considered capable of handling is 130 to 150°F. Increased lifts must be accompanied by a decrease in water temperature. When the water comes in contact with steam, the heat causes some of the impurities to drop out in the injector. This tends to scale up the nozzles, and the injector will fail to function properly.

Injectors are very inefficient pumping units. They are practical only on small boilers, and they are not entirely reliable. Moreover, modern power-plant practice favors high feedwater temperatures. Since an injector cannot handle hot water and also is unreliable, this method of feeding boilers has been largely discontinued. Injectors operate satisfactorily where load and pressure are somewhat uniform. With varying load conditions and fluctuating pressures, they become unreliable.

The injector, however, offers several advantages. It is very simple and has no moving parts to get out of order and require replacement. It is compact and occupies little room, and both the initial cost and the installation cost are low. It heats the feedwater without the aid of a heater, and thermally it is very efficient.

Causes of Injector Failure

If it fails to lift water, perhaps

Tubes and nozzles are plugged
Suction line is leaking

Suction pipe is corroded
End of suction pipe is not submerged
Suction strainer is stopped up
Overflow line is restricted
Steam is too wet
Supply water is too hot
Steam pressure is too low

If it fails to force water into boiler, perhaps

Tubes and nozzles are plugged
Delivery line is plugged
Suction line is leaking
Overflow is leaking
Check valve in discharge is not working properly
Steam is too wet
Steam pressure is too low

7.3 DUPLEX PUMPS. The duplex pump (Fig. 7.2) has two pumps mounted side by side. The operating medium can be water, compressed air, or steam; the pump described here is steam-actuated. The pump is direct-acting in that the pressure of the steam acts directly on a piston to move the second fluid, in this case, water. Action is secured through the motion of a piston or plunger reciprocating in a bored cylinder or a cylinder fitted with a liner. The action here is positive in that a definite amount of water is displaced per stroke; the quantity or volume delivered is reduced by leakage and slippage due to pump wear and the condition of the valves.

Pump dimensions are given in this manner: 3 by 2 by 3. The first figure refers to the diameter of the steam cylinder, the second to the diameter of the water cylinder, and the third to the length of the stroke; all dimensions are in inches.

Each of the two pumps of the duplex pump has a steam cylinder on one end and a water cylinder on the other. The rocker arm of

FIG. 7.2 *Exterior view of duplex pump. (American-Marsh Pumps, Inc.)*

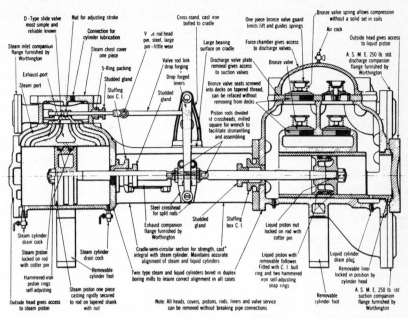

FIG. 7.3 *Horizontal duplex piston pump.* (*Worthington Corp.*)

one pump operates the steam valve of the opposite pump. A cross
section of a duplex piston pump is shown in Fig. 7.3; Fig. 7.4 shows
a plunger pump. The steam cylinders are fitted with pistons; they are
equipped with self-adjusting iron piston rings fitted to the cylinder bore.
Above each steam cylinder are four ports; the two outside are called
"steam ports," and the two inside "exhaust ports." A D slide valve controls the admission and exhaust of steam.

An inside guide is provided for the valve rod and a packing gland
for the piston rod where they enter the cylinder head and the valve
chest, respectively. The gland prevents leakage of steam at these
points. One rocker arm has a motion direct and the other indirect.
One valve rod moves in the same direction as the piston that is operating
it; the other moves in the opposite direction. Midway on the piston
rod is mounted a crosshead, to which is attached the rocker arm. On
the other end of the piston rod is located a piston (or plunger in the
case of the plunger pump) in the water cylinder. It has a renewable
liner and is actuated by the steam piston, which transmits its power
to the water side. The water piston is fitted with a removable follower
using fibrous or metallic packing rings and moves back and forth in
a bored cylinder. The ends of the cylinder are fitted with drain plugs
similar to those of the steam cylinder. Above the cylinder are located

two decks of valves, the lower group being the suction valves and the upper deck the discharge valves.

The seats of these valves are screwed into the decks on tapered thread and can be refaced without being removed from the decks. The lower valves are set over the suction inlet, and the discharge goes directly to the discharge pipe through the upper ones. An air cock is mounted on top of the chamber and is called a "vent." This vent is used to remove entrapped air from the pump when it starts.

The slide valve is very simple and is operated by direct-lever connection. Motion is imparted to the valve by a rocker arm actuated by the piston rod on the opposite cylinder. After admitting steam to the other cylinder, the piston completes its stroke and waits for its own valve to be operated on by the other piston, so that it may return on the next stroke. One of the valves is always open, and no dead centers are encountered.

The lost motion between the lugs on the back side of the valve and the nut does not permit the valve to move until the piston that actuates it (on the other side) has traveled some distance. The amount of lost motion permitted enables the valves to close slowly and quietly before the pump reverses the stroke. The flow of water, however, is not interrupted. While the one piston is being slowly brought to a stop, the other continues in action. This prevents fluctuations in the pressure

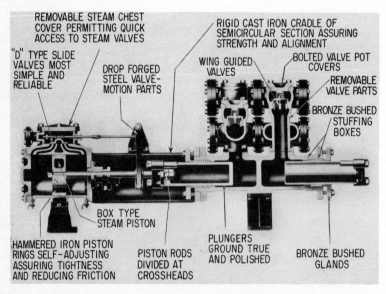

FIG. 7.4 *Horizontal duplex plunger pump.* (*Worthington Corp.*)

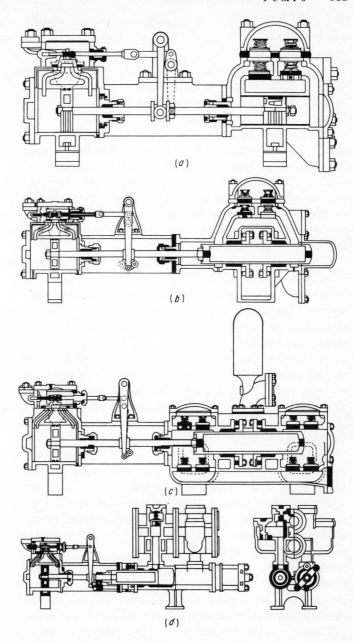

FIG. 7.5   *Duplex pumps   (a) Packed-piston pattern, valve-plate type.   (b) Outside-center-packed-plunger pattern, valve-plate type.   (c) Outside-center-packed-plunger pattern, turret type.   (d) Outside-end-packed pot-valve plunger pattern.   (American-Marsh Pumps, Inc.)*

and ensures a uniform flow.   The lost motion usually allowed in the valve mechanism permits the piston to travel one-half of its stroke before operating the valve.

Lost motion is provided in several ways.   In one model, the valve rod has threads on which is screwed a small block.   The block fits between the lugs on the back of the valve.   Lost motion is the distance from the edge of the block to the lug.   With this arrangement the amount of lost motion cannot be altered unless the block is replaced or the width changed.   Another model has the valve rod threaded, and the nuts are adjustable outside the lugs on the valve.   Lost motion can then be adjusted by moving the locknuts.   In a third type the lost-motion adjustment lies outside the steam chest, between the link connecting the valve rod to the rocker arm and the valve-rod headpin.   The advantage of this arrangement is that the valve can be adjusted for lost motion with the pump running.

Clearance space has been defined as that volume which the steam occupies when the piston is at the end of the stroke, that is, the space between the face of the piston and the underside of the valve.   The steam occupying this space does no useful work and hence a pump short-stroking increases clearance volume and as a result increases the steam consumption.

The duplex pump shown in Figs. 7.5a and b is designed to reduce this clearance and increase the economy.   The two large admission ports are eliminated, and a much smaller port is substituted.   Steam enters this port, slowly moving the piston forward until the main body of steam strikes it.   This reduction in clearance decreases the steam consumption and results in economical operation.

On some high-pressure duplex pumps a piston valve is employed instead of the usual slide valve.   This is a balanced valve; its advantages are that it is light, perfectly balanced, and very simple.   Valves on a duplex pump have neither lap nor lead.   As lap is necessary to cut off the steam before the piston reaches the end of the stroke in order to obtain expansion, it is evident that a duplex pump *must* take steam the *full length of the stroke.*

Some pumps are fitted with "cushioned valves." These are small valves fitted to duplex pumps, as in Fig. 7.6, and opening a connection from the steam port to the exhaust port at each end of the cylinder.   This pump is provided with a hand-operated regulating valve located between the steam and the exhaust port.   This valve provides a means for releasing the excess cushioning at each end of the stroke after the piston has closed the exhaust port.   The usual arrangement when the pump is being operated is to throttle or nearly close the valve at slow speed and completely open it at high speed.

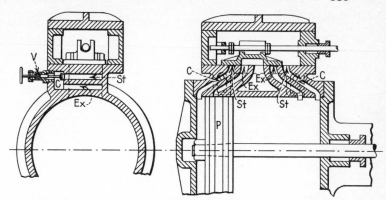

FIG. 7.6 *Duplex-pump steam-cylinder cushion valve.* (*Worthington Corp.*)

To set the valves on the duplex pump shown in Fig. 7.3, proceed as follows: Remove the steam-chest cover. Push the piston forward (this can be done by placing a bar behind the crosshead on the piston rod) until the piston strikes the head end. Make a chalk mark on the piston rod at the point where the rod goes through the studded gland. Move the piston rod in the opposite direction until the piston strikes the opposite head, and again make a mark on the piston rod (be sure to make the mark at the same end of the studded gland). With a pair of dividers, locate a *mark* midway between these two points, and place *this mark* at the end of the gland. The piston should now be located exactly in mid-position, and the rocker arm should be vertical. If the rocker arm is not in this position, it should be adjusted accordingly. The valve-rod headpin should now be removed from the valve rod of the opposite cylinder, and the valve placed in mid-position (so as just to cover the ports). The nut for adjusting the lost motion should be held exactly in the center of the space between the lugs on the back of the valve. The valve rod should now be screwed through the nut until the headpin again fits into the knuckle. The setting of one valve has now been completed, and it becomes necessary to repeat this operation for the other side. Before replacing the steam-chest cover, move one valve to uncover one of the ports; otherwise, the pump will not start. Once the pump has been started, it is impossible for it to stop on dead center.

When locknuts are employed to adjust the lost motion, the same general procedure is followed. However, it is not necessary to disconnect the rod from the valve-rod head. Hold the valve in mid-position and adjust the locknuts (Fig. 7.6) so that they will be at equal distances from the valve lugs, allowing about one-half of the width of the steam port for lost motion on each side. If the valve has outside adjustment, follow the same procedure as that just described. Owing to the fact that the jam nuts are on the outside of the steam chest, it is not necessary in this case to dismantle or stop the pump in order to adjust the valves.

Too much lost motion lengthens the stroke and may cause the piston to strike the head. *Too little lost motion* causes the pump to *short-stroke.* Short stroking results in an increase in steam consumption and a decrease in pumping capacity.

Valves on the water end are called "suction" and "discharge." The suction valves are for the purpose of admitting water to the cylinder and preventing a return of water to the pump suction. The discharge valves are for the purpose of discharging the water from the cylinder and preventing its return. Several different arrangements can be used to advantage:

1. The valve decks placed side by side over the cylinder (Figs. 7.4, 7.5d). This arrangement is used very frequently on high-pressure units.

2. The suction valves placed below the cylinder and the discharge above (Fig. 7.5c). This arrangement enables the flow to pass directly through the pump with a minimum amount of resistance being offered, as no reversing of flow is encountered.

3. The valve decks located one above the other. In this case both decks are above the cylinder (Figs. 7.3, 7.5 a, b). The discharge-valve plate is removable, giving easy access to both suction and discharge valves. The position of the suction valves over the cylinder ensures a full cylinder of water at all times and consequently a more uniform flow than is normally produced. Pumps of this design are especially adapted to low and medium pressures. The disadvantage of this valve arrangement is that water flowing through the pump must change its direction twice.

Many different kinds of valves are employed, the most common being the "flat-disk" and the "wing-disk" types. The flat-disk type is guided by a stem screwed into the valve seat (Fig. 7.3). Here the valve seats are screwed into decks on tapered thread and made of bronze. These can be refaced without removing them from the decks. Flat-disk valves are also frequently made of rubber or composition. Wing-type valves (Fig. 7.4) are also circular in form. They are usually fitted to bronze seats, although sometimes they are made of fiber, leather, or hard rubber.

For cold (low-pressure) and warm water, the usual procedure is to use rubber valves. For hot water and high pressure and temperature, bronze or steel should be used. (Practically all bronze and steel valves are made in the disk-poppet form.) Up to 100 lb it is general practice to use soft rubber or fiber material; for 200 lb, hard rubber or composition; above this, steel or bronze, the bronze valves being used more and more even for low pressure. For liquids other than water, special-type valves may be necessary. A number of small valves is preferable to one or two large ones since small valves are more positive in action, are easier to replace, are cheaper, do not warp, reduce leakage, and give less trouble from pounding. Moreover, they are usually accepted as being more reliable than large valves.

The difference between the theoretical and the actual displacement, expressed in percentage of the theoretical, is called "slip." Slip may be due to leaky valves (both suction and discharge) or to packing and piston leaks. In most pumps the slip varies from 5 to 10 per cent, as compared with a slip of 2 to 3 per cent for pumps in good mechanical condition.

An "air chamber" is a device located immediately ahead of the suction valves or directly behind the discharge valves. In the latter case it is placed directly above the water end of the pump. Air chambers on the suction side are recommended for short-stroke pumps, pumps with high suction lift, or those running at high speed. Air chambers on the discharge side are provided especially when the line is long. An air chamber absorbs shocks and surges in the discharge line and relieves the pump from excessive strains arising therefrom. Air chambers ensure a steady supply of water and reduce pounding.

The manner in which this is done is as follows: The upper section of the chamber is filled with air. As the water leaves the pump, some enters the chamber and compresses the air therein. When the piston momentarily hesitates in its stroke, the flow of water would be expected to remain stationary, but the compressed air expands and the flow continues. The air chamber thus forms a cushion to steady the flow. The volume of the air chamber is approximately $2\frac{1}{2}$ times the piston displacement.

All pumps require packing to prevent leakage of water past the piston or plunger. Different kinds of packing, based on pressures and temperatures of the liquid, are used on different pumps.

In piston pumps, the piston works back and forth in a bored cylinder (Fig. 7.3). Here the packing rings are securely fastened to the piston and move with it. The type of ring depends on the nature of the liquid to be pumped and on the pressure and temperature. Pistons used for general service employ soft canvas packing. This packing consists of a cotton fiber, square in cross section. Some pistons are fitted with grooved metallic packing rings. Others have snap rings like those employed in the engine.

In plunger pumps the plunger moves back and forth in a packing gland (Fig. 7.4) instead of a bored cylinder. The plunger type of pump can easily be distinguished from the piston type by this characteristic. Moreover, plungers are longer than pistons and long in comparison with the length of stroke. Plunger pumps are packed in different ways:

1. Outside-end-packed, Fig. 7.4 clearly showing the arrangement of stuffing boxes, glands, and packing.

2. Outside-center-packed, Figs. 7.5b and c showing how the plunger slides through the two packing glands. Access to the packing gland is had by removing the gland nuts, as it is outside-packed.

3. Inside-center-packed, in which the plunger slides back and forth in a packing gland. This gland consists either of a metal packing ring or of some form of hydraulic packing.

The inside-packed type has an advantage over the outside-packed in dusty and dirty locations, since the abrasive action of the dirt does not affect it as it affects outside-packed plungers. However, leakage in the inside-packed type is difficult to detect and repair, whereas with the outside-packed plungers leakage can be detected very easily and the packing replaced without dismantling the pump.

Several difficulties may be encountered when operating duplex pumps, particularly when handling hot water. Hot water, which comes to the pump by suction, is difficult to lift as it vaporizes when the pressure is reduced below atmospheric. As a result, the expansion of the vapor fills the suction chamber and pump cylinder. This vapor is compressed and reexpands, the pump thus failing to discharge a full cylinder of water. This results in considerable fluctuation in flow and pressure. Approximately 150°F is the maximum temperature at which water can be raised by suction. Hot water should *flow* to the pump and in this way ensure full pump cylinders. For pumping hot water a pump should be selected to run at low speed. A pump pumping hot water is usually placed 10 to 30 ft below the source of supply and the suction line made large enough to ensure ample delivery and the minimum interference with vapors formed.

Pumps generally should be provided with a strainer and a foot valve on the suction end of the line. The strainer prevents any foreign matter from clogging the line or lodging under the valves, while the foot valve prevents the water from draining out of the suction line. If there is no foot valve and water comes to the pump under suction, it becomes necessary to prime the pump. In this case, the pump is equipped with an aspirator which will exhaust the air in the suction line and draw water to the pump.

Thick liquids should always flow to the pump under pressure. Suction and discharge lines should be fitted with gate, in preference to globe, valves, as the latter offer too much resistance to the flow. The steam line to the pump should be fitted with a throttle valve and a drain at the pump inlet.

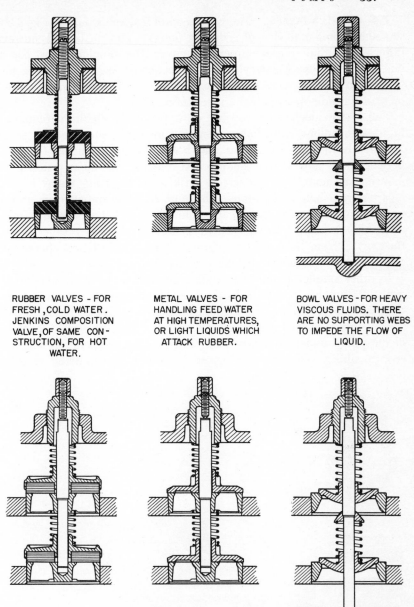

RUBBER VALVES - FOR FRESH, COLD WATER. JENKINS COMPOSITION VALVE, OF SAME CONSTRUCTION, FOR HOT WATER.

METAL VALVES - FOR HANDLING FEED WATER AT HIGH TEMPERATURES, OR LIGHT LIQUIDS WHICH ATTACK RUBBER.

BOWL VALVES - FOR HEAVY VISCOUS FLUIDS. THERE ARE NO SUPPORTING WEBS TO IMPEDE THE FLOW OF LIQUID.

FIG. 7.7 *Valve construction of single direct-acting pump.* (*Ingersoll-Rand Co.*)

7.4   SIMPLEX PUMPS.   The water end of a simplex pump (Fig. 7.8) is somewhat similar to that of a duplex pump.   The steam end, however, differs from that of the duplex; moreover, the simplex pump has but one steam and one water cylinder.   A sectional view of a simplex pump is given in Fig. 7.8, showing the pump interior with steam-valve arrangement and inside-operated valve mechanism.

This valve mechanism consists of a piston or auxiliary plunger F, fitted with hollow ends, holes extending into the piston ends as shown. Steam fills the valve chamber and hollow ends, balancing the piston between the ends of the steam chest.   When the main piston C has moved far enough to the left to strike the reversing valve I, steam escapes from behind F through the port E.   The unbalanced pressure on the auxiliary valve causes it to shift to the left and results in the main valve's being carried with it.   The steam which escapes through the port E from behind F is returned to the exhaust port through the exhaust cavity behind I to the exhaust chamber.

In other simplex pumps the actuating mechanism is an outside valve gear.   This valve gear actuates the auxiliary valve, which in turn throws the main valve.   In all simplex pumps the main valve is "steam-thrown."

The simplex valve mechanism is simple and ensures positive operation of the pump under all conditions.   The method of operation eliminates any possibility of short stroking, so frequently encountered in duplex

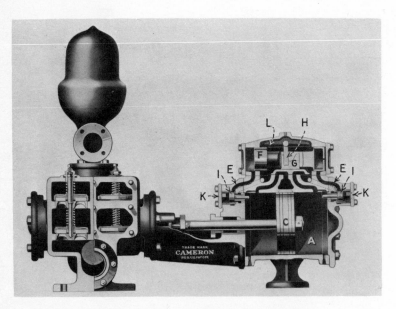

FIG. 7.8   *Sectional view of simplex direct-acting pump.* (*Ingersoll-Rand Co.*)

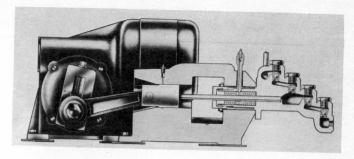

FIG. 7.9   *Volume control with step-valve design.*
(*Milton Roy Pumps.*)

pumps; since the simplex must travel the *entire stroke* in order to actuate the valve, short stroking is impossible.   The method of operation eliminates dead centers and ensures instant response when desired.

The simplex pump has but one port in place of two, as on a duplex pump.   This reduces the clearance volume and decreases the steam consumption.   The reduction in steam consumption results in an increase in economy.   This fact frequently accounts for the selection of a simplex in preference to a duplex pump.

7.5 POWER PUMPS.   A power pump is one in which the piston or plunger is operated by a unit not a part of the pump itself.   The motive power in the single and duplex pumps is the steam end of the pump. Power pumps, however, are geared, belted, or direct-connected to gas engines, electric motors, chain drive, etc.   Motors direct-connected by gears, belts, or chain drives are most commonly employed.   These pumps may have their cranks coupled at 90°, an arrangement which provides a steady and uniform flow.   The crank setting (degree) depends on the number of pistons or plungers employed.

The pump shown in Fig. 7.9 can be classified as a power pump. Pumps of this variety are used for chemical feed and offer a precise volume control by means of a variable-stroke adjustment mechanism. General methods of adjustment are provided, the one shown being by means of a screw.   The microadjustment is equipped with a vernier dial to permit quick and convenient resetting for any predetermined stroke length.   This pump employs the step-valve design shown.

Displacement power pumps are employed for low flow rates and high heads.   The efficiency of power pumps depends to a large extent on the character of the driving motor and the mechanism transmitting this power.   Their design is simple as well as practical; they can be located anywhere independently of belts or pulleys.   If, however, the power for the pump is an engine drive, it is necessary to have a steam generator, an item of expense not to be overlooked.

Mechanically driven pumps are very popular and as simple as the motor-driven units. The initial cost is lower. However, they occupy more floor space, have a poor system of regulation, and require an engine in order to operate. Steam-driven crank and flywheel pumps are also used. Steam can be used expansively on engines equipped with flywheels. The reciprocating motion of the pump piston is usually secured directly from the motion of the piston in the steam cylinder.

7.6   VACUUM PUMPS.   Vacuum pumps are provided for a variety of services.   The wet vacuum pump (Fig. 7.10) can be employed for heating systems and process requirements since it operates at vacuums up to 26 in. (as referred to a 30-in. barometer) and is available in a variety of sizes and capacities.   This pump is steam-driven and piston-packed; in operation it performs as a simplex valve.

Redi-Vac pumps automatically remove the air and condensate from return lines and discharge the air to the atmosphere, while at the same time returning the water to the boiler and maintaining the vacuum on the heating system. Removal of air and water from the heating system is accomplished by the vacuum created in delivering water through jets and into the telescopic draft tubes shown.

The Redi-Vac pump is divided into two sections.   On the left (Fig. 7.11) are shown the return line, the circulating-water chamber, and the vacuum chamber; on the other side are the accumulator chamber and the boiler feed chamber.   During normal operation the circulating pump draws water from the circulating-water chamber, discharges it through pipe 5 into the cord passage, which extends to pressure chamber 7 on top, and down through nozzles 8 into venturi throats 11 and draft tubes 12.   The air from the accumulator chamber is drawn in through ports 10 and carried by the jet water through the draft tubes 12 into the circulating-water chamber.   Here the air separates from the water, passing through space 13 and out through the air vent.

Exhausting the air from the accumulator chamber produces a lower

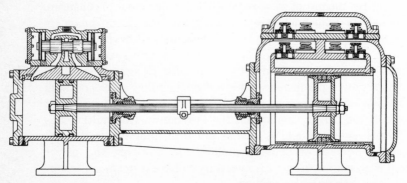

FIG. 7.10   *Wet vacuum pump.*   (*American-Marsh Pumps, Inc.*)

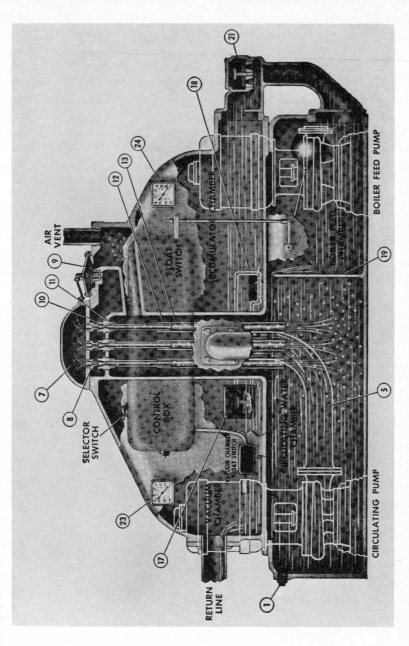

AIR VENT

FLOAT SWITCH

ACCUMULATOR CHAMBER

BOILER FEED CHAMBER

BOILER FEED PUMP

CONTROL BOX

VACUUM CHAMBER FLOAT SWITCH

CIRCULATING WATER CHAMBER

CIRCULATING PUMP

SELECTOR SWITCH

VACUUM CHAMBER

RETURN LINE

FIG 7.11 *Cut-open view of type AV single Redi-Vac pump.* (*American-Marsh Pumps, Inc.*)

pressure or partial vacuum in that chamber, resulting in a flow of air and condensate from the vacuum chamber through check valve 17 and rear manifold 18 into the accumulator chamber, the attenuated air continuing to pass into the jet stream through ports 10 and the water or condensate rising in the accumulator chamber until the unit is stopped by the vacuum switch, at which time the pressure of the circulating water in the pressure chamber 7 drops to zero, releasing pressure on the diaphragm 9 to allow the valve to open and admit atmospheric air through the valve into the accumulator chamber. This places this chamber under atmospheric pressure, closing check valve 17 and preventing a flow return into the vacuum chamber; this causes accumulated condensation to flow by gravity through check valve 21 to the boiler feed chamber.

The modified duplex water pump (Fig. 7.12) is used for boiler feed service; it is capable of handling water at high temperatures without vapor binding or cavitation. The duplex pumps are operated by two float switches set to start the water pumps at low- and high-water levels in the receiver. If one pump cannot handle an abnormally high flow of condensate, the second pump comes into operation. Of interest is the vacuum pump attached to this unit to remove entrained air from the system.

The vacuum pump (Fig. 7.13) is a centrifugal displacement pump consisting of a round multiblade rotor revolving freely in an elliptical casing partially filled with liquid. The curved rotor blades project radially from the hub to form with the side shrouds a series of pockets or buckets around the periphery. The rotor revolves at a speed high enough to throw the liquid out from the center by centrifugal force.

FIG. 7.12   Nash Manifold duplex vacuum heating pump.
(Nash Engineering Co.)

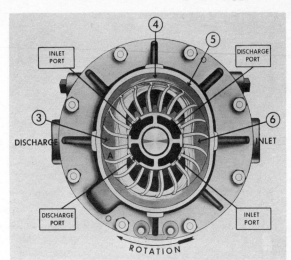

FIG. 7.13   *Nash Hytor vacuum pump.   (Nash Engineering Co.)*

This results in a solid ring of liquid revolving in the casing at the same speed as the rotor but following the elliptical shape of the casing.   This alternately forces the liquid to enter and recede from the rotor buckets, twice in each revolution.

The cycle of operation is as follows: The cycle starts at point A with the rotor bucket 3 full of liquid.   The liquid, because of the centrifugal force, follows the casing.   It withdraws from the rotor, pulling gas in through the inlet port, which is connected with the pump inlet.   At 4 the liquid has been thrown entirely from the chamber in the rotor and replaced with gas.   As rotation continues, the converging wall 5 of the casing forces the liquid back into the rotor chamber, compressing the gas trapped in the chamber and forcing it out through the discharge port, which is connected with the pump discharge.   The rotor chamber 6 is now full of liquid and ready to repeat the cycle.   The cycle takes place twice in each revolution.

7.7   ROTARY PUMPS.   The rotary pump is a positive displacement machine, similar in its characteristics to a reciprocating pump.   The rotary motion accomplishes positive displacement pumping by means of a rotating shaft or shafts, with rotors consisting of gears, vanes, screws. lobes, cams, etc., operating in a close-fitting casing.   Normal rotary-pump designs do not incorporate the use of valves or complicated water-ways, permitting the pump to operate efficiently on both low- and high-viscosity liquids with a low N.P.S.H. requirement (see Sec. 7.8 for an explanation of N.P.S.H.).

Since the rotary pump is a positive displacement machine, its theoretical displacement is a straight horizontal line when plotted against pres-

(a)

(b)

(c)

FIG. 7.14 Gear pump.
(a) Section through standard pump with packed stuffing box.
(b) Section through pump with mechanical seal and interchangeable stuffing box.
(c) Relief valve and bypass.
(Worthington Corp.)

sure with speed constant. When low-viscosity liquids are being handled, however, there is a loss in delivery due to slip at higher pressures. The actual pump capacity at any given speed and viscosity is the difference between the theoretical displacement and *slip*, slip being the leakage from the discharge back to the suction side of the pump through pump clearances. Low-viscosity liquids can short-circuit more easily as the pressure increases.

Rotary pumps have the following advantages: they are self-priming; are capable of high suction lifts; have low N.P.S.H. requirements; can handle high-viscosity liquids at high efficiency; have a wide speed range; and are available for low-capacity, high-head or high-capacity, high-head applications. There are many different designs, of which a few follow:

External-gear. This design consists of two meshing gears (Fig. 7.14) in a close-fitting housing. The gears can be spur, single-helical, or double-helical (herringbone type). In operation the liquid enters between the gear teeth and housing, on each side of the pump casing.

Internal-gear. This design consists of a single gear (Fig. 7.15) revolving between a cut gear (rotor) and an idle gear.

Vane-type. The rotor is slotted, and a series of vanes follows the bore of the casing; liquid is displaced between the vanes.

Since the rotary pump is a positive displacement machine, a relief valve (Fig. 7.14) is usually required. On small pumps, integral relief valves are sometimes provided; on larger pumps a relief valve should be located in the discharge piping and set approximately 10 per cent above the pump discharge pressure. Since rotary-pump performance depends upon the maintenance of close internal clearances and since many pumps of this type utilize internal bearings, it is important to prevent foreign matter from entering the pump. A strainer should be installed on the suction side of the pump; it should be of liberal size so as to avoid an undue friction drop.

Rotary pumps are equipped with a packed stuffing box, or the gland can be fitted with a mechanical seal as shown in Fig. 7.14b. A jacketed stuffing box for high pumping temperatures has two sets of glands, separated by a lantern gland, seal-lubricated. The inner packing gland is water-jacketed; since liquids are handled at high temperatures, cooling prolongs the life of the packing. The bellows-type shaft seal available with Buna bellows handles liquids to 240°F; for higher temperatures the seal can be made of other materials. Rotary pumps may be designed for forced-feed bearing lubrication.

The mechanical seal consists of a stationary lapped face, integral with the gland; a rotating assembly with a lapped carbon ring; synthetic rubber bellows on the shaft; and a spring to hold it in position. The mechanical seal is effective; its main limitation is that abrasives may get between the lapped sealing faces and cause leakage. On the basis of the type of liquid handled and its temperature and pressure, the bellows-type shaft seal is available in materials capable of handling the service for which it is designed.

The pump (Fig. 7.15) has two moving parts, the rotor (internal gear) and the idler gear. Each revolution of the internal gear means a definite rated output. The teeth separate at the suction port and mesh together at the discharge port, at the top. At position 1 the rotor and idler form a barrier; at position 2 the idler withdraws from the rotor and creates a suction opening to be filled with liquid; at position 3 spaces between the rotor and idler are filled; and at position 4 the rotor and idler come together, forcing liquid out through the discharge opening.

There are many applications for the rotary pump. In the power plant it is used chiefly for handling oil and lubricants. At times pumps of this type handle many forms of chemicals in solution.

7.8   CENTRIFUGAL PUMPS. A centrifugal pump is one which depends on centrifugal force and rotation of an impeller for its action. The type of pump employed depends upon the type of service for which it is intended and varies with the capacity required, the variations in suction and discharge head, the type of water handled (whether clean or dirty, hot or cold, corrosion effects, etc.), the nature of the load, and the type of drive to be employed.

End-suction vertical split-case centrifugal pumps have an almost unlimited range of applications; they are available in capacities ranging to more than 3,000 gpm and heads ranging to more than 500 ft. The

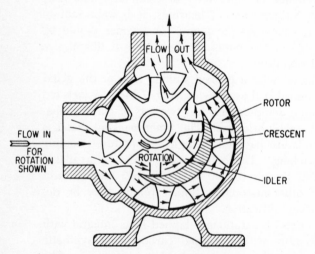

FIG. 7.15 *Cross section of internal-gear rotary pump. (Deming, Div. Crane Co.)*

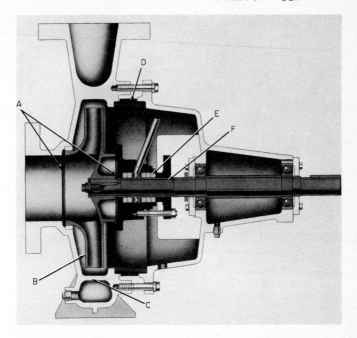

FIG. 7.16 *Sectional view of single-stage motor pump.*
(A) Casing ring. (B) Balanced, closed impeller.
(C) Dual-volute casing. (D) Self-centering stuffing
box cover. (E) Stuffing box. (F) Shaft sleeve.
(Ingersoll-Rand Co.)

single-stage double-suction centrifugal pump is used most often for general-service purposes. Sump pumps are most frequently vertically mounted. Boiler feed pumps are usually of the split-case variety, both single and multistage. Condenser pumps are of either horizontal or vertical design, depending somewhat on the suction lift and the headroom available for installation and removal; these pumps are of the low-head variety.

Centrifugal pumps may be classified as volute or turbine, depending on their construction; single- or double-suction, depending on the manner in which the water is made to enter the pump; single-, double-, or multistage, depending on the number of stages of impellers in the pump; open- or closed-impeller, etc.

A *volute* pump (Fig. 7.17) is one in which the impeller rotates in a casing of spiral design. The casing is designed to enclose the outer extremity of the impeller. The volute chamber changes velocity head to pressure head.

A *turbine* pump (Fig. 7.19) is one in which the impeller is surrounded by diffusion rings (Fig. 7.20). The diffusion ring takes the place of the spiral casing in the volute pump; it takes the water from the impeller, changing velocity head to pressure head. The quantity of water pumped depends on the size of the impeller and its speed.

A *single-suction* pump is one in which the water enters on only one side of the impeller (Figs. 7.16, 7.18). In Fig. 7.18 is shown a two-stage, hydraulically balanced opposed-impeller-design pump.

A *double-suction* pump (Fig. 7.17) is one in which the water enters on both sides of the impeller.

A *single-stage* pump is one in which but one impeller is mounted on a shaft (Figs. 7.16, 7.17).

A *double-stage* (two-stage) pump is one in which two impellers are mounted on a shaft (Fig. 7.18); this pump has opposed impellers.

A *multistage* pump (Fig. 7.19) is one in which two or more impellers are mounted on a shaft, the water passing from the discharge of one impeller to the suction of the next.

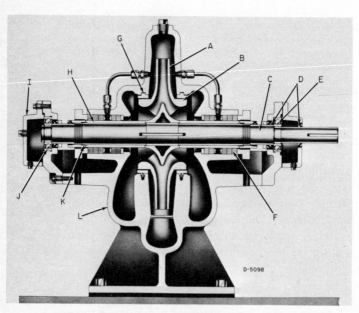

FIG. 7.17   *Cross section of double-suction volute pump.* (A) Impeller. (B) Impeller ring. (C) Shaft. (D) Seal rings. (E) Ball bearing (radial). (F) Packing stuffing box. (G) Casing ring. (H) Shaft sleeve. (I) Bearing end cover. (J) Ball bearing (thrust). (K) O ring. (L) Casing complete. (*Ingersoll-Rand Co.*)

Type HC pumps (Fig. 7.16) are cradle-mounted, flexible-coupled, and driven by standard motors. Pump and motor are mounted on a steel plate providing maximum rigidity, to prevent distortion and misalignment. The back-pullout design permits fast, easy access to stuffing box, impeller, and casing rings, without disturbing suction or discharge piping. When a spacer coupling is included, the pump can be completely inspected and serviced without moving the driver. These pumps are available in 400-ft heads and capacities to 1,200 gpm.

To protect the casing, replaceable casing rings are provided at both the front and back impeller clearances. When wear occurs they can easily be replaced. The closed impeller is dynamically balanced for vibration-free operation and hydraulically balanced to reduce stuffing-box pressure and to minimize thrust loads on the bearings. The impeller is keyed to the shaft, preventing the impeller from backing off if rotation is reversed. The dual-volute casing reduces radial hydraulic thrust on impeller and holds shaft deflection to 0.002 in. maximum.

The unit has a self-centering stuffing-box cover; the extra-deep stuffing box can accommodate either packing or mechanical seals; the shaft sleeve is replaceable; bearings are permanently lubricated. These units are designed for working pressures of 125 psi.

A cross section of a double-suction volute pump is shown in Fig. 7.17; this pump is of the horizontal split-case type, divided at the center line, with suction and discharge nozzles cast integrally with the lower half. The upper half of the casing can then be removed without disrupting the piping or the pump setting. The pump is designed to handle hot or cold liquid, capacities to 8,000 gpm, pressures to 250 psi, and temperatures to 250°F.

The advantages of this pump are: (a) less stuffing-box maintenance, sustained high efficiency, longer wear of rings and bearings; (b) prelubricated, grease-packed, or oil-lubricated and no routine lubrication required; bearings cannot be overloaded; (c) has a wide range of application; permits high suction pressure; (d) maximum packing or seal life; damage from reverse rotation, prevented; (e) a strong base plate and easy to grout with less vibration and long bearing life.

A cross-sectional view of a two-stage, hydraulically balanced centrifugal pump with opposed impellers is shown in Fig. 7.18; this arrangement reduces thrust to a minimum. The pump shaft has ball bearings; details of the oiled sleeve bearings are shown.

A cross section of a multistage centrifugal pump with forced-feed lubrication is shown in Fig. 7.19. It is designed for applications in which liquids must be handled at high pressures. This is a single-suction multistage diffuser-type unit featuring cylindrical "double-case" construction and "unit-type" rotor assembly. Such pumps are available for pressures to 6,500 psi, capacities to 12,000 gpm, and 70,000 hp.

The unit consists of two vertically split, concentric cylindrical casings, a high-strength outer casing built for full discharge pressure and a segmented inner casing formed by interlocking channel rings. The rotor (Fig. 7.20) is contained within the inner casing. The symmetrical design of both casings permits equalized expansion in all directions, thus eliminating any stress or distortion due to temperature changes. Since the space between the inner and outer casings is under discharge pressure and since this pressure acts on the discharge end of the inner casing assembly, interstage gasket sealing is assured by keeping the assembly under a compression force. The entire inner casing (and

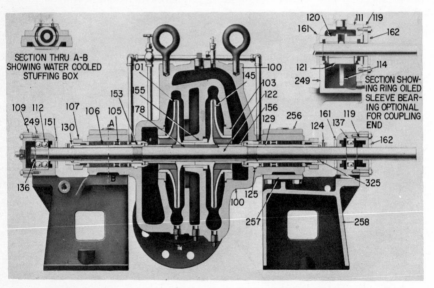

FIG. 7.18 Sectional view of two-stage, hydraulically balanced opposed-impeller pump. (100) Casing. (101) Impeller (suction). (103) Casing wearing ring (suction). (105) Seal ring. (106) Stuffed-box packing. (107) Stuffed-box gland. (109) End cover (thrust end). (111) Bearing housing cap. (112) Ball bearing (thrust end). (114) Oil ring. (119) End cover (coupling end). (120) Upper bearing shell. (121) Lower bearing shell. (122) Shaft. (124) Shaft-sleeve packing nut, right-hand. (125) Stuffing-box throat bushing. (129) Shaft sleeve (discharge). (130) Shaft-sleeve packing nut, left-hand. (136) Locknut (for ball bearing, thrust end). (137) Ball bearing (coupling end). (145) Impeller (discharge). (151) Liquid deflector (thrust end). (153) Shaft sleeve (suction). (155) Casing center bushing. (156) Spacer sleeve. (161) Liquid deflector (coupling end), inner. (162) Liquid deflector (coupling end), outer. (178) Impeller key. (249) Bearing housing. (256) Stuffing-box plate (upper). (257) Stuffing-box plate (lower). (258) Pump pedestal. (325) Shaft-sleeve-nut packing. (Goulds Pumps, Inc.)

rotor) can be pulled out without disturbing piping connections or driver.

The outer casing is made of forged steel proportioned to withstand maximum operating pressures; suction and discharge nozzles are welded in position.   Individual channel rings are interlocked with aligning rings; adjacent parts have ground joints which make metal-to-metal contact when the pump is assembled.   The outboard end of the rotor is equipped with a balancing drum (Fig. 7.21*b*) to take the axial thrust of the rotor; a thrust bearing is also provided (Fig. 7.21*a*).

Mounted at the outboard end of the pump is a gear oil pump located in the oil reservoir and driven through gears from the pump shaft.   Oil from the pump is made to pass through a filter and cooler before reaching the bearings.   An auxiliary motor-driven centrifugal pump is mounted on top of the oil reservoir for start-up purposes or for use if oil pressure falls below a safe level.

Impellers of the enclosed type are usually made of chrome steel.   Alloy steels are also used in impeller design; the materials employed are based on the water or fluid handled and conditions imposed by the liquids they are to pump.

The impellers have curved radial passages or vanes which connect with the hub.   Each impeller mounted on the shaft is provided with an impeller ring and casing ring, which reduce to a minimum the leakage from the discharge to the suction side.   The casing ring is held stationary in the casing, while the impeller ring is attached to the impeller and rotates with it.

Around the impeller (Fig. 7.20*b*) is placed a diffusion ring which is held stationary in the casing.   The ring contains slots or openings which receive the water from the impellers at a high velocity; by means of the gradually increasing area of the opening, the velocity head is converted

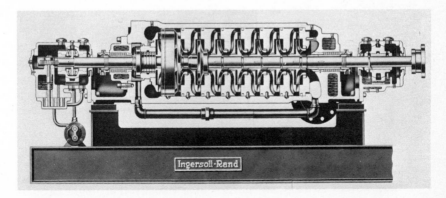

FIG. 7.19   *Cross section of multistage centrifugal pump equipped with forced-feed lubrication.   (Ingersoll-Rand Co.)*

(a)

(b)

(c)   (d)   (e) (f) (g)

FIG. 7.20 *Centrifugal-pump parts.* (a) *Inner assembly.* (b) *Channel ring and impeller.* (c) *Channel ring.* (d) *Impeller.* (e) *Key.* (f) *Retaining ring.* (g) *Split ring.* (Ingersoll-Rand Co.)

to pressure. This procedure then makes it possible for the water to advance from one impeller to the next, with little loss of energy.

In operation, the water enters at the right and proceeds to the first impeller, which is revolving at high speed. Immediately upon entering the impeller, it comes under the influence of the centrifugal force resulting from the rotation and is moved to the outer edge of the impeller at a gradually increasing speed, finally leaving at a high velocity. It then immediately enters the diffusion ring.

It enters the ring at a high velocity, which is to be changed into useful pressure and also reduced so that the turn of 180° at the top of the ring can be accomplished without serious friction loss and without the water's striking the surface at a velocity that might be destructive even if pure water were handled. The diffusion vanes accomplish both these results.

The water is now flowing at a reduced velocity (and at the pressure gained in the first stage) through the ring into the second stage. This process is repeated until the last stage has been reached, each impeller adding its pressure. In the final stage the diffusion vane is omitted and a volute chamber substituted. The vane is omitted here because it is not necessary for the water to make a turn, since it passes directly to the discharge line.

Centrifugal pumps have vanes or impellers, some of which are radial or inclined forward in the direction of rotation. With this type of vane, the head will increase as the delivery or output increases. If, however, the vanes are moved backwards, the head will remain relatively constant or fall off (Fig. 7.27) with the delivery. Naturally, best results are secured by basing the design on plant requirements.

Packing glands or stuffing boxes should be deep to facilitate packing and to ensure a minimum of leakage. The amount and kind of packing depend on the design of the pump and on the operating conditions.

Soft metallic packing is usually preferred. This is due to the fact that the centrifugal pump rotates at high speed and hard packing would score the shaft. Moreover, with hard packing it is impossible to make as tight a joint, with a minimum of pressure exerted, as with soft packing. Graphited packing made of soft asbestos is also used. Packing should be pulled up just enough to prevent air leakage and no attempt made to prevent some leakage of water past it. For the high-pressure pump (Fig. 7.19), the type of stuffing-box arrangement employed will depend on the service for which the pump is designed. Stuffing boxes are

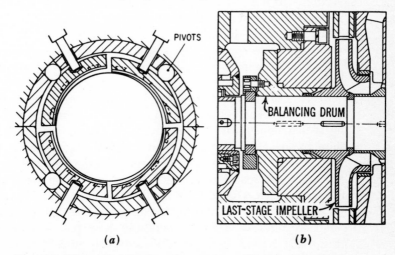

(a)   (b)

FIG. 7.21   (a) *Tilting-shoe bearing arrangement.*   (b) *Balancing drum.* (*Ingersoll-Rand Co.*)

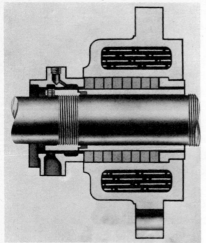

FIG. 7.22 *Water-cooled packed boxes.* (*Ingersoll-Rand Co.*)

usually of two types: (1) packed-water-cooled; and (2) floating-ring, arranged for cold-condensate injection. The latter is throttled between shaft sleeve and a self-adjusting floating ring. Mechanical seals frequently replace the conventional packing.

For water-cooled packed boxes (Fig. 7.22), the inbound and outbound boxes are water-jacketed to provide efficient cooling. They are designed to accommodate the packing arrangement that best meets the suction conditions under which the pump operates. For the normal range of boiler-feed application, the box is packed solid with eight rings of packing. At high suction pressures, a bleed-off bushing may be inserted to reduce the pressure on the packing.

The discharge stuffing box is contained in a removable extension, bolted to the casing with a gasket fit. This stuffing-box extension serves to hold the balancing-drum sleeve in place and permits its easy removal without opening the casing.

Mechanical seals are also employed; they provide a practical solution to the most difficult stuffing-box problems, for effective mechanical sealing is maintained without packing, and without wearing parts that require frequent attention and maintenance. Pump and seal design are integrated so that the seal can be serviced in the field without dismantling the pump. A spacer coupling with a taper fit on the shaft is recommended for all mechanical seal applications.

There are many varieties of mechanical seals; their design is based on pump pressures, capacity, temperatures, and rpm. The mechanical seal (Fig. 7.23) is employed for heavy-duty centrifugal process pumps with operating pressures to 720 psi, capacities to 7,500 gpm, temperatures to

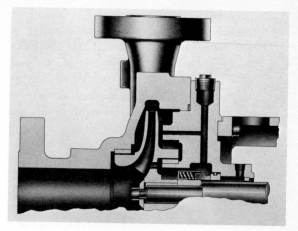

800°F, and 3,600 rpm. The type AS pumps have built-in seal alignment and built-in recirculation; no gland is required. Properly adjusted, the mechanical shaft-seal requires no further attention.

Thrust is eliminated or reduced in a number of ways. The pump shown in Fig. 7.17 has a double-suction inlet; while the water flow may not always be exactly equal on both sides, the inlet does go a long way toward neutralizing the unbalance which occurs with a single inlet. In the same pump is built a balanced port so that variations in pressure on either side of the impeller will be equalized, thus neutralizing the thrust which might be developed.

For the pump in Fig. 7.19, radial balance is obtained through multi-volute design, while axial balance is achieved by a hydraulic balancing drum (Fig. 7.21*b*). The axial thrust of the rotor toward the suction end, which is developed by the sum of the unbalanced pressure differentials across each impeller, is counteracted by the balancing drum located next to the last impeller at the outboard end of the rotor. Here the chamber to the left of the drum is connected to suction pressure. Discharge pressure from the last stage bleeds along the drum to act upon the face of the drum, and the difference in pressure on the two sides positions the rotor. Radial balance is secured by having the discharge from each impeller directed to the succeeding stage through a series of volutes (Fig. 7.20*c*) equally spaced around the entire circumference of the impeller; such an arrangement provides for radial thrust.

In addition to the balancing drum, stabilized bearings are usually furnished to prevent oil whip and possible vibration in high-speed pump operation (5,000 to 10,000 rpm). A Cameron-type Kingsbury thrust bearing (Fig. 7.21*a*) is incorporated with the outboard bearing to maintain longitudinal alignment of the rotor and to take up any thrust which may be set up by abnormal operating conditions. The thrust bearing contains a revolving collar attached to the shaft. The collar transmits any thrust that is developed to the stationary thrust shoes, which are made so that they lift slightly at one end (facing the direction of rotation), admitting a wedge-shaped film of oil under pressure between the surfaces. Four individual shoes centrally pivoted produce equally positioned oil wedges and stabilize any shaft movement.

To eliminate balancing devices, hydraulic thrust is frequently balanced by opposing single-suction impellers in equal groups, so that the thrust of one neutralizes that of another. The success of this method depends on the impeller arrangement and on the extent to which the various factors previously mentioned are satisfied. Although a great number of combinations and arrangements of stages are possible, only a few are worthy of consideration; Fig. 7.24 shows some of the common methods employed.

With opposed impellers, the most important factors are stuffing-box pressure, interstage leakage, and casing design. To satisfy these conditions, the first and second stages are placed next to the stuffing boxes in order to have the lowest pressure on them. To keep interstage leakage to a minimum and to make possible the simplest casing arrangements, the stages must be in series. Hydraulic balance requires that one-half of the stages face opposite to the other half. Since no stage arrangement satisfies all these conditions simultaneously, each arrangement must be a compromise that favors some particular point.

Staging has been resorted to because pumps with one impeller are limited as to the head against which they are capable of pumping. Volute pumps of a single stage are usually limited to about 300 psi.

A good system of lubrication is essential for the centrifugal pump. The reason is readily apparent when it is considered that these pumps operate at high speeds and very frequently handle water that is at an elevated temperature. Not only should the pump have an excellent lubricating system, but the lubricant should be of high quality. Bearings are both oil- and grease-lubricated; generally the grease-lubricated bearing is less expensive. If the grease-lubricated bearing has a disadvantage, it lies in the fact that it can be overlubricated, resulting in overheating. Since most pump manufacturers designate their preference on the basis of experience, indicating whether oil or grease is to be used and the quality of the lubricant, their recommendations should be rigidly adhered to.

A factor affecting the operation of a centrifugal or a rotary pump is the suction conditions. Suction performance (the relation between capacity and suction conditions) involves a quantity called N.P.S.H.

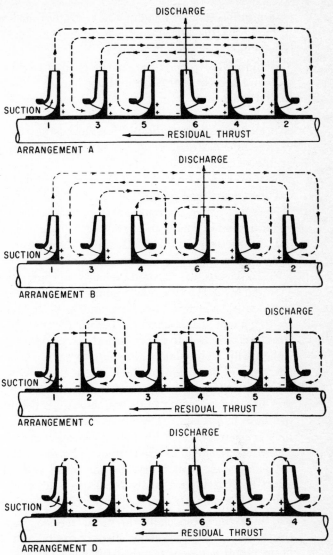

FIG. 7.24 *Four opposed single-suction impeller arrange-*
ments to obtain balance. With this, arrangements A
and B give the lowest pressure on the stuffing box.
(Ingersoll-Rand Co.)

(net positive suction head).[1]   Net positive suction head is the amount of energy in the liquid at the pump datum.   It is referred to as either "available" or "required" N.P.S.H.

Required N.P.S.H. is the energy needed to fill a pump on the suction side and overcome friction and flow losses from the suction connection to that point in the pump at which more energy is added.   The N.P.S.H. is a characteristic of the pump and varies with pump design, pump size, and operating conditions; it is supplied by the manufacturer.

Available N.P.S.H. is a characteristic of the system and is defined as the energy which is in a liquid at the suction connection of the pump, over and above the energy in the liquid due to its vapor pressure.

For a centrifugal pump, the required N.P.S.H. is the required amount of energy (in feet of liquid).   In a rotary pump, the required N.P.S.H. is the amount of energy in pounds per square inch required

1. To overcome friction losses from the suction opening to the impeller vanes, gears, etc.

2. To create the desired velocity of flow into the vanes or gears

So the N.P.S.H. varies with pump design, information based on conditions under which a specific pump is to be operated being provided by the pump manufacturer.   Before purchasing or installing a pump, therefore, it is well to know both the available and the required N.P.S.H. Abnormally high suction lifts cause a serious reduction in capacity and efficiency of the pump, which often leads to trouble in operating.

Should the static pressure at the impeller vanes fall below the vapor pressure corresponding to its temperature, a definite portion of the water will flash into steam.   These vapor bubbles, or "cavities," collapse when they reach regions of high pressure on their way through the pump. The obvious effect of cavitation is noise and vibration.   The bigger the pump, the greater may be the noise and vibration; vibration can cause bearing failure, shaft breakage, and other fatigue failures in a pump.   If a pump is operated under cavitating conditions for a sufficient length of time, especially on water service, impeller-vane pitting will take place.   The other major effect of cavitation is a drop-off in efficiency of the pump, apparent as a drop-off in capacity.

Piping to and from the pump is also important enough to warrant consideration.   The suction line to a centrifugal pump should be as short and straight as possible.   If at all possible, avoid long suction lines.   If a long suction line is unavoidable for one reason or another, make it several diameters larger than the inlet to the pump to reduce friction loss to a minimum.   Directly after the pump, place a check

[1] "Rotary and Centrifugal Pump Theory & Design." (Courtesy Worthington Corp.)

valve and a gate valve in the order mentioned; ahead of the pump place a gate valve. The check valve relieves the pressure from the pump when it is starting up and prevents waterhammer from injuring the pump. When the pump stops, close the gate valve to relieve the strain on the check valve.

Hot water requires an increase in suction head; the head varies with the temperature of water, sufficient head being required to compress the vapors and to prevent the pump from becoming steam-bound. For a given condition, increasing the water temperature decreases the head and capacity.

The output of centrifugal pumps can be controlled by throttling the discharge valve or by changing the speed of the pump (speed regulation). If the load is reduced to such an extent that the flow is almost completely stopped (throttled), the pump may become overheated. Where it becomes necessary to throttle the flow to such an extent, it is advisable to install a bypass orifice. Constant bypass orifice and piping size is provided by the pump manufacturer, who knows the characteristics of the pump.

The bypass line is installed on the discharge side of the pump and ahead of the check valve. The bypass line contains the orifice, with a gate valve ahead and behind the orifice, to permit removal for cleaning, if necessary. There should be a minimum of 3 ft straight run of pipe on either side of the orifice to reduce turbulence and noise. The recirculating line should enter the feedwater heater at a point below the minimum water level and as far away from the pump suction as possible.

If performance curves are provided (Fig. 7.27), they will show the relation of static head to capacity, power consumption, and the efficiency of the pump. Depending upon design, these characteristics will vary for each pump. The discharge velocity of water leaving the pump varies from 5 to 15 fps in ordinary practice.

Difficulties are frequently encountered in the operation of pumps. If no water is being delivered, the following conditions may exist: speed too low, discharge head too high, pump in need of priming, suction lift too high, and pump motor running in the wrong direction. If not enough water is being delivered, the following conditions may be the cause: air leaks in the suction line or stuffing box, speed too low, motor running the pump in the wrong direction, obstruction in the suction line or foot valve too small, pump in need of repair in that wearing rings are worn and impeller damaged, water too hot and insufficient head, and suction lift too high. If the pump is not developing enough pressure, the speed may be too low, the pump may be in need of repair, or air leaks may be present.

Comparison of Centrifugal Pumps with Other Pumps

Advantages

Absence of parts such as valves and packing
Simplicity
Application of motor drive
Application of synchronous motor for constant speed and improvement of power factor
Ability to maintain a uniform flow
Small floor-space requirement
Low initial cost
Ease of regulation
Economy of operation

Disadvantages

Lower efficiency than that of piston pumps
Frequent difficulty with the end thrust
Not suitable for high head requirements at low flow rates
Care necessary to align high-speed pumps properly
Possibility of overloading motor owing to certain load characteristics
Difficulty of regulating with wide fluctuations in load
Difficulty of operating at very low speed

Another major disadvantage that centrifugal pumps have is that of operating pumps in parallel. A series of pumps to be operated in parallel are selected with design characteristics which are similar. However, after they are in operation for a while, wear of the moving parts changes the design pattern so that the pumps no longer perform as they were designed to perform (for example, see Fig. 7.27). If the head at a given capacity will not match that of the next pump in line, the first pump will back the second pump off the line until it virtually shuts the second pump down. At any rate, the pump with the highest head will tend to hog the load. Pumps should, therefore, have the same, or very nearly the same, performance characteristics to avoid loss of load and difficulties in operation.

7.9 PUMP INSTALLATION AND OPERATION. Considerable attention must be paid to proper pump installation (Figs. 7.25, 7.26) to ensure trouble-free operation. The following factors must be taken into consideration: the pumps should be accessible for inspection and maintenance; they should be removed from areas that are subjected to flooding, water leakage, and corrosion; they should be located as close as possible to the source of supply; and headroom should be provided for lifting the rotor and casing. If the pumps must be located where the atmosphere is damp, it is possible to have the motors designed for this purpose.

The piping should run to a pump as directly as possible, avoiding, however, sharp bends; entrance to the pump should be provided with long-radius ells or bends to reduce inlet friction to a minimum. Piping must be supported to take the strain from the pump and provide for expansion and contraction to avoid pump misalignment. Expansion joints or loops should be used when hot liquids are being handled.

Short suction lines are recommended but are not always possible of installation; so to avoid problems experienced with long intake pipes and high suction lifts, provide an individual suction line for each pump. If this is impractical, make the suction-header size approximately 50 per cent larger than the pump connection and reduce it at the pump flange. The increase in suction size will reduce the pressure drop and permit the pump to operate to better advantage. Avoid tees or right-angle fittings and use long-radius ells or bends instead. Slope the suction header upward from the reservoir to the pump; avoid high spots in the header where air might collect. The suction end should be flared; the suction inlet should be well below the low-water level to eliminate the possibility of air entrainment. The suction should not be located close to the point of makeup or returning water, since eddy currents may interfere with pump operation. The velocity at the suction entrance should be less than 3 ft per sec. When supplying two or more pumps from a long intake header, employ a tapering header with Y branches to the pump suction. For pumps operating under high suction lift, stuffing boxes should be sealed, using water taken from the discharge header or from a separate source of supply.

Limit the suction lift to 10 to 15 ft, since only clean cold water can be raised this distance without experiencing operating difficulties. For hot water, it is necessary to bring the water to the pump under a positive head; the velocity should be low, and a suction head of 10 to 20 ft is desirable.

When starting up new boiler-feed- or high-temperature-water circulating pumps, a strainer is frequently installed to prevent foreign material from getting into the pumps (and lines) or clogging the suction intake; later the strainers can be removed when the system is clean. All strainers should be inspected periodically.

For a pump installation such as shown in Fig. 7.25, a foot valve should be installed in the suction connection below the normal water level; a check valve should be installed near the pump discharge, followed by a gate valve. This valve arrangement prevents the pump from running backward when shut down and facilitates start-up. For large pumps and dirty water, the installation of twin strainers is recommended; such an installation avoids shutdowns, loss of suction, and delays such as are experienced when a single strainer is used.

The installation of a small pump (Fig. 7.25) shows details of piping

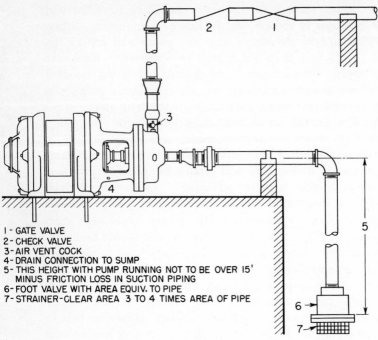

1 - GATE VALVE
2 - CHECK VALVE
3 - AIR VENT COCK
4 - DRAIN CONNECTION TO SUMP
5 - THIS HEIGHT WITH PUMP RUNNING NOT TO BE OVER 15'
    MINUS FRICTION LOSS IN SUCTION PIPING
6 - FOOT VALVE WITH AREA EQUIV. TO PIPE
7 - STRAINER - CLEAR AREA 3 TO 4 TIMES AREA OF PIPE

FIG. 7.25   *Installation of a small pump.*   (*Allis Chalmers Mfg. Co.*)

and fittings.   The pump strainer should not rest on the floor.   Turning the pipe down prevents objects from falling into the suction line.

A pump which has its suction line below the pump center line requires priming; that is, the air must be evacuated from the pump suction before water will enter the pump and fill the piping and casing.   Priming may be accomplished in a number of ways: (1) if the foot valve is tight, by filling the suction piping with water from some other source; (2) by using an ejector; and (3) by priming with a vacuum pump.   The ejector may be air-, steam-, or hydraulically operated; the injector is installed on top of the pump casing.   When water is ejected in a steady stream, it may be assumed that the pump is full of water and the pump may be started, after which the discharge valve may be opened slowly. If a vacuum pump is employed, it should preferably be a wet vacuum pump since water may damage a dry vacuum pump.

To ensure successful operation of a pump, it should be set on a foundation which is substantial and rigid enough to permit the pump to absorb

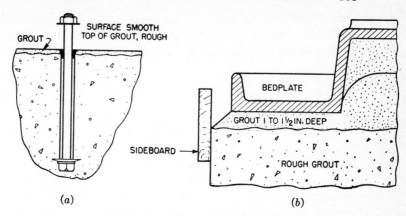

FIG. 7.26 (a) Method of installing foundation bolts. (b) Method of grouting in bedplates.

vibration. Good foundations are usually made of concrete. Before the concrete is poured, the foundation bolts are secured as in Fig. 7.26 and surrounded by a pipe sleeve several sizes larger than the bolt. This permits the bolt to be shifted or moved to meet bedplate drillings while the bolt is held securely in place by the washer, which is anchored down by concrete. The pipe surrounding the bolt must not be permitted to extend above the level of the concrete. The top surface of the foundation should be approximately 1 in. below the level at which the bedplate is to be set in order to allow for grouting in.

To prevent the bed plate from springing out of line, the pump must be carefully aligned. Pumps which come already mounted on a bedplate should be leveled up before they are placed in operation. Pumps which are to be set on a rough foundation should be set approximately in location and leveled with shims before the bedplate is grouted in position. Permit grout to be $\frac{3}{4}$ to $1\frac{1}{4}$ in. thick; the shims should preferably be tapered. Next check suction and discharge flanges to determine if they are level.

The driver and driven shaft should next be aligned by placing a straightedge across the top and side of the coupling and checking between coupling halves, using a feeler or thickness gauge to determine if the flanges are parallel. If the faces of the flanges are parallel, a feeler gauge will show the same measurement all the way around the circumference of the coupling. Any misalignment should be less than 0.005 in. Remember that each time a wedge is moved, the flanges should be checked again.

Build a wood form around the outside of the bedplate to hold the grout. Use 1 part cement and 2 parts sand and sufficient water to permit material to flow in and around the bedplate. This mixture should be permitted to set approximately 2 to 3 days, after which the holding-down bolts can be tightened. Be sure to recheck coupling for alignment.

After suction and discharge flanges have been bolted up, the alignment should again be checked. If connecting of piping has caused misalignment, repeat the procedure previously followed. Check pump and motor rotation and insert bolts in flanges and connect up. A short time after the pump has been in operation, the alignment should again be checked. After it is certain all is satisfactory, the unit can be doweled. The doweling is done with tapered pins usually provided with the pump. Dowel pins are located in the feet of the pump and driver as well as in the bedplate. Permit dowel pins to extend above the feet.

Before placing the pump in operation, make certain that bearings are lubricated. Rotate the motor by hand to make certain everything is free. Prime the pump by one of the methods previously explained. Fill the pump full of liquid, and with suction valve open and discharge valve closed start the pump, noting suction and discharge pressure. (All pumps should be fitted with pressure gauges on both suction and discharge.) After running the pump this way for a minute or two with the air vent open, close vent valves and open the discharge valve slowly. Observe gauges closely, and if anything unusual occurs or pressures seem unusual, stop the pump and check thoroughly.

Start and stop the pump several times and observe performance. If all looks satisfactory, continue in operation for $\frac{1}{2}$ to 1 hr, meanwhile observing bearing temperatures and watching gauges, lubrication, and general actions of pump and motor for overheating. Then shut pump down, recheck alignment, tighten all bolts, and give the entire installation a careful check.

Place the pump again in operation; observe packing gland, gland seals, lubrication, and overheating. Never run a pump for an extended period of time with the discharge valve closed unless the vents are open or the pump is equipped with a bypass to permit some water to circulate through the pump at all times.

7.10 PUMP TESTING AND CALCULATIONS. The pressure against which a reciprocating pump will operate depends on the dimensions of the pump and the steam pressure; that against which a centrifugal pump will operate, on design, speed, and the number of stages employed. The maximum head against which the pump will operate can be determined from the pressure.

Centrifugal pumps generate velocity in the liquid to move it from place to place, from one level to another, or to raise the pressure from the suction to the discharge of the pump. This difference in level (static head) and the difference in pressure (pressure head) must be taken into consideration in calculating losses. Then there are losses due to friction and velocity, which must also be converted to "head" in feet.

A column of water 1 ft high exerts a pressure at the base of this column of 0.433 psi. Assume that a pump operates against a head of water 275 ft high. The pressure in pounds per square inch at the base can then be determined in the following manner: as a column of water 1 ft high exerts a pressure of 0.433 psi, a column 275 ft high will exert $275 \times 0.433 = 119$ psi. If the pressure is known, it is possible to determine the head in feet or the height to which the water can be pumped. Suppose that the gauge at the pump reads 119 psi; the head to which this water can be pumped can be determined thus:

$$119 \div 0.433 = 275 \text{ ft}$$

The head to which a pump can raise water includes both suction and discharge. Water usually comes to the pump under suction. The distance that the water can be raised (vertical lift) is limited. The theoretical lift is approximately 34 ft. This height, however, is not possible in practice, as leaks in the suction line, packing leaks, etc., reduce the actual lift to about 24 ft. In fact, 10 to 15 ft is more usual because of friction in the line.

At atmospheric pressure of 14.7 psi (approximately equivalent to a 30-in. barometer), the theoretical lift is approximately

$$14.7 \div 0.433 = 33.9 \text{ ft}$$

As the temperature of the water is increased, the possible suction lift is decreased. At atmospheric pressure, the boiling point of water is 212°F. If the suction line is under a vacuum, the boiling point is reduced. As a result, vapor from the water would fill the space and partially destroy the vacuum. Water above 130°F should come to the pump under head. This ensures a full cylinder of water and prevents the pump from becoming steam-bound.

Water at its greatest density weighs approximately 62.5 lb per cu ft. As there are 1,728 cu in. in 1 cu ft, 1 cu in. weighs $62.5 \div 1,728 = 0.0361$ lb per cu in. Moreover, 231 cu in. is equivalent to 1 gal of water. Since 1 cu in. weighs 0.0361 lb, 231 cu in. will weigh $231 \times 0.0361 = 8.339$, the number of pounds in 1 gal. If 1 cu ft of water weighs 62.5 lb and 1 gal weighs 8.339 lb, it is apparent that there are $62.5 \div 8.339 = 7.5$ gal in 1 cu ft (approximately).

A pump delivering water requires a certain amount of power. The amount depends upon the rate at which the water is pumped and the height to which it is lifted. The number of pounds of water pumped per minute times the number of feet that the water is lifted equals the number of foot-pounds of work expended per minute.

Pumps with a given capacity may vary in characteristics to a considerable extent. A pump is said to have "steep" characteristics when the "shutoff" head is considerably above the "operating" head; a pump is said to have "flat" characteristics when the shutoff head is only slightly above the operating head. Figure 7.27 shows pump-performance curves for a double-suction single-stage pump operating at constant speed. The shutoff pressure is that shown by checking the head, or pressure, at zero flow or capacity. In Fig. 7.27*a*, this would be approximately 115 lb pressure.

It is obvious from the performance curves that as the pump capacity increases, the head decreases (drop in pressure); that an increase in pump capacity requires more horsepower input. The efficiency of the pump increases until it reaches some maximum point, after which it drops off.

A centrifugal pump will develop a given total head regardless of the weight (specific gravity) of the liquid, but by reason of the specific gravity (weight) of the liquid the discharge pressure will vary. For example, when hot water is being pumped, the head in feet will be the same as when cold water is being pumped, but the pressure will be less.

A centrifugal pump operates most efficiently under a head and a speed that approximate design conditions. For a given pump we find that (1) the power varies as the cube of the speed, (2) the head varies as the square of the speed, and (3) the quantity pumped varies directly with the speed.

The horsepower required to pump water is determined as follows:

$$\text{hp} = \frac{\text{gpm} \times 8.33 \times \text{head, ft}}{33,000 \times \text{efficiency of pump}} = \frac{\text{gpm} \times \text{head, ft}}{3,960 \times \text{efficiency of pump}}$$

$$\text{Theoretical hp} = \frac{\text{gpm} \times \text{total dynamic head, ft} \times \text{sp gr of liquid}}{3,960}$$

$$\text{Required hp} = \frac{\text{gpm} \times \text{total dynamic head, ft} \times \text{sp gr of liquid}}{3,960 \times \text{efficiency of pump}}$$

$$\text{kw input} = \frac{\text{gpm} \times \text{total dynamic head, ft} \times \text{sp gr of liquid}}{5,308 \times \text{pump efficiency} \times \text{motor efficiency}}$$

NOTE: 1 kw = 1.341 hp $1.341 \times 3,960 = 5,308$

Example 1 If we assume a pump handling 1,000 lb of water per min against a head of 200 ft, what is the theoretical horsepower required?

Solution 1 gal of water weighs approximately 8.339 lb; the pump handles 1,000 lb per min.

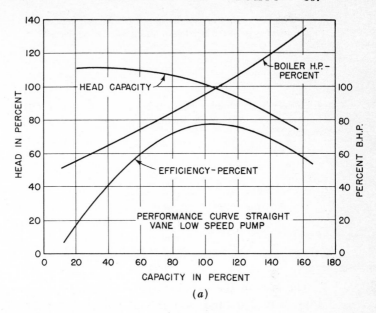

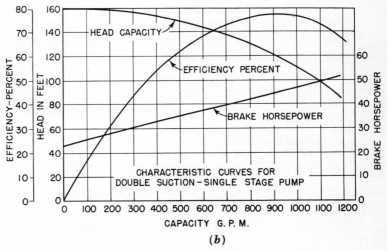

FIG. 7.27   *Performance curves for centrifugal pumps.*

$$\frac{1,000}{8.339} = 120 \text{ gpm}$$

$$\frac{120 \times 200 \times 1.0}{3,960} = 6.06 \text{ hp}$$

NOTE: To this must be added the horsepower required to overcome friction and losses.

Example 2   A pump is discharging 50 gpm against a head of 300 ft. What horsepower is required if friction and losses are neglected?

Solution

$$\frac{50 \times 300 \times 1}{3,960} = 3.79 \text{ hp}$$

Example 3   A simplex pump 6 by 4 by 5 is making 100 strokes per min against a pressure of 350 psi if friction and losses are neglected. What horsepower is required?

Solution

$$\text{Volume of cylinder} = 4 \times 4 \times 0.7854 \times 5 = 62.83 \text{ cu in.}$$
$$62.83 \times 100 = 6,283 \text{ cu in./min}$$
$$1 \text{ cu ft} = 1,728 \text{ cu in.} = 62.5 \text{ lb}$$
$$\frac{6,283}{1,728} \times 62.5 = 227.2 \text{ lb/min}$$
$$350 \div 0.433 = 808.3 \text{ ft}$$
$$\frac{227.2 \times 808.3}{33,000} = 5.56 \text{ hp}$$

The area of the steam piston of a boiler feed pump is usually made from 2 to 2½ times the size of the water piston. The steam piston is made larger than the water piston in order that the total pressure on the former may be sufficient to force the water into the boiler. If both pistons were equal in size, the pump would stop. The areas of the two pistons are unequal, but unit pressures are such that the total pressures are the *same*.

Example 1   Let us assume a simplex pump 3 by 2 by 3 supplied with steam at 100 psi. What pressure in pounds per square inch could be developed on the water side?

Solution   Consider the fact that the total pressure on the face of both pistons must be equal. The area of the steam piston in square inches times the pressure in pounds per square inch equals the total pressure on the piston:

$$3 \times 3 \times 0.7854 \times 100 = 706.86 \text{ lb total pressure}$$

The area of the water piston is

$$2 \times 2 \times 0.7854 = 3.1416 \text{ sq in.}$$

Therefore $\dfrac{706.86}{3.1416} = 225.00$ psi, pressure on water side

Although the area of the water piston is less than that of the steam piston, the total pressure distributed over this smaller area gives a higher pressure in pounds per square inch than that on the steam piston.

The steam pressure can be determined in the same way as the water pressure if the pump dimensions and the head against which the pump has to work are known.

Example 2 Assume a simplex pump 3 by 2 by 3 pump 250 ft. What steam pressure would be required?

Solution A column of water 250 ft high is equivalent to psi.

$$2 \times 2 \times 0.7854 \times 108.25 = 340.1 \text{ lb total pressure on}$$

As the area of the steam piston is $3 \times 3 \times 0.7854 = 7.0686$ s 340.1 lb pressure must be exerted on 7.0686 sq in., or

$$\frac{340.1}{7.0686} = 48.2 \text{ psi, steam pressure required}$$

A boiler feed pump should be large enough to handle approximately 2 to 2.5 times the actual boiler capacity; this will ensure adequate supply upon demand. For the older units, for which horsepower is determined on the basis of square feet of heating surface, we define a boiler horsepower as the evaporation of 34.5 lb of water per hr from and at 212°F. For this boiler the pump should be capable of handling 2.5×34.5, or 86 lb of water per hp calculated.

For the reciprocating pump, the capacity can be determined from the pump "displacement," which is the volume swept through by the piston or plunger for a specific number of strokes per unit of time. In practice the displacement of the pump must be corrected for the amount of slip.

Example 1 Suppose a simplex pump 6 by 4 by 8 makes 100 strokes per min. If slip is neglected, what is the capacity of the pump per minute in gallons, pounds, and cubic feet?

Solution

$$\text{gal/min} = \frac{4 \times 4 \times 0.7854 \times 8 \times 100}{231} = 43.5$$

$$\text{lb/min} = \frac{4 \times 4 \times 0.7854 \times 8 \times 100 \times 8.33}{231} = 362.5$$

$$\text{cu ft} = \frac{4 \times 4 \times 0.7854 \times 8 \times 100}{1,728} = 5.82$$

In the preceding examples the efficiency of the pump, the slippage, and the area of the piston rod were neglected. The area of the rod reduces the displacement on the crosshead end; slippage reduces the quantity pumped and the efficiency and increases the horsepower required for a given output.

Example 2 Consider a duplex pump 8 by 6 by 10, with a 2-in. piston rod, making 50 strokes (1 stroke of a duplex pump is both pistons moving once over their path) per min against a head of 150 ft. The pump is 75 per cent efficient and has a slip of 6 per cent.

Determine the capacity of this pump in gallons, pounds, and cubic feet per minute; also determine the horsepower developed.

$$6 \times 6 \times 0.7854 = \quad 28.27 \text{ sq in., area of head end}$$
$$6 \times 6 \times 0.7864 - (2 \times 2 \times 0.7854) = \quad 25.13 \text{ sq in., area of crosshead end}$$
$$28.27 \times 10 \times 25 = \quad 7,067.50 \text{ cu in., displacement of head end/min}$$
$$25.13 \times 10 \times 25 = \quad 6,282.50 \text{ cu in., displacement of crosshead end/min}$$
$$\overline{13,350.00} \text{ displacement of one side of pump, cu in./min}$$
$$13,350 \times 2 \text{ (duplex pump)} = 26,700.00 \text{ displacement of pump, cu in/min}$$
$$26,700 \times 0.06 \text{ (6 per cent slip)} = \quad 1,602 \text{ cu in., volume of slip/min}$$
$$\overline{25,098.00} \text{ actual displacement of pump, cu in./min}$$
$$25,098 \div 231 \text{ (cu in./gal)} = \quad 108.65 \text{ gal/min}$$
$$108.65 \times 8.339 \text{ (lb/gal)} = \quad 906.03 \text{ lb/min}$$
$$25,098 \div 1,728 \text{ (cu in. in 1 cu ft)} = \quad 14.52 \text{ cu ft/min}$$
$$\frac{906.03 \times 150}{33,000} = 4.118 \text{ hp}$$
$$4.118 \div 0.75 \text{ (efficiency)} = 5.49 \text{ hp}$$

The total head developed by the pump (Sec. 7.1) is composed of (1) static head, (2) pressure at the point of discharge, and (3) pipe friction.

Example 3 Find the total head required for a pump to operate under the following conditions: capacity, 40 gpm; discharge pressure, 15 psi; point of discharge, 30 ft above pump; suction lift, 5 ft; pipe and fittings, 2-in. size; pipe, 150 ft long; 6 elbows in line.

NOTE: Elbows offer resistance equivalent to running length of pipe; friction loss is given from tables not included in text.

Solution

$$30 \text{ ft (discharge head)} + 5 \text{ ft (suction lift)} = 35 \quad \text{ft}$$
$$15 \text{ psi (discharge pressure)} \times 2.31 = 34.6 \text{ ft}$$
$$6 \text{ 2-in. elbows} = 48 \text{ ft pipe (given)}$$
$$150 \text{ ft pipe plus } 48 \text{ ft elbows} = 198 \text{ ft pipe}$$
$$\text{Friction loss per 100 ft 2-in. pipe (given)} = 6.6$$
$$\text{Friction loss, pipe equivalent} = \frac{198 \times 6.6}{100} = 13.1 \text{ ft}$$
$$\text{Total head required} = \overline{82.7 \text{ ft}}$$

NOTE: Velocity head here is considered minor and is not included. If fluid comes to the pump from above or is delivered to the pump under pressure, this suction head or pressure is deducted, whereas in the example it is added.

7.11 PUMP MAINTENANCE. When a *reciprocating* pump is packed with canvas packing, the rings should be cut about ⅛ to ¼ in. short of meeting. The packing should be cut diagonally, and each joint lapped. The number of rings varies from one to four, and the width depends on the design of the pump. Before packing a pump with rings of canvas, the packing may be expanded to its working size by soaking it in water for several hours previous to packing. If this is not done, short stroking frequently results owing to swelling which occurs shortly after the pistons have been packed. Moreover, the packing should not be jammed or crowded into position. Allow enough room for expansion when the pump packing becomes wet.

Scored piston rods are a sign of improperly installed packing, of too tight packing, or of incorrect packing. Leaks around the packing glands are always difficult to detect in the early stages, as is valve leakage. Leakage past the piston is usually detected in the operation of the pump. A careful operator can detect any unusual conditions which arise from time to time. If it is required to run the pump considerably faster than usual to maintain the same service conditions, the packing or valves are evidently leaking. The first procedure, then, is to inspect the valves. After these have been examined, the actual capacity of the pump can be determined and compared with the displacement volume. In these calculations about 5 to 10 per cent should be allowed for slip.

Leakage in the water-suction valves can be determined by the appearance of the valves upon examination, by the erratic action of the pump, and by the failure of the pump to deliver the required amount of water or the desired water pressure (by leakage of water back into the suction chamber). At times this type of leakage may also be detected by the fact that the pump seems to speed up at points in the stroke rather than maintain a uniform speed. Leakage of the water-discharge valves also can be detected by their appearance upon examination, or the valves may be tested by blocking the piston in mid-stroke and closing the main suction valve on the pump. This will trap water between the head and the underside of the discharge valve. Now open the drain cock on the bottom of the pump on the side being tested. If after the water present in the cylinder has been drained out, water continues to pass through the cock, it is reasonably sure that the discharge valve or valves on that side of the pump are leaking, that is, if the piston packing is presupposed to be satisfactory.

The leakage of the steam valves can be detected by the erratic pump operation or by means of an indicator card taken in the same manner as that employed on engines. This latter is perhaps the quickest way to determine whether such leakage exists, although one can frequently tell from the general appearance of the valve whether or not it is apt to leak in service.

Steam valves wear in the seats, and after a while leakage occurs on both ends of the piston simultaneously. It is far cheaper to replace a valve which is thought to be giving trouble than to spend a great deal of time trying to determine how badly it leaks. Usually visual examination should suffice; when in doubt, replace the valve. Valves stick because of broken springs and foreign matter getting under them. When leakage occurs, it can sometimes be detected by the jerky action of the piston. If the piston can be moved back and forth with pump stopped and gate valve closed and discharge valves in good condition, leaking of the suction valves naturally is indicated.

Difficulty is frequently experienced with slippage of the crosshead. The crosshead can be reset without removing the steam-chest cover by disconnecting the rocker arm from the piston rod and pushing the piston to the head of the cylinder. Make a mark on the rod where it passes through the packing gland. Now push the piston until it strikes the opposite head of the cylinder. Again mark the piston rod at the packing gland. Find the center of these two positions marked on the rod. Mark and place this position at the packing gland. The piston is now centered up in the cylinder. Drop the rocker arm vertically, and fasten to the piston rod. The pump is now ready to run.

Another operating difficulty is the collection of solid deposit on water valves, causing leakage. This can be removed by scraping but more readily by cleaning with dilute acid solution. Also piston rings wear and must be replaced. Though difficulties are more frequently encountered on the water end of the pump, still anything and everything about the pump can give trouble. Therefore, when inspecting the pump, overlook nothing.

Many pump problems, such as vibration, bearing failure, shaft failure, and short coupling life, may be traced to misalignment between pump and driver. Nor does the flexible coupling make adjustment for such misalignment. It is important, therefore, that we read the manufacturer's instructions carefully before starting up the pump. The alignment record should become part of the permanent maintenance record for the pump.

Assuming alignment to be satisfactory on start up, this in no way assures that the pump will remain in that condition. It is therefore desirable to check alignment quarterly or whenever trouble is experienced. Alignment should be checked when pump and shaft are at room temperature and again when they are operating at elevated temperatures, since expansion (or shifting foundation) may frequently change the alignment.

Bearing life depends primarily upon *lubrication;* bearings should be kept free from dirt and foreign matter. Bearing life is extended in proportion to the care exercised in providing an excellent lubricant free from moisture and impurities. A *preventive-maintenance* schedule should see to it that the lubricant is drained and the bearing flushed at regular intervals.   Use oil recommended by the pump manufacturer. In all cases carry spare bearings; however, bearings and thrust shoes can be repaired.   If bearing linings are renewed, provide both lower and upper halves since if one face is worn, the new shoe will have to carry the entire thrust.   Ball bearings must be lubricated separately from the babbitt bearings; they should be lubricated with oil, not grease. There are so many different types of bearings that no effort will be made to cover them here.   Rather, it is important to read the instructions provided with each pump in which are presented the procedures to be followed for maintaining the bearings of that specific pump.

That a pump is operating with too much thrust in one direction is indicated by overheating or wearing out of thrust shoes and by damage to the impeller.   Overheating of the bearing can be detected by placing the back side of the hand on the bearing housing.   If it is impossible to keep the hand in contact with the bearing, the bearing is running too hot and the pump is out of alignment or has excessive clearance.

In single-stage pumps, the impeller should be in the center of the casing rings; this can be determined with the top half of the casing removed.

The life of a *centrifugal* pump as well as its reliable operation depends on the sealing or wearing rings and the thrust imposed.   As the sealing or wearing rings begin to wear between high and low pressure, the pump begins to lose efficiency; it is then time to renew the rings.   The amount of wear is influenced by the amount of clearance originally permitted, by the accuracy with which the pump is aligned, and by the nature of the liquid pumped. .

Packing used for a specific pump should follow the recommendation of the manufacturer.   Exercise care in installing this packing, first making certain that all old packing has been removed.   In adjusting the gland nuts, pull up evenly on the studs, preferably while the pump is running.   Permit a slight leakage to lubricate the packing.   Pulling up a gland too tightly will cause excessive shaft-sleeve wear and ruin the packing.   If the packing gland has been pulled up too tightly, this will be indicated by an increase in temperature and power consumption. If the packing is properly installed and the gland correctly adjusted, the shaft should turn freely by hand.   Should the packing gland leak excessively, do not pull up the gland nuts tightly to stop the leak; rather take the pump out of service and repack.   Neither should time be wasted in trying to pack a pump which has a scored shaft sleeve. Scored piston rods on reciprocating pumps and shafts on centrifugal pumps can be repaired by metallizing.

In packing against high suction pressures, the packing must fit the box exactly and the shaft sleeves must run true and not be scored. Likewise in using metallic packing, care must be exercised in order that the packing fit the box exactly. Furthermore, the metallic packing must be thoroughly lubricated, or difficulty with leaks and scored shafts will be experienced. Metallic packing should never be used where soft packing will do.

In some stuffing boxes, a sealing liquid is introduced through the lantern ring to the packing ring. This liquid may be employed for cooling or to prevent the escape of the liquid being pumped. In suction pumps water sealing is frequently used to keep air from leaking in while keeping packing lubricated.

In order to ensure continuous and uninterrupted service, it is necessary to carry in stock a sufficient number of spare parts, and in order to make certain that one secures the correct part, a record identifying both the pump and its component parts should be kept. For plants which contain many pumps and varieties, it is well to have a suitable storage place. The parts should be cataloged and classified (serially and with reference to catalog number and manufacturer) so that no time is lost in procurement and outage of equipment is prevented.

For reciprocating pumps, it is usually necessary to have only a number of valves and packing. Inspection will generally reveal any additional maintenance requirements before the parts are actually needed, so that corrections may be made after the necessary spare parts have been secured.

For centrifugal pumps it is necessary to carry such items as packing, a set of bearings, a shaft sleeve or a spare shaft, and a set of impeller wearing rings. If a number of similar pumps are employed, it might be well to carry two sets of the above, depending upon past experience and the importance of uninterrupted operation.

There is no standard rule which can be applied as to the number and variety of spare parts to be carried in stock. In normal times, the manufacturer may be able to supply the necessary spare parts and deliver them almost as quickly as they could be identified and located in the average stock room. The essential thing to remember is to keep the reserve or standby pumps in first-class condition at all times.

Just as high maintenance cost reflects poor operating or poor mechanical skill, so also delay in making necessary repairs is a sign of carelessness and indifference and poor management. Each pump should be repaired not immediately after it has failed or given trouble in operation but rather before failure or trouble occurs. This can be done by establishing a *preventive-maintenance* schedule wherein each pump is given a thorough going-over to determine, in advance of failure, whether maintenance is required. Preventive maintenance does not imply that the pump must be dismantled at a scheduled time, since there are many other ways of determining whether a pump is in good operating condition. Once it has been determined whether the pump is or is not in need of repair, it can be either dismantled or restored to service. If in doubt, dismantle the pump.

If a pump is set up on a preventive-maintenance schedule, the schedule should be determined less by time than by the number of operating hours. These figures are best secured by having the operators keep a log of the running time, from which the maintenance schedule can be fixed and lubrication needs satisfied. Too many oil changes are made when they are absolutely unnecessary.

Preventive-maintenance schedules are best established on a basis of experience; periods of maintaining the various pumps on location should be varied or staggered so that they do not occur all at the same time. If the operators work in shifts, it is well to spread the work by specifying certain types of pumps as the responsibility of a certain individual (by designation) and so to avoid the shifting of responsibility for carelessness in maintenance. The schedule should be so arranged that the proper work necessary to administer the schedule is held to a minimum.

In setting up such a schedule, it is well to write out all the various inspections that need be made together with the estimated time required to complete each. After this has been done, the total number of man-hours can be divided by the man-days over the extended period in which the work is to be completed. Otherwise the schedule may be so filled that should other difficulties be experienced at the same time, there would be little time to devote to this maintenance job.

When examination reveals that the pump requires maintenance, it is usually better to do a complete and thorough job rather than a temporary repair even if the scehdule does not demand it, since labor cost is a major item. On the other hand, the practice of replacing parts and making major overhauls, just because the schedule indicates them, is both foolish and costly. After the pump has been overhauled, the capacity should be checked against the head, or, if this is impossible, an approximate test of the pump's condition can be made by determining the static head at shutoff pressure and making a comparison with the design data.

There are many practical methods whereby it is possible to determine whether a pump is performing properly. If the pump is unusually noisy, that may be a sign of vibration, of water recirculation, of undue friction due to wear, or of packing trouble. If the pump casing gets unusually warm (for the liquid being pumped), this is another sign of trouble inside the pump. If the overflow in bearing-cooling water indicates low flow or hot cooling water, shut the pump down and inspect it.

A *preventive-maintenance* schedule might be initiated, such as this one for boiler feed pumps:

Weekly. Operate under normal load for ½ hr; inspect lubrication and cooling-water system; check for noise, unbalance, etc.

Monthly. Repeat above; check bearings and shaft; clean and inspect the motor; feel the bearing to determine heat due to friction, etc.

Yearly. Dismantle pump—overhaul if necessary; check wearing rings, impellers, packing, and alignment. (At the same time, inspect the motor and the turbine drive; inspect turbine buckets, nozzles, rotor, bearings, etc.) The pump should be checked for clearance of bearings, wear, pitting, corrosion, and alignment.

Since lubrication usually determines the life of the bearing and bearing failure frequently causes many other operating difficulties, lubrication is important. Not only should the lubricant be maintained at the proper level, but the correct lubricant for the particular service must be provided.

Since there are so many varieties of lubricants and special uses to which they can be applied, the manufacturer of the equipment involved should be consulted. Follow instructions implicitly and do not experiment with lubricants which have not been proved in the field. Lubricants must be free from foreign material, dirt, and water; when contamination exists, change the oil. Usual recommendations, depending on the grade of lubricant and the type of bearing, would be to change the lubricant after every 700 to 1,000 hr of operation—this is approximately once a month. Ball bearings are frequently grease-packed with a short-fiber soda-base grease, or a high-grade petroleum is used. Which to use depends on the type of bearing and the speed at which it operates. Follow manufacturers' instructions. Do not fill the bearing completely, since an excessive amount of lubricant causes overheating. Never run cooling water at such a rate as to keep the oil too cold. Lubricant should be maintained at approximately 120°F, or at such a temperature that one can place the back of his hand on the bearing at all times. When the pump has been stopped, shut off the cooling water to avoid condensation in the bearings. From time to time flush out the cooling system to make certain that it is in good operating condition.

QUESTIONS AND PROBLEMS

7.1. Mention a number of services for which pumps are used in a power plant.

7.2. Explain how an injector works.

7.3. What are the advantages and disadvantages of an injector?

7.4. What common difficulties are frequently encountered when ejectors are used?

7.5. How many ports has a duplex pump? Name each port and its location.

7.6. Why are the ports located in this manner?

7.7. What do the numbers 6 by 4 by 5 refer to in reference to a pump?

7.8. What is meant by lost motion?

7.9. What is the effect of too little or too much lost motion in duplex-pump valve gear?

7.10. What is meant by clearance volume?

7.11. What is a cushion valve, and what is its purpose?

7.12. What prevents the steam piston of a duplex pump from striking the cylinder head?

7.13. Why has one rocker arm a direct and the other an indirect motion?

7.14. How do you set the valves on a duplex pump?

7.15. Are all valves on the water end of the cylinder always the same?

7.16. What is meant by the percentage of slip of a pump?

7.17. What is the difference between a piston and a plunger?

7.18. What is meant by an inside-packed pump? An outside-packed pump? What are the advantages of each?

7.19. Does a duplex pump use steam expansively?

7.20. What is an air chamber?

7.21. What advantage does a number of small water valves offer over several large ones?

7.22. What kind of valves are used when hot water is being pumped?

7.23. When hot water is being pumped, why is it important to remember where to set the pump with respect to the water storage?

7.24. If the crosshead slips on the piston rod of a duplex pump, how will the pump act? How do you find the true position for it?

7.25. Will a duplex pump when short-stroking use more or less steam than when operating properly?

7.26. Why must we prime a pump?

7.27. How do you pack a reciprocating pump?

7.28. How can a pump be tested to determine the extent of leakage or slippage?

7.29. Explain how a simplex-pump steam valve works.

7.30. What are the advantages of a simplex over a duplex pump?

7.31. What is a power pump? Explain.

7.32. Mention the advantages and disadvantages of a power pump.

7.33. Which is the most efficient, a steam pump, a power pump, or an injector?

7.34. What is a vacuum pump? What is its purpose? Describe how one operates.

7.35. How does a rotary pump operate? What are its advantages?

7.36. What is a foot valve, and what is its purpose?

7.37. Would you prefer that a supply tank be located above or below the pump? Why?

7.38. What do you mean by a pump's becoming steam-bound?

7.39. What is a centrifugal pump? How does it operate?

7.40. Mention several types of centrifugal pumps and discuss their advantages and disadvantages.

7.41. What is the difference between a volute and a turbine pump?

7.42. What is the purpose of the diffusion ring on the centrifugal pump?

7.43. What is packing used for on a centrifugal pump?

7.44. What factors must be considered when selecting a centrifugal pump?

7.45. What type of packing is to be preferred for a high-speed centrifugal pump?

7.46. What is meant by a water-cooled packing box? What is the advantage of using one?

7.47. Some centrifugal pumps employ a mechanical seal. What advantage does this have over a packed pump?

7.48. What determines the type of mechanical seal to be employed for a centrifugal pump ?

7.49. Why is staging employed on centrifugal pumps?

7.50. What is meant by preventive maintenance? How would you go about setting up such a program?

7.51. What is meant by end thrust? How is end thrust taken care of?

7.52. Why is a good system of lubrication more important on a centrifugal pump than on a reciprocating pump?

7.53. What difficulties are ordinarily encountered in the operation of a centrifugal pump?

7.54. How can the output of a centrifugal pump be controlled without causing the pump to overheat?

7.55. When operating centrifugal pumps in parallel, what is the important thing to watch?

7.56. What is the significance of the shutoff pressure on a centrifugal pump?

7.57. How do you determine the capacity of a pump?

7.58. What is meant by cavitation? What are the results of cavitation?

7.59. Where is the check valve placed? Why?

7.60. Where is a foot valve placed? Why?

7.61. What important consideration should be given to suction and discharge piping attached to a pump?

7.62. How would you proceed to line up a pump?

7.63. A gauge is connected to the bottom of a column of water 100 ft high. What will it read, pounds per square inch?

7.64. A pressure gauge reads 125 psi at the pump discharge. How high can the water be lifted?

7.65. A tank 10 ft 6 in. long, 4 ft wide, and 20 ft high is filled with water. How many cubic inches does it contain? Cubic feet? Gallons? Pounds?

7.66. A pump 6 by 4 by 8 makes 111 strokes per min. How many gallons will it deliver per minute? Per hour?

7.67. How long will it take the pump in Prob. 7.66 to empty the tank in Prob. 7.65?

7.68. With reference to the pump in Prob. 7.66, how long will it take to empty a tank 15 ft in diameter and 20 ft long?

7.69. A pump 8 by 5 by 10 uses steam at 100 psi. What pressure will be developed on the water side? How high can this water be pumped?

7.70. A pump 3 by 2 by 3 runs 75 strokes per min with a slip of 15 percent. Calculate the capacity in gallons in 24 hr.

7.71. If the atmospheric pressure is 14.3 psi, what would be the theoretical lift?

7.72. A pump 4 by 3 by 6 delivers water at 200 psi. What steam pressure would be required to operate this pump?

7.73. A pump with a 5-in. steam piston operating with 150 psi delivers water at 300 psi. What is the size of the water piston?

7.74. What size of air chamber would be required for the pump in Prob. 7.70?

7.75. How fast will a duplex pump 5 by 3 by 8 run to supply 25,000 lb of water per hr?

7.76. A pump delivers 5,000 lb of water to a height of 100 ft in 1 min. How many horsepower are required?

7.77. A centrifugal pump delivers 300,000 gal of water per hr to a height of 250 ft. The pump is 65 per cent efficient. What horsepower is required to deliver this water?

CHAPTER 8

RECIPROCATING STEAM ENGINES

Heat engines operate on the principle that heat energy can be changed into mechanical energy. The steam engine makes use of this principle by converting the heat energy in steam into useful power. Consider a simple boiler with a pipe leading directly to a steam engine (Fig. 8.1). Heat is generated in the combustion process and then absorbed by the boiler for the generation of steam. The steam passes through the pipe and into the engine cylinder. The cylinder fills with steam, and the piston is forced to move, thus generating power. When the steam leaves the boiler, it must be replaced by other steam or the pressure will drop.

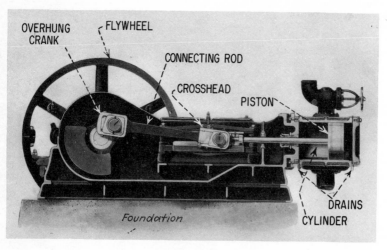

FIG. 8.1 *Cross section of a Skinner engine.*

Replacing this steam requires the application of additional heat to the boiler. Each time the piston moves, steam is required to fill the cylinder and some of the heat content of this steam is converted into mechanical energy. The rate at which the heat is supplied to the boiler must be changed to meet the demand upon the engine for mechanical power.

Internal-combustion engines and steam and gas turbines have replaced reciprocating steam engines for many applications. However, the steam engine is well adapted for variable speeds, for low rotative speeds, and for situations where great turning power is required. Low-speed steam engines are known for their reliability and low maintenance. In some plants steam engines are used to operate air compressors, pumps, and other auxiliaries. The exhaust steam from these engines is available for use in space heating and for industrial process work. When there is a need for this exhaust steam, the arrangement is economical as the power for driving the auxiliaries is a by-product of the normal steam requirements.

8.1 PISTONS AND CYLINDERS. Steam-engine cylinders (A, Fig. 8.2) are made of cast iron. The inside is accurately machined so that the piston and rings will fit properly. The piston rings exert pressure on the walls of the cylinder and cause a certain amount of wear. Because of the thickness of the piston and the clearance allowed between the piston and the cylinder at the end of the stroke, they do not travel the full length of the stroke. Naturally the cylinder wears faster where it is subjected to the pressure of the piston rings. This wear would, in time, cause a shoulder to form at the end of the travel of the piston rings. To prevent this the cylinder is bored slightly larger (counterbored) where it is not subjected to wear from the rings.

The cylinder is a complicated casting because, in addition to being accurately machined on the inside, it must contain steam passages for

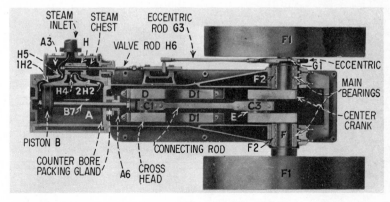

FIG. 8.2 *Sectional top view of a slide-valve steam engine.* (*Mobil Oil Corp.*)

the live steam going to the cylinder (1H2) and the exhaust steam leaving it (2H2, Fig. 8.2). In some engines the valves are also enclosed in the cylinder castings; in others they are located in the steam chest. [The steam chest (H, Fig. 8.2) is a compartment where the steam first enters the engine.] The end of the cylinder farthest from the flywheel is closed by a cylinder head which is bolted to the cylinder proper. The piston rod passes through the stuffing box in the other end of the cylinder. This arrangement applies only to the simple engines. When the steam is exhausted from one cylinder into another (in compound engines), the shaft sometimes extends through both ends of the cylinder. The cylinder, steam chest, and passages must be strong enough to safely withstand the steam pressure at which the engine is to operate.

There are two holes in each end of the cylinder: one in the side, the other in the bottom. These are tapped and fitted with cocks. The lower two are called "cylinder cocks" and are used to drain water from the cylinder when the engine is being started up. The two in the side of the cylinder are known as "pressure cocks" and are used for attaching an indicator.

The piston (B, Fig. 8.2) is cylindrical in shape and is made to fit the cylinder. It must move back and forth and at the same time must make an almost steamtight joint with the cylinder. It is not practical to have the piston fit the cylinder closely enough to prevent this steam leakage, since expansion due to heating would cause the piston to stick in the cylinder. For this reason the piston is fitted with "piston rings." These rings fit into slots in the piston. They press against the walls of the cylinder and prevent leakage yet are flexible enough to take care of expansion and to compensate, to some extent, for wear of the cylinder and rings.

The pistons of small engines are solid and have suitable grooves to receive the rings. In assembling, these rings are forced over the piston to these positions. This procedure is not possible with large pistons since the correspondingly large rings cannot be sprung enough to force them into the cylinder grooves. Therefore, large pistons are made up in sections, as shown in Fig. 8.3. The center of the piston which fits over the piston rod and supports the other parts is called the "piston body" or "spider." In some designs the spider is solid, but in others it is hollow, to make it as light as possible. The outer part of the piston is made up of "bull rings." These are divided at the piston-ring grooves so that the rings can be removed. The bullrings are supported from the spider either by springs or by adjusting studs. The "follower plates" on the ends of the piston hold the bullrings and assembly in place.

The common snap piston ring is made by machining a ring larger

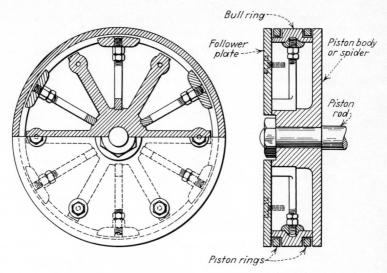

FIG. 8.3   *Built-up piston used in large engines.*

in diameter than the cylinder in which it is to be used. A slot is then cut in the ring, large enough so that when the ring is compressed, it will fit into the cylinder. When the ring is in place in the cylinder, there must be clearance between the ends to take care of expansion. The tendency of the ring to open to its original size holds it against the cylinder walls. Other rings are made in sections and are held against the cylinder walls by means of springs. The common snap ring is preferred for ordinary service.

When the engine is on dead center, there must be some distance between the piston and the cylinder head. This is referred to as "linear" or "piston clearance" and in some cases just as "clearance." This distance is referred to as "head-end" or "crank-end clearance" depending upon which end of the cylinder is being considered. Linear clearance is necessary to prevent the piston from striking the cylinder head and damaging the engine. The engine should be so adjusted that it will have equal clearance on the head and crank ends. The amount of clearance necessary depends upon the type of engine and steam temperature but is seldom less than ⅛ in.

The linear clearance is found by placing the engine on dead center and marking the position of the crosshead on the guides; the connecting rod is then disconnected and the piston pushed against the cylinder head. The distance between the position of the crosshead when the engine is on dead center and when it is striking the head is the linear clearance. This process is sometimes referred to as determining the "striking point" of an engine.

Another method of determining linear clearance is to insert a soft piece of solder in the cylinder through the indicator connection, after which the engine is turned over by hand and the solder flattened between the cylinder head and piston. The thickness of the flattened solder wire represents the linear clearance of the engine. The "volumetric clearance" of a steam engine is the volume of the steam passageways from the steam valves to the cylinder and the volume at the end of the cylinder when the piston is on dead center. This is usually expressed in percentage of piston "displacement volume." The displacement volume in cubic feet per piston stroke equals the length of the stroke in feet multiplied by the area of the piston in square feet. The clearance volume of engines varies from 9 per cent for slide-valve engines to less than 2 per cent for poppet-valve engines. Modern engines have volumetric clearances of 1.5 to 5 per cent. Reducing the clearance volume improves the economy of a steam engine.

The clearance volume is determined by placing the engine on dead center and filling the remaining space with water. A bucket of water is weighed, and water is poured from it into the clearance volume until it is filled to the valve. The water remaining in the bucket is then weighed. The amount of water poured into the cylinder is the difference between the two weights. The volume of water in cubic inches (clearance volume) can be calculated by dividing the weight of water in pounds by 0.036. Some of the water poured into the clearance volume will leak past the piston, and it is usually necessary to correct for this by noting the rate at which leakage occurs. The steam-engine cylinder is above room temperature by an amount that depends upon the steam pressure and degrees of superheat. This difference in temperature causes a continual flow of heat from the cylinder to the air in the room, which results in a direct loss in economy. This loss is reduced either by covering the cylinder with heat-insulating material or by making the cylinder double-walled (steam-jacketed) and filling the space with steam.

In the conventional engine steam enters and leaves through the same opening (port), resulting in the alternate heating of the port by the incoming steam and cooling by the exhaust. The result is a loss of a portion of the steam before it has had an opportunity to perform work in the engine.

8.2 PISTON RODS, CROSSHEADS, AND CONNECTING RODS. The force exerted on the piston is transmitted from the cylinder by the piston rod (B7, Fig. 8.2). One end of the piston rod is attached to the piston and the other end to the crosshead. During one stroke the piston rod is subjected to tension and during the other to compression. It must be mechanically strong enough to withstand these forces. The rod must be packed against full steam pressure where it goes through the end of the cylinder. For this reason it must have a smooth surface.

The piston rod enters the cylinder through a "stuffing box." The simplest type of stuffing box consists of a recess or cavity around the rod. The box is filled with soft packing and provided with a cover held on by bolts. The bolts are tightened so that the packing is drawn tight against the rod. The packing fits the rod so closely that steam cannot leak past.

Soft *braided packings* are used for low-pressure saturated steam. When high-pressure, highly superheated steam is used, the soft packing becomes inadequate. *Metallic packings* are used for these services; the first cost is higher, but they do not have to be replaced as often as the soft packings. There are two types of metallic packings: the flexible, which can be used in the same way as soft packing and does not require a special stuffing box; and the solid, made of rings of antifriction metal. The rings are held against the rod by springs. This packing requires a specially designed stuffing box.

The crosshead (D, Fig. 8.2) is rigidly attached to one end of the piston rod. It is connected to the connecting rod by means of a bearing. Since it is rigidly attached to the piston by means of the piston rod, it has the same motion as the piston. Its purpose is to guide the piston rod and keep it in line with the piston. The crosshead works between guides and thus ensures straight-line motion. The end ($C1$) of the connecting rod that is attached to the crosshead must move in a straight line, whereas the other end ($C3$), which is attached to the crank, has a circular motion.

The circular motion of the crank puts a thrust on the crosshead guides and causes wear. The crosshead is therefore fitted with slippers made of antifriction metal. These slippers are so arranged that they can be adjusted farther out from the body of the crosshead for lining up and to compensate for wear. In some designs this is accomplished by inserting thin strips of metal or paper (shims) between the crosshead proper and the slippers. Others have slippers that fit the body on a taper; in this case the adjustment is made by screws which move the slippers with respect to the crosshead (Fig. 8.4a).

The back-and-forth motion of the piston and crosshead is changed into circular motion by means of the connecting rod. The force developed in the cylinder is transmitted by the connecting rod from the crosshead to the crank. The connecting rod must have a bearing on each end. These bearings are made adjustable for wear.

There are two different types of connecting-rod bearings: the *closed-* or *solid*-end and the *marine*-end bearing. The closed- or solid-end type consists of a large rectangular hole in the enlarged end of the rod.

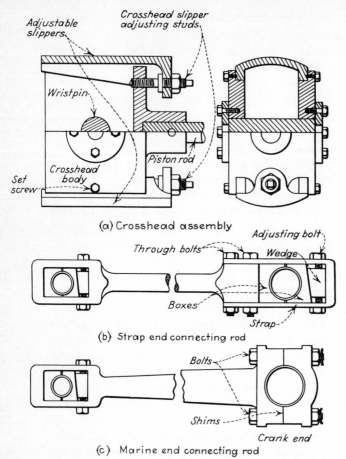

(a) Crosshead assembly

(b) Strap end connecting rod

(c) Marine end connecting rod

FIG. 8.4 *Crosshead and connecting-rod adjustments.*

This enlarged end may be an integral part of the rod, or it may be built up with bolted strips. The one-piece bearing can be used only on engines with overhung crank, but the built-up connecting rod can be used on any type of engine. The two halves of the bearing referred to as boxes or brasses fit into the enlarged end of the connecting rod. One half of the bearing is tapered and fits a tapered wedge. The bearing can be adjusted for wear by moving this wedge (Fig. 8.4*b*).

In the marine-end connecting rod (Fig. 8.4*c*) the two halves of the bearing are held together with bolts. Adjustments are made by removing shims from between the two halves of the bearing.

The distance between the piston and the cylinder head (linear clearance) when the engine is on dead center is determined by the length of the connecting rod and piston rod. When the connecting-rod bearings are adjusted for wear, their length is changed, and consequently the clearance. The effect upon the clearance depends upon the type of connecting-rod take-up. If the connecting rod is of the closed-end type and the wedge take-ups are on the *inside end* of the bearings, the connecting rod is lengthened when the bearings are adjusted for wear. In this case the crank-end clearance will be increased and the head-end clearance decreased. If, on the other hand, the adjusting wedges are on the *outside end* of the bearings, the connecting rod will be shortened, the crank-end clearance will be decreased, and the head-end clearance will be increased when the bearings are adjusted for wear. The closed-end-type connecting rod is sometimes built with the adjusting wedge for one bearing on the inside and that for the other bearing on the outside. If the wear on the bearings is equal, the shortening due to adjusting one will be compensated for by the lengthening of the other. When any adjustments are made to the connecting-rod bearings, the clearance should be checked before the engine is started.

8.3   CRANK, FLYWHEEL, AND VALVE MECHANISMS.   The crank holds the end of the connecting rod in a circular path and transmits the force from the connecting rod to the engine shaft. Several arrangements of main bearings, cranks, and flywheels are employed. Center crank engines have a main shaft bearing on each side of the crank. This arrangement distributes the forces equally on the two bearings. The center crank engine shown in Fig. 8.2 has two overhung flywheels. Overhung crank engines have a bearing on only one side of the crank. This arrangement provides accessibility for lubrication and maintenance, but too much overhung weight is undesirable. Outboard bearings are located on the end of the main shaft, outside the flywheel. The engine shown in Fig. 8.1 has an overhung crank and an outboard bearing. Small engines have a disk for a crank. The center of this disk is shrunk on the main engine shaft. The crankpin is shrunk in the disk. The disk is counterweighted on the side opposite the crank to compensate for the weight of the crankpin and part of the weight of the connecting rod. The cranks of large engines are made up in sections. The arm that holds the crankpin extends in one direction and the counterweight arm in the other. If the flywheel is on the right of the engine when viewed from the cylinder end, the engine is "right-handed." If the flywheel is on the left of the engine when viewed from the cylinder end, the engine is "left-handed." The engine shown in Fig. 8.1 is right-handed.

The distance from the center of the engine shaft to the center of the crankpin is called the "radius of throw of the crank." Twice the radius of throw of the crank is called the "throw of the crank," or "stroke of the engine." The piston travels once over its path when the engine makes a stroke. Two strokes are called a "revolution." However, the travel of the piston at intermediate points in the stroke varies, depending upon the proportional length of the connecting rod and the crank. The influence of the ratio of the length of the connecting rod to the crank, of a steam engine, upon the relative speed of the piston and crankpin is known as the "angularity effect" of the connecting rod. An engine having a connecting rod 9 ft long and a crank 2 ft long has a ratio of the connecting rod to the crank of $4\frac{1}{2}:1$.

Consider the diagram in Fig. 8.5, in which the engine is shown in two positions of a stroke: A, when the crankpin has moved a given angular distance from the head-end dead center, and B, when it is an equal distance from the crank-end dead center. This shows that for equal angular movements of the crank there is a greater linear movement of the piston near the head end than near the crank end. Since the flywheel and crank rotate at nearly constant speed, the piston travels faster on the head-end side of mid-position than it does on the crank-end side. The greater the ratio of the connecting-rod length to the crank, the less the effect of the angularity of the connecting rod.

From previous discussion of the operation of the piston in the cylinder it is evident that the force delivered to the engine shaft is by no means uniform throughout the stroke. This uneven application of force causes the shaft to run irregularly. To prevent any appreciable variation in speed a flywheel is attached to the engine shaft. The flywheel has a heavy rim and resists a change in the rate of motion. It takes a certain amount of power to increase its speed, but when the power applied to the shaft has decreased, this power is again available to prevent a serious change in speed. During part of the revolution the flywheel consumes power, and during another part it gives up power. The size of the flywheel is determined by the size of the engine, the speed, and the service for which it is to be used.

Flywheels for small engines are cast in a single piece. This is not

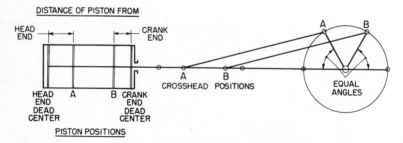

FIG. 8.5 *Diagram showing the effect of the angularity of the connecting rod.*

practical for large engines, not only because internal strains are set up in large castings, but because the large wheels cannot be transported in one piece. Large flywheels are cast in sections and assembled by means of links and bolts. The rim of the flywheel is made very heavy because its ability to compensate for irregularities in the supply of power depends upon the weight and linear velocity of the weight. The greater the distance at which the weight is located from the center of the wheel, the greater will be the linear velocity for a given rotative speed.

Any material resists being moved in a curved or circular path and always tries to move in a straight line. That is, the rim of a flywheel is continually trying to break away from the flywheel. The material of which the flywheel is constructed must overcome this force, or the flywheel will burst. A flywheel failure is disastrous and must be carefully guarded against. The general rule is that the rim of a flywheel shall not travel faster than 6,000 ft per min. All engines should be equipped with governors and overspeed tripping devices so that they cannot overspeed enough to become dangerous.

Engines are classified according to the direction of rotation of the flywheel. If when viewed from the cylinder end of the engine, a point on the upper part of the flywheel is moving away from the observer, the engine is said to be "running over." If when viewed from the same position, the point on the top of the flywheel moves toward the observer, it is said to be "running under." When an engine is running under, the crosshead is pushed against the upper guide on both forward and return strokes. When it is running over, the thrust of the crosshead is on the bottom guide during both strokes.

It is desirable to have an engine run over. When an engine runs under, the thrust is on the upper crosshead guide and at each end of the stroke; when the force is released, there is a tendency for the crosshead to drop, causing a bumping noise. This difficulty can be prevented by keeping the crosshead carefully adjusted at all times. When one is driving line shafting or machinery (located back of the engine) with a belt, it is best to have the engine run over. With this arrangement the lower side of the belt carries the load and allows the upper side to sag, increasing the arc of contact on the pulley.

The reciprocating engine must have a valve arrangement that will admit steam to the cylinder at the beginning and shut it off before the end of the stroke. The exhaust valve must open to discharge the exhaust steam at or slightly before the beginning of the return stroke. This exhaust valve must close at or a little before the piston has completed its return stroke. The valves on early steam engines were manually operated; the engines were slow and inefficient. One of the first improvements in the steam engine was to make the valves operate automatically. The engine was equipped with a mechanism that opens and closes the valves at the correct time with respect to the position of the piston.

Several different types of valves have been developed. They are explained in detail in Chap. 9. The valve motion, in practically all stationary engines, is obtained with the aid of an "eccentric." The simple eccentric consists of a circular disk larger in diameter than the engine shaft. In this disk a hole is bored equal in diameter to the diameter of the engine shaft. The center of this hole is not in the center of the disk. The distance from the center of the disk to the center of the hole is known as the "eccentricity," or "radius of throw of the eccentric." Twice the eccentricity is known as the "throw of the eccentric." This eccentric disk is placed on the shaft and attached so that it turns with the shaft. An eccentric strap goes around the eccentric. The valve rod is attached to the eccentric strap. When the engine shaft makes *one-half* of a revolution, the valve moves a distance *equal to* the throw of the eccentric, or twice the eccentricity. The simple slide-valve engine has but one eccentric. The more efficient engines have two, so that the intake and exhaust valves can be independently adjusted. The separate eccentric makes it possible for the governor to control the speed of the engine by adjusting the inlet-steam valve.

8.4   FRAMES AND FOUNDATIONS.   The engine must have a solid base and foundation to keep the stationary and moving parts in perfect alignment. Engines have been made with cylinder shafts and flywheels arranged in many different ways. In a "horizontal engine" the piston rod moves in a plane parallel to the floor. If the cylinder is located above the crank in a position so that the piston and piston rod travel up and down, the engine is called a "vertical engine." Sometimes the cylinder is inclined, and the engine is known as an "inclined engine." A "multiple-expansion engine" is one that has two or more cylinders, the steam exhausting from one to the other. With any arrangement it is necessary for the engine to have a solid cast-iron base. The base holds the main bearings, crosshead guides, and cylinder.

8.5   GOVERNORS.   In stationary practice, steam engines must operate at nearly constant speed. This is very important when the engines are being used to drive electric generators, since constant speed is required to ensure uniform voltage and frequency. When the load on an engine changes, the governor must adjust its steam supply to meet the change in demand.

Constant speed over the entire range of engine capacity is desirable, but some variation must be tolerated to obtain satisfactory governor operation. A governor mechanism must possess sensitivity, promptness of response, and stability. *Sensitivity* is the ability of a governor to detect, or be influenced by, small changes in engine speed. *Response* is the ability of the governor mechanism to make changes in the steam supply to the engine to compensate for small variations in speed. *Stability* is the characteristic of a governor which permits rapid changes in the load on the engine without fluctuations in speed.

Speed regulation is one measure of governor performance and is defined as the change in speed necessary to cause the governor to operate the engine from no load to full load. This change in speed is expressed in percentage of full-load speed.

Example An engine has a no-load speed of 205 rpm and a full-load speed of 200 rpm. What is the speed regulation expressed in percentage of full-load speed?

Solution

$$\text{Speed regulation} = \frac{205 - 200}{200} \times 100 = 2.5 \text{ per cent}$$

A governor which does not have adequate stability will cause the engine to run alternately fast and slow in rapid succession. This difficulty is known as "hunting," which may be overcome by installing a "dash-" or "gag" pot. This device improves the stability but reduces the sensitivity of the governor and must therefore be adjusted to provide stable operation with the minimum loss in sensitivity.

The dashpot consists of a cylinder which is fitted with a piston. The piston rod is operated by the moving arm of the governor, while the cylinder is attached to the flywheel. The cylinder is filled with oil. Movement of the governor arm produces movements of the piston, which are retarded by the oil. The amount of dampening or retarding action is controlled by regulating the leakage past the piston or by the choice of the viscosity of the oil. The dampening action gives the engine time to adjust its speed to the demands of the governor.

The simplest, although not the most efficient, way to control the speed of an engine is by means of a valve in the steam supply to regulate the pressure (Fig. 8.6a). Such a valve is of the balanced type, and it requires only a small force to open and close it. The governor moves the valve to vary the steam pressure to the engine, thus maintaining normal speed. When the engine's speed decreases, the governor opens the valve to increase the pressure. With an increase in speed, the steam pressure is reduced. Figure 8.7a is a series of indicator diagrams[1] showing the effect of the action of a throttling-type governor on the steam pressure in the cylinder for four different loads. The height of the diagram represents the pressure. Each time the throttle valve adjusts toward the closed position, the diagram becomes lower. In this way the average pressure on the piston (mean effective pressure) is reduced,

---

[1] See Sec. 10.6 for an explanation of indicator diagrams.

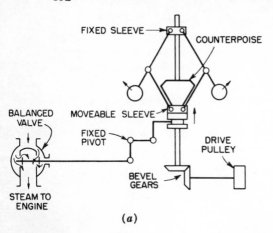

(a)

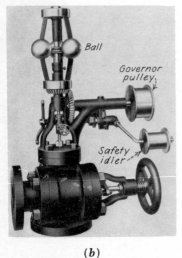

(b)

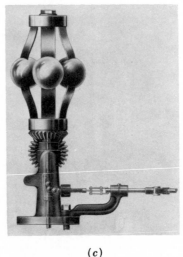

(c)

FIG. 8.6 *Flyball governor.* (*a*) *Schematic arrangement of flyball governor and balance valve.* (*b*) *Pickering flyball governor with safety stop.* (*c*) *Pickering governor speed-range adjustment.*

and the power output is lowered to meet the decreased demand. When more power is required, the valve opens to increase the pressure on the piston, thus enabling the engine to carry the extra load with only a small loss of speed. With this type of governor the steam-inlet valve always closes when the piston is at the same position in the cylinder (point of cutoff remains constant).

The output power of an engine can be controlled by changing the length of time the steam valve is allowed to remain open to admit steam to the cylinder. Engines which employ variable cutoff as a means of control are known as "automatic engines." Full steam-line pressure is admitted to the cylinder at the beginning of the stroke, but as the piston moves over its travel, the inlet valve is closed at a point that will allow only the required amount of steam to enter the cylinder.

The governor adjusts the valve-operating mechanism until the engine speed is normal. When the engine speed decreases, the governor operates the valve mechanism to close the steam-inlet valve later in the stroke; when the engine speed increases, the governor causes the steam-inlet valve to close earlier. Figure 8.7*b* is a series of indicator diagrams showing the effect of the action of a variable-speed cutoff governor on the steam pressure in the cylinder for four different loads. The longer the steam valve is open, the more steam admitted and hence the higher the average pressure exerted on the piston. When more power is required, the cutoff is later to increase the pressure on the piston, thus enabling the engine to carry the extra load with only a small loss in speed. With a decrease in power requirement the steam valve remains open a shorter time, reducing the pressure in the cylinder. With this type of governor the full steam-line pressure is admitted to the cylinder.

A *flyball governor* consists of two or more balls mounted on a vertical shaft. The shaft is usually driven from the engine shaft by means of pulleys and a belt. The balls are pivoted, permitting variations in their radial distance from the shaft, and this movement is transmitted through a system of linkages to control the steam supply to the engine. The rotating motion of the governor balls produces centrifugal force which reacts against the force of gravity, causing the balls to assume a position relative to the speed of the engine (Fig. 8.6). Springs are sometimes used to balance the centrifugal force created by the rotating motion of the governor weights. A decrease in engine speed and consequently in governor speed causes the balls to move through a smaller circle. This change in position of the balls increases the steam supply, thus restoring the speed of the engine. When the engine speed increases,

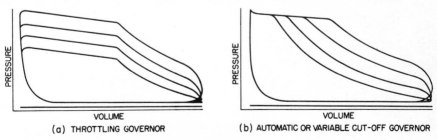

(a) THROTTLING GOVERNOR   (b) AUTOMATIC OR VARIABLE CUT-OFF GOVERNOR

FIG. 8.7 *Steam-engine indicator diagrams showing the effect of governor operation.*

the balls travel through a larger circle, effecting a decrease in the amount of steam supply and thus lowering the speed. If the belt breaks, the governor stops and the control mechanism responds just as if the engine were slowing down; that is, more steam is admitted, and the engine speeds up. A flywheel failure is likely to result. To prevent this condition "stop valves" are installed. The stop valve is held open by an arm which carries an idler on the belt (Fig. 8.6b). If the belt breaks, this arm is released and the stop valve closes, shutting down the engine.

The speed of the governor, relative to that of the engine, is determined by the size of the pulley on the engine and governor. The following formula applies:

$$V = \frac{vd}{D}$$

where V = rpm of engine
  v = rpm of governor
  D = diameter of engine pulley
  d = diameter of governor pulley

If, for example, the size of the governor pulley were increased, the governor speed would be lower for a given engine speed. The lower governor speed would mean that the governor would increase the steam supply, and the engine speed would increase; it would increase until the governor speed became the same as formerly. This is one way in which to change the speed of an engine equipped with a flyball governor. Another is to change the counterweight or the tension on the spring (Fig. 8.6c).

Shaft or *flywheel governors* also employ the principle of centrifugal force (Fig. 8.8). The weight or weights are mounted on pivoted arms and rotate with the flywheel. The centrifugal force created by the rotation of the weights pulls against a spring or springs attached to the weighted arms. The position of the weights with respect to the flywheel is determined by the rate of rotation of the engine and the pull of the spring. The speed of the engine can be adjusted by changing the tension on the spring. Increasing the tension requires more centrifugal force for a given position of the governor weights, hence an increase in engine speed. Decreasing the tension slows the engine down.

An increase in speed causes the governor weights to move against the action of the spring, reducing the supply of steam. It follows that breakage or failure of the governor spring will remove the restraining force and allow the governor to reduce the steam supply. This feature of shaft governors makes it unnecessary to use special safety devices to prevent overspeeding in case of governor-spring breakage.

The weighted arms of flywheel governors are connected to the eccentrics by linkages. Movements of the governor arms result in changes

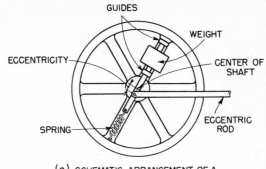

(a) SCHEMATIC ARRANGEMENT OF A
CENTRIFUGAL GOVENER

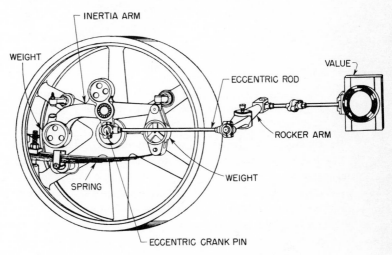

(b) SHAFT GOVERNOR ON SKINNER ENGINE

FIG. 8.8   *Centrifugal shaft governor.*

in the valve-operating mechanism to vary the point of cutoff (see Chap. 9 for a description of valve arrangement).

Inertia is defined as the property of all bodies which causes them to tend to remain at rest or in continued motion in a straight line. *Inertia governors* employ heavy weights held in place by arms which are pivoted to the flywheel (Fig. 8.9). The arms are held in an angular position with respect to the flywheel by means of springs. When the speed of the engine is reduced, the weights tend to continue at the same speed and run ahead of their normal position with respect to the flywheel. The valve mechanism is so arranged that when the governor weights run ahead of their normal position, the steam supply is increased; when the engine speed increases, the weights lag behind

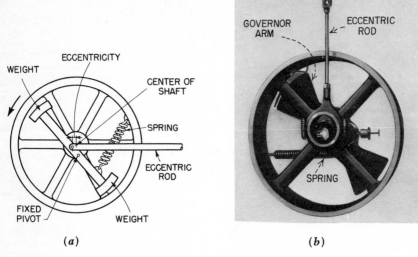

FIG. 8.9 *Inertia shaft governors.* (*a*) *Schematic arrangement of an inertia governor.* (*b*) *Troy-rites inertia governor.*

the flywheel and decrease the steam supply. The action of the governor upon the eccentric and valve mechanism is the same as that of a centrifugal governor.

The inertia principle is effective in quickly restoring the engine to normal but does not give a definite maximum engine speed. A governor which employed inertia alone would be very sensitive but would be so unstable that it would not be practical. Centrifugal force is employed to stabilize the engine at a definite speed. One governor weight is slightly larger than the other so that the governor assembly does not balance at its center. This results in an unbalanced portion of the weights, which produces centrifugal force. This force is balanced against the tension in a spring just as in a centrifugal governor. In reality this arrangement is a combination of centrifugal and inertia governors, but it is commonly referred to simply as an "inertia governor."

8.6   ENGINE SPEED.   The term "speed" as applied to an engine refers to the rotative speed measured in the revolutions per minute of the flywheel. The term "piston speed" refers to the rate of travel of the piston and is expressed in feet per minute. The length of the stroke in feet times the rotative speed in rpm times 2 equals the piston speed in feet per minute.

Two engines each having the same speed (rotative) will have different piston speeds if they have different lengths of stroke. Short-stroke engines have a piston travel less than 1.5 times the cylinder diameter, while long-stroke engines have a piston travel more than 1.5 times the cylinder diameter. Engines with rotative speeds of 125 rpm or less are classified as low-speed engines and those which exceed 125 rpm as high-speed. Piston speeds vary from 300 to 800 fpm, but speeds near 600 fpm have been found to be the best compromise in stationary practice. Engines having a long stroke and high piston speed usually have low rotative speed, while short-stroke low-piston-speed engines have high rotative speed.

High piston speed is desirable because more power can be developed in a given-sized cylinder.

High rotative speed is desirable because a smaller flywheel may be used, thus reducing the size and cost of an engine and electric generator for a given capacity. The high-speed operation permits the improved speed regulation essential to the operation of a directly connected electric generator. High-speed engines require that the valves be adjusted for more compression and lead than low-speed engines. This is necessary to fill the clearance space quickly, cushion the moving parts, and ensure full pressure behind the piston on the return stroke.

Valve mechanisms also impose speed limitations on engines. The D slide valve and releasing-type Corliss are usually restricted to engines operating at speeds of less than 125 rpm. Piston and nonreleasing Corliss valves can be applied to engines operating up to 500 rpm. Poppet valves are adapted to high-speed operation but in stationary practice are usually applied to large low-speed engines.

Low-speed engines are more desirable for continuous operation but require a larger, heavier, and more expensive framework than do high-speed engines. In addition, they must have more elaborate steam jacketing owing to an increase in cylinder condensation.

Characteristic	Type of engine			
	Slide	Piston	Corliss	Poppet
Steam pressure, psi............	50–175	75–225	75–150	125–225
Steam temperature, °F.........	450	600	525	600
Speed, rpm..................	50–350	100–500	50–200	75–175

Low-speed Engines

Advantages

Durability
Reduced maintenance
Possibility of regulating while in operation

Disadvantages

Heavier construction
Higher initial cost

Poor system of regulation
Increased cylinder condensation

High-speed Engines

Advantages

Coupled to generators
Compactness; small space
Excellent lubrication
Low initial cost
Better regulation
Reduced cylinder condensation

Disadvantages

Inefficiency
Increase maintenance; shorter life
Necessity for more attention

8.7 COMPOUND ENGINE. A "tandem-compound" engine (Fig. 8.10*a*) is one in which the cylinders are placed one behind the other. The pistons are both on the same piston rod, the engine having but one

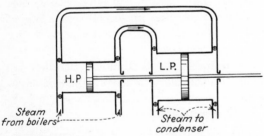

(a) Diagrammatic arrangement of a tandem compound engine

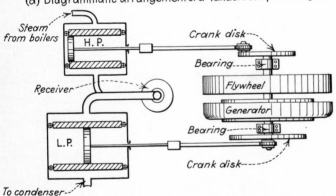

(b) Diagrammatic arrangement of a cross-compound engine

FIG. 8.10 *Compound engines.*

connecting rod and crank. The events in the cycle are identical for both units. It is not necessary to have a receiver on an engine of this kind, as both sets of valves open simultaneously. A "cross-compound" engine (Fig. 8.10*b*) is one in which the engine cylinders are set side by side. One is called the "high-pressure" cylinder; the other, the "low-pressure." Steam passes from the high- to the low-pressure cylinder. Each has a separate crank and connecting rod. The cranks here are set 90° apart, and as the cylinders do not work simultaneously, a receiver is interposed between the high-pressure discharge and the low-pressure admission for the purpose of storing the steam until the valves on the low-pressure cylinder are ready to admit steam.

In addition to a receiver, a reheater is frequently placed between the two cylinders. The exhaust steam from the high-pressure cylinder is reheated to the initial steam temperature before passing to the low-pressure cylinder.

Since the high-pressure engine crank is 90° ahead of the low-pressure engine crank, the high-pressure piston completes one-half of its stroke when the low-pressure piston is on dead center.

The low-pressure-cylinder steam valves are adjusted for more lead than the high-pressure. A lead of $\frac{1}{16}$ in. per ft of engine stroke has been found satisfactory.

Adjusting the low-pressure-cylinder steam valves for later cutoff reduces the horsepower developed in the low-pressure cylinder and increases that developed in the high-pressure cylinder. It follows that adjusting the low-pressure-cylinder steam valves for earlier cutoff increases the horsepower developed in the low-pressure cylinder and decreases that developed in the high-pressure cylinder.

Compound engines are considered more economical than ordinary engines. However, they are at a decided disadvantage with low-pressure steam and a changing load.

Compound Engine

Advantages

Less leakage and clearance
Reduced cylinder condensation
Equalized crank effort (cross-compound)
Increased range of expansion
Increase in economy

Disadvantages

High initial cost
Larger size and floor space
Complexity
Radiation losses

The greatest loss in the engine cylinder is that due to initial condensation and reevaporation. The compound engine reduces this loss to a considerable extent. A cross-compound engine is easy to start, the cranks being set 90° apart. Then, with one crank at dead center, the other reaches its maximum turning power. For any given size, compound engines require less steam per unit of output.

Owing to the arrangement of cranks on a cross-compound engine, the flywheel can be made considerably lighter than that for either the simple engine or the tandem-compound type. Tandem engines are inaccessible for repairs and inspection.

8.8   CONDENSING ENGINES. The push or thrust exerted upon a steam-engine piston depends upon the pressure of the steam on the one side and the exhaust on the other. Therefore, the thrust can be increased either by using higher steam pressure or by lowering the exhaust pressure. When the exhaust steam is used for heating or process work, the back pressure may be above that of the atmosphere. The simple noncondensing engine discharges the steam to the atmosphere; consequently, the thrust is caused by the difference between the steam-supply pressure and the atmospheric pressure. This is the lowest back pressure that can be obtained without the use of auxiliary equipment.

The back pressure on an engine may be reduced below atmospheric pressure by exhausting into a condenser. The exhaust steam is condensed by the use of cold water, which absorbs the latent heat of the steam. The volume of a given weight of steam is greatly reduced when it is condensed into water. If a space filled with steam is cooled until the steam condenses, the resulting water will occupy only a small portion of the volume and a vacuum will be produced. By continually condensing the exhaust steam the pressure is reduced below that of the atmosphere.

A given quantity of steam admitted to an engine cylinder will produce more work when operating condensing than when operating noncondensing (Fig. 8.11). The advantage to be derived from operating condens-

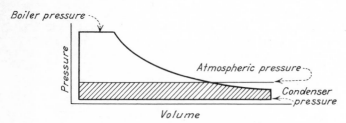

FIG. 8.11 *Pressure-volume diagram. Shaded area shows the gain due to operating an engine condensing over noncondensing.*

ing depends on engine design, valve arrangement, and the pressure employed. Simple engines lose much of the advantage of condensing operation owing to the increase in cylinder condensation (see Chap. 10 for a discussion of cylinder condensation). Condensers can practically always be used to advantage with uniflow and compound engines.

Steam engines that are to operate condensing must be designed to handle the larger volumes of steam encountered at the lower exhaust pressures. Cylinders, exhaust passages, and piping must be increased in size for a given horsepower output. More compression is required to give smooth operation. This may be taken into account in the design of the engine by decreasing the clearance volume and setting the exhaust valve to close earlier in the stroke, thus trapping a larger quantity of steam (to be compressed) into the clearance volume.

Several factors must be considered when choosing between condensing and noncondensing engines. Not only is the condensing engine more costly than the noncondensing, but it requires several auxiliary units which must be purchased, installed, operated, and maintained. (See Chap. 12 for a description of condensers.) The plant must be located where there is an available supply of water for condensing the steam. When the water supply is limited, spray ponds or cooling towers are required to cool the condensing water so that it can be recirculated. The pumps and other auxiliaries require additional power.

Surface condensers, in addition to increasing the efficiency, offer another important advantage. Since the condensate does not mix with the circulating water, it is available for reuse as boiler feedwater. This use of condensate lowers the required quantity of makeup, thus decreasing the cost of chemicals for water treatment. The resulting improved water condition decreases the blowdown and subsequent loss of heat and lessens the possibility of scale formation on the boiler heating surfaces and other difficulties encountered owing to impurities in the water supply.

8.9 UNIFLOW ENGINES. Steam enters at the ends of the uniflow-engine cylinder and leaves through ports located in the middle of the cylinder. (The uniflow-engine valve mechanism is explained in Sec. 9.7.) This uniflow principle avoids cooling of the intake valves and clearance surfaces by the cold expanded steam, making it possible to obtain, in one cylinder, economy comparable with that obtained with a compound engine. For the same reason the uniflow engine is suitable for operating condensing.

Uniflow engines are constructed with various cylinder and flywheel arrangements in the same way as counterflow engines. However, multiple-cylinder vertical uniflow engines are extensively used. Each cylinder receives high-pressure steam and exhausts to the condenser. When three cylinders are used, the cranks are spaced 120° apart, resulting in a distribution of effort and the smooth operation required for driving electric generators. Uniflow engines require relatively heavy pistons (see Fig. 9.15). The use of multiple cylinders distributes the weight of the pistons, and the vertical movement reduces the amount of wear and the necessity for the special piston supports sometimes used with horizontal engines.

QUESTIONS AND PROBLEMS

8.1. What is a steam engine?   Name its principal parts.

8.2. What keeps the piston and rings from wearing a shoulder in the cylinder at the end of the stroke?

8.3. How is leakage around the piston rod prevented?

8.4. What is the purpose of the piston rings?   What is a bullring?

8.5. What are meant by piston displacement, piston clearance, and clearance volume?   In what practical way could you find the piston clearance?

8.6. What is a steam jacket?   What is its purpose?

8.7. What is the difference between a cylinder cock and a pressure cock?

8.8. Find the area of pistons having respectively the following diameters: 4, 8, 11, 13.5, and 15 in.   Also find the total pressure on the face of the piston if the engine is operating at 100 psi pressure.

8.9. If the pistons in Prob. 8.8 have respectively piston rods with diameters of ½, 1, 1½, 1¾, and 2 in., find the total pressure on the crank side of the piston.

8.10. Calculate the volume of the following cylinders: 4 by 8 in., 8 by 12 in.

8.11. A piston is 8 in. in diameter, has a 1-in. piston rod, and uses steam at 150 psi pressure.   What is the total pressure on the crank end?   On the head end?

8.12. How is the crosshead adjusted to compensate for wear?

8.13. What precautions must be taken when adjusting the connecting-rod bearings for wear?

8.14. What effect would lengthening of the connecting rod have on an engine?

8.15. What is meant by the angularity effect of the connecting rod?

8.16. In keying up the crankpin brasses of a solid-end connecting rod with keys on the inside of the pin, what effect will be produced on the clearance?

8.17. Which crosshead shoe has the most wear when the engine is running over?   When running under?

8.18. How do you obtain the striking points of an engine, and what is the advantage of knowing them?

8.19. What do you mean by dead center?   What is the purpose of the flywheel?

8.20. What is meant by an engine's running over?   By running under?

8.21. A flywheel is 3 ft in diameter and is running 300 rpm. What is the rim speed in feet per minute?

8.22. A flywheel is 6 ft in diameter. According to the general rule, at what speed in rpm can it operate safely?

8.23. What is an eccentric, and what is its purpose?

8.24. What is the radius of throw of an eccentric?

8.25. What is meant by the throw of the eccentric?

8.26. The radius of throw of the eccentric is 3 in. How far does the valve travel in one direction?

8.27. What is the purpose of a governor? Name several different types.

8.28. Explain the operation of a flyball governor.

8.29. What is the difference between a throttling and a cutoff governor? Explain how each operates. State the advantages and disadvantages.

8.30. What is a centrifugal flywheel governor? How does it work?

8.31. What is an inertia governor? How does it work?

8.32. How could you increase the speed of an engine with a flyball governor?

8.33. If the spring broke on an automatic engine, what change would take place in the speed of the engine?

8.34. What is the purpose of a governor dashpot?

8.35. Explain the effect upon the governor and engine speeds of increasing the size of the governor pulley.

8.36. What is the meaning of the term "hunting" as applied to engine governors? How is it prevented?

8.37. What would happen to the engine speed if the belt broke on a throttling-type governor?

8.38. An engine pulley 10 in. in diameter makes 200 rpm; the governor pulley is 3 in. in diameter. How many revolutions per minute is the governor pulley making?

8.39. A governor pulley 4 in. in diameter is making 500 rpm, and the engine pulley is 18 in. in diameter. How fast is the engine running?

8.40. An engine is running 150 rpm, and the engine pulley is 14 in. in diameter. The governor is running 600 rpm. What is the diameter of the governor pulley?

8.41. What is the purpose and what are the requirements of engine frames?

8.42. An engine has a 10-in. stroke and makes 100 strokes per min. What is the piston speed in feet per minute?

8.43. An engine with a 14-in. stroke runs at a speed of 250 rpm. What is the piston speed in feet per minute?

8.44. An engine is to run at 200 rpm. What is the largest-sized flywheel allowed?

8.45. An engine operates at 150 rpm, and the crank throw is 10 in. What is the piston travel in feet per minute?

8.46. What is meant by piston speed?

8.47. How is the piston speed of an engine determined?

8.48. What are the usual piston speeds encountered in stationary practice?

8.49. What is the difference between high rotative and high piston speeds?

8.50. What are the advantages and disadvantages of high-speed engines?

8.51. What are the advantages and disadvantages of low-speed engines?

8.52. Which is given more lead, a high- or a low-speed engine?

8.53. What is the advantage of high piston speeds?

8.54. What is a cross-compound engine?

8.55. What is a tandem-compound engine?

8.56. Why is a receiver placed between cylinders on a cross-compound engine?

8.57. In a cross-compound engine, which cylinder is given the most lead?

8.58. What are the relative positions of the cranks on a cross-compound engine?

8.59. What are the advantages of running condensing over noncondensing?

8.60. Why are condensers practically always used with compound engines?

8.61. State the factors to be considered when deciding between condensing and noncondensing operation.

8.62. What is the advantage of a uniflow over a counterflow engine?

CHAPTER 9

VALVE-OPERATING MECHANISM

9.1 CYCLE OF EVENTS. The action of steam within an engine cylinder is best understood with reference to the diagrams in Fig. 9.1. The events are those occurring on the head end of the cylinder only, for a slide valve with outside admission.

The "point of admission" occurs at A, slightly before the piston reaches the end of the stroke. The position of the crank and piston is clearly shown in the diagrams. The pressure-volume diagram a indicates that this position on the indicator card would be as shown. In b the piston is shown near the end of the stroke A, the arrows behind the piston indicating that the piston has not yet reached the end of the stroke. In c the position of the crank A is slightly below dead center.

With admission of steam the pressure increases, as the vertical line from A shows. The pressure quickly reaches its maximum, which is usually a little below the boiler-pressure line KL. The variation in pressure between that of the boiler KL and that of the engine is due to the pressure drop through the lines, ports, and passages.

The period of admission is represented by the line from A to B. During this time the piston (Fig. 9.1b) has moved from A to B. The crank (Fig. 9.1c) has also moved from A to B. The valve in Fig. 9.1e has now assumed the position in Fig. 9.1f, having moved entirely to the right and returned to the position now shown. This period of admission continues from the time the valve opens, which is slightly before the piston reaches dead center, until the steam port has closed.

If the line rising from A in Fig. 9.1a is very nearly vertical, the steam has been admitted quickly. The slope of this line to the right or to the left is an indication of faulty valve setting. As we proceed along

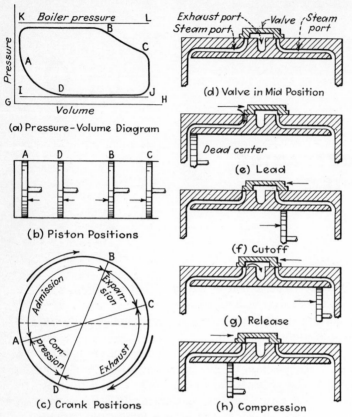

FIG. 9.1 *Outside-admission slide valve, showing piston, crank, and valve positions for cycle of events ocurring on the head end of the cylinder.*

the line *AB* (Fig. 9.1*a*), a slight reduction in pressure is noticeable before the point *B* is reached. This reduction in pressure is due to the restriction of the valve opening as the steam valve cuts off the steam.

The point and period of admission vary, depending on the type of engine, whether high- or low-speed, horizontal or vertical. Lengthening the period of admission *increases* the power developed but *reduces* the economy of the engine.

The "point of cutoff" is that point at which the admission of steam ceases. The valve is then over the ports, and the steam is trapped between the valve and the piston. Reference to Fig. 9.1*a* indicates that the point of cutoff occurs at *B*. It is sometimes very difficult to determine this point exactly, but it is usually located on the diagram by a change in the curvature of the sloping line along *BC*. This point is located where the curvature changes from convex to concave.

Figure 9.1*b* shows the position occupied by the piston (and its direction), and Fig. 9.1*c* shows the position of the crank. The valve (Fig. 9.1*e*) has moved from its position to its extreme-right-end position and returned to the position as shown in Fig. 9.1*f*. The port is now closed, and the piston is shown in position about one-third of the distance from the end of the stroke.

In order to use the expansive properties of the steam advantageously, it becomes necessary to shut it off at some point previous to the end of the stroke. This permits the remainder of the distance to be completed without the addition of more steam (steam used expansively).

The line *BC* (Fig. 9.1*a*) indicates the drop in pressure while expansion is taking place. During this period the steam valve must remain *closed,* and the steam in the cylinder becomes entirely *isolated.* Owing to the expansion of steam and the consequent forward motion of the piston, the volume increases. This increase in volume will result in a progressive drop in pressure.

Expansion continues until the valve (Fig. 9.1*f*) has moved to the position shown in Fig. 9.1*g*. This is called the "point of release" and is accompanied by the *opening* of the exhaust valve at *C* (Fig. 9.1*a*). The piston is then at *C*, as in Fig. 9.1*b*, and the crank at *C*, as in Fig. 9.1*c*. The exhaust valve opens slightly *before* the piston has reached the end of the stroke and remains open during the greater part of the return stroke. As the steam cannot be released at once, a drop in pressure usually accompanies the opening of the exhaust valve and continues until the back-pressure line has been reached.

The back-pressure line continues until the exhaust valve closes. This point is indicated at *D* (Fig. 9.1*a*), and likewise the position of the piston and the crank in Figs. 9.1*b* and *c*. In Fig. 9.1*h* the exhaust valve is seen closing the port and trapping the steam between it and the head end of the piston. The exact point cannot be located very definitely, as the change in pressure is first due to the gradual closing of the exhaust port.

The point *D* in Fig. 9.1*a* is the "point of compression." In Fig. 9.1*h*, the valve has now closed the exhaust port, trapping the steam between the piston and the head, while the piston is moving to the left. Compression continues until the valve in Fig. 9.1*h* uncovers the steam port and assumes a position as shown in Fig. 9.1*e*, completing a cycle. Corresponding positions for the piston and crank are shown in Figs. 9.1*b* and *c*.

The line *DA* is the compression curve and shows the rise in pressure due to the compression of the steam remaining in the cylinder *after* the exhaust valve has closed. Compression is desirable in order to cushion the piston on the return stroke and to fill the clearance volume with high-pressure steam before admission occurs. The latter increases the economy of the engine.

The line *IJ* is the atmospheric line and is made by the indicator when open to the atmosphere. The exhaust lines on noncondensing-engine diagrams fall on or above the line *IJ;* condensing-engine diagrams have their exhaust line below the line *IJ*, depending on the degree of vacuum secured. The line *GH* is the absolute-zero pressure line and is drawn to scale from the atmospheric line. *KL* is the boiler pressure. The difference between it and the admission line on the diagram represents the pressure drop through the piping, ports, and passages of the engine.

9.2   FUNCTION OF VALVES. The function of the valves is to open and close the passages to and from the engine at the proper time. There are various types and makes of valves, the purpose in each case being approximately the same.

One is the D slide valve, so called because of its cross-section resemblance to the letter D. This valve is box-shaped and oblong (Fig. 9.1*d*) and slides back and forth over smooth seats as shown. The valve seat has openings called "ports" which communicate with the interior of the cylinder. This valve controls both ports similarly. It receives its motion from an eccentric attached to the shaft. This eccentric operates the valve so as to regulate the events in the cycle.

The valve-rod length can be changed to position the valve with respect to the head end or crank end of the cylinder. Once the valve-rod length has been adjusted, the valve and eccentric always bear the same relation to each other. If a change in setting is required, we must alter the position of the eccentric with respect to the crank or change the valve.

"Lap" is the amount that the valve extends beyond the edge of the port when the valve is in mid-position. Without lap there could be no "cutoff" until the end of the stroke. Moreover, without cutoff, there would be no expansion, and consequently the efficiency would be lower.

In Fig. 9.2 is illustrated the operation of an engine employing a valve with lap and without lap. For the sake of simplicity the eccentric is shown above the crank. For the valve *without* lap, the piston is shown ready to start its stroke at *a*. The crank is on dead center with the eccentric 90° ahead of the crank; the crank and eccentric will turn to the right. The valve is now about to open the left steam port. At *b*, the piston is shown in the center of the cylinder, having completed half a stroke; the crank is at 12 o'clock and the eccentric's position at 3 o'clock; the valve has reached its extreme limit of travel to the right. In *c*, the piston has completed its stroke, and the valve has re-

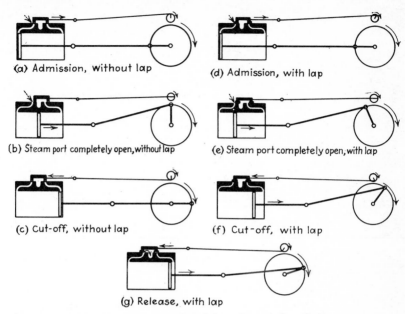

(a) Admission, without lap

(d) Admission, with lap

(b) Steam port completely open, without lap

(e) Steam port completely open, with lap

(c) Cut-off, without lap

(f) Cut-off, with lap

(g) Release, with lap

FIG. 9.2   *D slide valve without and with lap, showing the relative positions of the valve, eccentric, crank, and piston.*

turned to its original position with the left steam port now closed.   With this arrangement it is quite evident that the valve would admit steam for the entire stroke of the piston.   Expansion of steam would be impossible, since there is *no* lap on the valve to cut off the steam before the piston reaches the end of the stroke.

Now consider a valve *with* lap as shown in Fig. 9.2*d*.   Compare it with the valve arrangement in *a*.   In *d*, the eccentric has moved from its 90° position (12 o'clock) to approximately 115° (1:10 o'clock).   The valve is first moved far enough to the right to eliminate the lap; this is accomplished by shifting the position of the eccentric.   The valve (eccentric) is now set 90° *plus* ahead of the crank (1:10 versus 9 o'clock).   In *e*, the valve is completely open, and the piston has traveled to the position shown.   Note how the position differs from *b*.   At *f*, the point of cutoff, the valve has closed the left steam port, but the piston is still some distance from the end of the stroke and is still moving to the right.   With the steam valve closed, the piston travels the distance from *f* to *g*, using steam expansively behind the piston.

All references herein have been made with regard to the left side of the engine; action on the right side is identical to that described above.   This series of diagrams illustrates the necessity and desirability of lap.

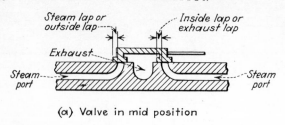

(a) Valve in mid position

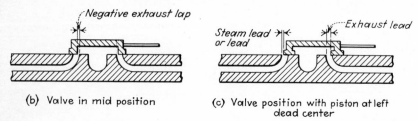

(b) Valve in mid position

(c) Valve position with piston at left dead center

FIG. 9.3 *Outside-admission D slide valve, showing lap and lead.*

There are different kinds of lap: *steam* or *outside* lap, *exhaust* or *inside* lap (Fig. 9.3*a*), and *negative exhaust lap* (Fig. 9.3*b*). Quite frequently, *inside clearance* is called "negative lap" (Fig. 9.3*b*). With a D slide valve and outside admission, the effect of increasing or decreasing the lap is as shown in the following table.

Lap	Increase	Decrease
Outside.........	Admission later; cutoff earlier	Admission earlier; cutoff later
Inside..........	Release later; compression earlier	Release earlier; compression later

The entire operation and performance as well as the efficiency depend to a considerable extent on the setting of the valve.

"Lead" is the amount that the valve is open when the engine is on dead center. It is often described as the distance between the edge of the valve and the edge of the port when the piston is at the end of its stroke. There are two kinds of lead: steam lead and exhaust lead (Fig. 9.3*c*). *Steam lead* is the amount that the steam port is open for the admission of steam when the engine is on dead center; *exhaust lead,* the amount that the exhaust port is open when the engine is on dead center.

Steam lead is given an engine to fill the clearance space and absorb the shock of the moving parts as well as to ensure full pressure against the piston on its return stroke. Exhaust lead is given to liberate the exhaust steam and thus avoid high back pressure.

The amount of lead is usually determined not by the valve and its dimensions but by the valve rod and mechanism. The lead varies with the engine speed, size, and type. The usual amount of lead ranges from $\frac{1}{32}$ to $\frac{1}{4}$ in. High-speed engines are given more lead than low-speed engines. Lead also varies for high- and low-compression engines and for vertical, horizontal, and compound units.

With the valve set in mid-position and the piston on dead center (Fig. 9.4a), the eccentric *leads* the crank by 90°. With this setting steam cannot enter the cylinder until the port has been uncovered by eliminating the lap; this is accomplished by advancing the position of the eccentric. Advancing the position of the eccentric (Fig. 9.4b) to remove the lap will bring the end of the valve in line with the end of the port. However, additional opening is necessary to give the valve lead. This is accomplished by advancing the eccentric from the position shown in Fig. 9.4b to that shown in Fig. 9.4c so that the engine may

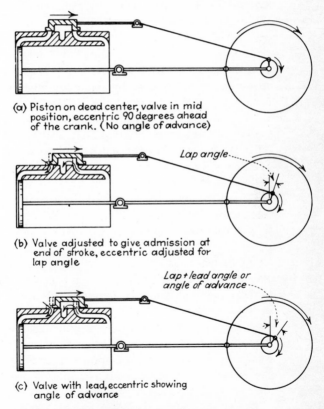

(a) Piston on dead center, valve in mid position, eccentric 90 degrees ahead of the crank. (No angle of advance)

(b) Valve adjusted to give admission at end of stroke, eccentric adjusted for lap angle

(c) Valve with lead, eccentric showing angle of advance

FIG. 9.4 *Outside-admission D slide valve, showing lead and the position of the eccentric, lap, lead, and angle of advance.*

operate. This distance is referred to as the "lead angle." The total distance that the eccentric must be moved to shift the valve from its mid-position when the engine is on dead center (Fig. 9.4*a*) to place it in its operating position (Fig. 9.4*c*) is a combination of lap plus lead angle referred to as the "angle of advance." The required angle of advance varies with the type of engine. NOTE: Advancing the position of the eccentric from *a* to *b* to eliminate the lap requires a change in angularity for the eccentric; such a change in angularity is referred to as the "lap angle." Similarly, reference is made to the lead angle, etc.

On engines with *direct*-connected valve gears (Figs. 9.5*a, b*), the eccentric *leads* the crank by 90° plus this angle of advance, the angle of advance normally being 20°. This places the eccentric 90 plus 20, or 110°, ahead of the crank.

On engines with *indirect*-connected valve gear (Figs. 9.5*c, d*) the eccentric *follows* the crank by 90° minus the angle of advance—the angle of advance being again 20°, placing the eccentric 90 minus 20, or 70°, behind the crank.

Setting the valves on a slide-valve engine can be accomplished in several ways: by trial and error, by the tram method, by the use of an indicator diagram, or by some combination of these. Several things should be known before valve setting is begun, namely, whether the valve has inside or outside admission, whether it is running over or under (the direction of rotation), whether the valve gear is direct or indirect, and whether the valve is to be set for equal lead or for cutoff.

Valve setting. Assume a simple D slide valve, direct-acting, outside-admission, to be set for equal lead, to run over by the trial method.

Place the engine on head-end dead center. Remove the valve cover. Loosen the set screw on the eccentric and rotate the latter in the direction of rotation (running over). Now place the valve directly over the ports. Rotate the eccentric and determine whether the valve travels as far to the head end of the cylinder as to the crank end. If it slides over its seat farther to the head end, *shorten* the valve rod; if the travel to the crank end is longer, *lengthen* the rod. Again rotate the eccentric in the direction of rotation and place the valve directly over the ports. Now advance the eccentric until the lead is equivalent to about $\frac{1}{16}$ in. for *each foot of piston travel*. Tighten the setscrew on the eccentric, rotate the engine until the opposite dead center is reached, and compare the leads. If the crank end has more or less than the head end, the difference must be equally divided.

To set the valves with the aid of a tram proceed as follows: First check the valve travel as in the previous setting. Next place the engine *almost* on dead center. Make a mark on the guide at one end of the crosshead. Place a tram against the flywheel, and inscribe a mark on the wheel for this position of the crosshead. Rotate the engine in the direction desired beyond the dead center until the crosshead *returns to the previous position on the guide* (as marked). Place the tram again in position, and make a new mark on the flywheel. With a pair of dividers locate the exact distance between these two marks on the flywheel. With the tram in position rotate the engine in the direction of rotation until this point and the tram coincide. This point should then represent the exact dead center on one end and may be called *x*. The other dead-center position is found in a similar manner and designated *y*. To obtain an accurate valve setting, all that is now required is to have *x* or *y* correspond to tram. Then set the valve as previously explained.

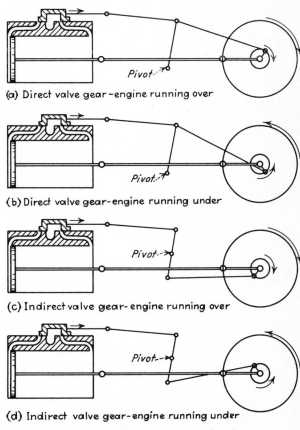

(a) Direct valve gear – engine running over

(b) Direct valve gear – engine running under

(c) Indirect valve gear – engine running over

(d) Indirect valve gear – engine running under

FIG. 9.5 *Direct and indirect valve gear, showing the relative position of the D slide valve, eccentric, and crank when the piston is on the head-end dead center.*

Always rotate the eccentric in the direction in which the engine is going to run. If the lead is incorrect, increasing or decreasing the angle of advance will remedy the difficulty. To reverse the direction of rotation, merely rotate the engine in the opposite direction and follow the procedure outlined for setting valves.

It sometimes happens that the eccentric slips and a hurried adjustment is necessary. To effect this it is not necessary to remove the valve cover. Place the engine on dead center and block the flywheel in position. Loosen the setscrew on the eccentric and turn the latter in the direction in which it is to run. Open the cylinder cock on one end. Admit steam to the chest by means of the throttle valve. Next, open the valve by moving the eccentric. When steam issues from the cock, tighten the eccentric, rotate the engine to the opposite dead center, and try again. If steam issues from this end, the setting is satisfactory and the engine is ready to run. Remember that this is only a temporary setup. Reference to an indicator diagram will disclose any errors, and corrections can be made.

Increasing the angle of advance makes all the events in the cycle *earlier; decreasing* the angle of advance makes them *later*. A valve is very seldom set for equal leads (to obtain the best economy) but is usually some compromise between lead and cutoff. The final setting of valves should always be checked by means of an indicator.

It frequently happens that the eccentric cannot be located on the shaft to bring in line the valve stem and the eccentric rod. This cannot be done when the valve rides on top of the cylinder without inclining the seat. For this and other reasons, such as flywheel interference, it may become necessary to use a rocker arm.

There are two kinds of rocker arms: direct and indirect. In the direct rocker arm (Fig. 9.5a) the valve stem and eccentric rod are both attached so as to lie on the same side of the fulcrum. In this case the direction of valve and that of the eccentric are the same. In the reversing rocker arm (Fig. 9.5c) the fulcrum lies between the points of attachment of the valve stem and eccentric rod. In this case the valve and eccentric have opposite directions. Figure 9.5 shows the eccentric setting and the relation of the crank for running over and under, direct and indirect.

In Fig. 9.5a we have a rocker arm with a direct motion. The eccentric leads the crank by 90° plus the angle of advance, which is the same as with an ordinary slide valve. The engine here is running over. In Fig. 9.5b the engine is running under, and the eccentric is again leading the crank by 90° plus the angle of advance. When the motion is indirect, as in c and d, the eccentric follows the crank by 90° minus the angle of advance.

Direction of rotation	Kind of rocker arm	Angle between crank and eccentric	Position of eccentric with respect to crank
Over...............	Direct	90 + angle of advance	Ahead
Over...............	Indirect	90 − angle of advance	Behind
Under..............	Direct	90 + angle of advance	Ahead
Under..............	Indirect	90 − angle of advance	Behind

The ordinary engine for stationary work is designed to run in one direction. It is sometimes required to run in the opposite direction. For the engine to do this, it becomes necessary to change the position of the eccentric. This is accomplished by loosening the eccentric and rotating it in the direction desired until the eccentric again leads the crank by 90° plus the angle of advance. To do this while the engine is running is very inconvenient and difficult. For this reason two eccentrics are frequently employed, one for running forward, the other for running in reverse. The two are interconnected by a link motion. With the link in a central position neither eccentric has any control over the valve. Quickly shifting one or the other into position gives the eccentric complete control, and the engine is ready for the particular direction desired.

9.3 THE SLIDE VALVE. In addition to the D slide valve, several other types of slide valves are employed on steam engines. Among these are the balanced valve, the multiported valve, the riding cutoff valve, and the piston valve.

The conventional slide valve is held against its seat by the pressure of steam exerted against the area of the valve. This pressure results in friction's being created between the flat surface of the valve and its seat, causing wear and some loss in efficiency. This is overcome by using a balanced valve which tends to relieve the pressure and reduce friction to a minimum.

The flat valve (Fig. 9.6) is recommended if the steam is likely to contain large amounts of water. The valve is balanced and stays steam-tight, lifting off the seat automatically to relieve the cylinder of excess condensate. The pressure plate is centered over the main valve and prevented from moving with the valve by lugs or ears, as shown. Springs are provided between the steam-chest cover and the pressure plate to hold the valve against the seat. Steam is prevented from getting on top of the valve by a seal located between the side of the valve and the pressure plate.

The multiported valve is similar to a D slide valve, except that steam

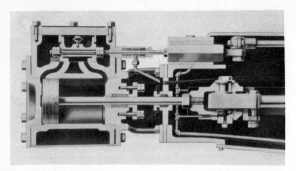

FIG. 9.6 *Balanced slide valve-flat type.* (*Troy Engine & Machine Co.*)

is admitted through several passageways at once. This design permits shorter valve travel with quick opening and closing of steam and exhaust ports.

The riding cutoff valve employs a double eccentric, one controlling the main valve, the other controlling the point of cutoff. This valve is used in connection with a governor on an automatic engine; its position is varied depending on the position that the governor occupies. The main valve controls the admission, release, and compression; the auxiliary valve can then be adjusted to vary the point of cutoff. The riding cutoff can adjust the point of cutoff beyond the range of the ordinary valve without altering the other events in the cycle. This auxiliary valve usually rides on top of the main valve and has a motion opposite to it.

The D slide valve is designed to take steam from around the outside edge and exhaust it past its inside edge. This valve has both steam and exhaust lap. The former controls the admission of steam and the latter the exhaust. Steam lap regulates admission and cutoff; exhaust lap controls release and compression. An increase in steam lap makes admission later and cutoff earlier; a decrease makes admission earlier and cutoff later. An increase in exhaust lap delays release and makes compression earlier; a decrease makes release earlier and compression later.

Regardless of whether the steam admission is inside or outside, the steam lap always affects admission and cutoff; likewise, exhaust lap affects release and compression. The amount of lap and lead required depends on the type of engine under consideration and on whether the engine is operating condensing or noncondensing.

The balanced, multiported, and riding cutoff valves are designed to overcome the shortcomings of the D slide valve. The balanced valve reduces friction to a minimum; the multiported valve overcomes the slowness of opening and closing; the riding cutoff provides a variable cutoff rather than a fixed cutoff point.

The ordinary D slide valve cannot cut off earlier than about five-eights of the stroke; hence the range of expansion is limited. High-pressure steam cannot be used satisfactorily, as excessive friction would result in damage to the valve and seat; high-temperature (superheated) steam would cause the valve to warp. For the D slide valve, the clearance volume is usually great; the valve does not lend itself to a wide range of adjustment for varying the events in the cycle. For these reasons the slide valve is used on small engines and where the exhaust steam can be used for process or heating; it is not normally used on engines over 100 hp.

The Slide-valve Engine

Advantages

Simple mechanism
Low maintenance
Low initial cost

Disadvantages

Large clearance volume
Unsuitability for high-pressure and high-temperature steam (usually)
High cylinder condensation owing to location of ports
Wiredrawn steam owing to valve design

9.4 PISTON VALVES. The piston valve (Figs. 9.7, 9.8) is a modified slide valve, the cylindrical construction and mechanism being essentially the same. The chief difference between an ordinary D slide valve and the piston valve lies in the construction of the steam chest, the valve, and the manner in which steam is admitted and in the fact that the piston valve is a balanced valve.

The piston valve (Fig. 9.8) is a cylindrical cast-iron plug, the ends K being larger in diameter than the central section H. The valve reciprocates within the sleeves $H5$ fixed in position in the steam chest. The sleeves are fitted with slots $H8$ located at each end and at the center $H7$ which coincide with passageways $1H2$ and $2H2$ to the ends of the cylinder A.

Steam is admitted at the center (balanced valve); as the piston valve moves to uncover the slots $H7$ in the sleeves $H5$, steam from the steam chest $A3$ enters the passageway $1H2$ (piston at the left). The exhaust steam in the cylinder A escapes through passageways $2H2$, passing through slots $H8$ and

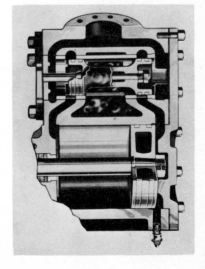

FIG. 9.7 *Piston valve.* (*Troy Engine & Machine Co.*)

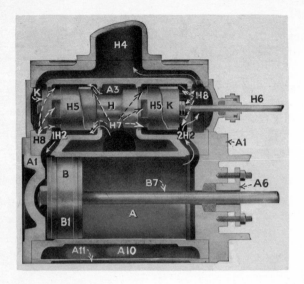

FIG. 9.8 *Part sectional view of cylinder and valve of a piston-valve steam engine.* (*Mobil Oil Corp.*)

around the end of the valve into the exhaust pipe *H*4. The piston valve in this manner controls the passage of steam in and out of the cylinder.

The amount of lap and lead given to the valve for proper operation is very much the same as that for setting the D slide valve. The admission of steam, however, occurs over the *inside* edge of the valve instead of the outside edge, as with the normal slide valve. Since the admission of steam is over the inside edge of the valve, the angular position of the eccentric is that for an indirect rocker arm (Sec. 9.2).

The piston valve is exceptionally light. It is considered to be perfectly balanced and is readily made double-ported. However, unless the valve is provided with special packing or means for adjusting the wear, leaks will result which in turn cause waste of steam.

The use of a piston slide valve on a vertical engine has an advantage in that less wear is experienced than with the horizontal type. If the valve seat consists of some type of replaceable bushing, the best procedure, if leakage is experienced, is to replace both the valve and the seat. This type of valve, however, is less economical than other valves since the area above the ports increases the clearance volume. The piston valve is frequently used with a riding cutoff valve on an automatic engine.

Piston Valve

Advantages

Can be used with high pressure and superheated steam
Cylinder condensation and radiation losses reduced with inside steam admission
Simplicity (equivalent to that of a common D slide valve)

Disadvantages

Wiredrawing due to slow opening and closing of ports
Leakage of valves difficult to detect or remedy
Only one eccentric, from which all events must be controlled
Impossibility of independent adjustment of valves, except when employing
riding cutoff valves

9.5   CORLISS VALVES.   The development of the Corliss valve resulted
from a desire to overcome the disadvantages of the common slide valve,
one major disadvantage being that all events in the cycle are controlled
by a single valve, the position of which is fixed.

The cylinder of the Corliss engine differs in design from that of other
engines discussed thus far.   The cylinder (Fig. 9.9) is fitted with four
valves, the two upper being steam "admission" and the two lower "exhaust" valves.   These valves are long and cylindrical and are usually
made of cast iron.   The center section of the valves is cut away; the
admission valves have double inlet edges, thus securing a large steam
opening with a small movement of the valves.   By reason of the small
arc through which the valves move, friction and wear are reduced to
a minimum.

In operation (Fig. 9.9), steam enters at the top left of the cylinder

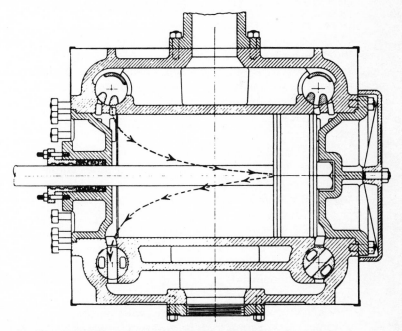

FIG. 9.9   *Four-valve engine cylinder with admission and exhaust
valves.*   (*Skinner Engine Co., subsidiary of Patterson Industries,
Inc.*)

FIG. 9.10   *Corliss-engine releasing valve gear with belt-driven governor with safety stop.*   (A) *Governor.*   (B) *Governor rod.*   (C) *Governor pulley.*   (D) *Governor safety stop.*   (E) *Wristplate.*   (F) *Eccentric rod.*   (G) *Steam-valve mechanism.*   (H) *Steam-valve rods.*   (I) *Exhaust valves.*   (J) *Exhaust-valve rods.*   (K) *Dashpot.*   (*Filer & Stowell Co.*)

through a two-ported valve to move the piston in one direction, as shown by the arrows; when the piston reaches the position shown, the exhaust valve opens (lower left) to discharge the steam.   Hence the steam changes its direction of flow from top to bottom, in contrast to the slide valve, in which steam enters and exhausts through the same passageway.

Corliss engines are operated with releasing or nonreleasing gears. *Releasing gears* or *trip gears* (Figs. 9.11, 9.12) employed on the Corliss engine control the point of cutoff by having the steam valve released from the eccentric's control.   The valve is then closed by the dashpot or some device such as a spring to obtain quick closing.   *Nonreleasing* or *positive-motion* valve gears (Fig. 9.13) are such as are directly controlled by the eccentric during the entire period of operation.   Nonreleasing valves permit relatively high speed and close speed regulation. High speed demands a valve gear with positive connection so as to provide quick opening and closing of the valves.

A single-eccentric Corliss engine with its component parts, including governor *A*, governor rod *B*, governor pulley *C*, governor safety stop *D*, wristplate *E*, and eccentric rod *F*, is shown in Fig. 9.10.   Steam valves *G* are on the upper right and left of the cylinder; the valve rods *H* are shown attached to the wristplate *E*. Exhaust valves *I* are located at the bottom of the cylinder and at each end, with valve rods *J* as shown. The dashpot rod is attached to the steam valve above and connected to the dashpot *K* below. The exhaust valves are operated directly from the wristplate and are of the nonreleasing type.

Keyed to the valve stem (Fig. 9.11) is a latch plate in which latch *M* can be positioned. The latch is free to move *around* the valve extension stem; it is positioned by the valve rod from the wristplate. The latch engages the latch plate to open the valve (and raise the dashpot); the length of time the steam valve is permitted to remain open is determined by the position of the governor rod *B*.

With reference to Fig. 9.11 and the upper steam-valve assembly, mounted on the steam-valve extension rod are three arms:

1. The outer arm, which is keyed to the valve extension rod at *C*. Attached to it are the latch plate *L* and the dashpot rod.

2. Latch *M*, which is attached to valve rod *B*. This arm swings *around* the valve extension rod but is not keyed to it. Latch *M* engages latch plate *L* to open the steam valve.

3. Governor rod arm (top). This also is free to swing *around* on the valve extension rod. On the back of this arm are the knockoff cam and the safety cam.

In operation, the governor positions the knockoff cam to trip latch plate *M*, disengaging the steam valve, which the dashpot closes. When the governor arm is completely advanced, the safety cam keeps the latch disengaged, preventing the steam valve from opening (this prevents the engine from racing).

The Corliss steam-valve mechanism (Fig. 9.11) is shown in three operating positions. In Fig. 9-11*a* the steam valve is closed, with the latch riding on top of the latch plate. The dashpot has dropped and closed the steam valve. In Fig. 9.11*b* the valve is shown ready to open; the latch will engage the latch plate while raising the dashpot. In Fig. 9.11*c* the steam valve is again closed as a result of the governor's positioning the safety cam to disengage the latch *M*, thus preventing the steam valve from opening.

The exhaust valves are directly connected to the wristplate by the valve rod and are nonreleasing. The steam valve in turn has its opening controlled by the latch positioned by the valve rod attached to the wristplate; the latter is operated from the eccentric rod. The length of time the steam valve is permitted to remain open is determined by the speed of the governor; the governor's speed varies the position of the knockoff cam to disengage the latch, permitting the dashpot to close the valve. Hence the steam valve is releasing and is not positive-acting.

(a)

(b)

(c)

FIG. 9.11 *Valve mechanism in various positions of operation.* (a) *Valve closed; latch riding on top latch plate.* (b) *Valve ready to open; latch engaging latch plate.* (c) *Valve closed; latch held in disengaging position by safety cam.* (B) *Valve rod.* (C) *Steam-valve mechanism.* (L) *Latch plate.* (M) *Latch.* (N) *Knockoff cam.* (O) *Safety cam.*

With changes in governor speed the position of the knockoff cam varies to make cutoff earlier or later in the stroke, thus regulating the engine speed.

The dashpots are used to close the steam valves quickly after the knockoff cam has disengaged the latch from the latch plate. Each dashpot consists of a cylinder fitted with a plunger. This plunger has packing rings fitted about its circumference to make an airtight joint. As the dashpot is raised, a vacuum is formed; when the steam valve is released, the atmospheric pressure forces it down. The air trapped in the chamber provides cushioning to prevent the plunger from striking the bottom. Too much compression or cushioning, on the other hand, will probably cause the plunger to rebound, reopening the steam valve.

The amount of lead given the engine depends on the relative proportions of the valve and the conditions under which the engine is to operate. In the slide-valve-engine setting, attention must be given the lap of the valve, as lap affects the point of cutoff and changes the expansion considerably. This is not the case with the Corliss engine. The cutoff here is a variable, being controlled by the governor through the medium of the knockoff cam.

The Corliss engine, owing to its many complicated parts, is essentially

FIG. 9.12 *Corliss-valve gear—releasing type.* (*Filer & Stowell Co.*)

FIG. 9.13   *Corliss-valve gear—nonreleasing type on uniflow engine.*   (*Filer & Stowell Co.*)

a slow-speed machine, ranging from 50 to 150 rpm.   This is about the maximum speed encountered with detaching Corliss engines using dash-pots.   With one eccentric the cutoff is limited to about one-half of the stroke, whereas with two eccentrics the cutoff can be extended to almost any length.

Positively operated valves may be operated at higher speeds than the releasing type.   With two eccentrics the steam and exhaust valves can be controlled independently.   This results in better steam distribution and a range of cutoff not possible with a single eccentric.

The Corliss engine was developed to overcome the disadvantages of controlling both admission and exhaust of steam by a single valve.   A single valve limits the degree to which events can be altered without affecting the other events in the cycle.   For this reason a Corliss is more flexible and considerably more economical than a simple slide-valve engine.

Corliss Valve

Advantages

Reduction in clearance volume due to valve's being set directly over the ports (which are very short)

Reduction of condensation due to separate admission and exhaust ports and
to reduction in clearance
Independent adjustment of valves
Increased economy of the engine due to range of possible cutoff

Disadvantages

Slowness of speed
Unsuitability for high pressure and high temperature (which causes valves
to warp and stick)
Complexity of mechanism

9.6   POPPET VALVES.   A view of the poppet-valve engine (Fig. 9.14)
shows details of cylinder and valve arrangement.   The poppet valve
$2H$ consists of a cast-iron disk or cylindrical body fitted to a valve stem.
The valve is designed to fit into an upper and a lower seat.   Steam-valve
chambers communicate through passageways $A3$ with the ends of the
cylinder A.   A cam or eccentric G, actuated by the crankshaft or dome
of the revolving shaft geared to the main shaft, raises valve rod $G2$
to lift the poppet valve from its seat.   Valve rod $G2$ raises the valve,
the opening being secured with a short, quick lift.   The valve is closed
by the action of the spring $1H$.   Distribution of steam is secured by
the independent timing of the steam and exhaust valves.

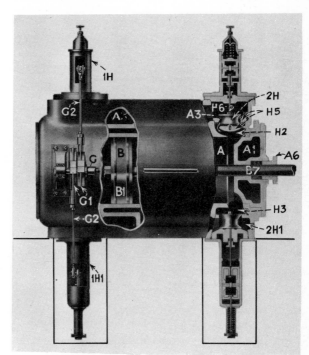

FIG. 9.14   *Side view of cylinder equipped with poppet
valves.   (Mobil Oil Corp.)*

Poppet valves are applied equally well to the conventional piston-type engine (Fig. 9.14) and to the uniflow engine (Fig. 9.16). The steam-valve eccentrics are controlled by a governor mounted on the lay shaft as shown. If required, the steam-valve gear may be controlled by a governor mounted in the flywheel; when so arranged, the exhaust valves are operated by a fixed eccentric mounted directly on the crankshaft.

The four-valve engine is more economical in the use of steam than is the simple slide-valve engine; in addition, it provides independent adjustment of the valves. The action of the valve is such that it opens and closes quickly with little throttling. Valves of this type are not

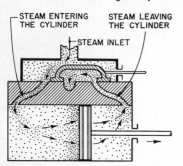
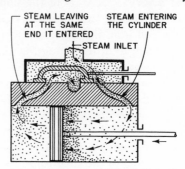

(a) COUNTER-FLOW STEAM ENGINE
FORWARD STROKE

(b) COUNTER-FLOW STEAM ENGINE
RETURN STROKE

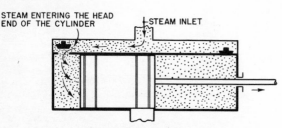

(c) UNIFLOW STEAM ENGINE
FORWARD STROKE

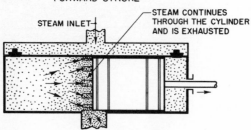

(d) UNIFLOW STEAM ENGINE
NEAR THE END OF THE FORWARD STROKE

FIG. 9.15 *Comparison of steam travel in the cylinders of counterflow and uniflow engines.*

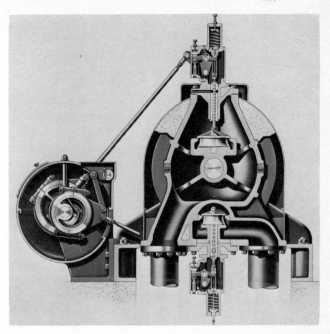

(a)

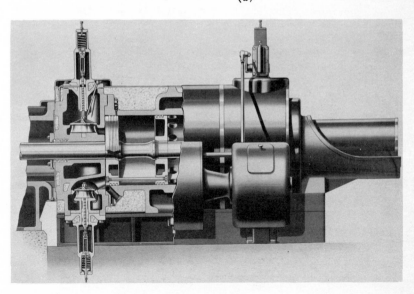

(b)

FIG. 9.16   (a) Section of poppet-valve uniflow engine and governor on lay shaft.   (b) Poppet valve in cylinder head on uniflow engine. (Filer & Stowell Co.)

subject to wear resulting from friction. Since they are set directly over the end of the cylinder, clearance volume is reduced. Poppet valves and operating mechanism are more complicated than are the slide valve and the Corliss, but they are extensively used for installations employing high-pressure and superheated steam.

9.7 THE UNIFLOW ENGINE. The conventional engine is classed as counterflow because the steam flows into the cylinder and reverses direction to flow out at the same end. For the simple slide-valve engine, steam enters and leaves through the same port (Figs. 9.15*a*, *b*). In the four-valve Corliss engine, steam enters through one valve and leaves through another, both on the same end of the cylinder. So, in each case, the steam has to reverse its direction of flow.

As the name "uniflow" implies, steam is always flowing in but one direction. The uniflow principle permits steam to enter at the end of the cylinder and leave through exhaust ports located in the middle of the cylinder (Figs. 9.15*c*, *d*). As the piston approaches the end of the stroke, it uncovers the exhaust ports, permitting the steam to be released shortly before the piston reaches its dead-center position. On the return stroke, steam remaining in the cylinder is compressed. Hence the piston must perform its normal function and also act as an exhaust valve. Note the length of the piston in comparison with that of the cylinder. The length of the piston is equivalent to approximately nine-

FIG. 9.17 *Longitudinal section through cylinder of four-valve nonreleasing Corliss uniflow engine.* (*Filer & Stowell Co.*)

tenths of the length of the stroke. The exhaust ports are greater in area than those of other engines.

The uniflow engine (Fig. 9.18) is intended primarily to eliminate the major loss encountered in steam engines, that of cylinder condensation. In the counterflow engine, the ends of the cylinder are alternately heated and cooled on each stroke, by contact first with the high-temperature steam entering and then with the exhaust. In the uniflow engine, the intake valves and cylinder heads are not cooled by the exhaust steam, and therefore the loss due to cylinder condensation is reduced. Uniflow engines are more costly to build than counterflow engines, since the cylinder must be larger to accommodate the larger piston. They are usually equipped with poppet valves, although Corliss valves (Fig. 9.17) are also employed. Since the piston is large and heavy, tail rods are used to support it. The poppet-valve uniflow engine is well adapted to the use of high-pressure, high-temperature steam and to operating condensing.

Since the uniflow engine was primarily intended to operate condensing, difficulty is experienced with excessive compression when it becomes necessary for the engine to operate noncondensing. This difficulty is avoided by equipping the engine with auxiliary exhaust valves (Fig. 9.19) near the end of the cylinder. These valves, also of the poppet type, are used to eliminate the excessive compression while the engine is operating noncondensing. The use of a uniflow engine should be considered for applications where only small quantities of exhaust steam will be required and for condensing operation. Increased steam economy usually warrants the added initial investment.

Characteristics of Steam Engines

Engine	Steam pressure, psi	Superheated steam	Speed, rpm	Clearance volume, per cent
Slide-valve.......	50–125	No	100–350	3–12
Corliss...........	100–175	Little*	50–150	3–10
Poppet...........	100–250	Yes	75–225	3–7
Piston...........	50–250	Yes	100–400	5–10
Uniflow..........	100–250	Yes	100–350	1–4

* Corliss poppet valve as high as 300°F.

In the conventional *condensing* uniflow, compression starts as soon as the piston covers the exhaust ports located in the center of the cylinder. This continues during the remainder of the stroke (approximately 90 per cent), and with small volumetric clearances of 2 to 5 per cent

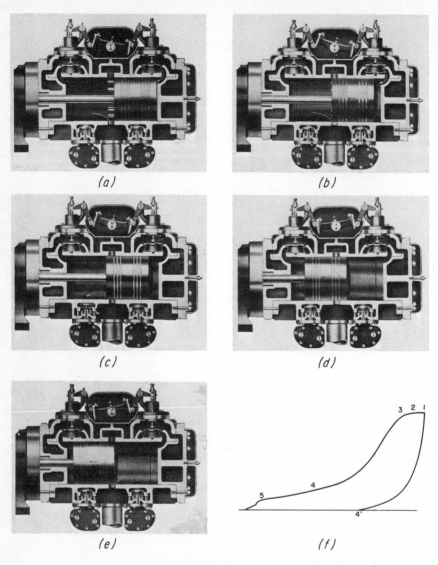

FIG. 9.18   *The Universal Uniflow engine operating noncondensing.*
(a) Head end—admission on dead center; frame end—exhaust through
central ports.   (b) Head end—full opening of valve; frame end—
exhaust through central ports about to close.   (c) Head end—closing
of admission valve and beginning of expansion; frame end—displacement
of trapped vapor through open auxiliary exhaust valve.   (d) Head
end—continuation of expansion; frame end—commencement of compres-
sion by piston covering auxiliary exhaust port.   (e) Head end—
beginning of exhaust through central exhaust ports; frame end—
continuation of compression.   (f) Noncondensing diagram.   (Skinner
Engine Co., subsidiary of Patterson Industries, Inc.)

the compression does not exceed the initial pressure. Should something happen to the vacuum, the compression would be so high as to prove detrimental to the engine.

If such an engine operates noncondensing and compression occurs during 90 per cent of the return stroke, large clearance volumes would be required (10 to 20 per cent) to prevent compression from exceeding the initial pressure. Such clearance chambers are controlled by hand-operated valves to relieve the high (compression) pressure by discharging some of the steam to the clearance pockets; this naturally reduces the efficiency of the engine.

The uniflow engine shown in Fig. 9.18 has largely corrected the condition mentioned above as occurring when running noncondensing without added clearance. This has been done by the application of auxiliary exhaust valves located at a point where compression usually begins in noncondensing operation. These auxiliary valves remain closed on the expansion stroke but open on the compression stroke. They are not located at the end of the cylinder so as to prevent exhaust steam from sweeping the length of the cylinder and causing initial condensation. The operation of the uniflow engine operating noncondensing can be explained by reference to Fig. 9.18:

a. Admission starting and exhaust continuing through the exhaust ports in the center.

b. Admission valve fully open; exhaust ports almost closed and auxiliary exhaust valve ready to open, though exhaust pressure is on both sides of the valve.

c. Steam valve closing; start of expansion. The auxiliary exhaust valve is wide open.

d. Expansion continued; the piston has crossed the auxiliary exhaust port with compression started. NOTE: Compression did not begin when the piston closed the exhaust ports in the center of the cylinder, but with the exhaust auxiliary valves this action was delayed and the efficiency increased.

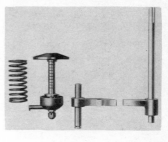

(a)

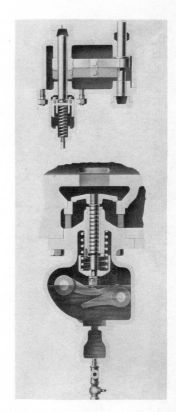

(b)

FIG. 9.19 (*a*) *Auxiliary-exhaust-valve gear.* (*b*) *Cross section.* (*Skinner Engine Co., subsidiary of Patterson Industries, Inc.*)

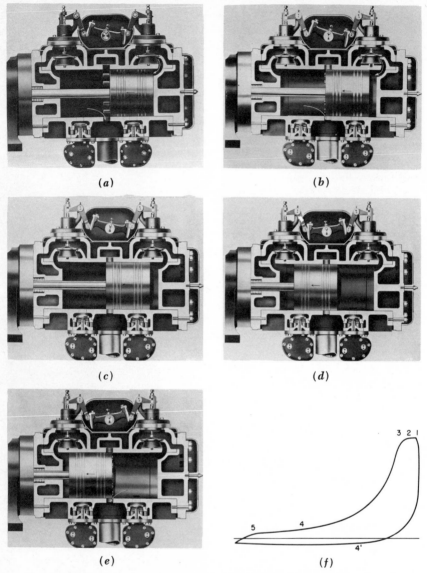

FIG. 9.20   *The Universal Uniflow engine operating condensing.*
(a) Head end—admission on dead center; frame end—exhaust through
central exhaust ports.   (b) Head end—full opening of valves; frame
end—exhaust through central exhaust ports about to close.   (c) Head
end—closing of admission valve and beginning of expansion; frame end—
commencement of compression caused by piston covering central exhaust
ports.   (d) Head end—continuation of expansion; frame end—continuation
of compression.   (e) Head end—beginning of exhaust through central
exhaust ports; frame end—continuation of compression.   (f) Condensing
diagram.   (Skinner Engine Co., subsidiary of Patterson Industries, Inc.)

e. Exhaust begun through the center ports on the side of admission (side where steam was previously admitted). The compression continues on the other side, and the exhaust auxiliary valve is almost ready to close.

The fact that the auxiliary exhaust valves open and close under no difference in pressure on either side of the valve permits operation with a single-seated poppet valve, requiring less clearance than a double-seated valve, which would ordinarily be necessary if valves were to open and close against the pressure. When operating condensing, the auxiliary exhaust valves are automatically disengaged and remain closed at all times unless the vacuum breaks. In that case valves are automatically placed in operation, and the engine becomes noncondensing in operation. This permits a plant to be operated condensing or noncondensing with the same relative efficiency, and the change from one to the other can be made automatically.

Details of the exhaust-valve gear are shown in Fig. 9.19 together with the automatic disengaging device. For condensing operation, when the vacuum reaches a predetermined point, it overcomes the tension of the spring and draws the shaft into the cylinder by means of the piston attached to the outer end of the shaft. This shifts the idler to the secondary position (shown in phantom), which is out of register with both the cam and the valve stem, so that while the cam still operates as before, it cannot lift the valve. The valve remains closed at all times when the vacuum exists in the exhaust line. The cam is so designed that the idler is always supported by it; the idler is held to the proper height so that if the vacuum fails, the auxiliary valve begins its normal functions and the engine becomes automatically a noncondensing engine.

Operation of the Universal Uniflow engine condensing is shown in Fig. 9.20. It is to be noted that the exhaust valves remain closed during the entire cycle of operation.

QUESTIONS AND PROBLEMS

9.1. What are meant by admission, cutoff, release, and compression?

9.2. How do you give an engine more compression?

9.3. What is the purpose of valves?

9.4. Make a sketch of a slide valve.

9.5. Mention several types of valves usually employed.

9.6. How does a slide valve work?

9.7. What is lap? Mention several different kinds.

9.8. What is the result of increasing the outside and inside lap?

9.9. What is lead? Why is an engine given lead, and how much lead is usually given?

9.10. What is the purpose of exhaust lead on a slide valve?

9.11. What do you mean by angular advance?

9.12. What is the result of increasing the angle of advance with no other change?

9.13. Does decreasing the angle of advance increase the power, and why?

9.14. What is the earliest cutoff that can be obtained with a common slide valve?

9.15. What keeps the piston ring from wearing a shoulder at each end of the cylinder?

9.16. How should you change the direction of rotation of a plain slide-valve engine?

9.17. How do you set the valves on a plain D-slide-valve engine?

9.18. What is the advantage of a riding cutoff valve over a plain slide valve?

9.19. What is meant by wiredrawn steam?

9.20. What is the advantage of a balanced valve?

9.21. How does a riding cutoff valve work?

9.22. What is meant by the term "positive cutoff"?

9.23. Which engine uses the more steam, one with a large, or one with a small, clearance volume?

9.24. Explain how a Corliss valve and valve gear operate.

9.25. What are the advantages of two eccentrics on a Corliss?

9.26. Give several advantages of a Corliss over the slide-valve engine.

9.27. What is the purpose of a dashpot? Is it used on all engines?

9.28. What is a wristplate? What is its purpose?

9.29. What is a poppet valve?

9.30. When is a poppet valve usually used? What are its advantages?

9.31. What is the chief difference between a uniflow and some of the other engines mentioned?

9.32. How do the pistons of a uniflow and a counterflow engine compare as to length?

9.33. Approximately how much clearance volume do the following types of engines have: (*a*) slide valve, (*b*) poppet valve, (*c*) uniflow engine?

9.34. What is meant by the term "negative lead"?

9.35. What two types of valves are best suited for high-temperature steam?

9.36. What are the advantages of a four-valve engine over a slide-valve engine?

CHAPTER 10

OPERATION
AND
MAINTENANCE
OF
STEAM ENGINES

The operators of steam engines must work to promote safety, assure uninterrupted service, and secure maximum efficiency and low maintenance costs. The principal safety hazards involve moving parts, overspeeding, water in the steam, and rupture of steam lines and other pressure parts. Consider these hazards and provide adequate protection in the form of guards, safety devices, and training of the personnel.

Continuous service and low maintenance are closely associated. When you provide satisfactory lubrication, close checks when operating, and periodic inspections and adjustments, both forced outages and maintenance will be kept at a minimum.

The possible maximum efficiency of an engine plant is determined by many factors, including steam pressure, superheat, and back pressure. It is the operator's duty to obtain the efficiency that was built into the engine plant by supplying the steam at the specified pressure and temperature, providing a minimum back pressure, checking and adjusting valves and governors, and maintaining the mechanical components of the engine in good operating condition.

10.1 INSPECTION OF ENGINES. When assuming charge and periodically when operating a plant, check the engines and auxiliaries to determine their condition, their need for adjustment and replacement of parts. The thoroughness necessary in these inspections will depend upon plant conditions and past experience.

Examine the main eccentric and connecting-rod bearings for wear; note any indication of faulty lubrication. Determine the "striking points" of the engine and mark them on the crosshead guides for future reference. Remove the steam-chest cover to observe the operation of the valves. Check the surfaces for wear and to see if they have been receiving sufficient lubrication. Turn the engine over and note the lead and point of cutoff at each end of the cylinder. The lead should be equal on both ends of the cylinder, and the point of cutoff should occur when the piston is in the same position on both ends of the stroke. For engines controlled by variable cutoff, it is necessary to block the governor in about mid-position before turning the engine over to check the operation of the valves. It may be found advisable to measure the lead and point of cutoff with the governor blocked in several different positions. Other things that can be checked during this observation are lost motion in the governor mechanism owing to wear and the amount of oil in the governor dash- or gag pot. Observe and periodically test the governor and all other devices provided to prevent the engine from overspeeding. In checking an engine equipped with a belt-driven fly-ball governor, remove the belt and see if the "safety" shuts off the steam supply. The final check on the valve and governor operation can be made by means of the steam-engine indicator after the engine is in operation. Observe the markings on the eccentric valve rods. If they are not marked, they should be, to establish the length and thus make it unnecessary to check the valve setting each time the engine is torn down.

Remove the cylinder head to inspect the cylinder walls for satisfactory lubrication and possible wear. Now is the time to determine whether or not to remove the piston and rings. If there has been no indication of leakage past the rings and the cylinder walls do not show wear, it will not be necessary to remove the piston or rings. When in satisfactory condition, the cylinder walls will be dark with no bright spots and if the engine has been operating recently will contain a very thin film of oil. This film will be sufficient to show on a paper towel rubbed over the surface. Faulty lubrication will be evident from scoring of the cylinder walls and excessive wear.

Examine the piston-rod and valve-rod packing. If soft packing is used and the gland is pulled up nearly all the way, it is best to replace the packing. Note the condition of metallic packing. Measure the shaft and packing box and arrange to have a replacement set of packing available at all times.

Check the lubrication system and make sure that all pipes are open and that all working parts are receiving oil. See if provisions have been made to atomize the cylinder oil as it is introduced into the steam lines. This is an advantage in providing satisfactory cylinder lubrication. Note the condition of the oil in the system; filter or replace if necessary. Fill all lubricators with the specified oil.

Now direct attention to the steam-supply and engine-exhaust system. The steam-supply line should have a separator ahead of the throttle valve with a trap for removing the accumulation of water and as a precaution against carry-over from the boilers. The steam chest and each end of the cylinder must be supplied with drain cocks to allow water to flow from the steam chest and cylinders during the starting-up period. The receiver between the high- and low-pressure cylinders on the compound engine must be provided with a means for draining the water. Examine the exhaust piping, including the atmospheric-relief valve and condenser connection. See that the engine has free exhaust to the low-pressure system, atmosphere and condenser depending upon the arrangement employed. Check the condenser exhaust piping for possible leaks, and determine the operating procedure to be used in connection with the condenser equipment. Note whether the engine is adaptable to either condensing or noncondensing operation and what changes are necessary when going from one type of operation to the other.

10.2 LUBRICATION OF STEAM ENGINES. Lubricants must be selected and applied to reduce the wastage of power and the wear on rubbing surfaces. Two types of lubricants are required: one for the valves and cylinders and another for the bearings.

The ideal steam-engine-cylinder oil must possess four qualities in balanced proportion to meet the needs of the specific plant: resistance to being washed away by condensation, ability to seal the piston rings against steam leakage, resistance to the action of heat, and adequate lubrication to reduce wear. The cylinder oil consists of petroleum compounded with animal fat in the form of acidless tallow and lard oil. Animal fat reduces the ability of a lubricant to withstand heat but improves its ability to cling to the cylinder walls and resist being washed off with wet steam. Engines using dry or superheated steam require lubricants with little or no compounding. The wetter the steam, the more animal fat (compounding) is required for satisfactory lubrication. Low-speed engines, in which there is likely to be considerable condensation, require large amounts of compounding.

There are two common types of lubricators for forcing the cylinder lubricating oil to the steam line against the pressure: the hydrostatic and the mechanical. The hydrostatic lubricator (Fig. 13.18) depends upon the weight of a column of water to force the oil into the cylinder. The pressure of the steam plus the weight of a column of water is greater than the pressure of the steam alone; hence a slight difference (equal to the weight of the column of water) is produced across the lubricator. The rate of flow is regulated by a needle valve, and the quantity is measured by counting the drops of oil as they rise in the glass. The oil is displaced with water, which must be drained out before the lubricator is filled with oil. This hydrostatic lubricator provides a simple means of supplying oil but is subject to some variations in the rate of feed as the viscosity of the oil changes, the needle valve clogs up, and the supply of oil becomes depleted. The mechanical lubricator (Fig. 13.19) consists of a pump driven by the engine. This pump forces the oil into the steam line. The stroke of the pump may be adjusted, or the angularity of the operating arm may be changed to vary the flow of oil and thus provide the necessary lubrication. The hand crank is provided for supplying extra lubricant to the cylinder when the engine is started up or an emergency arises. Lubricators of this type are furnished with multiple-piston plungers for supplying lubricant to several parts of the engine (Fig. 10.1). Sight feeds are furnished so that the operator can observe and adjust the feed to the various parts being lubricated. The oil lines are fitted with spring-loaded non-return valves. These hold the line full of oil and ensure a constant supply. Lubricators and lines should be located adjacent to the steam chest or other heated parts of the engine so that they will not be exposed to low temperatures which might cause the oil to congeal and restrict the flow. Each oil line should be checked to see that it is feeding the correct amount of oil. Oil feeds are usually specified in drops per minute.

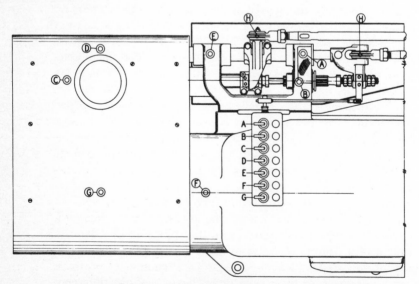

FIG. 10.1 *Lubrication of the steam cylinder and valve gear of an engine-driven compressor.* (*Ingersoll-Rand Co.*)

The application of a mechanical lubricator to the high-pressure steam cylinder of a steam-driven air compressor is shown in Fig. 10.1. The low-pressure cylinder has a similar but separate lubricator, permitting the use of a different grade of oil. Mechanical lubricators are driven by the reciprocating motion of the valve mechanism and provide seven pumps for automatically supplying oil to the cylinder and valve mechanism. The operator must fill the lubricator reservoir and check and adjust the feeders to the specified rate of flow for each of the seven points supplied. The rate of feed to the valve mechanism is correct when a small amount of oil appears at the bearing or on the rod. The quantity of oil required to lubricate the cylinder (points D and G) is discussed in a subsequent paragraph. The points designated by H have a low rubbing speed and are lubricated with grease.

The indirect or atomization method has been generally accepted for introducing oil into the cylinders of steam engines. This consists of injecting the oil into the steam before it reaches the valves. The steam is used to carry the oil and to deposit it on the friction surfaces of the valves and cylinders. The oil must be introduced into the steam in such a way as to ensure complete atomization and effective lubrication. Figure 10.2a shows oil being introduced into a steam line through a simple pipe. This method is unsatisfactory, as the oil runs down the side of the pipe and fails to distribute over the wearing surfaces. Better atomization is obtained by the quill arrangement shown in Fig. 10.2b. Here the oil is introduced into the center of the steam pipe,

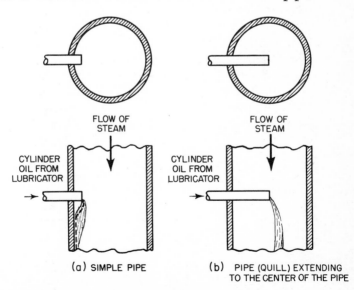

FIG. 10.2 *Atomizers used for introducing the cylinder oil into the steam line for engine-cylinder lubrication.*

and the steam velocity breaks up the oil into small particles and distributes it evenly over the friction surfaces. Further improvement in the atomization may be obtained by the use of patented devices which premix the oil with steam and condensate before it is discharged into the steam. These devices are useful, especially when it becomes necessary to introduce the oil near the steam chest. Normally the oil should be introduced into the steam line after the water separator but several feet ahead of the throttle valve.

Piston rods may not receive sufficient oil from the steam to provide adequate lubrication, in which case both packing and rod will wear rapidly. This is especially true when superheated steam is used or when it is difficult to atomize the cylinder oil. Satisfactory results are obtained by supplying oil directly to the packing gland from the lubricator (Fig. 10.1). When adequate lubrication is provided, it is unnecessary to draw the gland up tightly to prevent leakage; thus there is less possibility of overheating, rapid piston-rod wear, and packing failure.

Compound engines present some special problems in cylinder lubrication. Sometimes satisfactory results are obtained by introducing oil into the high-pressure cylinder and depending upon the steam to carry it through into the low-pressure cylinder. This does not always, however, meet the requirements because part of the oil is removed by the separator and because the lubrication requirements are different in the high- and low-pressure cylinders, a fact which makes it necessary not only to lubricate the cylinders separately but to use oil having different amounts of compounding.

Because of the many variables involved, it is impossible to give definite rules for the quantity of oil required for steam-engine-cylinder lubrication. The oil should be a minimum for satisfactory lubrication. Too much oil not only causes waste but is carried away by the condensate to give trouble in the boilers and other equipment. The following rule may be applied to obtain the approximate amount of cylinder oil required for speeds between 75 and 250 rpm and operating pressure between 100 and 250 psi:

$$\text{pt of oil/cylinder/hr} = D \times L \times \text{rpm} \times 0.00038$$

where D = diameter of cylinder, ft
L = length of stroke, ft
rpm = revolutions per min
0.00038 = constant[1]

[1] This constant is based on the use of 1 pt of cylinder oil per hr per 1 million sq ft of rubbing surface. For other rates of oil feed the constant must be changed proportionally before applying the formula.

$$\text{Constant} = \frac{2 \times 3.1416 \times 60}{\text{sq ft rubbing surface/pt of oil/hr}}$$

Example A steam engine with a cylinder 24 in. in diameter and a stroke of 48 in. operates on 150 psi steam pressure at 125 rpm. What would be the cylinder oil requirements in pints per hour?

Solution

$$\text{pt of cylinder oil/hr} = \frac{24}{12} \times \frac{48}{12} \times 125 \times 0.00038 = 0.38$$

When the atomization is good, less oil is required. For high-pressure superheated steam, high speed, or poor atomization, more oil is required.

Minimum oil requirements may be determined by reducing the amount of oil supplied to the engine cylinder and carefully observing its operation for indications of inadequate lubrication. First, determine the correct type of oil for the steam pressure, speed, and service; then reduce the rate of feed slowly. Observe the operation for any change in the sound of the engine. A slight tremor in the eccentric rods or, with a Corliss, a decided groaning in one or more of the valves is an indication of inadequate lubrication. During this trial period it is good practice to remove the valve cover and cylinder head and check for oil film by rubbing the friction surfaces with a soft piece of paper.

In addition to the valves and cylinders, the external rubbing surfaces or parts not exposed to the steam must also be lubricated. A wide range of lubricants and methods of application are used in the external lubrication of engines. The lubricant must be adapted to the method of application and the service.

Gravity lubrication systems employ overhead tanks from which the oil flows through sight-feed glasses to the parts of the engine to be lubricated. The flow of oil to each part may be observed and adjusted to meet the need. After passing through the bearings, the oil is collected in a sump, filtered, and pumped back into the overhead tank. Other engines have self-contained circulating-oil systems. The pumps are mounted on the engine frames and supply oil from a clean-oil compartment within the engine to the various parts to be lubricated. The oil drains from the bearings into a settling compartment, which overflows into the clean-oil compartment, where it is available for the pump. Circulating-oil systems require oils that are light and free-flowing, with the ability to withstand repeated use and to separate from water and solid particles. In addition to checking and adjusting the oil supply to the various lubrication points, the operator must maintain the level in the supply tank by adding new oil as required.

Splash lubrication systems consist of oil reservoirs or sumps in the bottom of the crankcase. The crank strikes the oil and splashes it over the parts requiring lubrication. Projections are provided on the crank for striking the oil, baffles for directing the splash, and openings and grooves for admitting the oil to the bearings. The oil drains from the bearings into the crankcase. Except for the addition of oil to maintain the level in the crankcase, the system is automatic. Pumps are sometimes employed in addition to the splash of the oil to provide lubrication to critical bearings. These pumps are driven from the main engine shaft.

Some engines are built with full-pressure lubrication systems (Figs. 10.3, 10.4). Bearings lubricated in this manner carry heavy loads at high speed and seldom require adjustment. Dirt is excluded, and an oil film is maintained between the bearing surfaces. The oil pump is supplied from the sump in the engine base and delivers oil under pressure to a hole drilled through the center of the crankshaft. The oil lubricates the main and connecting-rod bearings. A portion flows through a hole in the connecting rod to the crosshead pin and shoes. Another portion lubricates the eccentrics. Accurate bearing clearance permits a portion of the oil to be discharged from the bearings and returned to the sump, thus providing circulation and the necessary cooling. The high- and low-pressure frames of the engine have individual pumps, but both the sumps and the high-pressure lines are interconnected. Safety devices are provided either to sound an alarm or to shut the engine down if the oil pressure gets too low for safe operation. The operator must observe the oil pressure and maintain the oil level

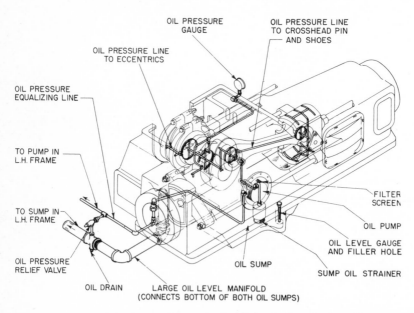

FIG. 10.3 *Isometric view of the oil-supply lines of the full-pressure lubricating system on a steam-engine–driven air compressor.* (*Ingersoll-Rand Co.*)

in the sump reservoirs.   Low oil pressure may be caused by worn bearings or by stoppage in the lines.   The pump-suction and discharge strainers should be examined for stoppage.   The oil is changed either after a given number of operating hours or when the analysis shows that it is no longer satisfactory for use.

When individual gravity sight-feed cups are used at each point that requires lubrication, a heavier oil is recommended.   With this method of application, the oil is not recirculated and should, therefore, stay on the surfaces longer.   These individual lubricators require frequent attention and refilling.

The conventional ring or chain method of oiling is applied to outboard pedestal bearings.   A small amount of oil is retained in a reservoir under the bearing.   A ring larger in diameter than the shaft or a chain hangs over the shaft and dips into this oil.   As the shaft rotates, the ring

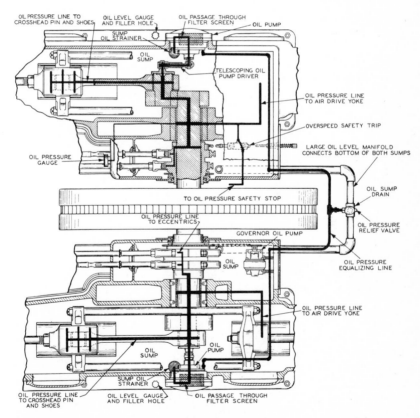

FIG. 10.4   *Cross-sectional view of high- and low-pressure units, showing oil holes in cranks, main shafts, and connecting rods on a steam-engine–driven air compressor.   (Ingersoll-Rand Co.)*

TABLE 10.1   Lubricating Oil Suitable for Steam Cylinders*

	Wet	Supt	Supt	Supt
Steam conditions:				
Pressure...................................	......	Below 250	Above 250	Above 250
Temperature..............................	Below 400	Below 500	Below 600	Above 600
Cylinder-oil specifications:†				
Flash point, minimum, °F..................	450	480	515	535
Viscosity, sec at 210°F (SSU):				
Minimum.............................	100	130	160	200
Maximum............................	135	165	200	
Carbon residue, percentage, maximum.........	3.0	3.0	3.0	3.0
Compounding, percentage (with acidless tallow oil):				
Minimum.............................	8.0	5.0	3.0	
Maximum............................	12.0	8.0	5.0	2.0

* Courtesy of The Ingersoll-Rand Co.
† See Sec. 13.9.

or chain carries the oil up onto the bearing. The oil must have lasting qualities and be free-flowing so that it will be carried to the bearing. The level must be maintained and the ring or chain inspected to make sure that the oil is being delivered to the bearing. The oil must be maintained at the specified level and changed at regular intervals.

Grease cups and pressure fittings are used to furnish grease to valve gears and miscellaneous moving parts. A good-quality high-melting-point cup grease is satisfactory for this service.

Study the method of application used on the engines in your plant. If in doubt as to the type of lubricant to use, consult your oil company for the recommended brand. Guard against contamination in handling or by contact with water and sludge in the engine. Packing that leaks is a source of contamination. Keep the filtering equipment in good operating condition. When in doubt as to the quality of oil in a circulating system, have it tested in a laboratory and replace or add new oil as required.

10.3   OPERATING A STEAM ENGINE.   Before attempting to start a steam engine, open the drains directly ahead of the throttle valve and allow them to remain open until steam has been discharged. Failure to drain the lines completely may result in waterhammer and damage to the engine. Now inspect the lubrication system, fill the lubricators and reservoirs, and lubricate with oil and grease as required.

If the engine operates condensing, start the circulating water pumps and vacuum pumps. In this way a vacuum can be maintained while the engine is being warmed up. Check the valves in the exhaust line to see that they are opened either to the atmosphere or to the condenser. Now open all drains, including those on the steam chest and engine cylinders. With the piston near one end of the cylinder open the throttle valve, quickly at first, and when the engine starts, close it down so that the engine will just continue to turn over. When the engine is operating condensing, observe the vacuum during the starting-up period. If there is a separate condensate pump, it should be operated to receive the condensate as it accumulates in the base of the condenser (hot well). Intermittent or continuous operation may be required, depending upon how rapidly the engine is brought up to speed and the load applied. Gradually bring the engine up to speed and then close the drains. It should require from 15 to 30 min to warm up an engine for service.

Slide, piston, and Corliss nonreleasing-type-valve engines can be started without changing the governor mechanism. This is true in the case of both throttling and automatic governors, as the governor mechanism does not function until the engine comes up to speed. When starting up Corliss engines of the detaching type, unhook the reach rod so that the latch may be operated by hand, independently of the eccentric. The admission valves may be lifted by means of a starting lever provided for moving the wristplate. In this manner each steam valve may be lifted in turn, thus permitting steam to move the piston from one side of the cylinder to the other. At the same time the exhaust steam is released from the cylinder.

Cross-compound engines may be started with the crank in any position by first noting whether the high- or the low-pressure piston side is off-center. If the high-pressure crank is on or near center, start the engine by admitting high-pressure steam to the low-pressure cylinder through the bypass. When the high-pressure side is in the starting position, admit steam to the high-pressure cylinder. Tandem-compound engines are started in the same way as single-cylinder engines except that the low-pressure cylinder must be drained and lubricated. These engines must be warmed up more slowly than simple engines owing to their greater length and corresponding greater expansion.

With the engine in operation, check the lubrication system and make sure that all moving parts are receiving sufficient lubrication. At this point a thorough knowledge of the methods of application and the kinds of lubricant is essential. The necessary frequency of inspection is determined by the method of application employed. These periods of inspection must include the condenser and other auxiliary equipment. Maintaining the highest possible vacuum is essential to efficient operation. Be on the alert for air leaks, dirty condensers, etc.

Bearings should receive attention during regular inspections while the engine is in service. If the main or crankpin bearings can be touched with the hands, they are in no immediate danger of overheating. Bearings may overheat because of insufficient lubrication, wrong kind of oil, dirt in the oil, overload, bearings out of line, too tight adjustment, or improper oil grooves. Temporary measures recommended for relieving overheated bearings are increase of the amount of oil to the bearing, cooling by the forced circulation of air, the application of cylinder oil, and the addition of a small quantity of graphite flour. Graphite must be added sparingly to avoid its stopping up the oil grooves. These procedures should be considered temporary measures; the cause of overheating should be determined and necessary corrective methods taken.

The operator may recognize loose bearings by the pounding or knocking noise produced, but difficulty may be encountered in determining which bearing is giving the trouble. It is advisable first to check the main and connecting-rod bearings as they are the first to become loose as a result of normal wear. Occasionally an engine will become noisy owing to wear or looseness in some unusual place. Loose crosshead shoes and lateral movement of the main shaft cause engine noises, the reasons for which are sometimes difficult to detect. Insufficient steam cushioning of the reciprocating parts when they reverse direction at the end of a stroke will cause the engine to become noisy. A cracking noise may indicate water in the cylinder. When this occurs, quickly open the drains and allow them to remain open until the noise disappears. If there is a recurrence, it is probably caused by priming of the boiler, and necessary corrective measures must be taken to produce dry steam.

When an engine is no longer needed for service, reduce the load so that it can be shut down without interfering with plant operation. If the engine is driving an electric generator, this will consist of shifting the load to the other units that are to be left in operation or to some other source of power supply. When the electric service in the building is going to be discontinued, assure yourself that shutting off the current will not cause inconvenience or personal hazard.

When the load on the engine has been reduced almost to zero and the governor has decreased the steam flow accordingly, close the throttle valve until the engine starts to slow down. Continue to close the valve slowly and try to prevent the engine from stopping on center (with practice this should be possible). After the engine has stopped, shut off the lubricators and open the drains from the cylinders, steam chests, exhaust line, etc.

In the case of a condensing plant the engine is stopped as explained above, and then the condensing equipment is shut down. After the engine has stopped, the vacuum and condensate pumps are stopped and the vacuum breaker opened. The circulating or condenser water pumps should remain in operation for a short time after the engine has been shut down.

When the engine is to remain out of service for an extended period of time, it should be conditioned to guard against corrosion. Remove the cylinder head and coat the cylinder walls with grease. Remove the packing and coat the piston rod, crosshead guides, etc., with grease. Oil companies supply specially prepared coatings for this purpose. They are easy to apply, prevent deterioration of bright metal surfaces, and may be easily removed when the engine is to be returned to service.

10.4 MAINTENANCE OF STEAM ENGINES. The cylinder of a steam engine must be strong enough to withstand the pressure of the steam. The inner surface must be smooth so that the piston rings can seal and prevent leakage and must be capable of resisting wear by the action of the piston rings. Wear on the cylinder walls occurs uniformly around the entire surface as a result of normal pressure exerted by the piston rings. The lower side of horizontal- or inclined-engine cylinders has additional wear as the result of the weight of the pistons. This is minimized by having the piston centrally located in the cylinder and the weight supported by the crosshead guides and, in some instances, by a rod extending through the packing gland in the cylinder head (tail rod).

Cylinder walls wear owing to a lack of lubrication, because of the piston not being centrally located, or as a result of long service. This wear progresses to a point where it becomes impossible to maintain a seal between the piston rings and the cylinder. With good care and normal operation, this will occur on an average of 7 to 10 years. Cylinders are made extra-thick so that they can be rebored two or more times to correct for this wear. When the cylinder walls become too thin for reboring, a sleeve or liner is pressed inside the cylinder to restore it to the original diameter. The counterbore must always be retained in the cylinder when repairs are made to prevent the rings from forming a shoulder in the cylinder bore at the end of the stroke.

Piston rings wear and fail to exert enough pressure on the cylinder walls to form the necessary seal. When this occurs and the cylinder walls are in satisfactory condition, the engine may be restored to operating condition by the installation of new piston rings. Under normal conditions and with continuous operation, this will be necessary every 3 to 5 years. One cylinder reboring or resleeving job will wear out two sets of piston rings.

When ordering new piston rings, specify the cylinder diameter, width and thickness of the ring groove, steam pressure, and temperature.   It is a good policy to order the rings from the engine manufacturer.   Check the ring dimensions before attempting final installation.   When the rings are placed in the cylinder, the clearance between the ends should be from 0.010 to 0.030 in., depending upon the particular installation.   Try the rings in the piston grooves and roll them around to make sure they are of the correct width.   When difficulty is encountered in getting the cylinder oil to adhere to the cylinder walls, it is sometimes advisable to round off the sharp, knifelike edge of the rings slightly, thus reducing their tendency to scrape the oil off the cylinder walls.

A solid piston must be removed from the cylinder in order to replace the rings.   The old rings may be lifted from the grooves by means of a hook and raised to a position where a piece of sheet metal or an old hacksaw blade may be inserted.   After metal strips have been inserted at several places around the piston, the rings may be removed easily.   This procedure is repeated until all rings have been removed.   A similar procedure, including the use of metal strips, is followed in passing the new rings over the pistons to their respective grooves.   A sheet-metal band clamped around the piston is useful in holding the rings compressed and in place while the assembly is being forced into the cylinder.   Consideration must be given to the counterbore arrangement during this operation.

The rings on a built-up piston may be replaced without removing it from the cylinder (Fig. 8.3).   After the cylinder head has been removed, the follower plate can be taken off the piston.   One of the piston rings can then be removed.   The adjusting screws, or centering bolts, are next disassembled and the bullring taken out.   This makes it possible to remove the second piston ring.   After checking the rings for correct diameter and thickness, the piston assembly can be replaced in the reverse order.   One ring is installed against the piston body, or spider.   Then the bullring is replaced and centered in the cylinder by means of the adjusting screws.   The other ring is then installed and the follower plate replaced.   Special attention should be given to see that the nuts which hold the follower plate are secure, for if they work off while the engine is running, serious damage will result.   Various safety devices are employed to reduce the possibility of studs' coming out of the piston.

Bearings must be adjusted to compensate for wear, but there are so many variables that it is impossible to predict how often this will be necessary.   Adequate lubrication is of utmost importance in reducing bearing difficulties.   With a full-pressure lubricating system and the correct oil, bearings will operate for a long period of time without adjustment.   The need for attention is indicated by a knocking noise or a tendency to overheat. With a totally enclosed crankcase, the operator must rely upon the sound of the engine and periodic inspection to determine the condition of the bearings.

Two general types of bearings are used on the connecting rods of engines (Fig. 8.4). The strap-end type of bearing is adjusted for wear by changing the position of the wedge by use of the adjusting bolt. The marine-end bearing is adjusted for wear by adding or removing shims between the two halves. Both the crosshead and the crankpins of an engine tend to wear oval owing to the variation in thrust at the different positions in the stroke of the engine. This fact must be taken into consideration when adjusting connecting-rod bearings. The bearings should be adjusted to allow clearance on the large diameter to prevent overheating. Another precaution must be taken relative to the change in the length of the connecting rod, which results in unequal mechanical clearance on the head and crank ends of the piston. After the bearings have been adjusted, the striking points of the engine must be checked. If they are found unequal, the length of the connecting rod or piston rod must be changed (see Sec. 8.2).

Crossheads are adjusted by wedge-shaped slippers, as shown in Fig. 8.4, or by adding or removing shims. Since the least wear on the crosshead guides is at the ends of the stroke, the engine should be on dead center when adjustment is being made. Care must be exercised to keep the crosshead centered between the guides. Check the level of the piston rod and make adjustments on both the top and the bottom guides to maintain the rod level at all positions of the engine. The lower crosshead guide will probably be worn more than the top one.

Main bearings of many engines are arranged with quarter boxes (a quarter section of bearing) that can be adjusted for wear. These quarter boxes are adjusted to the journals either by wedges or by screws and locknuts. These bearings must be adjusted with care to avoid getting them too tight and the main engine shaft out of square with the cylinder.

Occasions may arise when it is advisable to check the right-angle relation of the engine shaft to the center line of the cylinder without taking the engine apart. Since the end of the cylinder (with the head removed) is accurately machined at right angles to the center line of the cylinder, it can be taken as a reference. Place a straightedge across the end of the cylinder and support it in a level position. Measure from the straightedge to the engine shaft on both sides of the engine with a steel tape, taking into account differences in shaft diameter. If the engine shaft is at right angles to the center line of the cylinder, the two distances will be equal.

When bearings become unfit for further service as a result of overheating or wear owing to long usage, it is necessary to replace the babbitt-metal lining. Melt out the old lining, taking care not to burn the bearing boxes. Clean the interior of the bearing boxes and check to see that they are equipped with the necessary anchors to hold the new babbitt. Bolt the two halves of the bearings together with the necessary shims. Drive wooden plugs into the oil holes.

Secure a mandrel slightly smaller than the shaft or journal on which the bearing is to operate. Center the mandrel with respect to the bearing. Apply a light coating of graphite and cup grease to the mandrel to prevent the babbitt from bonding. Plug up all oil holes and openings to prevent the babbitt from running out. A mixture of asbestos and cylinder oil of a consistency of putty is useful in blocking small openings.

Before the babbitt is poured, the entire bearing should be heated to approximately 150°F. The babbitt metal must be supplied with a single pour. If the ladle does not hold sufficient babbitt, a second one must be ready and, once started, the pouring continued until the job is complete.

When the babbitt is cool, disassemble the bearing and hammer (peen) the metal into the recesses. Reassemble with the desired thickness of shims and bore to size in a lathe. With some types of bearings, such as solid-end connecting-rod bearings, it is necessary to make a gig to hold the boxes together while babbitting and boring. After doing this, cut oil grooves in both halves of the bearing in order that the oil will be directed toward the middle. Do not cut the oil grooves too wide, as this would reduce the effective bearing surface. Relieve the bearings ½ in. or more from the joint to prevent binding in the event of overheating. Now cover the shaft with prussian blue, assemble the bearings, and rotate. Disassemble and scrape off the high spots as indicated by the prussian blue on the bearings. Continue this operation until a good bearing surface has been obtained. Observe the bearing for overheating when it is placed in service. Operate at low speed for several hours.

In the case of small engines, for quick repairs in an emergency and when facilities are not available, satisfactory results may be obtained by using the shaft as a mandrel for rebabbitting bearings. The same general procedure is followed as that described above, but machining is unnecessary. Provisions must be made to hold the babbitt in place during the pouring operation. When the babbitt is cold, the oil grooves are cut and the bearings relieved at the joints. The bearing is assembled and adjusted reasonably tight and the engine run slowly. When heat has been developed, the bearing is disassembled and the high spots scraped off. If necessary, this operation must be repeated until the engine operates without the bearings' overheating.

Leaky valves are a source of much trouble and annoyance. They can usually be detected by placing the valve directly over the ports and blocking the engine in position. Both indicator cocks may then be opened. Steam is admitted to the valve chest, and any leakage past the valve is easily detected by the escapement of steam through the cocks. Another method is to use an indicator. With a little study, faulty valves can be detected very easily.

Slide valves must have their seats resurfaced and valves replaced at intervals. The frequency of replacement depends on the pressure, speed, and lubrication employed. Piston valves very frequently require replacement of both valve and seat, although some of them may be adjusted for wear. Corliss valves also have to be replaced occasionally. In this case the seat must be rebored and a new valve fitted in position. Poppet valves, for stationary work, very seldom need to be replaced. Occasionally, new springs are required. The valve is ground in, to a perfect fit, and again is satisfactory for service. (See Chap. 9 for procedure in setting valves.)

Examine the governor mechanism for binding of the joints owing to insufficient lubrication, lost motion as the result of wear, and faulty operation of the dashpot. All joints in the governor mechanism must operate freely and with very little friction. Reduce friction by lubrication or by "relieving" if the fits are too tight. Lost motion may be eliminated by replacing worn parts or by installing bushings. Dash- or gag pots are used to slow down the action of the governor and to prevent hunting. This action depends upon the viscosity of the oil and the adjustment of the bypass valve. The desired action of the dashpot can be determined by trial and error. Remember that increasing the viscosity of the oil and closing down the bypass slow down the action of the governor. Check the operation of the safety stop on belt-driven flyball governors.

10.5 ECONOMY OF STEAM ENGINES. An efficient steam engine is one that can deliver, as mechanical energy at its shaft, a relatively large portion of the heat energy available in the steam supplied. The heat energy available to an engine depends upon the initial steam pressure and temperature and upon the exhaust pressure. For example, if two engines exhaust at atmospheric pressure but one receives steam, dry and saturated, at 150 psig and the other, dry and saturated, at 100 psig, the one supplied with the higher pressure will have more heat energy available per pound of steam. However, if the engine using the 100-psig steam were more carefully maintained and operated, it might be more efficient in converting the heat available in the steam into mechanical energy. The part of the available heat which is converted into useful work depends upon the design of the engine and the skill employed in its operation. Owing to the various losses which are in part controllable, the actual engine converts only 50 to 80 per cent of the available heat into useful work. These losses may be classified as follows:

1. Cylinder condensation
2. Clearance volume
3. Wiredrawing of steam
4. Leakage past rings and valves
5. Moisture in the steam at admission
6. Friction in the mechanism
7. Incomplete expansion
8. Radiation

Cylinder condensation is by far the largest factor in preventing the actual engine from attaining ideal engine performance. In a counterflow engine the incoming steam comes into contact with the cylinder head and steam passages which have been cooled by the exhaust steam. When saturated steam is used, this cooling causes some of the steam to condense. When superheated steam is used, the superheat is reduced by this cooling. The effect of condensation is reduced by insulating the cylinder, steam-jacketing, superheating, compounding, increasing speed, decreasing clearance volume, and employing the uniflow principle.

Clearance volume is that portion of the cylinder which is not displaced by the travel of the piston plus the volume of the steam passageways between the valves and cylinder. The clearance volume varies from 2 to 10 per cent of the total piston displacement. It increases cylinder condensation and thereby reduces the efficiency of the engine. In theory the loss in efficiency resulting from large clearance volume can be partially removed by increasing the compression. However, in counterflow engines the possible increase in efficiency resulting from an increase in compression is lost owing to the fact that a larger quantity of steam is compressed in the clearance volume to be expanded on the next stroke.

"Wiredrawing" is a term applied to the reduction in steam pressure which occurs when steam passes through a partially closed valve. This occurs owing to the throttling action of governor valves and owing to the closing of the cylinder steam-admission valves. The resulting pressure drop wastes energy that could otherwise be used by the engine. The wiredrawing loss is reduced by the use of an automatic-type governor in place of a throttling type and by the use of a riding cutoff, releasing Corliss, or poppet valve in place of the conventional slow-closing D-slide or piston valves.

Leakage past the piston rings and valves allows the high-pressure steam to enter the exhaust without performing work in the engine cylinder. Adequate cylinder lubrication and maintenance of cylinders, pistons, and rings will reduce this loss to a minimum.

Moisture carried by the steam does no work in the cylinder. Tests show that the consumption of dry steam per horsepower is practically constant, the water acting as an inert quantity. There is great danger of the moisture's carrying along impurities that will seriously damage the equipment. It is best to solve this problem by reducing carry-over from the boiler. The use of a separator in the steam lines to the engine should be considered as a secondary measure.

The difference between the horsepower developed in the cylinder, called "indicated horsepower," and the brake horsepower is due to the power consumed in overcoming friction. This loss of power varies from 4 to 20 per cent of the indicated horsepower. Maintaining correct engine alignment and bearing adjustment and supplying adequate lubrication will reduce the friction losses.

The perfect engine is assumed to work on a complete expansion cycle; the steam is expanded to the existing back pressure. With the actual engine this is not practical. The increase in mean effective pressure by complete expansion is small. The cylinder required would be so large that its expense would not be justified. The exhaust valve of the actual engine opens when the pressure in the cylinder is above the exhaust pressure, giving an incomplete expansion cycle. As the load on an engine increases, the pressure at the time the exhaust valve opens increases, allowing a great amount of heat to be discharged to the exhaust. Preventing overloads on an engine reduces the loss due to incomplete expansion. Adjustment of the time at which the exhaust valve opens is also important in expanding the steam to the lowest pressure possible before releasing it to the exhaust.

Radiation is the heat loss from the engine cylinder to the surrounding air. This is effectively reduced by insulating the cylinder.

10.6   STEAM-ENGINE INDICATORS. James Watt invented the steam-engine indicator and applied it while experimenting with engines that used steam expansively. By use of the indicator Watt was able to make many improvements in the steam engine. Engineers continue to use this remarkable instrument to assist in adjusting engine valve mechanisms for maximum performance and efficiency.

The indicator (Fig. 10.5) records both the steam pressure in the cylinder and the position of the piston in the cylinder. The resulting graph is a "picture" of the conditions inside the engine cylinder and is known as an "indicator diagram." The paper on which the diagram is made is termed an "indicator card." The horizontal distance locates the

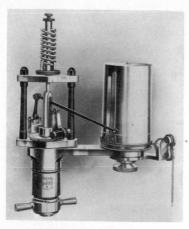

FIG. 10.5 *Crosby steam-engine indicator.*

position of the piston, while the vertical distance indicates the corresponding pressure. This means that the indicator pencil must be actuated by two mechanisms: one is operated by the steam pressure in the cylinder; the other, by the back-and-forth motion of the piston. The indicator diagram is referred to as a "pressure-volume diagram" because it shows the relation between the pressure and volume of steam in the cylinder.

The pressure-recording mechanism consists of a small (usually ½ sq in.) cylinder accurately fitted with a piston. A spring counteracts the force exerted by the steam on the piston. Indicators are supplied with a series of springs, rated and marked 20, 40, 60, etc. The spring is selected compatible with the steam pressure involved. For example, a 40-lb spring subjected to 40 lb variation in pressure within the engine cylinder will cause the pencil to move 1 in. vertically on the chart.

The piston position is recorded on the indicator card by having the cylinder rotate in relation to the movement of the piston. This motion is accomplished by a cord which connects the crosshead to the indicator and causes the cylinder to rotate. It is impractical to make the indicator cylinder diameter large enough, and hence the indicator card correspondingly long enough, to accommodate the entire length of the engine stroke. It therefore becomes necessary to use a reducing mechanism between the point where the cord is attached to the crosshead and the indicator. For example, on an engine with a 24-in. stroke the reducing mechanism would restrict the oscillating motion at the indicator to 6 in. These reducing mechanisms may be either built into the indicators or attached to the engine frame, and reduce the motion of the cord before it is attached to the indicator.

Procedures and precautions to follow and observe in taking engine indicator diagrams:

The spring strength must be matched with the steam pressure involved.

Reducing mechanism must be selected and adjusted for maximum length of the indicator card.

Cord length must be adjusted so the indicator does not strike the stops at either end of the card.

Check to see that the position of the crosshead is accurately reproduced on the diagram: ¼ position of crosshead, ¼ position on card; ½ position of crosshead, ½ position on card; etc.

Card must be securely held on the indicator drum by the clips.

Open the indicator cock to the atmosphere, hook up the cord, and with light but firm pressure on the pencil draw the atmospheric line.

Open the indicator cock to the engine cylinder and allow the indicator to reach temperature before "taking" the card.

Open the indicator cock for actuating the pressure element of the indicator, attach the cord to actuate the piston motion mechanism, and with light but firm pressure on the pencil draw the diagram.

Disconnect the cord, close the cock, and remove the card which now shows the performance of the engine.

When a force is exerted through a distance, work is said to be accomplished. For example, when a 10-lb weight is lifted from the floor to a table 3 ft high, $3 \times 10 = 30$ ft-lb of work is done. The steam pressure in the cylinder of a steam engine exerts a force on the piston, and as a result it moves over its stroke doing work. The force or pressure on the piston times the distance that the piston moves is the amount of work done. The intake valve of most engines closes when the piston has partly completed its stroke. Full pressure is exerted on the piston while steam is being admitted, but after the steam valve has closed, the pressure drops. This means that there is a variation in pressure on the piston during the stroke. The average or *mean effective pressure* (mep) is one of the factors that determine the power output per stroke. The indicator diagram shows the pressure at any position of the piston during the stroke. The average height of the diagram represents the mean effective pressure. The area of the diagram represents the work done during the stroke.

The importance of proper valve setting is explained in Chap. 9. By means of the indicator diagram it is possible to see just how the valves operate when the engine is carrying load. Careful study and adjustment of valves by means of the indicator diagram are effective not only in improving engine economy but also in increasing the capacity.

The conventional indicator diagram (Fig. 10.6a) is a diagram which represents the greatest possible work that can be obtained with a given set of conditions. It is made not by an indicator but from a known set of existing conditions. There is no compression, the exhaust valve does not close until the end of the return stroke, and the intake valve opens as soon as the exhaust valve closes. The pressure in the cylinder reaches the maximum as soon as the steam valve opens. The steam is admitted to the cylinder without loss of pressure until the point of cutoff is reached. The steam expands from this point without giving up or receiving heat. This is called "adiabatic expansion," and the line is a hyperbolic curve. At the end of the stroke the exhaust valve opens, and the pressure immediately drops to the pressure at which the engine

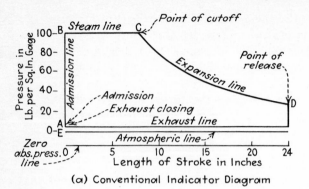

(a) Conventional Indicator Diagram

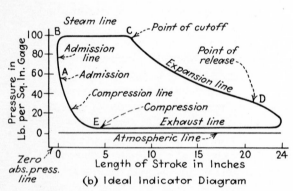

FIG. 10.6 *Conventional and ideal indicator diagrams.*

(b) Ideal Indicator Diagram

is exhausting. The back pressure remains equal to the exhaust pressure during the entire return stroke; that is, if an engine were exhausting at atmospheric pressure, the back pressure on the conventional diagram would be atmospheric pressure.

The actual indicator diagram, even if ideal, varies from the conventional in several respects, depending upon the engine and especially upon the valve arrangement used. Figure 10.6b is an ideal indicator diagram, or a diagram in which the events occur in such a manner that best practical results are obtained. In practice the steam valve cannot be opened instantaneously. It is therefore opened *before* the piston gets to the end of the stroke (given lead) so that enough steam is admitted to exert full pressure on the piston when it starts the power stroke. The pressure in the cylinder is *less* than the pressure in the steam header because of the drop through the valves and ports. The diagram is *rounded off* at the point of cutoff because the valve throttles "wiredraw" the steam as it closes. At the beginning of expansion the actual curve falls *below* the conventional curve because some heat is transferred to the cylinder walls. During the last part of expansion the actual curve is *above* the conventional curve because the steam is cooler and now receives heat from the cylinder walls.   The point of release must occur *before* the piston reaches the end of the stroke so that the valve will be fully open when the return stroke begins.   On the return stroke the cylinder pressure is *above* the pressure in the exhaust header because of the pressure drop through the ports.   Before the piston completes the exhaust stroke, the exhaust valve *closes* and the steam remaining in the cylinder is compressed in the clearance volume.   This increased pressure *slows* up the piston, and it is brought to a stop without causing a knock in the engine.   There is some steam left in the clearance volume when the piston is at the end of the stroke; this steam reexpands on the return stroke.   A large clearance volume lowers both the efficiency and the capacity of the engine.

Five of the many possible indicator diagrams are shown in Fig. 10.7. These reveal the effect of various faulty valve adjustments.

Figure 10.7a shows an ideal indicator diagram.   Admission at *A* takes place early enough so that full pressure is exerted on the piston when it starts on its power stroke.   The passages and ports are large enough to prevent appreciable pressure drop when the engine is taking steam. Quick closing of the steam valve at the point of cutoff reduces the amount of wiredrawing.   Expansion continues to *C* near the end of the stroke, but the release is early enough to relieve the steam and ensure minimum back pressure when the piston starts to return.   The exhaust ports are large enough to prevent the pressure from greatly exceeding the exhaust-header pressure.   The exhaust valve closes early enough at D to give sufficient compression to cushion the piston and provide smooth operation.

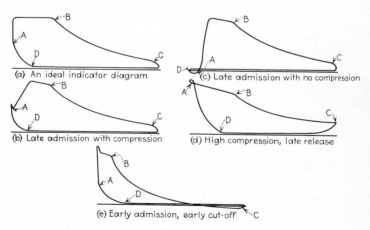

FIG. 10.7   *Indicator diagrams showing the effect of faulty valve setting.*

Figure 10.7b shows the effect of late admission and slow opening of the steam valve. Here the exhaust valve has closed at the proper time to give sufficient compression. But the steam valve opens so late that the compression pressure has started to drop before the live steam is admitted. The valve is slow in opening, and as a result the piston has completed part of its return travel before full pressure is obtained in the cylinder. The steam valve should be given more lead.

Figure 10.7c is an example of an indicator diagram obtained from an engine with no compression and late admission. The exhaust line runs almost straight to the end of the stroke. The piston starts its power stroke before the steam valve opens; that is, at the beginning of the stroke both valves are closed, the pressure is reduced below the exhaust pressure, and a negative work loop is formed. The area of this loop must be subtracted from the positive area of the card.

In Fig. 10.7d compression begins so early in the exhaust stroke that the compression pressure exceeds the throttle pressure before the piston reaches the end of the stroke. When the steam valve first opens, steam flows from the cylinder into the steam chest until the movement of the piston reexpands the steam. The sloping "steam line" in this diagram shows that the steam pipeline or ports are too small. Release does not occur until the end of the stroke. The steam does not have time to escape. During the first part of the exhaust stroke the back pressure on the piston is high.

Figure 10.7e shows too much lead and the steam admitted so early that it is pushed back into the steam chest by the piston. This causes the cylinder pressure to exceed the header pressure. The pressure quickly drops when the piston starts on its power stroke. This diagram was taken on an automatic engine. Owing to the fact that the load on the engine is very light, the cutoff is early. The steam in the cylinder expands until the pressure in the cylinder is lower than the exhaust pressure. This kind of loop in a diagram represents negative work. Power was actually used in moving the piston over this part of the stroke.

10.7   ENGINE RATING AND EFFICIENCY. The full throttle pressure is not available for pushing the piston during the entire stroke because of cutoff, compression, and other conditions previously explained. The indicator diagram makes it possible to determine the actual average or mean effective pressure exerted on the piston.

A planimeter is an instrument used to determine the magnitude of irregular-shaped areas and may, therefore, be used to determine the area of an indicator diagram.[1] A conventional planimeter has two arms and a wheel with a graduated scale. The two arms are hinged together;

---

[1] The ordinate method of determining the approximate mean effective pressure from an indicator diagram is explained in Appendix F.

one end of the assembly is held in a fixed position, while the other end contains a pointer which is moved around the perimeter of the area to be measured. The wheel, which is attached to one of the arms, runs on the paper. The scale indicates the area encircled by the pointer.

To determine the area of a diagram with a planimeter place the tracing point at some position on the outline. Note and record the reading on the graduated cylinder, and then move the point around the outline of the diagram until it returns to the starting point. Again read the graduated cylinder. The difference between the two readings is the area of the diagram expressed in the units in which the planimeter is made to read, usually square inches. The area of the diagram divided by the length equals the height. The height multiplied by the spring scale equals the mean effective pressure. The planimeter automatically compensates for the area of a negative loop. It is advisable to make several area determinations and take the average as final. Some planimeters are made especially for determining the mean effective pressure of indicator diagrams and read direct.

The heat energy in steam is converted into mechanical energy in the steam-engine cylinder. The capacity of engines is expressed in horsepower (hp). A *horsepower* is 33,000 ft-lb of work done in 1 min. The mean effective pressure p in pounds per square inch multiplied by the area of the piston A in square inches equals the total force exerted on the piston in pounds. The total force in pounds exerted on the piston multiplied by the length of the stroke L in feet equals the number of foot-pounds of work done by the engine per stroke. The number of foot-pounds per stroke times the number of strokes per minute n equals the total foot-pounds developed per minute. This quantity divided by the constant 33,000 gives the indicated horsepower (ihp) developed. That is to say, the number of foot-pounds of work done in the engine cylinder per minute (hence the indicated horsepower) depends upon the mean effective pressure, the length of the stroke, the area of the piston, and the number of revolutions per minute.

$$\text{ihp} = \frac{PLAn}{33,000}$$

where P = mean effective pressure, lb per sq in.
   L = length of stroke, ft
   A = area of piston, sq in.
   n = number of strokes per min (rpm $\times$ 2)

The effective area of the piston is reduced by an amount equal to the area of the piston rod on the crank end of simple engines and on both ends of engines in which the rod extends through the head. This means that a different area of the piston must be used in calculating the power developed by the crank end of the engine from the one used in calculating the power developed by the head end. The mean effective pressure, as calculated from the indicator diagram, will seldom be the same for both ends of the cylinder.

Example A double-acting steam engine has a cylinder 12 in. in diameter. The stroke is 18 in., and the engine operates at 150 rpm. The mean effective pressure on the crank end is 90 psi; on the head end, 95 psi. The piston rod is 1½ in. in diameter. Calculate the horsepower developed by the engine.

Solution

Head end:

P (mean effective pressure) = 95 psi
L (length of stroke) = $18/12$ = 1.5 ft
A (effective area of piston) = 12 × 12 × 0.7854 = 113.098 sq in.
n (number of strokes) = 150 per min

$$\text{ihp (head end)} = \frac{95 \times 1.5 \times 113.098 \times 150}{33,000} = 73.26$$

Crank end:

P (mean effective pressure) = 90 psi
L (length of stroke) = $18/12$ = 1.5 ft
A (effective area of piston) = (12 × 12 × 0.7854)
  − (1.5 × 1.5 × 0.7854) = 111.331
n (number of strokes) = 150 per min

$$\text{ihp (crank end)} = \frac{90 \times 1.5 \times 111.331 \times 150}{33,000} = 68.32$$

$$\text{ihp of engine} = \overline{141.58}$$

The horsepower as calculated by using the mean effective pressure determined from the indicator diagram is known as the "indicated horse-power." Not all this power is transmitted to the engine flywheel, because part of it is lost in overcoming friction in the engine. The power actually delivered to the flywheel of the engine is known as the "brake horsepower" (bhp).

$$\text{Mechanical efficiency} = \frac{\text{bhp}}{\text{ihp}}$$

One method of determining the brake horsepower of an engine is by use of the Prony brake shown in Fig. 10.8. The Prony brake consists of a steel band that goes around the flywheel. The band is lined with material that will produce friction and yet resist heat and wear. Wooden or asbestos blocks are frequently used. The band is attached to an arm that prevents it from turning with the flywheel. The end of the arm is supported on scales so that the force *tending* to cause it to turn can be measured. The band has an adjusting wheel by means of which the tension can be adjusted. In this way the load on the engine can be varied. The energy developed by the engine is converted

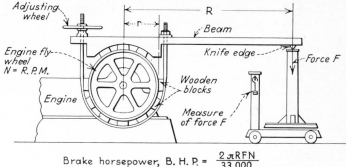

Brake horsepower, B. H. P. $= \dfrac{2\pi RFN}{33,000}$

FIG. 10.8   *Prony brake as arranged for measuring the brake horsepower of an engine.*

into heat by the brake.   For this reason some method of water cooling is frequently employed to keep the brake cool.   The formula for calculating the brake horsepower, from the data obtained by use of the Prony brake, is as follows:

$$\text{bhp} = \frac{2\pi RFN}{33,000}$$

where R = distance from center of engine shaft to knife edge on scales, ft
$\quad F$ = force exerted by arm as shown by scales, lb
$\quad N$ = rpm

Part of the force exerted on the scales is due to the weight of the brake arm.   The scale reading or value of F must be corrected for this force.   The correction is found by loosening the band and turning the engine flywheel first *over* and then *under,* noting the reading of the scales in each case.   The sum of these two scale readings divided by 2 gives the average.   This is the correction which must be deducted from all scale readings obtained during the test.

If the engine is connected to an electric generator and the efficiency of the generator is known, the brake horsepower can be calculated from the known electric output of the generator.   The output of the generator in kilowatts divided by its efficiency gives the engine output in kilowatts. This output of the engine in kilowatts multiplied by 1.34 will give the brake horsepower of the engine.

It is common practice to state the performance of an engine in terms of its "water rate."   The water rate of an engine refers to the pounds of steam that it utilizes per horsepower-hour or per kilowatthour of energy produced. The water rate of an engine may be expressed in pounds of steam per indicated horsepower-hour, per brake-horsepower hour, or per kilowatthour, depending upon whether the energy referred to is that delivered to the cylinder, the flywheel, or the generator terminals.

The water rate is not the true measure of the economy of an engine. An engine utilizes heat, and the heat that must be supplied to generate a unit of energy is the true measure of economy. An engine may receive steam at a low pressure and exhaust against considerable back pressure. The heat per pound of steam available for use in the engine is small, and the steam consumption per unit of energy generated is correspondingly high. An engine operating on high-pressure steam and exhausting into a partial vacuum should have a low water rate. The water rate is useful in comparing the daily performance of an engine operating under constant steam and exhaust conditions. The operator can tell how his engine performance compares with the manufacturer's guarantee. In this case it is usually necessary to make corrections for the variation between actual steam and exhaust conditions and those given in connection with the guarantee.

The water rate of an engine is used in determining the size of a boiler which will be required to furnish the necessary steam.

Following are the approximate steam requirements in pounds per indicated horsepower-hour (water rate) of engines utilizing saturated steam at 100 psig pressure.

Single-valve noncondensing............. 28
Single-valve condensing................. 23
Four-valve noncondensing............... 22
Four-valve condensing.................. 16

The steam consumption of small engines is determined by condensing the exhaust steam and weighing or measuring the condensate. This procedure accounts for the term "water rate." In the case of the simple unit the weight of condensate discharged is equal to the weight of steam supplied to the throttle. This is not always true of turbines and compound engines because part of the steam is discharged at intermediate points and does not appear in the final exhaust. Meters are now available for measuring the flow of steam to the engine or turbine. These meters will measure the flow of steam, water, or gases. They make it possible to check the daily performance of an engine and detect any decrease in economy.

The Btu of heat required by an engine to produce a unit of energy is the true measure of the economy of operation. The heat supplied to a simple engine in Btu per indicated horsepower-hour may be found as follows:

$$\text{Btu/ihp} = W\,(H - h)$$

where W = steam per ihp-hr as determined from test

H = Btu per lb of steam supplied to engine throttle

h = heat of liquid as found in steam table, corresponding to pressure of engine exhaust

This formula takes into account not only the steam supplied but also the heat utilized by the engine.

The thermal efficiency of an engine is found by comparing the actual Btu required to produce a unit of energy to that theoretically required. The heat equivalent to a horsepower-hour is 2545 Btu.

$$\text{Thermal efficiency} = \frac{2545}{\text{Btu supplied/ihp}}$$

The Rankine-cycle ratio is another method of expressing the performance of a steam engine. In this case the heat units required by the actual engine are compared with the heat units that would be required by an ideal engine working on a Rankine cycle. The Rankine cycle assumes an engine taking steam at constant pressure to point of cutoff, adiabatic expansion to exhaust pressure before release occurs (complete expansion), exhaust at constant pressure and temperature (no compression), and feedwater returned to the boiler at a temperature corresponding to the exhaust pressure. This ideal engine is assumed to operate without throttling, radiation, and friction losses.

The Rankine-cycle ratio which can be expected depends upon the valve mechanism employed, upon whether saturated or superheated steam is used, and upon whether the engine is operating condensing or noncondensing. Simple single-valve engines operating on saturated steam have a Rankine-cycle ratio of 50 to 63 per cent when operating noncondensing and 40 to 50 per cent when operating condensing. Multivalve engines operating on saturated steam have a Rankine-cycle ratio of 65 to 75 per cent when operating noncondensing and 45 to 60 per cent when operating condensing. Compound and uniflow engines have Rakine-cycle ratios from 5 to 10 per cent higher than multivalve engines. The use of superheated steam increases the ratio by 5 to 15 per cent. The application of steam tables and calculation of cycle efficiencies are presented in detail in advanced work on this subject.

QUESTIONS AND PROBLEMS

10.1. What are some of the measures that should be taken to ensure continuous operation and low maintenance of steam engines?

10.2. List the items that should be included in the inspection of a steam engine.

10.3. Explain the two types of lubricators used to supply cylinder oil to the engine cylinders.

10.4. What is a compounded cylinder oil, and under what conditions should it be used?

10.5. State the precautions necessary in lubricating a compound engine.

10.6. How do you determine the amount of cylinder oil required? What are the effects of using too much cylinder oil?

10.7. Discuss three methods of lubricating the external parts of steam engines.

10.8. What precautions are necessary before starting an engine?

10.9. Outline the procedure that should be followed in starting a small condensing engine.

10.10. What are some of the parts of an engine that might cause noisy operation?

10.11. What conditions will cause rapid wear of the cylinder walls?

10.12. How would you proceed to replace the piston rings on a built-up piston?

10.13. What precautions would you take when adjusting connecting-rod bearings if the crankpin were a badly worn oval?

10.14. How would you proceed to rebabbitt a solid-end connecting-rod bearing?

10.15. How are leaking Corliss valves repaired?

10.16. How is engine economy improved by the use of superheated steam?

10.17. How does wiredrawing of steam at the point of cutoff lower engine efficiency?

10.18. Why is it impractical to build engines that will utilize the complete expansive power of steam?

10.19. Mention some methods used to reduce the loss caused by cylinder condensation.

10.20. What is a steam-engine indicator?

10.21. What benefits may be derived from taking indicator cards on a steam engine?

10.22. What precautions are necessary when taking indicator cards?

10.23. Draw an indicator diagram in connection with a cylinder, piston, and valve. Mark the important events on the diagram.

10.24. If an indicator diagram shows good operation of a slide-valve engine except for too high back pressure, what should you do?

10.25. What does the area of an indicator diagram represent?

10.26. Give two ways of finding the area of an indicator diagram.

10.27. An engine has a cylinder 14 by 24 in. and runs at a speed of 90 rpm. The average mean effective pressure is 65 psi. Calculate the indicated horsepower.

10.28. An engine with a cylinder 10 by 12 in. operates at 250 rpm. The mean effective pressure is 45 psi. What is the indicated horsepower?

10.29. Find the indicated horsepower of an engine with a cylinder 10 by 24 in. and a piston rod 2¼ in. in diameter. The engine operates at 100 rpm. It has a head-end mean effective pressure of 72 psi and a crank-end mean effective pressure of 68 psi.

10.30. If the engine mentioned in Prob. 10.29 has a mechanical efficiency of 90 per cent, what would be the brake horsepower delivered to the engine shaft?

10.31. A Prony brake was fitted to an engine and on trial showed that the correction for the weight of the arm should be 25 lb. The Prony-brake arm was 60 in. long. On test the scales read 135 lb, and the engine operated at 225 rpm. What was the brake horsepower of the engine?

10.32. A 50-hp simple-valve engine operating at full load requires 1,375 lb of steam per hr. The steam as it enters the engine is 100 psi gauge. The exhaust is at atmospheric pressure. Find (*a*) the water rate of the engine, (*b*) the thermal efficiency.

CHAPTER 11

STEAM TURBINES
AND
AUXILIARIES

11.1 TURBINES: GENERAL. A heat engine is one that converts heat energy into mechanical energy. So the steam turbine is classed as a heat engine, as are the steam and internal-combustion engines. The turbine makes use of the fact that steam when issuing from a small opening attains a high velocity. The velocity attained during expansion depends upon the initial and final heat content of the steam. This difference in heat content represents the heat energy converted into kinetic energy (energy due to velocity) during the process. The kinetic energy or work available in the steam leaving a nozzle is equal to the work that the steam could have done had it been allowed to expand (with the same heat loss) behind a piston in a cylinder.

The fact that any moving substance possesses energy, or the ability to do work, is shown by many everyday examples. A stream of water discharged from a fire hose may break a window glass if directed against it. When the speed of an automobile is reduced by the use of brakes, an appreciable amount of heat is generated. In like manner the steam turbine permits the steam to expand and attain high velocity. It then converts this velocity energy into mechanical energy. There are two general principles by which this can be accomplished. In the case of the fire hose, as the stream of water issued from the nozzle, its velocity was increased, and owing to this impulse it struck the window glass with considerable force. A turbine that makes use of the impulsive force of high-velocity steam is known as an "impulse turbine." While the water issuing from the nozzle of the fire hose is increased in velocity,

466

a reactionary force is exerted on the nozzle. This reactionary force is opposite in direction to the flow of the water. A turbine that makes use of the reaction force produced by the flow of steam through a nozzle is a "reaction turbine." In practically all commercial turbines a combination of impulse and reactive forces is employed; both impulse and reaction blading on the same shaft utilize the steam more efficiently than does one alone.

A simple *impulse* turbine is illustrated in Fig. 11.1*a*. Here water is seen striking a flat plate *P* and being scattered so that any energy remaining is splashed, lost, or dissipated. Only impulse force is employed.

Another simple illustration of an impulse turbine is shown in Fig. 11.2*a*. A unit of this type, using steam, was made in the seventeenth century by an Italian named Branca and is the first impulse turbine of which we have any record. Here pressure causes the steam to issue with high velocity from a small jet or nozzle. This steam is directed against the paddle wheel, and rotation results. All pressure drop takes place in the nozzle or stationary elements, and the moving paddles absorb the velocity energy in the steam issuing from the nozzles.

For best economy the moving element should travel at *one-half* of the velocity of the steam from which it is receiving its energy. To meet this requirement the rotating element of large turbines, using high-pressure steam, would have to operate at excessive speed. The required velocity of the moving parts is reduced by applying the principle of "pressure staging." Pressure staging consists of allowing only a limited pressure drop in one set of nozzles. After the steam from a set of

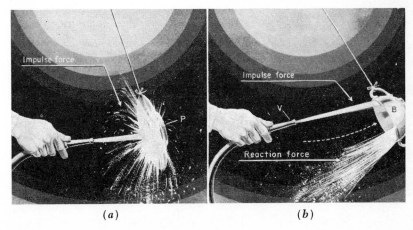

(*a*)   (*b*)

FIG. 11.1   (*a*) *Simple illustration of impulse force only.*   (*b*) *Simple illustration of impulse and reaction forces.* (***Tidewater Associated Oil Co.***)

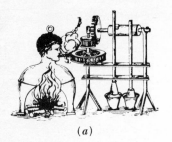

(a)

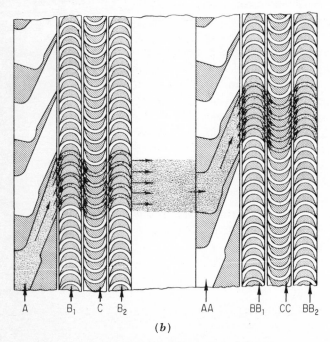

A   B₁   C   B₂   AA   BB₁   CC   BB₂

(b)

FIG. 11.2 *The impulse turbine.* (a) *Early impulse turbine* (*Branca, seventeenth century*). (b) *Diagrammatic impulse turbines.* (*Mobil Oil Corp.*)

nozzles has passed the rotating element (blades), it is expanded in another set of nozzles. Figure 11.2b shows how the steam expands through nozzles A and after passing the blading is again expanded farther through nozzles AA. A turbine may have many sets of nozzles (pressure stages), each increasing in size to accommodate the increased volume of the steam.

The principle of "velocity compounding" is also employed in the operation of impulse turbines. This means applying the velocity energy in steam coming from the nozzles to two or more sets of moving blades. The operation makes use of a set of stationary blades, which reverse the flow of steam, between each set of rotating blades. In Fig. 11.2*b* the steam in the first pressure stage is expanded through nozzle *A*. Then it strikes, in the following order, the moving blades B_1, the stationary blades *C*, and the moving blades B_2. From B_2 the steam enters the second stage and expands through the nozzles *AA*. It then goes through moving blades BB_1, stationary blades *CC*, and moving blades BB_2 and into the exhaust or next pressure stage. Large high-pressure turbines usually employ many pressure stages and in addition use velocity compounding in the high-pressure stages.

Figure 11.3 shows an application of velocity compounding in a Terry turbine. The steam is expanded in the nozzles and strikes the single row of buckets on the rotor. It then enters the steam-reversing passages in the casing and is again directed against the rotor buckets. This procedure is repeated several times until the velocity of the steam has been reduced. Such an arrangement is known as a "reentry-type turbine."

If one set of nozzles is employed and all the pressure drop occurs in this group of nozzles and if all the energy is directed against a single wheel, we have a single-stage simple velocity turbine.

The *reaction* turbine is one in which the pressure drop takes place in the rotating element. Figure 11.4*a* shows a reaction turbine of the simplest type. A turbine similar to this type was made by Hero of Alexandria about 150 B.C. but was never put to any practical use. Steam generated in the boiler passes through trunnions to the hollow sphere. From here steam is discharged through nozzles attached to the sphere. The steam leaving the nozzles at high velocity causes them to move in a direction opposite to the direction of the flow of steam; this causes the sphere to rotate on the trunnions.

The illustration in Fig. 11.1*b* shows a combination of impulse and reaction forces in action. Note that, with the same stream of water, the bowl *B* moves farther than the plate *P*. The water from the jet produces an impulse force in the direction in which the jet is moving, and on leaving the bowl the water produces

FIG. 11.3   *Action of steam in a terry wheel and reversing chamber.   (The Terry Steam Turbine Co.)*

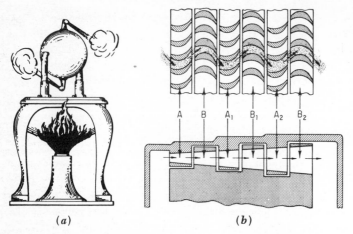

(a)   (b)

FIG. 11.4   *The reaction turbine.   (a) Earliest reaction*
steam turbine.   (b) Diagrammatic illustration of section of
blading of reaction turbine.   (Mobil Oil Corp.)

a reaction force against the bowl.   All turbine blades are shaped to
give somewhat the same effect as the water in the bowl.

The modern reaction steam turbine consists of stationary and moving
blades.   The blades are similar, each being arranged so that the area
through which the steam leaves is less than that through which it enters.
Pressure drop occurs in both the stationary and the moving blades.   The
restricted area at the outlet of the blades causes the steam to increase
in velocity as it leaves the blading.   This same reaction principle is
employed today in launching rockets into space; the modern airplane
uses the same jet reaction to propel its jet engines.

Figure 11.4b is a diagrammatic arrangement of the blading in a typical
reaction turbine. A, A_1, A_2, etc., are the stationary blades mounted
in the casing; B, B_1, B_2, etc., are the moving blades mounted on the
drum or spindle.   The steam in passing through the turbine expands
alternately through the stationary and moving blades; the stationary
blades are designed to direct the steam against the next row of blades.
Note that the area of the steam passages at the outlet of the blades
is less than that at the inlet.   As a result of this change in area of
the steam passage, the velocity increases and the pressure decreases
as the steam leaves each set of blades.

The reaction turbine is in reality an expansion nozzle in which the
pressure of the steam is used to increase its velocity.   This velocity
is converted into mechanical energy by the rotating element.   During
this process of expansion, the steam increases in volume.   The area
of the steam passage through the blading must increase, from the high-

pressure to the low-pressure end of the turbine, in order to accommodate this larger volume of steam.

Turbine Design and Construction. An elementary turbine consists of a shaft on which is mounted one or more disks. On the circumference of the disks are located blades or buckets to receive the steam and convert it into useful work. The rims of the disks have dovetail channels for receiving the blades. The ends of the blades are made to fit these dovetail channels. This turbine requires bearings for support, a suitable housing or casing to enclose the moving wheel, a system of lubrication, and a device known as a "governor" to maintain control over the speed.

The economy and satisfactory performance of the turbine depend to a great extent upon the design and construction of the blading. The blades must therefore be made to withstand the action of the steam and the centrifugal force caused by the high speed at which the turbine must operate. In designing turbine blading a compromise must be made between strength and economy. The stationary and moving elements have very little clearance, and any vibration of the moving element will cause them to rub. For this reason extreme care must be exercised to design the wheels so that they will not vibrate. The length and size of the blades must be increased as the steam pressure drops to accommodate the increase in volume. Large condensing turbines have large wheels and long buckets in their last stages. This reduces the velocity of the steam leaving the turbine and as a result improves the economy.

Turbine blades are drop forgings made of steel or alloys, depending upon the conditions under which they are to operate. Very highly superheated steam requires blades especially designed to prevent warping and deterioration. Steam in the last stages of a turbine becomes very wet, and this moisture erodes the turbine blades. Special materials are employed to lengthen their life.

The blades are assembled on the disk, and a "shroud ring" is placed around their outer ends (Figs. 11.5, 11.6f, 11.8). The tips of the blades pass through holes in the shroud ring. The ends are then riveted so that they are held securely by the ring. When the blades are very long, extra lacing is used to tie them together (Fig. 11.8).

Rotors for small turbines consist of a machined-steel disk shrunk and keyed onto a heavy steel shaft. The shaft is rust-protected at the gland zones by a sprayed coating of stainless steel. The element is statically and dynamically balanced to ensure smooth operation throughout its operating range.

The small steam turbine (Fig. 11.5) employed for mechanical drive has a number of essential features; in Fig. 11.6 are shown component parts of this turbine. The steam chest is bolted to the base and is made

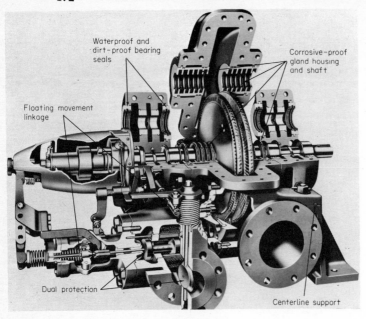

FIG. 11.5 *General-purpose turbine for mechanical drive.*
(*Westinghouse Elec. Corp.*)

of iron or steel.   It contains the governor valve, strainer, and operating hand valve for manual adjustment and to secure maximum efficiency. Regardless of whether or not a hand valve is provided, the steam is made to pass through the governor-controlled admission valve contained in the steam chest.

The governor valve located in the right end of the steam chest is a double-seated balanced valve operating in a renewable cage (Fig. 11.6c); the valve and cage are made of noncorrosive material and the valve stem of stainless steel.   Packing is provided where the valve stem passes through the chest cover.

Speed regulation is secured by means of the governor (Fig. 11.6b); this governor is of the centrifugal-weight type and is connected to the admission valve by the arm and linkage shown in Fig. 11.5.   Movement of the governor weights is opposed by the compression spring to transmit its motion from the rotating spindle to a stationary sleeve through a self-aligning ball-thrust bearing.   If the speed of the turbine exceeds a safe limit (10 per cent above normal speed), an overspeed trip device (Fig. 11.6d) is provided.   This trip is eccentrically mounted and restrained by a spring to trip simultaneously the governor valve (Fig. 11.6c) and the butterfly valve (Fig. 11.5).   Hand reset is provided when the speed has been reduced to normal. In normal operation the butterfly valve is held open by the trip linkage against the force of the coil spring shown. In some cases the butterfly trip valve is omitted, and overspeeding closes merely the governor valve.

Located in the casing are the steam-admission nozzles (Fig. 11.6e), which are cut into a solid block of bronze or alloy steel, depending

(a)　　　　　　　　　　　　　　　　　(b)

(c)　　　　　　　　　　　　　　　　　(d)

(e)　　　　　　　　　　　　　　　　　(f)

FIG. 11.6 Component parts of general-purpose turbine. (a) Steam chest. (b) Speed-regulating governor. (c) Governing valve. (d) Overspeed trip. (e) Nozzles. (f) Rotors and blades. (Westinghouse Elec. Corp.)

on steam conditions. Nozzles are so proportioned as to be contributory to efficient operation and are made of corrosion- and erosion-resistant materials. This nozzle block is bolted to the steam chest, which in turn is bolted to the base of the turbine casing. The entire assembly of nozzles for one stage is called a "diaphragm." The casing assembly with the stationary blading or nozzles is referred to as the "turbine cylinder." The cylinder of an impulse turbine is frequently referred to as the "wheel casing." The turbine blades (Fig. 11.6f) are made from rolled and drawn sections of stainless steel. Dovetail roots are obtained by broaching. The blading, shrouding, and rotor rim are contoured to approximate the steam-expansion characteristics. The rotating blades are secured in dovetail grooves cut into the rotor disk and are regularly spaced by soft-iron packing pieces.

Around the outer rim of the blades or nozzles is provided a shroud ring (Fig. 11.6f) to stiffen the blades against vibration and confine the steam to the blade path. Each group of blades is tied together. Round tenons at the blade ends are machine-spun to secure shroud bands to the blades.

Bearings support the rotor; they are horizontally split and lined with high-grade tin babbitt. Access to the bearings is possible without raising the cylinder cover or rotor. The governor-end bearing is babbitt-faced at both ends and acts as a combined thrust and journal bearing. Rings suspended in oil roll on the shaft to provide lubrication to the bearing. Water jackets are provided for cooling the oil.

To prevent leakage of steam where the rotor shaft extends through the turbine casing, carbon-ring-type glands are provided. These consist of segmental rings held around the shaft by garter springs. Between the two outer rings of each gland is a leak-off connection which prevents steam that may pass the inner rings from leaking into the engine room. The casing proper (Fig. 11.7) is bolted together at the horizontal joint. Flanges are frequently finish-ground, making possible a steamtight joint without use of a gasket.

On units slightly larger than that shown in Fig. 11.5, oil for governing and flood lubrication is supplied by a shaft-driven gear-type pump. If lubrication requirements exceed the capacity of a single pump, a separate gear-type pump is furnished to supply lubricating oil. The direct-acting oil governor with hand-speed changer from 3:1 maximum-speed adjustment is employed for variable-speed applications, such as driving fans, blowers, etc. For the mechanically driven turbine, the speed is frequently automatically controlled by the draft, air pressure, feedwater pressure, and other means. NOTE: Design standards for materials and governor performance are covered in publications of the NEMA.

[1] National Electrical Manufacturers Association.

(a)

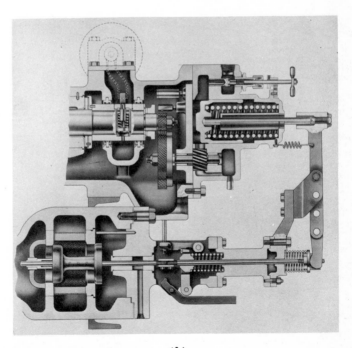

(b)

FIG. 11.7   (a) Turbine casing.   (b) Hydraulic orifice
governor.   (Westinghouse Elec. Corp.)

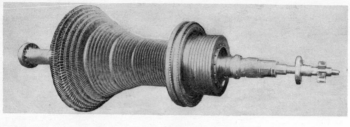

(a)

(b)

(c)

(d)

FIG. 11.8   Rotors for various types of turbines.   (a) Rotor
for condensing turbine.   (b) Rotor for noncondensing
turbine.   (c) Rotor for noncondensing single-extraction
turbine.   (d) Rotor for condensing double-extraction turbine.
(Westinghouse Elec. Corp.)

On large turbines there is normally a high- and low-pressure section with the steam chest integral with the high-pressure section. High-pressure castings are made of steel, whereas exhaust sections are made of cast iron; cylinders are split along the horizontal plane. Blades may be assembled in separate blade rings or directly in the cylinder, depending on the turbine size and pressure and on temperature conditions. In order to minimize misalignment and distortion turbines are so designed as to permit expansion and contraction in response to temperature changes.

Larger turbines have their rotors (Fig. 11.8) formed from a single-piece forging, including both the journals and the coupling flange. Thrust-bearing collar and oil impeller may be carried on a stub shaft bolted to the end of the rotor. Forgings of this type are carefully heat-treated and must conform to specifications. Rotors are machined; after the blades are in place, they are dynamically balanced and tested.

As with the small turbine, the efficiency and life of larger turbines are influenced chiefly by the form or shape of the blades, the manner in which they are fastened in place, and the materials from which they are made. *Rotating* blades are secured by a T-root fastening with lugs (Fig. 11.9) machined on the shank, straddling the blade groove. Blades are held against a shoulder in the groove by half-round sections calked in place at the bottom. *Stationary* blades are anchored in straight-sided grooves by a series of short keys which fit into auxiliary grooves cut in the blade shank and in the side of the main groove. If high pressure and temperature are employed, steam leakage across impulse stages is controlled by thin sealing strips (Fig. 11.9a) which can be set with close running clearance. On reaction turbines of the larger sizes where high temperature and pressure are used, *shrouded* blades (Fig. 11.9b) are employed, with radial seal strips to control leakage between stages. Seal strips are made very thin, permitting close running clearances.

Two types of nozzles are employed (Fig. 11.9a): round nozzles with holes drilled and reamed in a solid block of steel and curved-vane nozzles. The interstage diaphragms are located in grooves which are accurately spaced and machined into the casing. The upper halves are attached to, and lift with, the casing cover. Labyrinth seals (Fig. 11.12) minimize steam leaks along the shaft where it passes through the diaphragm. The seal rings are spring-backed and are made of a material that permits close running clearances with complete safety.

Main bearings (Fig. 11.11) consist of a cast-steel shell, split horizontally and lined with high-grade tin-base babbitt. They are supported in the bearing housing by steel blocks. Between the supporting blocks and the bearing shell are steel liners by means of which the bearings

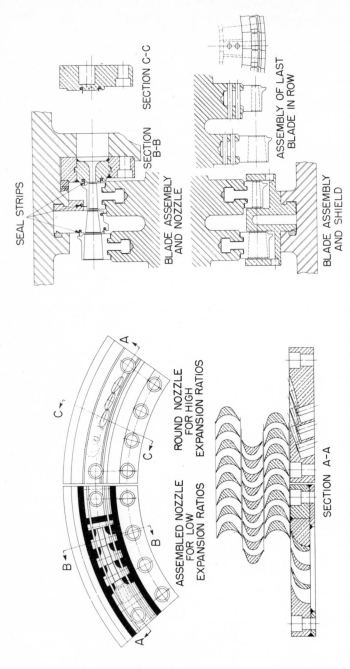

SEAL STRIPS

SECTION C-C

SECTION B-B

BLADE ASSEMBLY
AND NOZZLE

ASSEMBLY OF LAST
BLADE IN ROW

BLADE ASSEMBLY
AND SHIELD

ASSEMBLED NOZZLE
FOR LOW
EXPANSION RATIOS

ROUND NOZZLE
FOR HIGH
EXPANSION RATIOS

SECTION A-A

FIG. 11.9 (a) Nozzles and impulse blading. (Westinghouse Elec. Corp.)

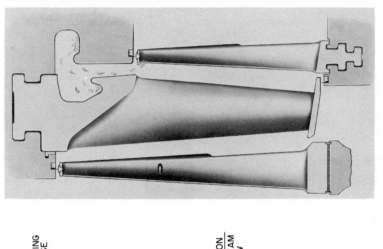

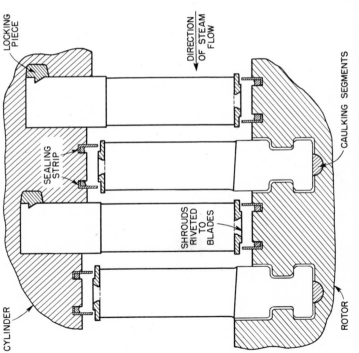

FIG. 11.9 (b) Low pressure reaction blading showing moisture-catcher, erosion-protection shields. (Westinghouse Elec. Corp.)

can be moved both vertically and horizontally to align the rotor accurately within the cylinder.

The conventional turbine bearing consists of a cylindrical shell divided into halves so that it can be assembled on the shaft. The outside of the shell has a spherical section at the middle. This fits into a similarly shaped seat in the bearing support pedestal and is an aid in properly aligning the bearing. In most cases the top half is grooved on the trailing side of the bearing in such a manner that there is a tendency to draw or aspirate oil into the bearing. The bottom half of a split bearing should be relieved on each side of the joint by scraping it clear of the shaft. The clearance of the bearing can be measured by placing a soft lead wire on the shaft and bolting the top half down tight. The flattened wire can be removed and the thickness measured by a micrometer. The clearance in the bearing should be 0.002 in. per in. of shaft diameter.

Oil is supplied through a hole drilled in the bearing housing that matches a similar hole in the lower supporting block. For large bearings, orifice outlets are provided for positive control.

The pressure drop through the rotating blades of the reaction turbine combines to produce a force which tends to push the turbine rotor toward the low-pressure end. In some turbines this force is balanced by what is known as the "double-flow principle." In this case, steam enters the middle of the turbine and flows in both directions. This produces two forces which balance each other. Another method is to employ balancing or dummy pistons as shown in Fig. 11.10.

Thrust Bearing. The thrust bearing (Figs. 11.10, 11.11) consists of a collar rigidly attached to the turbine shaft rotating between two babbitt-lined shoes. The clearance between the collar and the shoes is small. The piston is attached to the spindle; steam pressure is exerted on one side and atmospheric pressure on the other. The difference in pressure produces a force which balances the thrust exerted on the rotating blades. If the shaft starts to move in either direction, the collar comes into contact with the shoes and the shaft is held in proper position. Larger thrust bearings have several collars on the shaft and a corresponding number of stationary shoes.

The *Kingsbury thrust bearing* (Fig. 11.11) is used when a large thrust load must be carried to maintain the proper axial position in the turbine cylinder. (The one shown in Fig. 11.11 is a combination of the Kingsbury and collar types.) The thrust collar is the same as that used in the common type of thrust bearing. The thrust shoes are made up of segments which are individually pivoted. With this arrangement, the pressure is distributed equally not only between the different seg-

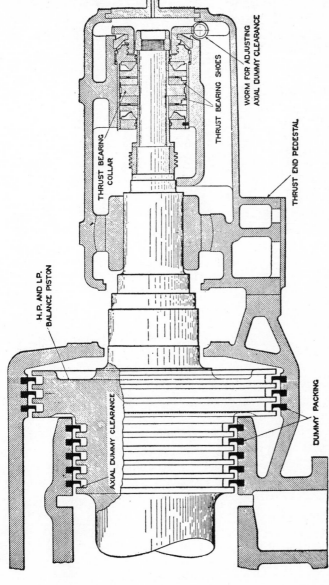

H.P. AND L.P.
BALANCE PISTON

THRUST BEARING
COLLAR

THRUST BEARING SHOES

WORM FOR ADJUSTING
AXIAL DUMMY CLEARANCE

THRUST END PEDESTAL

AXIAL DUMMY CLEARANCE

DUMMY PACKING

FIG. 11.10   *Turbine thrust end showing balance piston and thrust bearing.*   (*Allis Chalmers Mfg. Co.*)

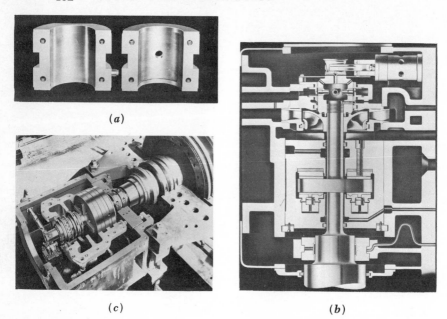

FIG. 11.11   *Main and thrust bearings.* (*a*) *Main bearing.* (*b*) *Section of thrust bearing and housing.* (*c*) *Thrust-bearing cage in place.* (*Westinghouse Elec. Corp.*)

ments but also upon the individual segments. The openings between the segments permit the oil to enter the bearing surfaces. Almost 10 times as much pressure per square inch can be carried on the Kingsbury-type bearing as on the ordinary-type thrust bearing. Axial position of the bearing and turbine rotor may be adjusted by liners, located at the retainer rings, on each end of the bearing. The bearing is lubricated by circulating oil to all its moving parts.

The impulse turbine does not require so large a thrust bearing as the reaction turbine because there is little or no pressure drop through the rotating blades. However, the thrust bearing must be used to ensure proper clearance between the stationary and rotating elements. Reaction turbines that do not have some method of balancing the force caused by the drop in pressure in the rotating blades must be equipped with large thrust bearings.

Turbine bearings are subjected to very severe service and require careful attention on the part of the operator. Most turbines operate at high speed and so are subjected to the heat generated in the bearing itself as well as that received from the high-temperature steam. These conditions make necessary some method of cooling. In some cases the bearings are cooled by water jacketing; in others the oil is circulated through a cooler.

Packing. The shaft at the high-pressure end of the turbine must be packed to prevent leakage of steam from the turbine. The one at the low-pressure end of a condensing turbine must be packed to prevent the leakage of air into the condenser. Here, again, because of shaft speed, steam pressure, and steam temperature, conditions are different from those encountered in steam-engine practice.

Labyrinth packing is used widely in steam-turbine practice. It gets its name from the fact that it is so constructed that steam in leaking must follow a winding path and change its direction many times. This device consists of a drum which turns with the shaft and is grooved on the outside. The drum turns inside a stationary cylinder which is grooved on the inside (Figs. 11.10, 11.12). There are many different types of labyrinth packing, but the general principle involved is the same for all. Steam in leaking past the packing is subjected to a throttling action. This action produces a reduction in pressure with each groove that the steam passes. The amount of leakage past the packing depends upon the clearance between the stationary and the rotating elements. The amount of clearance necessary depends upon the type of equipment, steam temperatures, and general service conditions. The steam that leaks past the labyrinth packing is piped to some low-pressure system or to a low stage on the turbine.

A water-packed gland consists of a centrifugal-pump runner attached to the turbine shaft. The runner rotates in a chamber in the gland casing. Water is supplied to the chamber at a pressure of 3 to 8 lb and is thrown out against the sides by the runner, forming a seal. Water seals are used in connection with labyrinth packing to prevent the steam that passes the packing from leaking into the turbine room. Such a seal is also used on the low pressure end of condensing turbines. In this case the leakage to the condenser is water instead of air.

Figure 11.12 shows water-sealed glands and labyrinth seals as used on the high-pressure end of condensing turbines. They are used singly or in combination, depending upon the service required. Each labyrinth consists of a multiplicity of seals to minimize steam leakage. The seal rings are spring-backed and made of material that permits close running clearances with safety. The glands are usually supplied with condensate for sealing to prevent contamination of the hot-well water.

Carbon packing is composed of rings of carbon held against the shaft by means of springs. Each ring fits into a separate groove in the gland casing. When adjustments are made while the turbine is cold, carbon packing should have from 0.001 to 0.002 in. of clearance per in. of shaft diameter. The width of the groove in the packing casing should

(a)   (b)

FIG. 11-12   (a) Water-sealed glands and labyrinth seals as used on the high-pressure end of condensing turbines.   (b) Labyrinth-type gland as employed on noncondensing turbines.   (Westinghouse Elec. Corp.)

exceed the axial thickness of the packing ring by about 0.005 in.   Carbon packing is sometimes used to pack the diaphragms of impulse turbines. Steam seals are used in connection with carbon packing.   This is essential when carbon packing is used on the low-pressure end of condensing turbines, because if there is a slight packing leak, steam instead of air will leak into the condenser.   In operating a turbine equipped with carbon packing, a slight leak is desirable because a small amount of steam keeps the packing lubricated.

Flexible metallic packing is used to pack small single-stage turbines operating at low back pressure.   In most cases the pressure in the casing of these turbines is only a few pounds above atmospheric pressure.   The application is the same as when this packing is used for other purposes except that care must be exercised in adjusting.   Owing to the high speed at which the shaft operates, even a small amount of friction will cause overheating.

Governors.   A close control of turbine speed is essential from the standpoint of safety and satisfactory service.   The same theory of centrifugal force that applied to engine flywheels also applies to the rotating elements of turbines.   If the turbine runs at a speed far above that for which it was designed, the blading will be thrown out of the wheels. When a turbine is thrown apart (bursts) in this manner, the resulting damage may be as great as, or even greater than, that caused by a boiler explosion.   Turbines operating a-c generators must operate at constant speed.   Some electrical appliances are seriously affected by a slight change of frequency in the power supply.   The speed of the generator determines the frequency of the electric-current generator. Even with d-c generators a small change in speed will affect the voltage.

There are two ways of changing the turbine supply of steam to meet the load demand. One consists of throttling the steam, by means of a valve, in such a manner that the pressure on the first-stage elements is changed with the load demand. Small turbines, like that shown in Figs. 11.5 and 11.13, have a main governor of the throttling type. The operating mechanism is driven directly by the main turbine shaft. Overload is taken care of by means of hand-operated valves which admit additional steam to the turbine. The other method employs several valves, governor-operated, that are opened separately to supply steam to secondary nozzles as the load increases. This arrangement is shown in Fig. 11.14; it is used only on large turbines.

Economical partial-load operation is obtained by minimizing throttling losses. This is accomplished by dividing the first-stage nozzles into several groups and providing a separate valve to control the flow of steam to each group. Valves are then opened and closed in sequence, and the number of nozzle groups in service is proportional to the load on the unit. Valve seats are of the diffuser type to minimize pressure drop; valve seats are renewable.

The multivalve steam chest (Fig. 11.14) is cast integrally with the cylinder cover with a cored passage from each valve to a nozzle group. Single-seated valves are used, arranged in parallel within the steam chest and surrounded by steam at throttle pressure. The governor

FIG. 11.13 *Cross section of type G Terry steam turbine.* (*The Terry Steam Turbine Co.*)

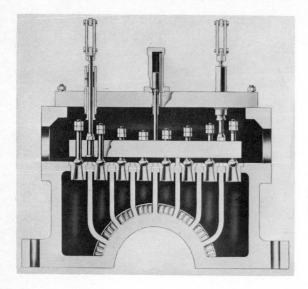

FIG. 11.14 *Simplified steam chest with multiple valves.* (*Westinghouse Elec. Corp.*)

mechanism raises and lowers the valve-lift bar in a horizontal plane, opening the valves in sequence with an unbalanced force tending to close the valves.

In the *oil-relay* system the governor operates a small valve which admits oil to, and allows it to drain from, a cylinder. This oil cylinder contains a piston which is connected, by means of a rod, to the steam-valve mechanism. The governor admits oil either above or below the piston, depending upon which way the load is changing. There is no connection between the governor and the governor-valve mechanism except by means of the oil cylinder. The movement of the governor is transmitted to the governor valves by the oil pressure. Oil is supplied to the governor system by the same pump that supplies oil to the turbine bearings. If this oil supply should fail, the governor valve would close and stop the turbine, thus preventing injury to the bearings.

The hydraulically operated throttle valve (Fig. 11.15) is used to control the flow of steam when starting a turbine and in addition functions as an automatic stop valve in case of overspeed. It cannot be opened nor the turbine started until after normal operating pressures for the turbine oiling system have been established. If oil pressure falls to an unsafe degree, the valve automatically closes and the flow of steam to the turbine is interrupted. A strainer is located within the valve to protect both valve and turbine.

The emergency overspeed governor (Fig. 11.15) is separate and independent of the speed governor. It functions to protect the unit from excessive speed by disengaging a trip at a predetermined

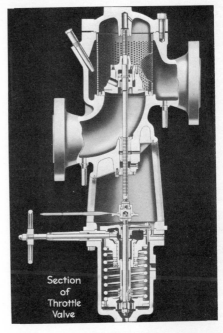

(*a*)

FIG. 11.15   (*a*) *Section of throttle valve.*   (*b*) *Overspeed trip assembly.*   (*Westinghouse Elec. Corp.*)

(*b*)

speed, permitting the throttle valve to close.   This trip may be reset and the throttle valve reopened before the turbine speed returns to normal.

Large turbines are arranged frequently for extraction of steam at various points along the path of steam.   One turbine uses a grid-type extrac-

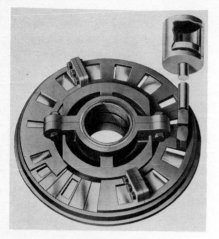

FIG. 11.16 *Grid-type extraction valve for lower extraction pressure.* (*Westinghouse Elec. Corp.*)

tion valve (Fig. 11.16) at normal pressure and temperature. This valve consists of a stationary port ring and a rotating grid; the rotating grid turns against the stationary ring and opens the ports in sequence. Thus there are obtained simple hydraulic interconnections between the several components of the control system and accurate and positive control of speed or load carried by the unit and extraction steam pressures.

Where electric-power generation is the prime consideration, condensing turbines are used. Openings are provided in the turbines for the extraction of steam for heating the feedwater of boilers. A condensing turbine with 13 stages and 1 extraction stage is shown in Fig. 11.17.

The change in speed of a turbine from no load to full load divided by the full-load speed is known as the "speed regulation." A turbine with a full-load speed of 1,764 rpm and a no-load speed of 1,800 rpm would have a change of 36 rpm, or regulation of 2 per cent.

The governors of turbines operating a-c machines are supplied with synchronizing springs. These are arranged to aid in moving the governor weights and lowering the turbine speed or to work against the governor weights and increase the turbine speed. The tension on this spring is varied by means of a small motor (synchronizing motor). By adjusting the tension on this spring the operator can at will change the load on the a-c turbogenerator.

Dashpots are frequently employed in connection with turbine governors. They are used to keep the governor from overtraveling because of the weight of the governor parts. When a governor adjusts the speed of a turbine and makes so great a change that an adjustment in the opposite direction is immediately necessary, the governor is said to "hunt" or overtravel. In other cases dashpots are used to prevent the governor from operating too quickly. Friction or lost motion in the valve mechanism or, in some cases, too heavy or improperly installed dashpots will cause a turbine to hunt. Rapid changes in speed and the corresponding variation in load are referred to as "surges."

A governor mechanism must be adjusted as follows: the valve should be closed when the governor is in closed position and open the correct amount when it is in open position, the speed adjustment must be such

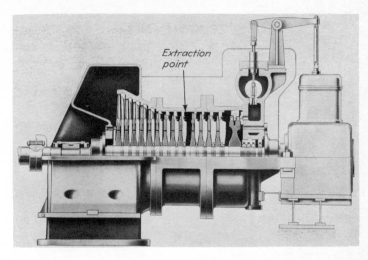

Extraction
point

FIG. 11.17 *Condensing turbine.* (*Westinghouse Elec. Corp.*)

that the governor will be in open position when the turbine is at rated speed, and the speed regulation (that is, the change in speed from no load to full load) must be adjusted.

Turbine governors require very little attention on the part of the operator. They must, however, be kept oiled and the joints working freely. With the oil-relay system, oil leaks must be stopped as soon as possible. The overspeed trip must be checked at regular intervals. It is good practice to operate the overspeed every time that the turbine is placed in service.

Lubrication. Proper lubrication is of utmost importance in the operation of a steam turbine. High journal speed, heat conducted from the steam to the bearing, and the possibility of water's leaking into the oil are some of the conditions that make lubrication difficult.

There are two methods of lubricating turbine bearings. One utilizes oil rings in supplying oil to the bearings; the other consists of a pressure system which circulates the oil to the bearings.

The oil-ring system is used on small turbines. There is an oil well under each bearing. The oil level in the bearing is kept below the bottom of the shaft. The oil rings have a larger diameter than that of the shaft. They hang over the shaft, and a section extends into the oil. When the shaft rotates, the rings also turn and the oil which adheres to them is carried to the bearing. The bearings are grooved in such a manner that the oil is properly distributed to the bearing surfaces.

A typical self-contained oil-pumping system for a large turbine generator is shown in Fig. 11.18. This system is designed to supply bearing

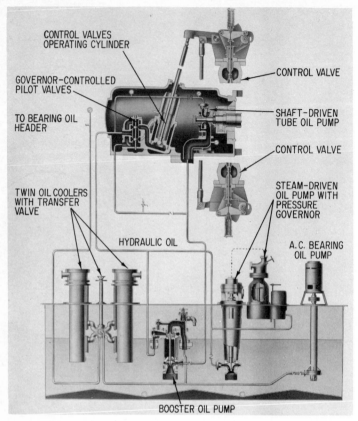

FIG. 11.18 *Schematic diagram of oil-pumping system for steam turbine. (General Electric Co.)*

oil at 25 psig and hydraulic oil at 200 psig, approximately. It operates as follows: During the start-up of the turbine, the steam-driven oil pump provides hydraulic oil to the control system, a portion of the hydraulic oil being fed to the booster-pump controls. Part of this oil pressure is reduced as the oil passes through the oil cooler to the bearing header for journal lubrication (low pressure). The remainder of the oil to the booster pump drives an oil turbine coupled to the booster pump itself. The function of the booster pump is to provide suction pressure to the shaft-driven oil pump.

When the main turbine reaches near-rated speed, the shaft-driven pump supplies the hydraulic oil to the controls as well as hydraulic oil to the booster pump. At this point, the steam-driven oil pump can be put in standby service. The a-c bearing-oil pump is used to provide bearing oil only when the turbine is shut down and when it is operating on turning gear; during this period oil is required for the bearing journals.

With the governor in operation, the governor controls the pilot valve to position the control-valve piston that operates to open or close the steam-admission valves, thus regulating the speed of the turbine. Oil drained from journals and operating mechanism is, returned to the oil reservoir, repeating the cycle.

A variation of the foregoing comprises replacing the steam-driven oil pump with an a-c motor-driven oil pump and, for emergency or standby service, the use of a d-c emergency oil pump operating in parallel with the a-c pump. The d-c emergency unit is employed when the turbine is operating below rated speed or when a-c power is lost.

A somewhat more complicated hydraulic system, for large turbines, is shown in Fig. 11.19. This is a combined lubricating and control system designed for 325 psig; reducing valves are provided as required. The main oil pump provides oil, basically, for two systems: (1) lubricating oil to bearings and auxiliary pumps, as shown at the left; and (2) control oil for the various control points, as shown at the right. Oil coolers and filters are installed at required locations.

Erection of the steam turbine. Steam turbines up to 50 or 60 kw are usually shipped completely assembled and are installed by the purchaser. These units are assembled on their baseplates or bedplates and may be placed in position by means of rollers or a crane. In making hitches with a crane, care must be exercised to prevent appreciable deflection. The foundation must be roughed and swept clean to receive the grout. The turbine is set in position on iron blocks and tapered wedges. These must be arranged to hold the baseplate from ¾ to 1½ in. from the foundation to allow space for the grouting. The correct level is obtained by placing a sensitive spirit level across machined bosses on the base. The level should be checked at all points.

Small turbine units are assembled complete on one bedplate, but in the case of medium-sized units the turbine is separate from the machine which it is to drive. When the machines are separate, they must be aligned. The shafts must be central and parallel. The central relation of the two shafts may be checked by placing a straightedge parallel to the shafts and across the rims of the coupling flanges. This check should be applied at four places about the flanges. The parallel relation of the two shafts can be checked by measuring the distance between the coupling-flange faces with a feeler gauge. If the two shafts are parallel, these measurements will be equal at all points on the coupling. Patent couplings which will compensate for a certain amount of misalignment are used. Even if these couplings are used, however, it is advisable to have the alignment as close as possible.

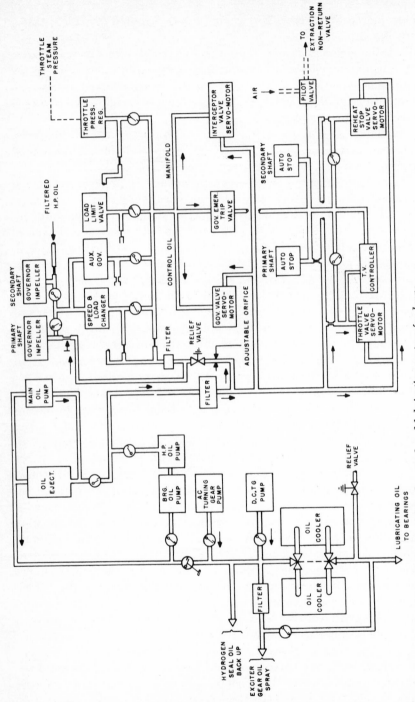

FIG. 11.19 *Hydraulic system—combined control and lubricating system for large turbine.* (*Westinghouse Elec. Corp.*)

The turbine changes in temperature more than the generator does, and if the shafts are lined up cold, they must be set to allow for expansion. Even if they are properly lined when cold, this does not necessarily mean that they will be in proper alignment when in operation.

The foundation bolts can now be tightened up and the level rechecked. If the level is found to be correct, the machine is ready to be grouted. A dam of boards must be built around the foundation to keep the grouting in place. A suitable place must be provided for admitting the grout. Some foundations have holes for this purpose. The grouting mixtures vary from a pure cement to 2 parts of sand and 1 part of cement. The foundation should be wet before the grouting is poured. The mixture should be thin enough to flow readily, so that it will find its way under the base. It is customary to leave the wedges in place, but in some cases they are removed after $1\frac{1}{2}$ or 2 days.

Many medium-sized and large turbines are shipped completely disassembled. Their erection should be supervised by a thoroughly trained factory representative. They require careful adjustment, and each different type presents different problems. The bearings are aligned by means of a fine steel line (tight line) which is stretched through what is to be the center of the shaft. Laser beams are also employed on the large turbine to secure proper alignment. The axial clearance between stationary and moving parts must be checked with a taper gauge. The packing clearance must be carefully checked and adjusted.

The piping to a steam turbine must be large enough to handle the steam flow at full load without excessive pressure drop. Stem velocities of 6,000 fpm were at one time considered the upper limit, but large high-pressure piping is now designed for velocities of 10,000, or even 12,000, fpm.

The piping must be arranged so that it will not produce strains on the turbine. This precaution is especially important for medium-sized and large turbines. Expansion is taken care of by means of loops or bends. The piping near the throttle valve is supported by springs which give with the expansion produced by the change from hot to cold condition yet hold the pipe firm enough to prevent vibration. A valve is placed in the steam line near the main header. This makes it possible to close off the line to the turbine, allowing work on the throttle valve while there is pressure on the main header. A strainer is placed in the line to the turbine just ahead of the throttle valve. This prevents solid particles from entering the turbine.

A noncondensing turbine with the exhaust line connected to an exhaust header must have a shutoff valve located near the header. The exhaust line must also have an expansion joint to prevent strains on the turbine or exhaust header. There must be a relief valve between the

FIG. 11.20 (a) Rupture disc—reverse buckling unit (KBA-ZAP). (Continental Disc Corp.)

turbine and the shutoff valve. This reduces the possibility of a turbine explosion if by mistake the turbine should be operated with the shutoff valve closed.

When a turbine is operated condensing, an atmospheric-relief valve must be placed in the exhaust line between the turbine and the condenser. If the vacuum should fail, this valve would open and the turbine would exhaust to the atmosphere. It is necessary to keep a water seal on the atmospheric-relief valve to prevent the leakage of air into the condenser. Atmospheric-relief valves are costly; a substitute is to be found in the use of "rupture discs."

A rupture disc (Fig. 11.20a) is a prebulged membrane made of various metals based on the service for which it is intended. The disc may be used in lieu of an atmospheric-relief valve or installed ahead of the atmospheric-relief valve provided: (a) The valve has ample capacity. (b) The maximum pressure range of the disc designed to rupture does not exceed the maximum allowable pressure of the vessel. (c) The area is at least equal to the area of the relief valve. (d) The disc unit has a specified bursting pressure at a specific temperature and is guaranteed to burst within 5 per cent (plus or minus) of its specific bursting pressure. If the pressure is greater than 15 psi, the ASME code states that either a rupture disc or safety valve or a combination of the two may be used (Fig. 11.20b); or the disc can be used in parallel with the safety valve, so that the safety valve maintains all normal over relief protection. Then the rupture disc is set at a higher pressure (approximately 20 per cent above safety relief valve). We can then take care of excessive pressures and assure a safety system in case the safety relief valve fails.

FIG. 11.20   (b) *Rupture disc installed in series with relief valve.   (Continental Disc Corp.)*

The rupture disc is fail safe and assures a seal-tight system; no moving parts and therefore nothing to malfunction, stick, or corrode shut; hence a true safety fuse.   A rupture disc cannot reclose itself as can a safety relief valve.

The rupture-disc assembly installed beneath the safety valve includes a Tell-Tale assembly consisting of a pressure gauge and an excess-flow valve; leakage is immediately apparent.   Jack screws lift discharge piping or outlet flange and provide for quick and easy rupture-disk replacement.

Automatic vacuum breakers are also employed to prevent the turbine from overspeeding.   When the overspeed trip closes the throttle, there is a possibility of enough steam's leaking past to cause the unit to overspeed.   The emergency governor operates the vacuum breaker, which eliminates any possibility of enough steam's being drawn through the turbine to cause overspeeding.

The piping and turbine must be supplied with drains at all places where water can accumulate.   The steam piping must have a drain located between the header valve and the throttle.   This drain is used to remove moisture from the line before placing the turbine in service. When saturated steam is used, a steam separator is placed in this line to remove the condensation.   The drain is then connected to the separator.   These drains are supplied with traps that are left open when the turbine is in operation and bypassed when it is being started.

The turbine casing must be drained, as an accumulation of water while the turbine is out of service would cause the blades to rust, and this would result in the rotor's becoming unbalanced. A drain must be connected to the exhaust line of a noncondensing turbine between the turbine and the shutoff valve. Drains must be supplied to carry the water away from the gland seals. These lines must be of sufficient size to handle an excess of water, because it is advisable to have this water discharge into funnels so that the flow can be observed by the operator. The lines must be piped to a place where there is no back pressure.

Mechanically driven turbines operating below 250 psig are frequently employed to start up (automatically) emergency electrical generators in case of a power failure; they are also employed to start up automatically pumps, air compressors, etc. For the automatic operation, the throttle valve opens to place the turbine in service. The exhaust line remains open; casing and other low points are provided with drain traps.

Bearing-cooling water is supplied either to an oil cooler, in the case

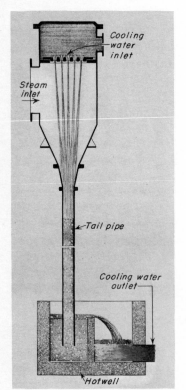

FIG. 11.21   *Ejector-jet barometric   condenser. (Ingersoll-Rand Co.)*

of large turbines, or to water-jacketed bearings in the case of small turbines. The cooling water for bearings should be as free from scale-forming material as possible. The water entering the drain lines should be visible to the operators. The discharge lines should be large enough to handle the discharge.

The grouting of a new turbine must be allowed to set until sufficiently hard before the turbine is tried out. The piping and drain should be carefully inspected to see that the valves are properly arranged. It is good practice to blow out the steam lines with compressed air to remove mill scale.

11.2 CONDENSERS.   A steam condenser is a vessel in which exhaust steam is condensed by contact with cooling water; in condensing the steam, a vacuum is created. The vacuum reduces the back pressure on the engine or turbine; reduction in back pressure increases the horsepower and efficiency of the unit.

There are two principal types of condensers: the jet barometric type, in which water and steam come into direct contact; and the surface type, in which the cooling water and steam remain separate.

Jet barometric condensers.   These condensers are also referred to as "ejector-jet" or "disk-flow" types. In Fig. 11.21, water enters a series of water-injection nozzles at the top. It meets steam entering on the side, and together they drop into the hot well, overflowing by gravity. The steam-inlet connection may be either on the top or at the side of the condensing chamber; for the latter, space requirements are reduced. Brackets are attached directly to the shell to support the weight of the condenser.

In Fig. 11.22 steam enters the chamber at the "exhaust inlet," and the injection water enters at the point shown. Upon entering, the steam meets numerous jets of cold water. The water and condensed steam drop to the tail pipe and then to the hot well. If the condensing chamber is placed at a high elevation, about 35 ft, the water is discharged from the tank without the aid of a removal pump. (Low-level jet condensers are similar to barometric condensers, except that the water must be removed by a pump.) As the steam is condensed, air is released and passes through a baffle before being liberated at the top of the condenser by means of a vacuum pump or air ejector.

In the disk-flow condenser (Fig. 11.23), cooling water enters at the top, cascading down over disks, to condense the steam entering at the point shown. The condensate and cooling water mix, to pass down the tail pipe to the hot well. The large-diameter tail pipe makes the

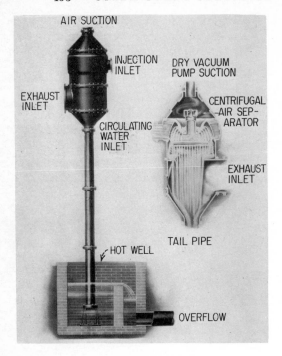

AIR SUCTION

INJECTION INLET

DRY VACUUM PUMP SUCTION

EXHAUST INLET

CENTRIFUGAL -AIR SEP- ARATOR

CIRCULATING WATER INLET

EXHAUST INLET

TAIL PIPE

HOT WELL

OVERFLOW

FIG. 11.22 *Barometric condenser.* (*Worthington Corp.*)

condenser self-supporting and at the same time provides a vent adequate for the escape of steam to the atmosphere in case the water fails.

Air and noncondensable gases pass up through the water curtain to the air-vapor outlet and from there to a two-stage ejector system. The first-stage ejector employs steam to pull the air vapor from the condenser to a barometric intercondenser; the intercondenser employs water to condense the steam, condensate and water flowing to the hot well. Steam injection is provided for the second stage; the noncondensable gases are aspirated from the intercondenser to discharge to the atmosphere.

For the condenser shown in Fig. 11.21, access to the nozzle plate is secured by removing a manhole cover; this simplifies cleaning and inspection. The entire nozzle tray can be raised or lowered by turning a single nut to locate the tray in its best operating position.

Jet condensers are made of cast iron, welded steel plate, or other noncorrosive materials to meet specific conditions. Bronze throat pieces are supplied with cast-iron condensers, and cast-iron throat pieces with steel-plate condensers. Impingement baffles are made of extra-thick plate to withstand erosion; center baffles and plates are cast in one piece and securely bolted together in place. This is accomplished by installing lugs to support the first tray, whereas the supporting member for other trays is all in one piece. The disk-flow condenser and its auxiliary equipment require less cooling water and at lower pressure than does the ejector-jet condenser. For the ejector-jet type, the water quantity need not be controlled so accurately.

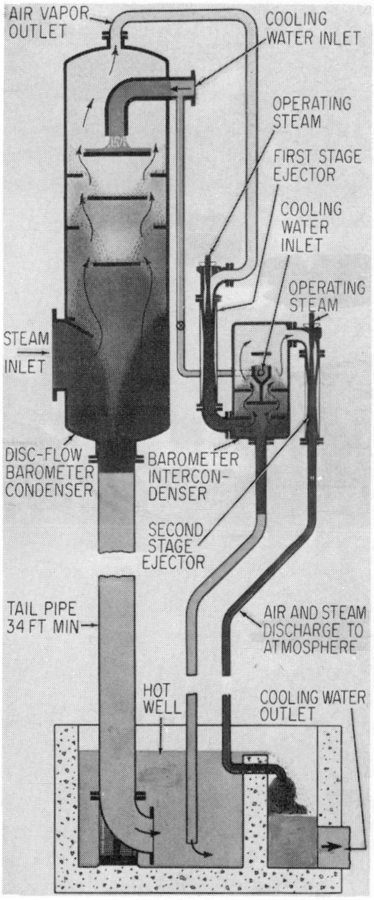

Surface condensers. The surface condenser is a closed vessel filled with many tubes of small diameter; condensing water flows through the tubes with steam on the outside of the tube bank. The water flowing through the condenser may be once-through or single-pass; or it may be made to reverse one or more times before being discharged.

Exhaust steam (Fig. 11.25) from the turbine enters at the top of the condenser, passing down and around and between the tubes in the tube bank. Cold condensing water flows through the tubes in sufficient quantity to condense the steam. Depending on the design, the circulating water makes one or more passes through the tubes before being discharged to the river or cooling tower. In condensing a large

FIG. 11.23 *Flow diagram of disk-flow condenser and two-stage condensing ejector. (Ingersoll-Rand Co.)*

volume of steam into a small volume of water, a vacuum is created, reducing the back pressure for the turbine. The condensate passes to a hot well, where the steam (water) is recovered and pumped back to the boiler. Noncondensable gases and air finding their way into the condenser are removed by ejectors or other suitable air-removable devices. The removal of air and the oxygen which it contains not only reduces the back pressure but prevents its reentry into the system, thus reducing the possibility of corrosion in the piping and boiler.

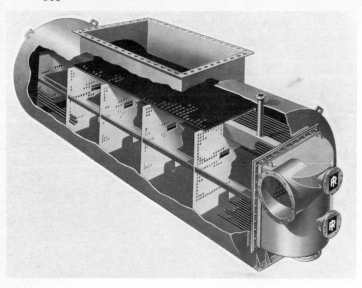

FIG. 11.24 *Surface condenser for industrial and utility stations.* (*Ingersoll-Rand Co.*)

Condensers such as shown in Fig. 11.24 are adaptable to a wide range of turbine-condenser arrangements, for industrial and utility stations. Some features of the Class II condenser are: (*a*) Shell—including the tube support plates, hot well, and connecting piece; is fabricated from heavy steel plate welded into one integral unit. Liberal flow area above and around the tube bundle allows complete distribution of steam to all tubes. (*b*) Waterboxes—are deep with large nozzles to keep water velocities low; they can be furnished with a suitable lining for corrosive water applications. (*c*) Hotwell—has ample capacity for storage requirements. Reheating features keep condensate depression to a minimum and oxygen contents below 42 ppb. (*d*) Tubes—are rolled and expanded into the tube sheets at both ends, forming a leak-proof and mechanically strong tube-to-tube sheet joint. (*e*) Expansion—a diaphragm-type shell-expansion element provides for thermal expansion and contraction of the tube bundle under all operating conditions. (*f*) Air cooler—provides full-length scavenging of noncondensables. (*g*) Impingement plate—the large perforated baffle plate is provided directly below the steam inlet to protect the tube bundle against moisture impingement.

Details of the heart-shaped condenser construction are shown in Fig. 11.25. Steam distribution through the condenser is such as to control the path of steam by making it converge as closely as possible to the tube nest at the air-vapor outlet; steam is brought into intimate contact with the cooling surface.   Longitudinal distribution is secured by changing the tube area at the ends to produce more condensation at this point and also by regulating the terminal pressure in each section (this is accomplished by using an orifice in the air-vapor outlet).

Attached to the bottom of the condenser are a multipass air cooler and a reheating hot well.   Maximum cooling of the air before ejection is desired since a lower vacuum is produced, resulting in a decrease in the volume of air, which is thus easier and less costly to remove.   The air and noncondensable gases are made to pass to the air-vapor outlet and then to the air ejector.   Condensate collects in a long trough which has staggered rows of holes along each side; condensate sprays from the trough through these openings, passing through steam previously bypassed from the turbine exhaust.   In this manner the condensate reaches a maximum possible temperature on its return to the steam-water cycle.

To accommodate the large steam generators presently being installed, very large condensers are required; at times twin units are employed. Flanges of the shell are made of quality steel plate, and all seams are welded.   Shells are flanged or butt-welded for air take-offs, condensate pump suction, return vents, drains, etc.   Tube heads are made of muntz metal, aluminum bronze, silicon bronze, copper nickel, or steel.   Sheets are machined and drilled to receive the tubes.   If the tubes are welded

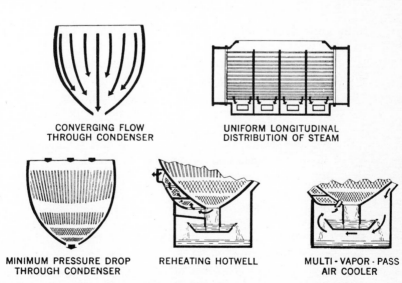

CONVERGING FLOW
THROUGH CONDENSER

UNIFORM LONGITUDINAL
DISTRIBUTION OF STEAM

MINIMUM PRESSURE DROP
THROUGH CONDENSER

REHEATING HOTWELL

MULTI - VAPOR - PASS
AIR COOLER

FIG. 11.25   *Details of surface-condenser design.*   (*Ingersoll-Rand Co.*)

FIG. 11.26 *Shell section of condenser under construction, shoding tube sheets and support plates.* (*Foster Wheeler Corp.*)

to the tube sheets, the tube sheets are made of silicon bronze to facilitate welding.

Tube supports are provided across the width and length of the condenser to hold the tubes in place and avoid the tube damage which may result from vibration. The support plates are made of steel; the interior bolting and cap nuts exposed to the circulating water are made of corrosion-resisting composition or steel. The plates in turn are welded to the shell. The necessary drains and vents must be provided for removing condensate, for venting pumps back to the condenser, etc. Braces, drain plates, baffles, and supports in the steam space are welded in place. Water boxes are fabricated of steel for fresh water; for unusual conditions cast iron is used. For salt water, the boxes are made of cast iron or suitably lined fabricated steel. In Fig. 11.26 is shown an end view of the partially shop-assembled condenser for Paradise Steam Plant Unit No. I of the TVA. This condenser contains 300,000 sq ft of surface and is of the modified double-flow design with divided single-pass water boxes. There are 21,910 Admiralty Metal $7/8$-in.-diameter tubes 60 ft long.

Condenser tubes are made of Admiralty Metal, aluminum alloys, and various stainless-steel compositions. Aluminum alloys are less expensive than the other materials, but experience has indicated some difference in fouling characteristics with performance falling off more rapidly than for copper-base alloys. Although this fault is corrected by using a larger condensing surface and more circulating water, this practice has the disadvantage of increasing condenser and power costs. Where circulating water is corrosive and no reasonable life can be obtained with copper-base alloys, stainless steel is used. Stainless steel appears to have more favorable fouling characteristics than either Admiralty Metal or other copper-base alloys when used with acidic waters.

Tubes are small, usually $\frac{1}{2}$ or 1 in. in diameter, and fit into holes drilled and machined to receive them. Condensers may have stationary tube sheets, or they may be constructed with a floating tube sheet on one end. The floating-tube-sheet design takes care of tube expansion. Various methods are employed for holding the tubes in place: (1) They may be held in position by packing both tube ends with fiber or metallic rings held in place by a ferrule. The ferrule is an open-end brass plug with a flared end on one side and a threaded end on the other; it is fitted with lips to prevent the tube from creeping. The packing rings are placed around the tube end; the ferrule is pulled up against the packing, thus holding the tube securely in place. The inlet end of the tube is flared over to provide a smooth water entry. (2) The tubes are packed at the water outlet, the inlet ends being expanded into the tube sheet. As in the first method, the packing rings fit into a threaded stuffing box. (3) Tubes are rolled on both ends, the inlet end being flared over. The tubes can be welded if required.

In large condensers tubes are installed in the form of an S-shaped bow, the deflection curve of a uniformly loaded beam, to provide for unequal expansion and set. One end is higher than the other to eliminate any pockets where corrosive gases could collect and to afford positive drainage of circulating water during shutdown periods.

For the large condenser, the shell is manufactured in a minimum number of sections to minimize field erection. The maximum size of each prefabricated section is dictated by shipping clearances. The shell and exhaust connections have the various edges of the various sections chamfered, as though they were to be finished welded. Sections are then assembled in the shop, temporary lugs and bolts being used to fit the sections together. Tube support plates and struts are welded in place to act as braces. When the unit has been completely assembled, it is marked for future erection, dismantled, and shipped. When the unit is in its final location, all seams are welded to form a rigid structure free from air leaks, after which the temporary holdings are removed.

Condensers are frequently designed to serve as a complete foundation for the turbine, generator, and exciter. Such a unit is installed in the power plant of the General Electric Company; this condenser (Fig. 11.27) is a two-pass unit, containing 7,150 sq ft of surface and using tubes $\frac{7}{8}$ in. in diameter. All auxiliaries are mounted on the side of the condenser, which provides a compact installation.

Condensers require large quantities of circulating water to dissipate the heat released when the steam is condensed (see Sec. 12.2). So the large power plants must be located on rivers or lakes where water is available.

In passing through the condenser and depending on station design and capacity, the temperature of the circulating water increases from 6 to 10°F. The nuclear power plant requires (for an equivalent power output) more circulating water than that for the fossil-fuel–fired plant.

In instances where the supply of these natural waters is limited, the temperature of a portion or all of the water is increased, giving rise to the term "heat or thermal pollution." This condition adversely affects aquatic and marine life, and environmental legislation has established limits on the use of natural waters receiving the rejected heat from power plants. It therefore becomes necessary to increase plant investment by installing spray ponds or cooling towers which reject this heat to the atmosphere.

11.3   SPRAY PONDS AND COOLING TOWERS. The simplest type of cooling-water system is the pond or spray pond. The condensing or circulating water is pumped into the pond or basin; the basin, in impounding the water, provides storage in addition to the cooling.

Spray ponds. The cooling pond is converted to a spray pond by locating a series of sprays above the surface of the water. Sprays are mounted on a series of piers (concrete or timber) and normally set at a height of 1 to 5 ft above the base of the pit. The height is determined by the depth of the pit and the quantity of water it contains. The sprays are supplied from a series of pipes located on top of the supporting piers, their size and the number of headers depending upon the quantity of water to be handled. The piping header (Fig. 11.28) is tapered, thus reducing the pressure drop; this arrangement also decreases the installation cost. Large headers are supported by rollers to permit expansion and contraction. Individual branch lines are equipped with valves to regulate the flow of water and also to permit taking a header

FIG. 11.27 *Condenser designed to serve as a complete foundation for turbine, generator, and its exciter. (Ingersoll-Rand Co.)*

out of service. The pipe may contain a series of nozzles, or the header may be equipped with a series of radial pipes extending from the pipe in a cluster arrangement as shown. The water issues vertically from the spray to form a cone; such spraying provides uniform water distribution, increasing the area of exposure and improving the efficiency.

As the water is sprayed into the air, particles in varying amounts are carried away by the wind. This results in a loss of water and creates a nuisance which is a source of complaint in congested areas. This loss of water and nuisance are reduced by the installation of a louvered fence around the spray pond.

Cooling towers. An atmospheric spray cooling tower is a cooling pond equipped with louvered walls and arranged in one or more decks. Sprays located on top of the structure discharge the water to be cooled, the water dropping to the pond below. Air passes through the louvers from the side, up through the descending water spray; this type of cooling tower is called a "natural-draft tower." Some atmospheric spray

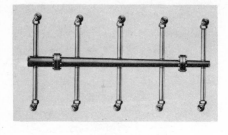

(a)

A TYPICAL ASSEMBLY,
ILLUSTRATING ONE
MANNER OF COMBINING
MARLEY NOZZLES, SPRAY
ARMS, END PIECE AND
TAPERED PIPE

MARLEY CAST IRON
TAPERED END PIECE

MARLEY BRONZE
SPRAY NOZZLES

SPRAY ARMS OF
GALVANIZED STEEL
OR CAST IRON

RUNNING FLAT BOSS
ALONG EACH SIDE PERMITS
LOCATING TAPS FOR SPRAY
ARMS ON ANY CENTERS
DESIRED

MARLEY CAST IRON
TAPERED PIPE

FIG. 11.28 *Spray system and
tapered piping header.* (a)
Tapered piping header. (b)
*Spray nozzles with cluster ar-
rangement.* (*The Marley Co.
Inc.*)

(b)

towers are filled with decks of timbers over which the water is made to
cascade, exposing more water surface to the air passing through and
thus increasing the evaporation and lowering the water temperature.
Louvers are inclined to permit ease of air circulation through the tower
and prevent the discharge of water.

A natural-draft hyperbolic cooling tower for large-volume cooling is
shown in Fig. 11.31. The unit is designed for the Kentucky Power
Company Big Sandy power station; it is capable of handling 120,000

FIG. 11.29   *Double-flow cooling tower—induced type.*   (*The Marley Co. Inc.*)

gpm and cooling water from 109.7°F on entering to 87°F on leaving. It has a minimum diameter of 130 ft and a maximum diameter of 245 ft and is 320 ft high; it serves a 265,000-kw generator.

The reinforced-concrete towers have a cooling capacity comparable with that of a multicell installation of induced-draft cooling towers, but they require considerably less ground area. Since no fans are needed, power cost and auxiliary equipment are eliminated; operating and maintenance costs are consequently reduced. Operation is much like that of other natural-draft spray cooling towers, with hot water cascading over timber splash-type filling through which cooler air is moving. For the tall tower there is little chance of air recirculation. Approximately the same consideration must be given to the design of the hyperbolic unit as to that of other natural-draft towers.

Mechanical-draft cooling towers do not depend on natural wind movement, since the fan has the required capacity to effect air cooling based on the outside wet-bulb temperature. They are of two general types: forced-draft and induced-draft towers. In the former, the fan is located at the base of the tower and air is blown by the fan up through the descending water. In the latter, the fan is located at the top of the tower; the air enters through louvers located on the tower's sides and is drawn up and discharged through the fan casing to the atmosphere.

The double-flow-principle mechanical-induced-draft tower shown in Fig. 11.29 (transverse cross section in Fig. 11.30) serves a 100,000-kw turbine generator, cooling 50,000 gpm. This unit features 28-ft cells,

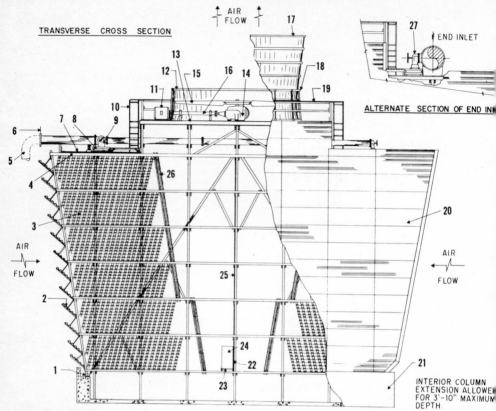

TRANSVERSE CROSS SECTION

ALTERNATE SECTION OF END INLET

FIG. 11.30 *Transverse cross section of double-flow cooling tower.* *(1) Perimeter anchorage.* *(2) Corrugated asbestos-cement louvers.* *(3) Splash bars in fiberglass grid supports.* *(4) Diffusion decks (above fill).* *(5) Side-inlet pipe system located as shown.* *(6) Pipework stops at face of inlet flange.* *(7) Open distribution basin (removable nozzles).* *(8) Covered distribution box.* *(9) Flow-control valve.* *(10) Ladder.* *(11) Motor mounted on transverse center line of each cell.* *(12) Vertical laminated fan cylinder.* *(13) Unitized steel mechanical equipment support.* *(14) Gear reducer.* *(15) Multiblade fan.* *(16) Driveshaft.* *(17) Velocity recovery cylinder.* *(18) Access door in fan cylinder opposite motor for gear-reducer removal.* *(19) Handrail around fan deck.* *(20) Horizontal corrugated asbestos-cement board end-wall casing.* *(21) Concrete basin by purchaser. Interior column extension allowed for 3 ft 10 in. maximum depth.* *(22) Access opening through longitudinal partition at each cell (no doors).* *(23) Walkway (one side only).* *(24) Partition and end-wall access door.* *(25) Longitudinal partition.* *(26) Herringbone drift eliminators.* *(27) Flow-control valve.* *(The Marley Co. Inc.)*

24-ft fans, 6-ft fiber-glass fan cylinder bases, and a 12-ft laminated-wood velocity-recovery stack extension. Special mechanical equipment-handling devices include removal davits and dollies and a monorail system for transporting heavy equipment to tower end walls.

The double-flow tower consists of two identical sections divided at the tower center by a partition which extends from the water level in the basin to a point close to the fan-inlet housing. The housing consists of vertical columns made from sturdy timbers that are spaced on close centers for added strength; the vertical members are supported mechanically by bracing to provide rigidity where required. Transverse bracing as well as longitudinal bracing is provided to give maximum strength.

To avoid the corrosive influences to which cooling towers are subjected by water conditions and atmospheric contamination, moldings are made of glass-reinforced polyester. Structural-ceramic rings are applied in conventional connector-ring joints where values exceeding those available in bolted joints are required. The ceramic material is a complex porcelain oxide of silicon and magnesium. This construction and use of materials eliminate the possibility of bearing failure in the wood under the inroads of rot in a rust-deposit area. Asbestos-cement board covers much of the exterior of the tower, the covering including end-wall casings and board louvers that form the tower sides. This adds to fire safety and structural sturdiness.

The material under each "section" is filled with splash bars set in fiber-glass supports, which are impervious to all corrosive conditions. These high-strength grids are on close centers so that there can be no sagging fill or channeling of water flow. Splash bars are securely retained without nails or other corrosive fasteners and can elongate or shrink without distortion or cracking. Between the filling and the center longitudinal partition are herringbone drift eliminators for removing water entrained in the air. The drift eliminators also function as effective diffusers, equalizing pressure through the cooling chamber.

For large fans, blades of glass-fiber–reinforced polyester are used to eliminate corrosion; aluminum blades are also available for other models. The gear reducer contains an enclosed lubrication system with a renewable cartridge filter.

The use of glass-fiber–reinforced grid supports and molding together with the method of construction eliminates corrosion or degradation effects resulting from contaminants in the circulating water; hence there is no galvanic corrosion or rusting, as occurred with ferrous parts in the past. Structural-ceramic rings are applied in conventional connector-ring joints; the ceramic material is porcelain. The combination of the foregoing reduces deterioration by corrosion regardless of acid, caus-

FIG. 11.31 *Hyperbolic natural-draft cooling tower.* (*The Marley-Mouchel Co.*)

tic, salt, or other contaminants in the circulating water; there are no metallic salt deposits to be absorbed in the wood, which is subject to biological attack.

Directly over the center of the tower are located the induced-draft fans. The motors are removed from the airstream and set outside the fan housing. In the illustration shown are seven individual towers comprising one cooling-tower unit. The louvered openings are on the side through which the air is drawn. Across the top of the tower are the open-distribution hot-water basins; the bottom of each basin contains a series of porcelain nozzles so located as to provide uniform water distribution to the "filling" below. Flow-control valves are provided for each tower and are used to vary the flow if desired. In the center of the fan housing is the fan assembly with gear-reducer drive and flexible drive shaft; two-speed motors are available.

The level is held constant by automatic-control float makeup valves to replenish the water supply, by automatic control of the blowdown from the basin, and by the overflow pipe to the drain. If well designed, the water-distribution system should be readily accessible for regulation of flow and for cleaning and inspection. With the double-flow tower one-half of any cell may be shut down while the other half remains in service. At low wet-bulb operation, sufficient cooling can be obtained without operating the fan.

Large cooling towers are usually provided with concrete basins; these basins should be watertight and deep enough to store an adequate amount of water. Sufficient space and clearance should be permitted for ease in cleaning and painting. Access doors should be conveniently located and large enough to permit equipment necessary for repairs and maintenance to be moved into the tower's interior.

In designing a cooling tower, consideration should be given to the prevailing wind direction and to any obstructions surrounding the tower, since any interference will reduce the efficiency of operation and influence the performance. Cooling towers must be designed and built to support their own weight, together with the weight of the water and the force produced by the wind. Standard wind-pressure design is 30 psi wind load, though in certain areas where windstorms prevail additional safeguards must be taken to withstand loading. Wood towers are to be recommended over steel towers to avoid corrosion. If bolts are used, they should be bronze or galvanized; nails are never used to carry a load. Consideration should be given to fire prevention, to proper access to facilities, and to walkways for adjustments, servicing, and maintenance.

11.4 CONDENSER AUXILIARIES. Various auxiliaries are required for the proper operation of a condenser. Circulating pumps must provide water to condense the steam and produce a satisfactory back pressure or vacuum; condensate pumps must remove the steam thus condensed; a vacuum pump or ejector must remove the air and noncondensable gases; to avoid pressure in the condenser, an atmospheric-pressure-relief valve is required. A typical piping installation of a large surface condenser and auxiliaries is shown in Fig. 11.32. The figure shows the arrangement of auxiliaries on a two-pass surface condenser.

Circulating pumps used for surface condensers are of three types: (1) The impeller type is used where high suction lift or pumping head is specified. This type of pump is usually double-suction and horizontally mounted. (2) The propeller type is suitable where a moderately large quantity of water is required at a relatively low head. This pump is mounted vertically and operates with the pump submerged; no priming is required. (3) The axial type is used where moderate pumping head and suction lift are required. This pump too is set submerged and does not require priming.

Since the impeller pump is not much different from that described in Chap. 7, reference here is made only to the mixed- and axial-flow pumps as illustrated in Fig. 11.33. These are vertical pumps built to meet specific requirements. Both operate with submerged suction and hence are self-priming; they are made adaptable to float-controlled operation, start and stop. The axial-flow pump is a single-stage, low-head,

FIG. 11.32 *Typical piping arrangement of a two-pass surface condenser with auxiliaries.* (*Ingersoll-Rand Co.*)

large-capacity unit and is made in sizes up to 50,000 gpm at 25 ft. If variable head and capacity conditions are encountered, an adjustable-vane impeller can be furnished. The mixed-flow pump combines axial- and centrifugal-flow characteristics; it is made in sizes up to 100,000 gpm at 50 ft. Pumps of this type can be designed for higher heads if desired.

The shaft of the vertical pump is held in alignment by column bearings spaced at approximately 9 ft or less. If water lubrication is provided, rubber bearings are used. If oil lubrication is used, bearing linings are either bronze or iron. Shaft sections are coupled with heavy threaded sleeves, self-locking even when flow through the pump is reversed.

The impeller is located in its most efficient position with respect to the casing by means of an adjusting nut in the top of the motor housing. The impeller is keyed and locked on the shaft to prevent movement relative to the shaft.

The vertical pump, as normally installed, can be pulled out for inspection after the pipe has been disconnected. This must be done when the water level is such as to permit removal. Complete removal facilitates inspection and repairs.

Condensate pumps remove the water from the hot well and maintain a continuous water supply through the extraction heaters to the boiler feed pumps. The head against which the pump is required to operate

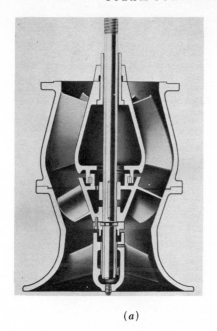
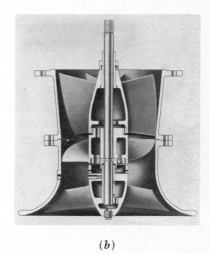

(a)   (b)

FIG. 11.33   (a) *Mixed-flow pumping element.*   (b) *Axial-flow impeller pumps.*   (*Ingersoll-Rand Co.*)

depends on the type of installation.   As a result some condensate pumps are single-stage while others are two-, three- or four-stage.   Pumps of this general description have been explained in Chap. 7.

High vacuum requires the removal of the noncondensable gases. Sometimes the pump which removes the condensate also removes the air.   More often the condensers are provided with separate *vacuum pumps.*   Vacuum pumps may be of the reciprocating type or of the hydraulic-vacuum type.   In the former we have a positive displacement unit exhausting the air from the condenser at a point close to the water line in the hot well.   In the hydraulic-vacuum unit, water is recirculated from a hurling water tank through a revolving jet wheel, causing air entrainment.   Both water and air are discharged through the jet to the tank, where the air is liberated.

The steam-air ejector (Fig. 11.34) consists of a steam nozzle, a suction chamber, and a diffuser.   Steam enters the nozzle and discharges a jet of high-velocity steam across a suction chamber into a venturi-shaped diffuser.   Air or gases to be evacuated enter the ejector suction, become entrained by the moving jet of steam, and are then discharged through the diffuser, where the velocity energy is converted into pressure, thereby compressing the mixture to a lower vacuum.   The steam pressure is a variable depending upon the size of the nozzle employed and vacuum to be produced.

Steam-jet ejectors are used with barometric condensers as well as for air removal on surface condensers. They are simple in operation and rugged in construction, can employ a wide range of steam pressures, and will handle wet or dry mixtures of air, gases, or vapors at a near-perfect vacuum. However, several ejectors may be required to secure the desired vacuum.

Where lower vacuums are desired or where it is not economical to do the entire job of evacuation in one stage, multistage ejectors are employed. Here the discharge of the primary stages enters the suction chamber of the succeeding stages either directly or through an inter-cooler placed between the stages. Precoolers are sometimes placed ahead of the first-stage ejector to remove condensable vapors before they enter the ejector, thereby reducing its size and steam consumption. Single-stage ejectors are suitable where moderate vacuum is desired, after which two-stage ejectors are employed. Ejectors of this type require little maintenance and are more or less trouble-free in operation, and they have consequently been universally accepted for this operation.

The steam-air ejector has virtually replaced all other forms of air-re-

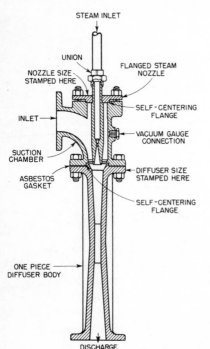

FIG. 11.34 *Steam-jet ejector.*
(*Ingersoll-Rand Co.*)

moval devices. In most condenser installations, the vacuum to be maintained is greater than can be secured economically by a single-stage ejector. Two-stage ejectors usually have some type of condenser between them to condense the steam and condensable vapors from the first stage.

When large quantities of water are being handled, difficulty is frequently encountered with foreign matter clogging up the tubes and openings of surface and jet condensers as well as interfering with the proper operation of circulating and condensing pumps. In some localities, leaves and fish prove especially troublesome during certain seasons of the year. To avoid such operating difficulties, large revolving or *traveling screens* are placed in the entrance to the intake water tunnel. The screen consists of a string of baskets joined together to form a continuous chain. Baskets or screens are made of brass or galvanized wire to resist corrosion.

In operation, one end of the chain revolves about a sprocket placed at the bottom of the intake tunnel; the other end, the driving motor, and the sprocket are above the water level. As water flows through this revolving or stationary screen, it deposits the foreign matter on the wire basket. As the screen revolves on its way to the bottom of the pit, it encounters a spray of water directed against it from the inside which then washes off the foreign matter into a trough, from which point the foreign matter is carried away for disposal.

Relief valves are placed in a branch line leading from the main exhaust pipe to the atmosphere to avoid the possibility of having a pressure on the condenser. To prevent air leakage into the condensers, the valve is provided with a water seal. A relief valve is similar to a check valve; it is held closed by the atmospheric pressure on top of the valve. As long as a vacuum is maintained in the condenser, it remains closed. If the vacuum is lost, the valve promptly opens to discharge the steam to the atmosphere. The valve continues to remain open until the vacuum has been restored, at which time the valve closes automatically. To prevent pounding during intermittent exhaust on the engine such valves are frequently fitted with dashpots. Leakage of air is prevented at the seat by placing a water seal at the raised lip. These valves usually have attached an arrangement to hold the valve wide open if it is found necessary.

QUESTIONS AND PROBLEMS

11.1. What is a steam turbine?

11.2. Explain how it operates.

11.3. What is the difference between an impulse and a reaction turbine?

11.4. How does steam impart energy to a turbine to develop work or horsepower?

11.5. Mention the essential parts of a small turbine.

11.6. Of what material are turbine blades made?

11.7. What precautions must be taken in designing turbine blading?

11.8. How is speed regulation on a turbine secured?

11.9. What is the purpose of the hand reset valve?

11.10. What is meant by a diaphragm?

11.11. How is leakage of steam along the rotor prevented?

11.12. Are large turbine rotors made of one or more pieces? Explain.

11.13. How are the blades fastened and held in place?

11.14. What is a thrust bearing, and how does it work?

11.15. Which type of turbine requires the largest thrust bearing, an impulse or a reaction turbine?

11.16. How is steam leakage prevented on large turbines?

11.17. What is meant by an extraction turbine?

11.18. How does an oil-relay governor operate?

11.19. Mention two different methods of lubricating turbine bearings.

11.20. Mention four or five essentials in lining up a new turbine.

11.21. What is a condenser?

11.22. What is the purpose of a condenser?

11.23. What is a jet condenser? A barometric condenser?

11.24. What are the advantages of each? Discuss them.

11.25. What is a surface condenser, and what are its advantages?

11.26. What auxiliaries are placed in the exhaust line from a turbine to the condenser?

11.27. What is the purpose of an atmospheric-relief valve on a condenser?

11.28. What is meant by parallel and counterflow?

11.29. Of what are condenser tubes made?

11.30. How are the tubes fastened in the tube sheet?

11.31. What is a rupture disc? What is its purpose? How does it operate?

11.32. What is a cooling pond, and what is its purpose?

11.33. Explain the difference between a spray pond and a cooling pond.

11.34. How does a cooling tower compare with a spray pond? What is its purpose?

11.35. What is the advantage of a hyperbolic cooling tower over the mechanical-, forced-, or induced-draft type?

11.36. Mention several important turbine auxiliaries and their use.

11.37. What is a steam-air ejector, and what is its purpose?

CHAPTER $\overline{\underline{}}$ 12

OPERATING
AND
MAINTAINING
TURBINES
AND
AUXILIARIES

12.1 TURBINES. Turbines may be classified as to (1) steam supply, that is, whether they are low-, medium-, high-, or mixed-pressure units; (2) exhaust arrangement, which may be of the extraction, condensing, and noncondensing or back-pressure types; (3) physical arrangement of turbine shaft or shafts, that is, as single, tandem, or compound; (4) driven equipment, consisting of mechanical or electric generators; and (5) connection to drive unit, that is, as direct or geared.

Turbines designed for steam at 150 psig or less are called "*low-pressure*" units; those for steam from 151 to 450 psig, "*medium-pressure*" units; and those for steam over 450 psig, "*high-pressure*" units. Mixed-pressure turbines are designed to admit steam at two or more subsequent openings in the casing; here the low-pressure steam combines with that which entered the first stage. This arrangement provides a means of using steam at different pressures without dissipating the heat energy available by passing the steam through a reducing valve. The temperature and pressure of the steam supplied are an important factor in determining ultimate efficiency; materials used in the construction of the turbine play an important part in the overall performance.

High-pressure, high-temperature turbines are used primarily in large industrial and utility stations; several types and applications are shown

517

in Fig. 12.1. Pressures for these turbines range from 400 to 3,500 psig with steam temperatures to 1050°F. Most of the large units operate with reheat; here the steam from the high-pressure turbine is taken back to a reheater (boiler) where the steam is reheated to its initial temperature. High-pressure turbines are sometimes employed as topping units. This arrangement consists of installing a high-pressure unit the exhaust of which enters a low-pressure turbine (installed earlier and operating at some lower pressure). In essence the high-pressure turbine acts as a reducing valve and, while so doing, generates power; the exhaust steam to the low-pressure unit produces the same amount of power as previously, provided steam conditions on entering and leaving remain the same.

The turbine in Fig. 12.2 is a tandem-cross-compound unit. The upper portion shows a high- and intermediate-pressure (tandem) turbine on a single shaft. The lower portion is the low-pressure unit. Mounted on the right of each (not shown) are the generators.

In operation initial steam enters the high-pressure turbine at 3,500 psig and 1000°F through two inlets (top and bottom). It passes through the turbine to exit at the left (and below) at approximately 600 psig and 550°F and then passes to a reheater, where the steam is again heated to 1000°F. On leaving the reheater, steam enters the intermediate unit (at bottom center) and passes through the double-flow turbine, exhausting at the top through two outlets. This steam is at approximately 170 psig and 710°F when it passes to both sections of the low-pressure unit, finally exhausting to the condenser.

This is an extraction turbine, steam being extracted from one stage of the high-pressure turbine and four stages of each low-pressure turbine; it is used for feedwater heating and other purposes. Note that the intermediate-pressure unit and both low-pressure units have double-flow arrangements. The stators are water-cooled; hydrogen cooling is provided for generator windings and magnetic core. This unit is operating in the Bull Run station of the TVA and has a capacity of 900,000 kw.

Steam from a turbine may be exhausted into a condenser to obtain the maximum amount of energy in the steam, or it may be removed at any intermediate pressure by the use of a noncondensing or back-pressure turbine. While the latter reduces the amount of energy available to the turbine, it provides steam for process or space heating. A compromise between the condensing and noncondensing arrangements is the extraction turbine. Here steam, having passed partway through the turbine, is removed from the casing at a point or points with the desired steam conditions. The pressure at a given turbine stage (that is, connection to the casing) varies with the load. For this reason two

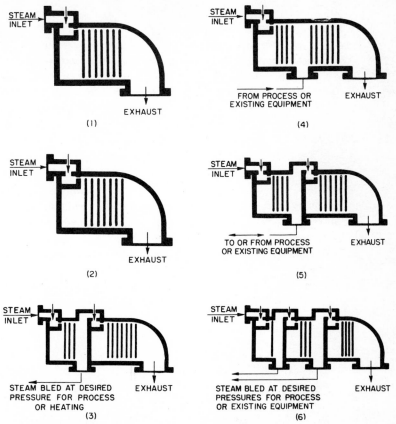

FIG. 12.1   *Steam-turbine types and applications.   (1) Condensing; used when exhaust steam from the turbine cannot be utilized and power must be generated on a minimum amount of steam. (2) Noncondensing; used when all or practically all the exhaust steam from the turbine can be used for process or heating.   (3) Single-extraction; used when process steam requirements are variable or intermittent.   (A noncondensing extraction turbine can be used when process steam is required at two different pressures.)   (4) Mixed-pressure; used when excess steam is available at lower-than-inlet pressure and when this supply is intermittent.   (5) Mixed-pressure extraction; used to supply process steam when necessary and to utilize a surplus of process steam when available.   (6) Double-extraction; used when process steam is required at two different pressures (or at three different pressures if noncondensing).*

methods of extraction are employed: (1) The uncontrolled extraction method consists of connection to the casing with varying pressures, increasing as the load increases and decreasing as the load decreases. This arrangement is used extensively in central power stations for extraction feedwater heating. (2) The controlled extraction method maintains constant pressure by regulating the flow of steam through the turbine on the downstream side of the extraction point. This arrangement is employed to secure steam at constant pressure for process or heating. The amount of superheat (or steam quality) delivered at the point of extraction depends upon the initial steam conditions and load on the turbine. The turbine is adaptable to the use of steam at various pressures and temperatures; it can supply maximum power in the condensing operation or a lesser amount of power in addition to steam at a reduced pressure. Several types of turbines together with their applications are shown in Fig. 12.1.

Noncondensing turbines are employed where process steam at only one pressure is required and where no condensing water is available or the cost of providing it is prohibitive. An example of such a turbine

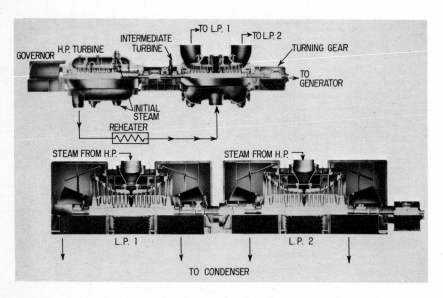

FIG. 12.2 *3,600-rpm cross-compound four-flow reheat steam turbine; 900,000 kw, with steam at 3,500 psig and 1000/1000°F.* (*General Electric Co.*)

would be the topping unit referred to previously.   Noncondensing turbines may be designed for one or more stages of extraction; such turbines lend themselves to the mixed-pressure extraction-type unit.   Extraction steam can be used for process and to drive plant auxiliaries.   Mixed-pressure extraction turbines are used in paper mills and other industries where large quantities of steam at varying pressure are required at one time while at other times an excess amount of steam at this same pressure is available.

In a single turbine, the steam expands from the initial condition to exhaust in one unit.   The tandem-compound turbine consists of two separate units, arranged in line with their shafts connected end to end; steam passes from the high- to the low-pressure turbine.   The cross-compound unit also consists of two separate turbines; these are high- and low-pressure turbines mounted side by side with their shafts parallel.

Steam-turbine drive is employed for pumps, air compressors, fans, and other mechanical equipment.   To be efficient a turbine must operate at high speed.   When the driven unit can also be operated at high speed, the turbine shaft is connected by a coupling to the driven shaft. When, however, a driven machine is required to operate at a lower speed than that of the turbine, reducing gears are used to transmit the power.

When steam expands in a turbine, it imparts energy to the rotating elements.   This energy is taken from the steam, and as a result of this heat loss part of the steam is condensed.   When saturated steam is expanded in a turbine, the loss of heat quickly results in the formation of moisture.   The water thus formed can do no work in the turbine; in fact it increases friction and actually hinders the flow of steam, resulting in a lowering of turbine efficiency.   The formation of moisture is delayed by superheating the steam before it enters the turbine.   Superheating improves the economy of the turbine in two ways: the additional heat increases the energy available for conversion into work, and it also reduces friction.   With superheats up to 100°F, the saving in steam may be estimated as 1 per cent for each 10°F.   At higher superheats, the savings are slightly less.

The greater the pressure range through which the steam is expanded, the more superheat is required to prevent the formation of excessive moisture in the last stages of the turbine.   The steam temperature is limited to about 1050°F by the metals available.

Some turbines employ a reheat cycle; this consists of passing the high-temperature steam through a turbine, then returning it to a special super-heater, and resuperheating it before it is expanded further in the low-pressure turbine.   This system makes it possible to have dry steam in all but the last few stages of the turbine. It would be possible, but is seldom economically advisable, to employ more than one stage of reheating.

Turbines are well adapted to the utilization of high-pressure steam. At 100 to 150 psi pressure, there is a decrease in water rate of 1¼ per cent for each 10 lb increase in pressure. From 150 to 250 psi pressure, this decrease becomes ¾ per cent for each 10 lb increase in pressure. At higher pressures, the reduction in water rate is lower.

Condensing turbines are operated effectively and efficiently with low back pressure. Surface condensers are generally used in connection with large turbines. The use of these condensers decreases the water rate by approximately 5 per cent for each inch of mercury improvement in vacuum within the range of 25 to 29 in. Hg.

Turbines are especially adapted to the use of superheated steam since there is no complicated valve mechanism to come into contact with such steam, as in the case of reciprocating engines. Neither does the use of superheated steam introduce lubricating difficulties, since the turbine lubricating oil does not come into direct contact with the steam and hence the condensate is free from oil. This is very important because oil in the condensate causes serious boiler-operating and maintenance problems.

For equal capacity the turbine requires less floor space than does the engine. Turbine units can be built for almost any desired capacity. Owing to the high speed at which the turbine operates, the direct-connected electric generator for a turbine is smaller and cheaper than a generator of equal capacity driven by an engine. When a turbine is used to operate a machine which must run slowly, it is necessary to use a reducing gear. In these cases, the steam engine is often more desirable.

Turbine operation. Before starting the turbine, the operator should familiarize himself with the general piping layout and the operating characteristics of the unit, after which he will proceed as follows:

Starting a small noncondensing turbine (like that shown in Fig. 11.5):

1. Fill all grease cups and oil the governor and other miscellaneous parts.

2. Open drains on the header, separator, casing, and exhaust lines. If these are equipped with traps, open the bypass line.

3. Slowly open the exhaust-line shutoff valve.

4. If the turbine has a pressure lubricating system with an auxiliary oil pump, start the pump now.

5. Inspect the bearings for ample flow of oil; check the oil pressure and see if the pump is operating properly; 3 to 5 lb pressure is sufficient.

6. Turn the cooling water on the bearings or oil cooler.

7. Open the throttle quickly to start the turbine. During this time

observe the turbine carefully for signs of rubbing (a small pipe or rod placed between the ear and the points to be checked will aid in detecting unusual noise and rubbing).

8. Slowly bring the turbine up to speed, approximately 300 rpm, and operate it at this speed for a period of 15 to 30 min.

9. Trip the emergency valve by hand to see that it closes properly.

10. Open the throttle wide and allow the governor to regulate the speed; keep a close check to prevent overspeeding. NOTE: If the governor will not control the speed at no load, the hand throttle may be used until the load is on.

11. Inspect the bearings to make certain they are getting oil. Oil should be up to the level recommended by the manufacturer. Usually the level drops because of oil's filling all the various cavities in the system.

12. The drains, which were opened before the turbine was started, should now be closed or arranged to discharge through the trap.

13. Gradually increase the load on the turbine while keeping close check on the oil, cooling-water, and bearing temperatures.

Starting a medium-sized or large condensing turbine:

1. Inspect the governor mechanism, fill all grease cups, and oil where necessary.

2. If the boiler stop valve is not open, open it so as to permit as much heating of the steam line as possible and avoid condensation in the line.

3. Open the following drains: header, separator, throttle, and turbine-casing.

4. Open the stop valve in the steam supply to the auxiliary oil pump. NOTE: The operation of this pump is controlled by a governor to shut off the supply of steam after the main oil pump has delivered oil at normal pressure; it opens when the pressure fails.

5. Adjust needle valves to secure 10 to 15 lb oil pressure on the main bearings and 15 to 20 lb oil pressure on the thrust bearing; make sure gauges are in operating condition and have been calibrated.

6. Start the condensing equipment, circulating pumps, and dry vacuum pump; operate the condensate pumps as found necessary to remove water during the warming-up period.

7. The turbine steam or water seal should be turned on and the vacuum maintained at 24 to 26 in. of mercury during the warm-up period; maintain approximately 1-lb pressure in the packing chamber.

8. Turn on the water to the generator air cooler, and see that water flows properly to this and other points requiring water.

9. If the drain ahead of the throttle valve has been closed for any reason, open it again and keep it open until all water of condensation has been removed.

10. Now open the throttle valve quickly to set the rotor in motion.

11. As soon as the turbine is rolling, trip the overspeed by using the hand lever; this is to determine if the tripping mechanism operates properly and to prevent the turbine from accelerating too rapidly.

12. Reset the emergency overspeed valve, and before the turbine comes to rest, adjust the throttle so that the turbine will operate between 200 and 300 rpm.

13. While the rotor revolves slowly, use a metal rod or listening device to determine rubbing or mechanical difficulty.

14. When the oil leaving the bearing reaches a temperature of approximately 110 to 120°F, start the circulating water through the oil cooler to maintain these temperatures. At this time, the bearing-oil pressure should again be checked.

15. Gradually increase the speed. The rate depends on the size of the turbine. Follow the manufacturer's instructions.

16. Adjust the water seal on the turbine and the atmospheric-relief valve.

17. When normal operating speed has been reached and the machine is under the control of the governor, test the emergency governor by opening the valve in the oil line to it. See that all valves controlled by this tripping mechanism close promptly. Reset, open throttle valve, and restore speed to normal.

18. If high-pressure packing is of the water-seal type, adjust the water to 15 lb pressure and shut off the steam.

19. Close drains mentioned in No. 3.

20. Open leak-off from high-pressure packing so that any excess steam may flow to the feedwater heater or to one of the lower stages of the turbine.

21. Synchronize the generator and tie it in the line.

22. With the throttle valve wide open the speed is controlled by the governor. The turbine is now ready for load and is regulated from the turbine control panel.

NOTE: Coordination of effort on the part of the boiler, turbine, and switchboard operators is essential in placing the turbine in operation.

It must be remembered that a large turbine has close clearances and that expansion or improper operation is likely to cause more damage than in the case of a small unit. Large turbines are provided with instruments, including oil-pressure gauges and thermometers. These instruments should be observed at frequent intervals and the readings recorded on the log sheet. In addition to the turbine, the condenser and other auxiliaries require attention, and this phase of the operation must not be forgotten.

NOTE: Always read and follow the manufacturer's instructions, to the letter.

The turbine in motion

1. Apply the load gradually

2. Observe the oil level; check to see if an ample supply of oil is going to the bearings and the hydraulic cylinder. This can best be observed by watching the pressure gauge and sight indicator on the oil discharge.

3. Watch the oil-bearing temperature. This is always a good indication of overheating and mechanical trouble. Temperatures of approximately 140 to 150°F are desirable; above 175°F, serious operating difficulties may be experienced.

4. Observe the turbine for any unusual noise, vibration, etc.

Shutting the turbine down

1. Gradually reduce the load to zero.

2. Start the auxiliary oil pump, and make certain that the proper pressure is maintained while the turbine is coming to a stop.

3. Trip the emergency valve. In most cases this valve also operates the vacuum breaker.

4. Close the leak-off from the high-pressure packing; admit steam to chamber at approximately 1 psi and shut off water.

5. Shut off supply of cooling or condensing water.

6. Shut down condensing equipment and open drains on turbine piping and casing.

7. Continue auxiliary oil pump in operation until the turbine rotor has stopped.

8. If the turbine is to be left idle for a period long enough for it to cool to room temperature, operate the condenser air pump to dry it out. In this way corrosion can be avoided.

While a turbine is operating, it is good practice to keep a log sheet and record the hourly readings of the instruments. These readings should be taken by the operator while making a regular inspection. Such a procedure prevents the operator from neglecting some important inspection, and the data make a valuable record for future reference. The log sheets will have to be made specially for the particular turbine for which they are to be used. The number of readings to be taken depends upon the size of the turbine and the number of instruments installed. Some of the readings which might prove valuable are load on the generator in kilowatts, throttle steam pressure and temperature, exhaust pressure, temperature of cooling water entering and leaving the cooler, bearing-oil pressure and temperature, and perhaps the throttle steam-flow rate.

The proper application of oil to the bearings and a continuous flow of cooling water are the main requirements of a turbine while it is operating. The operator detects trouble on the inside of the turbine by noise or vibration, or both. An appreciable drop in oil pressure can be corrected by operating the auxiliary oil pump until there is an opportunity to investigate the cause of the trouble.

Some small turbines have a number of hand-controlled individual nozzles for admitting the steam. The operator can improve the economy by having a minimum number of these nozzles open for the load at which the turbine is operating.

Turbine maintenance. Proper maintenance is essential to the continuous and efficient operation of the turbine. Items requiring maintenance should not be permitted to accumulate but rather be taken care of as soon as the trouble arises. The exception, of course, is the general overhaul, which is scheduled in advance and during which careful inspection and examination are made of the complete machine. Breakdown or emergency maintenance is best avoided by employing a preventive-maintenance schedule. Maintenance is costly not only from an equipment and labor standpoint but more so because of the outage of the equipment, which in many cases represents the investment of many thousands of dollars on which a return is not forthcoming while the unit is out of service.

Since each and every turbine is basically different, maintenance requirements must be discussed here in a rather general way. Specific recommendations are to be found in the literature and catalogs supplied by the turbine manufacturer, where peculiarities in the particular design are explained and described together with the proper maintenance of the unit. It should be remembered that timely replacement of parts and adjustment for wear may prevent a shutdown and save costly repairs.

Many operators find it economical and advisable to dismantle their turbines completely and give them a thorough internal inspection once a year. Although this may be desirable, experience has demonstrated that a turbine properly operated and maintained can run for years without dismantling. The merit of such inspections, however, together with their frequency, is a matter of operating experience and depends to a large extent on the age of the equipment and the service-hour record if it is assumed that proper operation, etc., has been provided.

An annual inspection of a turbine would consist of examination of blading and nozzles for wear, erosion, cracks, and scale deposits; examination of main bearings and thrust bearing for wear; inspection of clearance between stationary and moving parts and packing clearance; and thorough inspection of the oil pumps and the various governor mechanisms.

Since turbine rotors rotate at high speed, one of the difficulties encountered is vibration. Vibration in many cases is caused by unbalanced weight in the rotating parts and misalignment of the shafts. Vibration due to an unbalanced condition changes with the speed at which the machine is operating. It may be caused by broken, eroded, or corroded blades, by a bent shaft, by distortion due to unequal heating, or by scale deposits on the blading.

Improper alignment of turbine and rotor shafts causes a vibration which does not change with the turbine speed but which increases as the load on the turbine increases. It might be caused by improper alignment of a new turbine. Foundation settling or strains due to pipe expansion could cause vibration in a turbine that had been in operation for some time.

Turbine vibration is also due to water coming over from the boiler with the steam, which will cause the turbine to become noisy and, in severe cases, will cause vibration; electric-generator troubles, such as an unequal air gap or a loose coil in the stator; internal rubbing of parts caused by warping of bladings or diaphragms, improper adjustment, or worn thrust bearings; too much clearance in the main bearings; and overheating due to faulty lubrication or lack of cooling. Also vibration may result if the shaft becomes distorted as a result of heat generated by having the packing rub.

The remedy for most of these difficulties immediately suggests itself: the different parts of the equipment must be supported rigidly and aligned properly, the bearing clearance must be within proper limits, and the bolts between caps and standards must be well tightened.

The rotor of the turbine is held in correct axial clearance by the thrust bearing (Figs. 11.10, 11.11) adjusted with shims. The space provided for the shims is greater than the internal clearance between the wheel and the nozzles on one side and the stationary buckets on the other. With shims removed, the striking limits of the rotor can be determined and measured at the bearing. If we assume a rotor with two rows of buckets on the first wheel and with a total clearance of 0.160 in. or less, the thrust must be shimmed so that the clearance is divided equally and the rotor is in mid-position between striking limits when washers are set. If wear occurs, a complete thrust bearing should be ordered as an assembly; a spare should always be carried in stock.

Carbon packing rings (Fig. 11.12) are used at the high- and low-pressure ends of turbines that operate condensing, on those turbines exhausting against high back pressure, and where high initial temperature and pressure are employed. Metallic packing rings are placed at the shaft bore of diaphragms to minimize steam leakage along the shaft due to the difference in pressure between the stages.

Packing (carbon) clearance for 2- to 3-in.-diameter shafts is recommended as 0.002 to 0.005 in. for clearance of the bore of the ring in excess of the diameter of the shaft. When diaphragm packing is used, the bore of the teeth of the packing ring must be 0.010 to 0.013 in. larger than the shaft.

Proper maintenance of the emergency governor is likewise important. To ensure proper operation, the governor should be tested at frequent intervals. All parts should be kept clean and pins and bearings oiled. Alteration in the tripping speed on most governors is accomplished by altering the number or thickness of shims. The emergency governor should be tested weekly or when being placed in operation after an extended outage.

Lubrication is of extreme importance to the successful operation of the turbine. Proper lubrication of a steam turbine depends upon the proper application of the oil to the bearings, the selection of a suitable oil, and the maintenance of the oil in good condition after it is in the turbine. The proper method of application of the oil to the bearing is the duty of the designer. The operator is responsible for the selection of the proper oil and for keeping it in good condition after it is in service in the turbine.

Oil having the proper viscosity must be selected. A thin oil is recommended, but it must not thin out and lose its lubricating properties when heated to operating temperature. It should have the ability to separate readily from water. Some oils mix with water and form sludge, which is very objectionable in the lubricating system. Oil must be free from acid so that it will not corrode the highly polished surface of the bearings. It must not give off volatile gases when heated to temperatures of 325 to 350°F. Oils that have their gases distilled off at low temperatures lose their lubricating properties after they have been in use for some time. Some oils leave a sludge deposit in lines, tanks, etc.

Oils in service pick up particles of dirt mixed with water and tend to become acid. Medium-sized and large turbine installations have filters and cleaners for keeping the oil in good condition. Several different methods are employed as follows: *bypass filtration,* in which part of the oil is continually circulated through a filtration system; *continuous filtration,* in which the entire amount of oil returning from the turbine is filtered; *makeup system,* in which quantities of oil are drawn off at intervals and replaced with new or reconditioned oil; and *batch treatment,* in which all the oil in the turbine is removed and replaced with new or reconditioned oil. These systems are explained in Chap. 13.

The oil pump, as well as the oil cooler, strainer, and other appliances through which the oil passes, should be kept in first-class condition. The worm and worm-gear teeth of the pump should be inspected. The relief valve should be adjusted for 50 to 55 lb. To ensure the best lubrication, the oil should be withdrawn from the tank and the tank cleaned thoroughly.   All openings, pipes, etc., should be blown out with air or steam.

Oil coolers require attention; they should be removed from service for the necessary maintenance.   This can best be accomplished by having available a spare tube bundle.   After removing the heads, clean the tubes by passing a brush through them.   The cleaning of the outside of the tube is accomplished by boiling it in a solution of hot water or cleansing agent, the kind of solution depending on the type of deposit on the tube.   The entire tube bundle is then immersed in a tank with sufficient capacity to house the bundle conveniently.   The task can be accomplished to better advantage if a pump is used to circulate the cleaning solution.

Preventive maintenance, which includes inspections at frequent intervals, will do much to keep the turbine in first-class operating condition. The purpose of the above-mentioned inspections is to find possible sources of trouble before they have had an opportunity to cause serious damage.   Testing of the emergency governor and checking the operating governor are two points worth remembering.

Weekly.   Inspect the turbine exterior.   Keep control-line valves free from dirt and grease at the guides.   Clean and oil the spindle of the throttle valve and connections between levers of the governing mechanism, using a light oil to prevent gumming; repair all oil leaks.

Monthly.   Inspect the oil reservoir to determine if any sludge is accumulating or any water is getting into the oil.   Check the oil strainer.   Check the operation of the governor and automatic tripping devices, vacuum breakers, etc.   Observe the governor operation to determine how it acts and whether it will carry the maximum turbine load.   If the turbine is operating condensing, inspect the condensing equipment and check the vacuum to see if it is the best obtainable with the equipment and circulating-water temperatures.

Yearly.   Dismantle turbine and auxiliaries; check for corrosion, dirt, scale, and fouled passages and for encrusted, corroded, and eroded blading.   Inspect oil and steam atomizers.   Check all valves and seats.   Check governor knife edges and bearing blocks, emergency tripping device, etc.   Check vacuum pumps and ejectors.   Check for air leaks.

At this time a test should be made to check the spring calibration and to determine the speed above normal at which the oil-tripped emergency governor will operate.   In operation, the emergency governor should be tested weekly.   Testing the emergency governor should be done after the turbine has been placed back in service or approximately once a month.

1. Attach a vibration tachometer to the turbine so that it can be read by the man operating the throttle valve; the man at the other end of the turbine should be provided with a hand tachometer.

2. Operate the throttle valve slowly to bring the turbine up to the required speed, where the governor takes over.

3. Disconnect the governor beam from the vertical connecting linkage; hold it firmly to prevent it from closing the controlling valve. Lift it slowly, permitting the control valve to open and speed to increase slowly until the tripping point is reached on the emergency governor. Do not permit the speed to exceed the safe limit established.

4. Take tachometer readings to check the speed at which the emergency governor operates.

5. If and when the emergency governor acts, check to see that the throttle valve trips and speed decreases. If the emergency governor fails to operate, repeat all the previous check points to determine what the problem might be.

6. Trip the emergency governor two or three times to make certain speed check is correct and all parts are in working order.

7. Test the oil-tripped governor when shutting down the turbine by

a. Running the turbine without load, throttle valve wide open and turbine under control of the operating governor.

b. Open the oil cock in the pipe supplying oil to the nozzle and see to it that the governor operates instantly when the oil jet is applied and the throttle valve trips, at which time the speed should drop, indicating the throttle valve has closed.

c. When the governor trips, shut the oil cock and determine if the governor returns to normal; then reset the tripping mechanism. Never keep a turbine in service unless you are certain that the emergency devices are reliable. A complete record should be kept of the changes and adjustments made on the occasion of these tests.

12.2 CONDENSERS. The condenser reduces the back pressure, thereby increasing the horsepower and efficiency of the turbine; the condensed steam becomes excellent feedwater. The vacuum created in the condenser depends on the type of condenser, the effectiveness of the air-removal devices, and many other factors.

Comparison of Low-level Jet Condensers and Barometric Condensers

Low-level Jet Condensers

Advantages

Low initial cost
Low maintenance (no tubes to leak or repair)
Small floor space required

Disadvantages

Possibility of flooding the turbine
High pumping cost for air and water removal

Barometric Condensers

Advantages

Less power for pumping required
No danger of flooding
Has most of the advantages of low-level condensers

Disadvantages

High cost of piping installation (because of height of setting, especially if vertical space is not immediately available and the system has to be removed some distance from the plant)
Difficulty of keeping piping and system tight

A *jet condenser* is a unit in which steam is condensed by direct contact with the cooling water. Jet condensers may be classified as low-level or barometric (Figs. 11.21, 11.22, 11.23). In the low-level jet condenser both cooling water and condensate are removed from the same hot well by a pump, whereas in the barometeric-type jet condenser the cooling water and condensate can be removed by installing a tail pipe and permitting the condensate to be removed by gravity. Circulation of condensing water in the low-level condenser is effected by lifting the water by the vacuum created; the barometric type, on the other hand, requires a separate condensing pump. Low-level condensers usually operate with medium vacuum unless a separate ejector or pump is provided to remove the entrained air. Steam-jet condensers are limited to facilities where the water contains no foreign matter which might clog the nozzles (if the water is dirty, a strainer must be installed), where the quantity of noncondensable gases and entrained air is low, and where feedwater and makeup are of little importance. Water having corrosive characteristics may be used in jet condensers, as the parts are easily and cheaply replaced. However, if the water is to be reused for boiler feed, the circulating water must be suitable. If the condensing water is not suitable, the condensate must be discarded and the entire boiler feed treated.

Barometric condensers are used because of their simplicity in design and operation. They are used in process work as well as for the power plant. They find ready acceptance for many industrial applications.

In the *surface condenser,* recovery of condensate is made possible, and condensate contamination is avoided. When condensate is used for boiler feed, the makeup required is but a small percentage of the total and the cost of chemical treatment of minor consideration. Surface condensers maintain a better vacuum than can be secured with jet condensers. The main disadvantages of these condensers are the initial cost, the huge quantities of circulating water required, the maintenance due to leaks, cleaning of the cooling surfaces, and the power required to operate the auxiliaries.

Condenser operation. In operation of the ejector-jet barometric condenser (Fig. 11.21), water enters at the top of the condenser and passes through a series of hydraulic nozzles, forming high-velocity jets of water through which steam is made to pass. As the steam is condensed, air and noncondensable gases are entrained by the converging stream and carried to the throat piece, where the high-velocity jet is converted into pressure. At this point the air and gas become entrained in the high-moving jet of water, and the entire mixture discharges to the hot well, permitting the air and gas to escape.

To operate, fill the system with steam under a slight pressure before turning on the water. This will permit the air to be cleared from the system immediately when the water flow is established. The water should be clean and slightly above atmospheric pressure. If it is necessary to produce a partial vacuum in the system before admitting steam to the condenser, the water should be turned on first and the condenser then acts as its own evacuating apparatus.

All that is necessary normally to place the condenser in operation or to take it out of service is to open or close the valve in the waterline. There are no auxiliaries, and the starting procedure and operation are extremely simple. From time to time it may be necessary to inspect the nozzles for cleanliness and to check to determine if the system is tight; this is done by checking the vacuum.

The disk-type barometric condenser (Fig. 11.23) has steam and air entering at the lower section of the shell with condensing water admitted at the top. As water cascades down and over the series of trays, it meets the steam rising through the descending spray of water and condenses it; the water drops to the tail pipe. This condenser requires no pumps, vacuum breakers, or atmospheric-relief valves; it handles dirty water and is more or less foolproof in operation. To improve the operation of the condenser, it is sometimes equipped with an air-removal device.

The air-removal device (air ejector; Fig. 11.23) can be a single- or a two-stage ejector. In the first stage, steam pulls the air and noncondensable gases from the top of the barometric condenser and passes them to a small barometric intercondenser, which is mounted at the same elevation as the large condenser and is similar in design but smaller. Attached to this small intercondenser is a second-stage ejector pulling the air and noncondensable gases from this unit and discharging this entrainment to the atmosphere.

To place in operation a barometric condenser equipped with one air ejector, proceed as follows: Open the water valve to start the water through the condenser. Next operate the valve on the steam ejector to maintain the vacuum. Turn the steam into the condenser, and the unit is in operation. Water is removed by gravity due to the head of water above it, which is sufficient to offset the vacuum created.

Since the barometric condenser is located approximately 40 ft above the water level in the hot well and since the vacuum will raise the condensing water by only about 15 to 20 ft, a pump, known as an "injection pump," must raise the water the remaining distance.

In the operation of the surface condenser (Figs. 11.24, 11.25), steam passes over the bank of tubes through which the condensing water circulates. The steam is condensed and drops as condensate to the hot well, being removed by the condensate or hot-well pump. Air and noncondensable gases are removed by ejectors. The condenser is so designed that double curtains of condensate spray from the trough, condensing the steam bypassed from the turbine exhaust. In this way a large amount of condensate is exposed to the steam, heating the condensate.

To place a surface condenser in operation:

1. Open the discharge and inlet valves to and from the condenser and start the circulating pump.

2. Open the steam-jet air-ejector valves and slowly bring up the vacuum.

3. Operate the turbine throttle to admit steam and permit steam to pass to the condenser.

4. Start the condensate pump; operate it to remove the water as it condenses. This can be accomplished by operating the pump on an intermittent basis.

5. As the turbine comes up to its operating speed, increase the vacuum by operating the steam valves to the air ejector.

6. If the load increases, speed up the circulating pump if a variable-speed motor is provided; operate the condensate pump on a continuous cycle to remove the water from the hot well.

The quantity of water required to condense the steam can readily be computed. Since the heat absorbed by the circulating water in passing through the condenser must equal that given up by the exhaust steam, if leakage and radiation are neglected, we have

$$Q = \frac{H - (t_0 - 32)}{t_2 - t_1}$$

where Q = weight of water, lb, to condense 1 lb of steam

H = heat content of exhaust steam above 32°F

t_0 = temperature of condensate, °F

t_1 = temperature of condensing water entering, °F

t_2 = temperature of condensing water leaving, °F

Example   Assume the heat in the exhaust to be 1100 Btu; the temperature $t_0 = 100$, $t_1 = 80$, and $t_2 = 90$.

$$\frac{1,100 - (100 - 32)}{90 - 80} = \frac{1,032}{10}$$

$$= 103.2 \text{ lb, amount of water required to condense 1 lb of steam}$$

Condenser maintenance. The ejector-jet barometric condenser requires little or no maintenance. Joints need to be checked occasionally for leakage. The nozzle plate (Fig. 11.21) can be inspected by loosening the nut and lowering the plate; the nozzles can then be inspected and cleaned. Condensers of the disk-flow type are in the same category as the ejector-jet type and require little maintenance other than stopping leaks and routine inspection.

The greatest disadvantage of the surface condenser is the large amount of maintenance which may be required. As a result of expansion and contraction, we may encounter tube leakage that requires rerolling or replacement of the tube.

Tube leakage can be determined in a number of ways. The condensate continues to exceed the normal or average requirements, causing the hot-well level to rise. This may be due to any one of a number of things: leaks around the tube ends, split tubes, or excessive leakage into the system from the sealing water. The unit must be inspected to determine the cause and location of the leak.

This can be done by pulling a vacuum on the condenser and exposing the flame of a candle to each tube. This is a slow and tedious job but is usually effective. Or the inspection can be made with the condenser out of service and with circulating water passing through the tubes. By entering the space below the tubes, it is possible to detect drips or leaks. At times it is difficult to detect actual leakage and the location of the leaks because the tubes tend to "sweat," giving one the impression that a leak exists where there is none. This sweating is due to condensation of moisture in the air which is in contact with the cold surface through which the circulating water is passing. Still another method frequently employed for locating leaks is as follows: With the condenser out of service, fill the steam space with water. The tubes can be inspected for leakage by entering the water boxes. Tubes found to be leaking can be replaced and repairs made where necessary.

The quantity of water leaking per unit of time can be determined with the condenser out of service by pulling a vacuum on the condenser while noting the rise of water level in the hot well. Or a condition may arise in which the leakage is not so easily noted as in the former case. Small amounts of water leaking into the hot well are best determined by running a chemical analysis on the condensate from time to time or by testing the condensate to determine the resistance or conductivity provided by a given sample of water.

If a tube is split, temporary repairs may be made by driving a wooden plug into the end of the tube or by replacing the ferrules with a special brass cap. Permanent repairs are made by replacing the tube. In the case of tubes expanded in the tube sheet, care must be exercised to secure a tight joint without bending the tubes by overexpanding them. When packing is used to prevent leakage between the tubes and the tube sheet, the specified amount of packing is placed in the box and the ferrule is screwed down tightly but with care to prevent crushing (necking) the tube.

Air leaks which develop are difficult to detect and locate because of the large surface area of the condenser and the many valves, fittings, and auxiliaries attached to it. These leaks are detected by (1) examining all the joints for leaks, using a candle; (2) filling the entire condenser with water and maintaining a slight pressure on the system; and (3) using low-pressure air and checking for leaks, employing a soap solution on the joints. The leaks must be checked around all auxiliaries connecting the turbine seals, bleeder connections, etc.; this is very often a slow and tedious job.

A convenient method of determining the extent of leakage is to pull a vacuum on the system and close down, permitting the unit to stand for a while. If the vacuum drops more than 2 to 4 in. Hg per hr, it will be necessary to hunt for leaks. Another method frequently used to determine whether the condenser is leaking air is to check the pressure and see if the pressure in the condenser corresponds to the condensate temperature. Should the pressure be higher, air leaks or noncondensable gases are present in the condenser. As an example, suppose the pressure in the condenser is 1.932 in. Hg abs (0.9847 psia) and the temperature in the hot well is 96°F. Referring to the steam tables, we find that at 96°F the pressure should be 1.711 in. Hg abs, or 0.8403 psia. Hence, the added or excess pressure is due to air or noncondensable gases which have leaked into the condenser.

Condensers are sometimes equipped with metering devices to measure the air removed from the condenser. These meters are usually calibrated in terms of cubic feet of free air, at a given temperature, leaking into the condenser per unit of time.

Condensers are frequently troubled with scale and algae growth. These reduce the heat transfer and at times become so serious as to interfere with the circulation of water. Backwashing is sometimes effective in removing foreign material. Condensers frequently have their piping so arranged that this can be accomplished while the condensers are in service. Two-pass condensers often have a divided water box with horizontal decks; one-half of the surface can be cut out of service for cleaning while the remainder carries the load, at reduced vacuum. This avoids interruption to operation and, if done at light load, causes little inconvenience.

Scale can be removed by washing or turbining the tubes. Sediment inside the tubes can be removed by "shooting" the tubes, using rubber plugs and compressed air. Algae growth is inhibited or destroyed by using various chemical solutions, depending on the nature of the algae growth; at times chlorine can be used very effectively.

When replacing tubes or making repetitive adjustments, it is well to keep a continuous record for future reference. Continued maintenance and repeated adjustments indicate trouble which must be taken care of before the situation becomes serious.

12.3 COOLING TOWERS AND SPRAY PONDS. Power plants are not always located adjacent to a plentiful supply of condensing water. At other times the source of supply is found to be contaminated by the presence of foreign matter which would prove injurious to the pumps, condensers, and circulating systems through which the water must pass. The cost of placing this water in a satisfactory condition for use would prove prohibitive. Yet if the turbines are to be operated condensing, cold water is required.

Hence *cooling towers* (Figs. 11.29, 11.30) and *spray ponds* (Fig. 11.28) are provided where sufficient water is not available, their purpose being to discharge the heat absorbed in the condenser. In order to do this in the shortest possible time, the water must be broken up into very fine spray to present to the atmosphere as much water surface as possible.

Both the spray pond and the cooling tower enjoy advantages depending upon the specific installation for which they are intended and must be evaluated accordingly. The factors to be considered in choosing between a cooling tower and a spray pond are the amount of water to be cooled, the amount or degree of cooling desired, the cost of water makeup, piping, space available for installation, pumping cost, water-treatment cost, fixed charges, maintenance, and operation.

Cooling ponds and spray ponds are usually less expensive to install and maintain than are cooling towers. However, better results are obtained with a properly designed cooling tower, since it can provide water at a lower temperature, which in turn can produce a better or lower vacuum in the condenser. The cooling tower is very compact and can be located almost anywhere that space permits, such as the top of buildings, etc. It is usually installed where land is expensive. By comparison, a spray pond requires considerably less area than a cooling pond.

The principle of cooling applied in both cooling pond and tower is that of transferring sensible heat in the condensing water to the atmosphere, thereby lowering the water temperature. Most of the cooling is due to exchange of latent heat resulting from evaporation of a small portion of the water. In cooling (evaporation) some water is lost; it amounts to approximately 1 per cent for each 12 to 14°F of actual cooling.

The capacity of the air to absorb additional water vapor depends on the wet-bulb temperature, the wet-bulb temperature being an indication of the total heat content of the air. The dry-bulb temperature is an indication of only the sensible heat contained in the water or substance and not the total heat. When the wet- and dry-bulb temperatures are the same, the relative humidity[1] is 100 per cent and the air's capacity to absorb additional water vapor is zero. Hence the ability of a spray pond or cooling tower to give off heat contained in the water depends on the wet-bulb temperature or relative humidity of the air with which it comes into contact. Influencing factors which must be taken into consideration are wind direction and velocity. The extent of cooling is also dependent upon the area of the water exposed to the atmosphere and the temperature of the water when it enters the basin. Cooling is brought about by evaporation and conduction.

The degree to which the temperature of the water approaches the air wet-bulb temperature is a measure of performance for the pond and cooling tower.

The wet-bulb temperature of the surrounding atmosphere is the theoretical point to which water can be cooled. Usually it is not economical to approach closer than 5°F of the wet-bulb temperature. The temperature drop (cooling range) will vary from 10 to 12°F for a spray pond versus 12 to 17°F for a mechanical-draft cooling tower. Since the wet-bulb temperature of air varies geographically, it follows that the performance of a cooling tower will also vary with its location. So the wet-bulb-temperature record for a given location must receive consideration when a tower is being designed.

The cooling pond is employed where space adjacent to the power plant is plentiful and where land is relatively inexpensive. Cooling ponds are also employed in certain parts of the country where atmospheric conditions are favorable. As an example, a cooling pond located in an area where the relative humidity is usually low would prove more effective than one installed in an area where the humidity approaches 100 per cent.

---

[1] Relative humidity is the ratio of the amount of water vapor contained in the air to the amount which the air could contain if saturated.

Occasionally the decision to use a cooling or spray pond is based on the fact that it is possible to take advantage of a natural water hole or depression in which water can be stored and cooled and which could be used at little or no expense.

The advantages of the *forced-draft* tower are that it is suitable for corrosive waters; that the fan can be mounted, firmly, close to the ground; and that the fan is accessible. Its disadvantages are recirculation of air with reduction in tower efficiency (air leaving the tower exits at low velocity), higher fan power requirements and hence increased operating cost, and increased maintenance. Also since the size of the fan is limited, more fans and foundations may be required.

The *induced-draft* tower (Fig. 11.29) discharges the air at a high velocity, thereby avoiding recirculation; induced-draft fans are usually quiet in operation and permit more uniform air distribution. They have, however, several disadvantages: they require a heavy superstructure, they must be perfectly balanced, and they are somewhat difficult to service when major repairs are required.

Operation of cooling towers and spray ponds. In the spray-pond system operation is extremely simple. Water from the condenser is pumped to the system of piping and laterals, where it is ejected through nozzles, falling to the surface of the pond in a spray; it is then returned to the condenser, being used over and over again. A combination of wind movement and draft created by the spray provides the evaporation necessary. A very fine spray or mist is undesirable since, during periods of high wind, too much water is lost. A fine spray, in addition to becoming objectionable, increases the cost of makeup water and water conditioning. Loss of water (drift) varies, depending on the design of the spray and on the wind velocity, and ranges from 1 to 5 per cent. Pressure on the nozzles is usually low, 3 to 10 psi.

In the operation of the induced-draft cooling tower, water from the condenser is passed to the top of the tower (Fig. 11.29) to a flow-control valve at each section or cooling chamber. The flow-control valve permits the water to fill the overhead water-distribution basin, in which are located nozzles for uniform distribution of water. Interior decking spreads the water uniformly, permitting intimate contact with air passing up and through the descending spray of water. The water finally arrives at the bottom of the tower, where storage is provided; from this point it is returned to the condenser as required. This cycle is repeated again and again.

Air enters the side louvers, passing up and over the interior decking and cooling the water; the air then is made to pass through the drift eliminators, where moisture carry-over is reduced to a minimum. Figures 11.29 and 11.30 show a double-flow tower operated by means of an induced-draft fan. Air and water pass in counterflow, so that the air is in contact with the hottest part of the water immediately before it leaves the tower. Power costs are kept low by holding the flow of air to a minimum. Merely shutting off the valve to the water-distribution system reduces pumping costs, and air regulation is secured by operating the fan on a low-speed motor.

To place an induced-draft cooling tower in service, open the makeup valves on the tower to provide a sufficient amount of water storage. Open the valve to the tower outlet and inlet as well as valves on the condenser outlet. Now slowly open the inlet valve to the condenser to permit water to fill the condenser and discharge piping. Next close the valve on the pump discharge and start the circulating pump, after which the valve on the pump discharge can be slowly opened. This is to avoid the shock and vibration that might result were the pump to be placed on the line suddenly. Adjust the gates on the top of the tower to secure equal distribution of water to the various sections. Now start the induced-draft fans. Make a routine inspection of fans and motors to make certain that they are operating properly.

If the water temperature leaving the tower is lower than required or if the turbine load has been reduced so that not all the tower capacity is needed, reduce the fan speed if two-speed motors are provided rather than take a cell out of service. The reason for this is that it is less expensive to operate cells at half capacity or rating than to shut them down and permit other cells to operate at top fan speed, because reducing the load to 50 per cent capacity reduces the horsepower or energy required to operate the fan. This is so because the horsepower requirements vary with the cube of the speed of the fan and thus reducing the fan speed by 50 per cent drops the fan horsepower to approximately one-eight of full-load power requirements. This is an appreciable power saving and the reason why two-speed motors are employed to take care of seasonal changes in load.

Should the load be further reduced and more than one circulating pump be provided, take one pump out of service. Further load reductions will necessitate taking towers out of service or shutting fans down in multiple units and permitting water to cascade from top to bottom.

Cold-weather operation introduces a few hazards which must be carefully watched. To prevent icing, maintain the circulating-water temperature as high as is practical. For induced-draft towers, several recommendations are provided.

1. Shut off the fans to the tower or run the motor at low speed.
2. Shut down cells and run the water over fewer compartments.

3. Bypass some of the water to the tower and reduce the number of cells operating.

4. If ice forms on the deck filling or louvers, shut down the fans temporarily while continuing to circulate the water.

As a further precaution against the accumulation of ice, the fans are sometimes made reversible; also the sump beneath the cooling tower is provided with heating coils.

In the operation of cooling towers, an important item to consider is water treatment. High water concentrations will deposit scale in the condenser heating surfaces, in the piping transmission system, and the tower. Low water concentrations result in higher operating cost since they waste water. Since most cooling towers are installed in locations where water is a scarce or costly commodity, it is very easy to see why excess blowoff is uneconomical.

Water concentrations build up because part of the original water is evaporated in the heat transfer, while additional water is lost as drift or as entrained moisture in the air passing through the tower. Evaporation accounts for a loss of approximately 1 per cent of the water for each 10°F range of water circulated. In a good cooling tower, drift is approximately 0.1 per cent. The amount of blowdown water wasted depends on the hardness of the circulating water, the type of treatment used, and the drift loss. To control the scale-forming solids at a satisfactory level, blowing down is resorted to. This is done by testing the hardness of the water and by blowing down to keep the water below the point where it would produce scale or where it would become corrosive. Other factors that must be considered are the amount of dirt in the water or other foreign material (depending on the location). The pH value is important in controlling the chemical properties of cooling-tower water. Neutral water has a pH value of 7; below this value the water is acid and tends to be corrosive, while above 7 the water is alkaline and scale-forming. If the pH value is maintained too high, serious damage may result to the wooden cooling tower because of delignification. This results in the carbonates' dissolving the lignin which binds the wood fibers together, a condition which reduces the structural strength of the wood. This condition first appears at points where the timbers are alternately wet and dry because here the water concentration is increased owing to the rapid evaporation. Sulfuric acid, zeolite, and sodium hexametaphosphate are the chemicals most frequently employed for treatment. If the concentration of solids increases, the effective wet-bulb temperature will be raised.

Algae formation is also frequently experienced. This growth collects in the tower and circulating system, reduces the heat transfer, and is a source of considerable trouble and annoyance. It can be controlled by a number of chemicals, such as chlorine, copper sulfate, potassium permanganate, and others. The type of algae growth determines the treatment to be employed.

Inspection of equipment should be made at least once a shift. Usually mechanical failures give first warning by excessive noise or vibration. Overheated motors can be detected by feeling the motor; vibration in fans is detected by peculiar noises and by observing the vibration after placing a listening rod between the point under observation and the ear.

Maintenance of cooling towers. Mechanical maintenance is reduced first of all by keeping the equipment clean and adjusted after frequent inspections. Grease the motor bearings; add oil to the gearbox and speed reducers as recommended by the equipment manufacturers. Keep all the clamps and bolts tightened. Inspect fans, motors, and housing to avoid undue vibration; they should be painted yearly, although the necessity for painting is most frequently determined by experience. Redwood does not require painting for protection but for appearance only.

Remove all debris, scale, etc., from the decking and distribution system. Drain and wash down the storage basin and overhead deck from time to time. Keep all parts of the tower in alignment. Inspect nozzles for clogging (weekly). Constant vigilance is important; observe anything unusual in the form of high water temperatures which may indicate faulty operation, scale in the system, etc. Check gearboxes weekly; locate any unusual noise or vibration immediately and correct it. Check the oil level weekly and change the oil after 90 days' operation. At least once a year give the entire installation a thorough going over: dismantle the motors, gearboxes, etc.; check for structural weakness while tightening the bolts.

Do not permit the tower to remain out of service (drained) for any length of time if avoidable. To avoid fires in the tower, keep it wet. If the tower is kept out of service "dry" for any length of time, fire extinguishers should be made available for emergency use; they should be located for easy accessibility.

The best guarantee of trouble-free performance is the adoption of a preventive-maintenance schedule with a continuous record of performance. The life of a cooling tower depends on the care and attention it is given.

12.4 AUXILIARIES. A *traveling screen*, placed in the intake tunnel supplying water to the plant, is made to operate at variable speed. This permits the screen to be rotated fast or slowly, continuously or intermittently. Operating costs can be reduced if the screen is run only when necessary. The frequency of operation can best be determined by installing an indicating or recording gauge to notify the operator when the pressure drop (differential head or water-level drop) across the screen has increased, indicating that the screen needs to be cleaned. Allowance must be made, however, for the head of water on the inlet side because the pressure drop across the screen will vary with the area exposed to the inlet water.

Usual operation is to run the revolving screen at intervals frequent enough to expose a new set of screens or buckets to the water flowing to the tunnel to avoid pressure drop across the screen. Since the pressure drop is occasioned by foreign material on the screen, the simplest way to determine whether screens are clean or dirty is to operate as follows:

Assuming the water is 20 ft deep and each screen is 1 ft wide,

1. Turn the screens over slowly until all 20 ft (20 buckets) have been exposed. Check the screens for foreign material.

2. If the screens are dirty, continue to run them until they begin to run clean.

3. Determine the frequency of running necessary to keep them in this condition. Without a gauge, this is best determined by operating experience.

4. Once the frequency of rotation has been determined, a schedule can be set up to rotate or raise the screens to the necessary height once each hour or less frequently, depending on the condition of the water.

When starting the screen, operate at the slowest speed so that if interference exists in the form of logs, etc., fouling will not occur and damage result which would necessitate pulling the assembly from the well or sending a diver down to make an inspection.

In certain seasons of the year more difficulty is experienced than at others, due to seasonal rains, floods, etc., which increase the foreign matter carried along by the moving stream. It is best under these conditions to run the screen continuously.

Care should be exercised to prevent freezing and damage to the screen if ice should form in the intake during cold weather. To prevent this, screens are run continuously, but even at best difficulty is encountered. In some plants, all or part of the condensing water, after leaving the condenser, is discharged at the mouth of the intake. This maintains the entire mass of water 1° or more above freezing temperature and prevents ice formation on the screens.

Most of these drives are equipped with automatic lubricators. If they are not so equipped, lubrication must be applied while the screen is running. The motor should be properly lubricated, and a coating of grease should be applied to the rollers on the rear of each basket.

Preventive maintenance is the first step toward lower maintenance costs.

Weekly. Check the entire installation for rubbing, holes in the baskets, overloaded motor, etc. Lubricate where necessary.

Monthly. Check the bearings on the sprocket drive; check rollers and bearings. Replace damaged baskets.

Yearly. Replace all worn bushings and bearings and make a thorough and complete check of all moving parts. Overhaul the motor, gear reducer, etc.

Pumps. Of all the auxiliaries employed, the circulating pump requires the most power because of the large quantity of water handled. Most condensers are equipped with two circulating pumps (Fig. 11.32), both operated at maximum load, while at partial load only one is required to maintain the vacuum desired. Some circulating pumps are provided with two-speed or variable-speed motors to reduce operating cost to a minimum.

Circulating pumps (Fig. 11.33) are operated at relatively low speed and normally require little attention once alignment has been correctly made. Properly lubricated, operated, and maintained, such pumps will run for extended periods of time without difficulty.

Some pumps are set above the normal water level and have to be primed to be placed in service. Priming is accomplished by evacuating the air in the pump until water is ejected from the siphon. At this point the motor is started, after which the discharge valve is opened slowly so as not to lose the vacuum. The priming is then discontinued.

If strainers are provided ahead of the pump, they must be inspected at frequent intervals. If twin strainers are provided, the cleaning can be accomplished without taking the pump out of service. If not, one condenser must be taken out of service for this inspection. The cleanliness of the strainer can be determined by placing a set of pressure gauges before and after the strainer and noting the pressure drop (loss) across the strainer.

Condensate (hot-well) pumps and other auxiliary pumps used on heater drains and sumps require the same type of service and maintenance as that outlined for pumps in Chap. 7.

Steam-jet ejectors and vacuum pumps. In order to maintain the highest possible vacuum, air and noncondensable gases must be removed from the condenser. Sometimes the pump which removes the condensate also removes the air. This type is called a "wet vacuum pump." More often, condensers are provided with ejectors (dry vacuum pumps) to remove the air or noncondensable gases and a separate pump to remove the water. The design of wet vacuum pumps is similar to that of reciprocating pumps. Dry vacuum pumps may be either of the hydraulic type or steam ejectors. In the former, jets of water are hurled at high velocity and are made to pass through a revolving wheel which rotates because of the water passing through it. The water rushing through the discharge cone and diffuser in the form of a helix encloses the vapors which enter around the rotary transforming wheel between the suction and discharge of the pump. The air and vapors, in addition to the hurling water, are discharged into the hurling-water tank, from which the gases are liberated by means of a vent. In operation, the hurling-water temperature should be below the temperature in the condenser; this is accomplished by adding sufficient makeup water or by using a heat exchanger to reduce the hurling-water temperature. In the latter, the steam ejector (Fig. 11.34) entrains the air and non-condensable gases in very much the same way as the hydraulic-vacuum pump. Ejectors are designed for the vacuum and steam economy desired. The type of steam ejector usually used for large steam-driven turbine generators consists of two-stage ejectors in series provided with surface inter- and aftercondensers, with the steam being returned to the main condenser.

Before placing the ejector (Fig. 11.23) in service, it might be well to fill the ejector and piping system with water to test for air leakage. The steam pressure to the ejectors should be at least 10 lb above the minimum for which the ejector was designed to operate.

Turn cooling water into the barometric condenser and open the steam valve on the second-stage ejector. As the turbine gradually comes up to speed, vacuum can be maintained with the second-stage ejector. When the turbine is ready for load, the steam valve to the first-stage ejector is admitted to maintain the desired vacuum.

If a two-stage ejector is used on a surface condenser with an intercondenser between stages, proceed as follows, placing water pressure on the system to check for leaks before placing in service:

1. Open evacuator suction and discharge valves.

2. Open vacuum-trap drains and suction valve to first-stage ejector.

3. Turn on steam and maintain pressure at least 10 psi above the minimum pressure required.

4. With turbine starting to roll and steam on the seals, air leakage is reduced.

5. After introduction of steam to the condenser, supply condensate to inter- and after condensers.

6. When pressure in the main condenser reaches 8 or 10 in. Hg abs and with condensate in intercooler, open steam valve on first-stage ejector to secure best possible vacuum.

To close down proceed as follows:

1. Close the evacuator suction valve before shutting off the steam; then close the steam valve to the evacuator.

2. Close the evacuator discharge valve.

If vacuum cannot be maintained, carefully check the steam pressure to the ejectors; the orifice in the supply line or the nozzles may be stopped up.

Low or no vacuum may result from lack of water or an insufficient flow of water in the inter- or aftercoolers, from improper operation of the traps, etc.

Vacuum gauges. Pressures below atmospheric are determined by means of a vacuum gauge. The gauge may be of the dial type actuated by a bourdon tube similar to the conventional pressure gauge. Or it may consist of a mercury tube, with the height of the mercury column indicating the difference in pressure between the condenser and the atmosphere. In either case the gauge is calibrated to read in inches of mercury. Every inch of mercury corresponds to a pressure of 0.491 psi. A vacuum, then, of 27.25 is equivalent to $27.25 \times 0.491 = 13.38$ psi pressure below atmospheric.

As pressures in the steam tables are given in pounds per square inch absolute, it becomes necessary to convert gauge pressure into terms absolute.

Example Assuming a barometer with a pressure of 29.50 in. Hg and a vacuum of 27.25 in., determine the absolute pressure in pounds per square inch.

Solution

$29.50 \times 0.491 = 14.48$ psi, pressure of atmosphere
$27.25 \times 0.491 = 13.38$ psi, pressure in condenser
$14.48 - 13.38 = 1.10$ psi, pressure absolute in condenser

QUESTIONS AND PROBLEMS

12.1. Why is superheated steam employed in turbine operation?

12.2. What are the disadvantages of using wet steam in a turbine?

12.3. What are the most essential things to watch in bringing a turbine on the line? After it is on the line and operating?

12.4. How would you start and place in service a small noncondensing turbine?

12.5. How often should a turbine be dismantled to receive a thorough inspection and overhaul?

12.6. What are some of the common causes of turbine vibration?

12.7. How would you give a turbine a complete internal inspection?

12.8. What are some of the reasons for high turbine maintenance?

12.9. Discuss the following: low-pressure, mixed-pressure, and bleeder turbines.

12.10. What difficulties are usually encountered with surface condensers that require maintenance?

12.11. How would you proceed to determine whether a condenser was leaking?

12.12. How would you determine the reason for low vacuum which had suddenly occurred on a turbine operating condensing?

12.13. How does the relative humidity of the atmosphere affect the operation of a cooling tower?

12.14. What precautions should be taken relative to the operation of cooling towers?

12.15. The vacuum gauge reads 26.3 in. Hg and the barometer 30.03 in. Hg. What is the absolute pressure in pounds per square inch?

12.16. The pressure of the atmosphere is 14.36 psi, and the vacuum gauge reads 28.23 in. Hg. What is the back pressure in pounds per square inch absolute?

CHAPTER 13

AUXILIARY
STEAM-PLANT
EQUIPMENT

This chapter considers some of the more important general auxiliaries found in steam plants. These devices are required to make it possible for the major equipment to perform its function. Satisfactory boiler operation depends upon the feedwater being heated and conditioned. It is essential that steam and water flow from one part of the plant to the other through pipelines of adequate strength and size. Condensate must be automatically removed, from steam lines to prevent the possibility of waterhammer and from steam-heating coils to keep the water from blanking the heating surface and decreasing the rate of heat transfer. Pumps, filters, and various feeding devices must be used in the process of lubricating machinery. This auxiliary equipment warrants careful manipulation and the necessary maintenance to keep it in working condition.

13.1 OPEN FEEDWATER HEATERS. An open feedwater heater is one in which the steam and water come into direct contact and the effluent is a combination of the inlet water and supply steam. Under design conditions the temperature of the outlet water approaches the saturation temperature of the supply steam. This is in contrast to the closed feedwater heater, in which the steam and water are separated by tubes through which the heat is transferred.

Open heaters are widely used to heat feedwater, and in this application they provide several additional functions. The combined makeup water and condensate is introduced into the heater through sprays, to provide thorough mixing of the water and steam. This results in rapid heating of the water and removal of the oxygen and other noncondensable gases. These gases are vented from the heater. The heated and deaerated water is collected in a storage section of the heater or a separate tank. The level of water in this storage compartment controls the flow of the makeup water to the heater. When the level of water in the storage drops, a valve opens to supply the required makeup water to the heater. It is advisable to use stainless steel in the section of heater supply piping which carries the mixture of condensate and makeup water. The hot condensate will cause the dissolved oxygen to be released from the relatively cold makeup water and this results in rapid corrosion of steel piping. The interior of the deaerating section of the heater is made of corrosion-resistant material. Condensate which is above the temperature and pressure in the heater may be returned uncontrolled through a separate nozzle. In this way all of the heat content of this high-temperature condensate is reclaimed in the heater. The heater and water storage section is elevated to provide the necessary suction head on the boiler feed pumps. (See Sec. 7.8.) Sodium sulfite or hydrazine is frequently added to the water in the heater storage tank to remove the last trace of dissolved oxygen.

Open feedwater heaters must be provided with safety devices and instrumentation. Steam is supplied to these heaters from turbine exhaust or other low-pressure systems. Pressure-reducing valves provide steam from high-pressure sources to supplement the low-pressure steam. Feedwater heaters are selected to operate in a range of from 5 to 15 psig, depending upon the pressure of the available exhaust steam. The heaters are guaranteed to reduce oxygen content of the water to 0.005 ml per liter and the carbon dioxide to zero. However, to maintain these guarantees, the heater must be adequately vented and the pressure must be maintained at that for which the heater was designed to operate.

Relief valves are provided to make sure the safe pressure limits are not exceeded. The storage section must have an overflow valve or loop seal to limit the level. Owing to the critical nature of the level in the storage section, it is advisable to have a level indicator on the control panel and an annunciator alarm, to warn the operator when the level is too high or too low.

The steam pressure in the heater and the water level in the storage tank are automatically regulated. Nevertheless the operator should keep a close surveillance and know what action to take in event of an emergency. The available quantity of feedwater in the storage section will normally last only a few minutes if the supply fails. It is generally assumed that if the steam pressure is maintained in the heater the water will be deaerated satisfactorily. However, it is good practice to occasionally run a dissolved oxygen analysis on the effluent water. Should

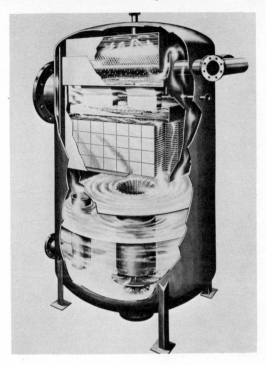

(*a*)

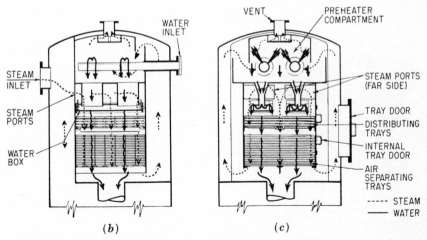

FIG. 13.1 *Deaerating heater.* (*a*) *Section view.* (*b*), (*c*) *Flow diagrams.*
(*Cochrane Division—Crane Co.*)

the oxygen content be found excessive the necessary corrective measures can be taken and thus prevent the possibility of corrosion in the boilers.

It might, at first, appear that an open feedwater heater is a cure-all, but this is not the case. The water and steam mix, and a part of the impurities appears in the makeup water and is introduced into the boiler.

The deaerating-type open heater shown in Fig. 13.1a consists of an external shell of welded low-carbon-steel plate designed for the operating pressure. An inner chamber of corrosion-resistant material is provided to prevent the gases from coming in contact with the carbon-steel shell. The perforated preheating spray-distribution pipes are also made of corrosion-resistant material. Spring-loaded spray nozzles are provided, when required for a special application, to control the flow of water into the preheater. The tray stacks are made of stainless steel or cast iron. Tray access doors are provided in the shell and tray compartment.

The deaerator utilizes two heaters in series: the preheater, where the incoming water is sprayed into the steam space; and the secondary section, where the water flows down over the trays.

The operation of this heater can be explained by reference to the flow diagrams in Figs. 13.1b and c. Steam enters the shell and flows into the preheater and through ports into the tray compartment. Water enters the preheater through the perforated distribution pipe, comes into contact with the steam, and is heated to nearly the saturation temperature of steam. This heating releases most of the noncondensable gases, which, along with a small amount of steam, are discharged through the vent. The water flows from the preheater through the seals to the tray compartment. The steam and water come into contact as they flow downward through the distribution trays. The heated and deaerated water falls into the storage space. The steam flows up between the tray compartment and the shell to the preheater compartment.

FIG. 13.2 *Package feedwater unit, including deaerating heater and pumps. (Cochrane Division—Crane Co.)*

Feedwater systems are available in factory-assembled package units, as shown in Fig. 13.2. These systems include deaerating heaters, condensate receivers, and booster and boiler feed pumps. The components are shop-assembled on skids and include piping, motors, wiring, and controls. The use of these units reduces the fieldwork of installation to connecting the supply and discharge piping and providing the electric power.

The heating and conditioning of feedwater are performed by a hot-process water softener. A typical installation is shown diagrammatically in Fig. 13.3*a*, and a section through the heater and softener unit is shown in Fig. 13.3*b*. There are no provisions for the introduction of condensate into this unit. It is intended for use where there is 100 per cent makeup water or where the condensate returns are delivered to an independent deaerator. The installation includes a system which proportions the chemical feed to the rate of flow of makeup water, a combination heater, softener, and settling tank, a sludge-recirculating pump, filters, and wash pump. The flow of cold makeup water to the softener is regulated by means of a flow-controlled valve to maintain a constant level in the treated-water chamber. An orifice-type flowmeter actuates the chemical feed in proportion to the rate of water flow. The raw water is sprayed into the steam-filled sections at the top of the softener. The oxygen and other noncondensable gases are discharged through the vent to the atmosphere. The sludge-recirculation pump transfers the sludge from the bottom of the softener into the chemical-reaction portion of the softener. This recirculation of sludge accelerates the chemical action, thereby increasing the effectiveness of the reaction chamber. Treated water is discharged from the inverted-funnel–shaped baffle to the pressure filters, and then to the boiler feed pumps. The filter-backwash-pump suction line is also located below this funnel-shaped baffle. After passing through the filters, the wash water is discharged to the upper section of the softener, thus reclaiming the filter-backwash water.

These open feedwater heaters and hot-process softeners are essentially automatic, but there are some points that require the attention of the operator. Adequate steam must be supplied to maintain the specified pressure; this steam is usually obtained from the exhaust of plant auxiliaries, and when this exhaust is not sufficient, steam from the high-pressure lines is admitted through a reducing valve. These heaters and softeners may be designed for any desired steam pressure to satisfy plant conditions, but they are usually operated at 3 to 10 psig. Some softeners are equipped to receive and blend condensate with the makeup water. When the condensate returns are not sufficient to meet the demands for boiler feedwater, the level in the heater drops and makeup is added. The controls should be adjusted to supply a proportional amount of makeup and condensate to assure uniform boiler feedwater conditions. The water-level control is important since failure to maintain the level will result in the plant's running out of feedwater, while flooding may cause the water to enter the steam-supply lines. This may cause waterhammer or injury to the unit that is supplying the exhaust steam. Alarms are frequently installed to warn the operator when the water is too high or too low in the heater. Noncondensable gases are liberated in open heaters and hot-process water softeners.

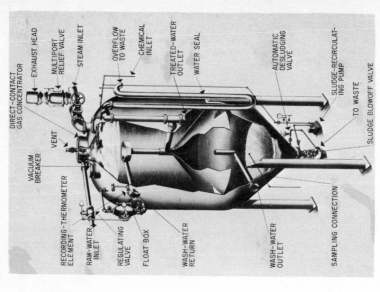

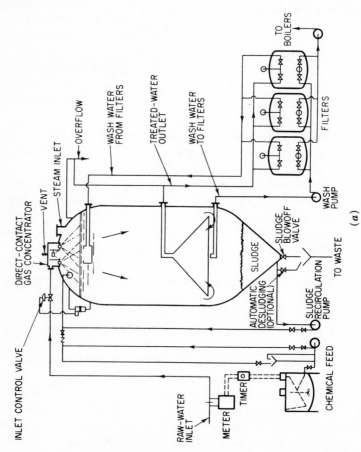

FIG. 13.3 *Deaerating hot-process softener.* (*a*) *Diagrammatic arrangement of equipment.* (*b*) *Downflow softener unit.* (*Cochrane Division—Crane Co.*)

These gases must be vented to prevent them from accumulating and partially blocking the flow of steam to the heater.

The softener-type open heaters provide a means of sludge removal during operation. Materials known as "coagulants" are added to the softener to speed up the settling action and thus decrease the load on the filters.

The control of the chemical treatment of water in a lime-soda softener is accomplished by testing or titrating[1] a sample of the effluent water for P (phenolphthalein), M (methyl orange) alkalinity, and for H (hardness). These values are expressed as parts per million (abbreviated ppm)[2] as calcium carbonate.

The lime control of a hot-process softener, $2P - M$, should normally be between 5 and 10 ppm as calcium carbonate. If this value is too low, increase the lime; if it is too high, decrease the lime.

The soda-ash control of a hot-process softener, $M - H$, should normally be between 30 and 40 ppm as calcium carbonate. If this value is too low, increase the soda ash; if it is too high, decrease the soda ash.

This control should result in a hardness H between 15 and 25 ppm as calcium carbonate. Solids deposited in the bottom of the softener sedimentation compartment must be removed by blowing down once each 8-hr shift. The filters must be backwashed before the pressure drop through them becomes excessive. A pump is usually provided, and the water used to backwash the filters is recirculated through the softener, effecting a saving in both water and treating chemicals. The

---

[1] Titration for alkalinity consists of adding a few drops of a dye, phenolphthalein, to the sample. The sample turns pink, indicating the presence of hydroxides and carbonates. A standard-strength acid solution is then added in sufficient quantity to just remove the pink color. The quantity of acid used is the P alkalinity. Then another dye, methyl orange, is added and the sample becomes a straw color, indicating the presence of hydroxides, carbonates, or bicarbonates. As more standard-strength acid is added, the sample color will change from straw to pink. The quantity of acid added during this second titration is the M alkalinity.

One method of determining hardness is to add a standard-strength soap solution to the sample, and recording the amount required to produce a lather. A second method is by titration using an organic dye which causes the sample of hard water to turn red. Then a standard-strength sequestering (separating) agent is added in sufficient amount to change the color of the sample from red to blue. The ml (milliliters) of the titrating agent added multiplied by a constant gives the hardness H of the sample. The constant corrects for the size of the sample and strength of the titrating solution.

[2] Parts per million refers to a concentration of 1 in 1 million, that is to say, 1 lb of salts in 1,000,000 lb of water. Another way of expressing solids in water is grains per gallon, a grain being 1/7,000 lb. To convert from grains per gallon to parts per million multiply by 17.1. The P and M readings denote the alkalinity of the water. The M quantity is known as the "total alkalinity." The alkalinity and hardness are expressed as calcium carbonate so that they may be added and subtracted directly. This procedure simplifies the calculations used in determining the necessary changes in treatment as indicated by the control test.

chemical-mixing tank must be charged with lime, soda ash, and coagulants as indicated by the control analysis.   A record should be kept of the total weight of chemicals and the amount in pounds per inch of water in the mixing tanks.   The rate of chemical feed can be varied by changing either the strength of the solution in the mixing tank or the rate at which the solution is fed to the softener in proportion to the water.

13.2   CLOSED FEEDWATER HEATERS.   A closed feedwater heater is one in which the heat in the steam is transferred through tubes which separate the water from the steam.   It consists of a shell made of either steel or cast iron which holds the steam for heating and houses the tubes that supply the heat-transfer surface (Fig. 13.4).   The tubes are made of either copper or brass to resist the corrosive action of the water. Closed feedwater heaters are classified as one-, two-, three-, or four-pass, depending upon the number of times the water traverses the length of the unit before being discharged.   The heater shown in Fig. 13.4 is baffled to provide four passes for the water.   A good closed feedwater heater will heat the water to within a few degrees of the steam temperature.

The shell and tubes of closed feedwater heaters are made of dissimilar metals and therefore have different rates of expansion.   Several methods are used to compensate for this difference in the expansion rates of the tubes and shell.

One of the first methods of constructing closed feedwater heaters consisted of packing each tube so that it could expand independently of the shell.   In this type the tube sheet at each end of the shell is drilled to receive the tubes and provide a packing gland.   This arrangement requires two packed joints for each tube, which are a possible source of trouble and necessitate maintenance.

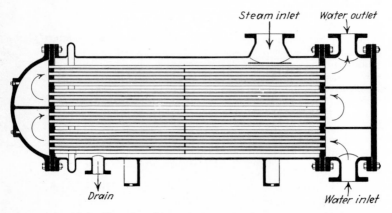

FIG. 13.4   *Closed feedwater heater.*   (*Schutte and Koerting Co.*)

The floating-head-type heater consists of a separate tube, sheet, and head or cover which is free to move inside the heater shell. The tubes are expanded in both the fixed and movable tube sheets; the movement of the "free" head compensates for unequal expansion. To gain access to the tubes for inspection and cleaning it is necessary to remove the head of the shell as well as the cover over the tube sheet.

Another method of building a heater without using packed joints between the tubes and tube sheet is to bend the tubes into the form of the letter U; these tubes are frequently referred to as "hairpin tubes." Both ends of these tubes are expanded into the same tube sheet, and the water box is baffled to direct the flow of water through the tubes from one pass to the other. Although both ends of these tubes are rigidly attached to the tube sheets, their shapes permit free expansion. Owing to their U shape, these tubes cannot be readily cleaned by mechanical means.

The heater shown in Fig. 13.4 has an expansion joint in the shell which compensates for variation in the length of the tubes and shell owing to the expansion and contraction caused by temperature change. This type of shell construction permits the use of straight tubes expanded in both tube sheets. The straight tubes facilitate cleaning and repairs.

The tube sections of closed feedwater heaters must withstand the pressure of the water, and the shell must withstand the pressure of the steam. The water is forced through one or more heaters by a single pump. High-pressure steam plants employ these heaters between the boiler feed pumps and the boilers, exposing the water side of the heater to the full boiler-feed-pump pressure. Safety measures in the form of relief valves or rupture disks (see Sec. 11.1) must be installed to prevent the possibility of either the water or the steam pressures exceeding that for which the heater was built.

In the closed feedwater heater, the steam and water do not mix; consequently oil in the steam is not introduced into the feedwater. One pump may be used to force the water through several heaters and thus utilize the steam from several extraction points in the steam turbine. These closed feedwater heaters are not suitable for heating hard water, as deposits form on the tube surfaces, retarding heat transfer and restricting the flow. The steam space must be vented, but there are no provisions for removing the noncondensable gases liberated from the water by the application of heat.

When a closed feedwater heater fails to produce the expected outlet-water temperature, observe the steam pressure, note the gauge glass to make sure that the water has not accumulated in the shell, check the vent system, and if in doubt increase the amount of venting. If all is in order and the outlet temperature is still low, the heater must be given an internal inspection for dirty or plugged tubes. Cleaning is accomplished by the use of rotary cleaners or by means of chemical solutions.

Tube leakage may be detected by an excessive amount of water being discharged from the shell. A large leak may overload the drainage system, causing the heater to become flooded. Leakage may occur at the packed or expanded joints where the tubes enter the tube sheet or as a result of a defective tube or tubes. A heater may be tested for leakage by taking it out of service, removing the heads, filling the steam space with water, and then observing the point of leakage. If a considerable number of packed tubes leak, it is best to repack. Pulling the packing too tightly reduces the diameter of the tubes, making it impossible to secure a tight joint and necessitating tube replacement. Minor leaks, where the tubes are expanded in the tube sheet, may be stopped by reexpanding. Care must be exercised not to overexpand, as this may split the tube ends or enlarge the holes in the tube sheets.

The greatest fuel saving is obtained by heating feedwater with exhaust steam from turbines, engines, or reciprocating pumps. For approximately every 10°F rise in feedwater temperature, there is a 1 per cent saving in fuel.

An example will show the possible saving that results from heating feedwater. The heat supplied to the boiler is 1190 Btu per lb of steam. The temperature of the water entering the heater is 55°F, and that of the water leaving is 195°F. The saving that results from preheating would then be

$$\frac{195 - 55}{1190 - (55 - 32)} \times 100 = 12.00 \text{ per cent}$$

Closed heaters may be arranged to utilize steam at any pressure, either above or below that of the atmosphere. When they are functioning properly, the outlet-water temperature will be 2 to 5°F below that of the steam. The effectiveness of the heater is reduced by deposits on the tubes, the accumulation of noncondensable gases in the shell, or flooding of the shell with condensate. These heaters are not satisfactory for use with hard water owing to the rapid accumulation of scale on the tube surfaces. When they are operating with a positive pressure in the shell, noncondensable gases may be vented to the atmosphere. When the shell is below atmospheric pressure, it can be vented either to a condenser or by a steam jet or other vacuum-producing auxiliary. Venting of the noncondensable gases is essential, but opening the vents more than is required results in wastage of steam. Condensate may be discharged through a trap to a vessel having a pressure sufficiently lower than that of the shell. If a low-pressure vessel is not available, a condensate pump is required. Gauge glasses located at the bottom of the heater provide a means of checking the performance of the drainage system.

13.3 ION-EXCHANGE WATER CONDITIONERS. One of the simplest methods of removing hardness from a water supply is by use of a sodium zeolite water softener (Fig. 13.5). Softeners of this type employ zeolite, an insoluble granular material having the ability to exchange hardness in the form of calcium and magnesium ions for sodium ions which do not produce hardness. The total amount of chemicals dissolved in the water is not reduced, but they are changed to non-hardness-producing chemicals, thus softening the water.

The exchange material is contained in a steel tank adequate to withstand the working pressure. The raw water is introduced above the bed, and the exchange takes place as the water flows downward through the zeolite. These softeners are equipped with meters and the necessary valves and accessories for controlling the regeneration procedure.

Normally the outlet water from these softeners shows zero hardness. When a sample of the outlet water shows hardness, the capacity of the exchange material has been exhausted and it must be restored by regenerating with a brine solution (sodium chloride).

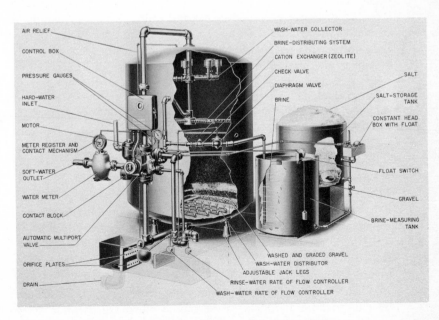

FIG. 13.5 *Zeolite-water-softener installation.* (*The Permutit Co.*)

The regeneration procedure involves three steps:

1. The softener is taken out of service, and a swift current of water is passed upward through the zeolite bed. This flow of water agitates and regrades the zeolite and at the same time washes away dirt that may have been deposited during the softening operation.

2. After the backwashing has been completed, the brine is introduced by means of a water-actuated eductor. The brine is evenly distributed above the zeolite bed by a system of piping, assuring an even flow through the bed and regeneration with a minimum of salt. During the regeneration process calcium and magnesium are removed as soluble chlorides and the zeolite sodium content is replenished.

3. After the salt has been introduced, the calcium and magnesium chloride together with the excess salts must be washed from the zeolite bed by means of a relatively slow flow of rinse water. After this rinsing operation has been completed, the softener may be returned to service. The hard water enters the top of the softener tank and flows downward through the bed, then through the meter, and out to service.

The regeneration requires a total of 35 to 65 min, and the soft-water requirement plus the backwash and rinse must be provided either from a stored water supply or by another softener. It is customary to install two or more units to furnish adequate water during the regenerative cycle.

These softeners can be regenerated either by hand manipulation or by automatic control. Softeners may be equipped with the necessary hand valves for backwashing, supplying the salt solution, and rinsing. However, the operation is simplified by the use of a multiported valve with a single control handle which is moved from the run position through the backwash, salt, and rinse positions and back to run, allowing the correct amount of time in each position. The regeneration is per-formed automatically by a power-driven multiported valve and timers which allow the valve to remain in each position the predetermined amount of time.

When the raw-water hardness remains nearly constant or when the variation takes place slowly, a meter provides an adequate means of determining the time to regenerate. After a predetermined amount of water has passed through the meter, a signal warns the operator that it is time to regenerate. In the case of an automatic unit the signal from the meter may be used either to alert the operator or actually to initiate the regeneration cycle.

When the hardness of the raw water varies widely, the effluent must be analyzed for hardness to determine when regeneration is required.

This determination may be made with an automatic hardness tester and the resulting signal used to warn the operator or initiate the regeneration cycle. When two or more automatically controlled softeners are installed, they are interlocked so that only one can be out of service for regeneration at any given time.

The dissolved solids are not removed by sodium zeolite softening; therefore, when the makeup water contains large quantities of impurities, the concentration of dissolved solids and alkalinity in the boiler water may become excessive. The concentration of the boiler water can be reduced by increasing the amount of blowdown, but supplemental treatment of the makeup water provides a more satisfactory solution. The alkalinity can be reduced by the addition of acid but its use is limited, owing to the close control which must be maintained and the precautions necessary in storage and handling. Another option is the use of a dealkalizer which is physically like a zeolite softener but contains chloride anion resin. The effluent from the sodium zeolite softener flows through the dealkalizer. The alkaline sodium compounds formed in the zeolite softener are converted to chlorides in the dealkalizer. The alkalinity is reduced, but the solids remain in the water to build up the concentration in the boiler water. The dealkalizer, like the zeolite softener, is regenerated with sodium chloride and operated in the same manner.

Hot-lime–soda softeners are used in connection with sodium zeolite softeners. The makeup water first enters the lime–soda softener where it is heated; the dissolved solids, including silica. are partly removed; and the hardness is reduced. This hot treated water is then pumped through a sodium zeolite softener, where the hardness is reduced to zero. The ion-exchange material in sodium zeolite softeners must be selected for use with hot water.

Sodium zeolite softeners are relatively low in first cost and are simple to operate. The regeneration cycle including backwash admission of brine and rinse can be made fully automatic. The salt requirement varies from 0.3 to 0.5 lb per 1,000 grains of hardness removed. The salt can be stored in a tank into which water is added to provide a ready supply of brine for regeneration. Where the raw-water-supply pressure is adequate, it may be utilized to force the water through the softener, thus eliminating the use of extra pumps.

The use of the ion-exchange principle is not limited to the removal of calcium and magnesium hardness from water. By selection of exchange material and the solution used for regeneration these units can be employed to provide makeup for use in high-pressure boiler plants and in applications requiring water having a high degree of purity.

An arrangement of ion exchangers that removes practically all the dissolved solids in water is referred to as a "demineralizer" or "deionizer" (Fig. 13.6). The raw water passes through the cation unit, where the calcium, magnesium, and sodium ions are removed. The cation unit is regenerated with acid. From the cation unit the water cascades over fill material in the degasifier tank. Air is blown up through the degasifier tank in counterflow to the water. By this action the carbon dioxide gas is removed from the water and discharged from the tank with the air. The water discharged from the degasifier tank is next pumped through the anion exchanger and into the service mains. The anion unit contains an exchange material which will remove silica, carbon dioxide, chloride, and sulfate. The anion unit is regenerated with sodium hydroxide. The demineralizer will operate without the degasifier tank, but use of the latter reduces the size of the required anion unit and the amount of sodium hydroxide needed for regeneration.

A demineralizer is capable of providing an effluent water having a total-dissolved-solids content of less than 3 ppm calcium carbonate, less than 0.5 ppm silica, less than 1.0 ppm carbon dioxide, and a specific conductance of approximately 0.3 micromho.[1]

When the supply of water to a demineralizer is obtained from city water mains precaution must be taken since municipal water frequently contains residual chlorine. Chlorine deteriorates the exchange material, necessitating costly replacement. The chlorine can be removed by passing the water through a carbon filter before it reaches the demineralizer.

This ion-exchange principle is adaptable to many possible arrangements for different types of raw water and conditioned-water requirements.

13.4   EVAPORATORS.   Impurities in water are effectively removed by evaporators. Heat is applied, and the steam or vapor produced leaves the impurities to concentrate in the remaining water. When the steam or vapor is condensed, the resulting condensate has a high degree of purity.

---

[1] Pure water has a high resistance to the flow of electricity; therefore, its conductivity, the reciprocal of resistance, is low. The purity of the water is frequently determined by measuring its conductivity, using a dip cell and conductivity bridge. The results are expressed in micromhos of specific conductivity. The relation between micromhos of specific conductivity and parts per million of dissolved solids varies to some extent with the kind of dissolved solids in the water. The presence of carbon dioxide and ammonia in the water also affects the conductivity, and the readings must be corrected accordingly in order to determine the total dissolved solids.

In samples which do not contain carbon dioxide or ammonia or when corrections for them have been made, the approximate purity in parts per million of dissolved solids can be determined by multiplying the conductivity in micromhos by 0.55. In addition to checking relatively pure water, this method may, with suitably calibrated dip cells and conductivity bridges, be used to determine the concentration of boiler water.

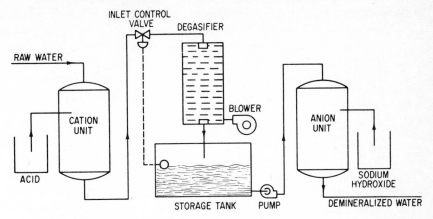

FIG. 13.6 *Diagrammatic arrangement of a demineralizer.*

The typical evaporator consists of a steel shell with the necessary outlets and a bank of tubes which comprise the heating surface. Steam supplied to the inside of the tubes gives up heat to the surrounding water and is thereby condensed. The condensate is removed through a trap which prevents the tubes from becoming flooded with condensate and at the same time prevents steam from being discharged and wasted. The heat received from the steam causes the water surrounding the tubes to evaporate, forming vapor which leaves the unit through the outlet while the impurities remain in the shell of the evaporator. The vapor is at a lower pressure and hence a lower temperature than the steam supplied to the evaporator. The rate of evaporation depends upon this temperature difference between the steam in the coils and the water surrounding them.

In the design and operation of an evaporator, consideration must be given to several details to ensure service and performance. Both the tube bundle and the shell must be designed to be strong enough to withstand the pressure involved. The shell must have a safety valve to prevent damage as a result of accidental overpressure. Priming and foaming, as experienced in boiler operation, may also occur in evaporators. This fault is overcome by designing the unit for sufficient water surface to prevent excessive agitation by steam bubbles and by installing moisture eliminators in the top of the shell at the steam outlet. The operator must keep the concentration of the liquid low enough to prevent foaming and to see that the water is at the specified level in the gauge glass. Careless operation results in solids' being carried over with the vapor, thus diminishing the possible advantages derived from the use of the evaporator.

Except for radiation and blowdown losses the heat supplied by the steam is returned to the condensate by the vapor condenser. The operator must remove the solids from the water in the evaporator by blowing down in order to prevent priming and foaming since this would cause solids to be carried over with the vapor, thus defeating the purpose of the evaporator. The vapor must be checked to make sure that the water being produced is of the desired quality. The output of the evaporator may be controlled by varying the steam pressure to provide the necessary quality of distilled makeup water. Scale deposits on the evaporator tubes lower the maximum capacity but do not affect the efficiency or cause tube failures. Some scale is permissible as long as sufficient distilled water can be produced. Evaporators should be large enough to provide the necessary water with some scale on the tubes. Some plants prevent the formation of an excessive amount of scale by pretreating the water or by adding chemicals to the water in the evaporator. Another method of control consists of allowing the scale to deposit for several hours and then taking the evaporator out of service and subjecting the tubes to a rapid temperature change which causes the hard scale to chip off the tubes. Evaporators are built to give a considerable tube movement with a change of temperature to cause the scale to chip off with this method of operation. When cold water is supplied to evaporators, oxygen, carbon dioxide, and other gases are driven off in the vapor, causing it to be corrosive. This difficulty is being overcome by the installation of hot-process softeners and deaerators to condition the water before it is fed to the evaporators.

When it is necessary to obtain a large quantity of distilled water, multiple-effect evaporators are employed. These units consist of a number of successive stages or effects through which the vapor generated in one effect flows to the tubes of the next. This multistage operation increases the amount of distilled water that can be obtained from a given amount of steam.

13.5 BOILER BLOWDOWN. In order to limit the concentration of solids in boiler water it is necessary to discharge (blow down) some of the concentrated boiler water and replace it with makeup water (see Sec. 6.2). This procedure results in a loss of heat because the water blown down had been heated to the temperature corresponding to the boiler pressure. When blown down it must be replaced by makeup water at a lower temperature. Furthermore the cost of treating this added makeup must be charged against the blowdown operation.

Several arrangements of flash tanks and heat exchangers are available for reclaiming the heat from boiler blowdown water. The system shown in Fig. 13.7 shows a schematic blowdown heat-recovery system.

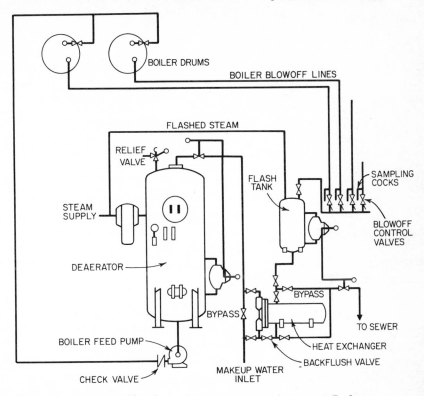

FIG. 13.7 *Continuous-blowdown heat-recovery equipment.* (*Cochrane Division—Crane Co.*)

The flash tank removes heat from the blowdown water by decreasing its pressure to that of the heater steam supply and then using the resulting flash steam in the heater. Boiler blowdown water flows from the flash tank through the heat exchanger (closed type) to the sewer. Heat is transferred from the blowdown water to the makeup water as it flows through the heat exchanger to the feedwater heater. The system provides valves for controlling the rate of blowdown and cocks for obtaining samples of this water. Consider the following example, to estimate the possible saving that can be made by application of blowdown heat recovery.

Example A plant generates steam at an average rate of 150,000 lb per hr; pressure, 400 psig; blowdown, 10 per cent of the boiler feedwater; fuel consumption is 8 tons of coal per hr; heating value of coal, 12,000 Btu per lb. What fuel saving in tons of coal per year will result from the installation of a blowdown heat-recovery system which will reduce the blowdown water temperature to 100°F?

Solution

$$\text{Boiler feedwater} = \frac{150,000}{1.00 - 0.10} = 166,700 \text{ lb per hr}$$

Blowdown water $= 166,700 - 150,000 = 16,700$ lb per hr

Enthalpy in blowdown water leaving the boiler $= 427.88$ Btu
  (400 psig, 414.7 psia)

Enthalpy in water to sewer (100°F)	$= 67.97$ Btu
Heat reclaimed per lb of blowdown water	$= \overline{359.91}$ Btu

Savings:

$$359.91 \times 16,700 = 6,010,500 \text{ Btu per hr}$$
$$6,010,500/12,000 = 500.9 \text{ lb coal per hr}$$
$$500.9 \times 24 \times 365/2,000 = 2,194 \text{ tons per yr}$$

13.6 PIPING SYSTEM. Piping is employed to transmit water, steam, air, gas, oil, and vapor from one piece of equipment to another. In this way the piping provides an essential steam-plant function. An adequate piping system must provide for the necessary size to carry the required flow, sufficient strength to withstand the working pressure and temperature, expansion resulting from temperature change, the necessary support to hold the pipeline in place, drainage of water from steam and compressed-air lines, and insulation to reduce heat losses.

In determining the size of a pipeline, one must consider the velocity and pressure which will occur at the designed rate of flow. It is standard practice to limit the velocity in steam lines that supply reciprocating steam engines and pumps to 6,000 fpm. High-pressure systems that supply steam turbines are designed for velocities of 10,000 to 15,000 fpm.

The best-designed piping systems employ the minimum sizes which will result in velocities and pressure drops that can be tolerated. The smaller pipes are lower in initial cost and have less radiation loss. However, it is false economy to select piping which is too small, since the pressure drop incurred may reduce the capacity of the equipment served by the piping.

The diameter of pipe required for a given velocity and rate of steam flow may be determined from the following formula:

$$D = \sqrt{\frac{144Q}{0.7854V}}$$

where D = diameter of the pipe, in.
  Q = cu ft of steam per min
  V = velocity of steam, fpm

Example   Consider 4,800 lb of steam per hr flowing through a pipe at 100 psia pressure.   Assume a velocity of 5,280 fpm.   What size of pipe is required?

Solution   Saturated steam at 100 psia has a volume of 4.432 cu ft per lb (see steam tables in Appendix D).

$$Q = 4.432 \times \frac{4,800}{60} = 354.56 \text{ cu ft of steam per min}$$

$$D = \sqrt{\frac{144 \times 354.56}{5,280 \times 0.7854}} = 3.509 \text{ in.}$$

A 4-in. pipe is therefore required.

Piping must be selected to withstand the internal pressure, the strains resulting from expansion, and the external forces to which it will be subjected.   The safe working pressure of a pipe may be calculated in the same manner as that for a boiler drum (see Chap. 2), except that the formula must be modified to provide for threading and other factors. One method of designating the wall thickness of piping is by the terms "standard," "extra-strong," and "double-extra-strong" (see Table 13.1). As steam pressures and temperatures increased, these designations became inadequate, and they have been supplemented by a system which employs schedule numbers[1] to designate the wall thickness of pipe.

Pipe sizes 12 in. and under are designated by their nominal inside diameter.   Since the outside diameter of any nominal size of pipe remains the same, the extra-strong pipe has a smaller inside diameter than a standard pipe having the same nominal-size designation (see Table 13.1).   This also applies to piping when the thickness is specified by schedule numbers.   Pipe sizes above 12 in. and all tubing are designated by the outside diameter.

---

[1] These schedule numbers are related to the steam pressure and strength by the following:

$$\text{Schedule numbers} = 1,000 \times P/S$$

where P = steam pressure, lb per sq in.

S = working stress of pipe material (usually taken as 10 to 15 per cent of ultimate strength)

Pipe having a schedule number of 40 corresponds approximately to standard pipe, and a schedule number of 80 to extra-heavy.   The recommended values for S for steel and alloys at various temperatures are given in the American Standard Association (ASA) Code for Pressure Piping B31.1.

In addition to wall thickness, piping materials and method of manufacture must be considered. The material used in the manufacture of piping is designated by American Society of Testing Materials (ASTM) numbers.[2]

In order to assemble piping systems and make connections to equipment it is necessary to have joints and use fittings of various types. Screwed, flanged, and welded joints are used for steel piping.

Screwed fittings are made of cast or malleable iron. It is advisable to limit the use of screwed fittings to sizes of 2 in. and smaller. The larger sizes of screwed fittings are bulky and joints difficult to make up. Unions must be used in connection with screwed pipe assemblies to facilitate maintenance and replacement.

Standard fittings are satisfactory for pressures of 125 psi and 450°F. Extra-heavy fittings are required for pressure from 125 to 250 psi and 450°F. For pressures above 250 psi and temperatures above 450°F, steel fittings are selected in accordance with specific code requirements for the design conditions.

Flanges and flanged fittings are used in connection with larger piping. The flanges may be attached to the pipe by screwed joints, slipped over the pipe and seal-welded (slip-on flanges), or by butt welding (weld-neck flanges). The use of screwed joints is restricted to low pressures. Welding is preferred for medium pressures and required for high pressures. Asbestos gaskets are used between the pipe flanges for water and saturated steam, while steel or stainless-steel gaskets are used for superheated steam. Flanged joints are subject to leaks owing to failure of the gaskets but provide accessibility for replacement and maintenance.

Piping systems can be assembled by butt or socket welding. Butt welding is successfully used for all piping above $1\frac{1}{2}$ or 2 in. in diameter.

---

[2] The chemical analysis and manufacturing procedure for steel to meet these code requirements are fully explained in "ASTM Standard for Ferrous Metal."

These ASTM specifications cover pipe for all applications, from the ordinary low pressure to steels containing various percentages of molybdenum, chrome, nickel, and other materials for high-temperature applications.

ASTM Specification A-120 seamless or resistance-welded steel pipe is satisfactory for steam pressures not exceeding 125 psig and temperatures not above 450°F except where close coiling or bending is required. This pipe is also suitable for ordinary usage where high pressures and temperatures are not involved.

ASTM Specification A-53 is available in seamless or resistance-welded steel pipe and in Grades A and B. This pipe is suitable for use with steam up to 750°F. Grade A must be used for close coiling, cold bending, or forging.

ASTM Specification A-106 is a seamless steel pipe. This specification requires close control of the chemical composition and is suitable for use up to steam temperatures in excess of 750°F with specified reduction in S values. The Grade A pipe must be used for close coiling, cold bending, and forging.

TABLE 13.1   Wrought-steel-pipe Data

Size, in.	Standard				Extra-strong			
	Diameter, in.		Wall thickness, in.	Weight, lb/ft	Diameter, in.		Wall thickness, in.	Weight, lb/ft
	External	Internal			External	Internal		
⅛	0.405	0.269	0.068	0.244	0.405	0.215	0.095	0.314
¼	0.540	0.364	0.088	0.424	0.540	0.302	0.119	0.535
⅜	0.675	0.493	0.091	0.567	0.675	0.423	0.126	0.738
½	0.840	0.622	0.109	0.850	0.840	0.546	0.147	1.087
¾	1.050	0.824	0.113	1.130	·1.050	0.742	0.154	1.473
1	1.315	1.049	0.133	1.678	1.315	0.957	0.179	2.171
1¼	1.660	1.380	0.140	2.272	1.660	1.278	0.191	2.996
1½	1.900	1.610	0.145	2.717	1.900	1.500	0.200	3.631
2	2.375	2.067	0.154	3.652	2.375	1.939	0.218	5.022
2½	2.875	2.469	0.203	5.793	2.875	2.323	0.276	7.661
3	3.500	3.068	0.216	7.575	3.500	2.900	0.300	10.252
3½	4.000	3.548	0.226	9.109	4.000	3.364	0.318	12.505
4	4.500	4.026	0.237	10.790	4.500	3.826	0.337	14.983
4½	5.000	4.506	0.247	12.538	5.000	4.290	0.355	17.611
5	5.563	5.047	0.258	14.617	5.563	4.813	0.375	20.778
6	6.625	6.065	0.280	18.974	6.625	5.761	0.432	28.573
7	7.625	7.023	0.301	23.544	7.625	6.625	0.500	38.048
8	8.625	8.071	0.277	24.696				
8	8.625	7.981	0.322	28.554	8.625	7.625	0.500	43.388
10	10.750	10.192	0.279	31.201				
10	10.750	10.136	0.307	34.240				
10	10.750	10.020	0.365	40.483	10.750	9.750	0.500	54.735
12	12.750	12.090	0.330	43.773				
12	12.750	12.000	0.375	49.562	12.750	11.750	0.500	65.415

Fittings are available with chamfered ends for butt welding, suitable for all pressures and made of materials for all temperatures.

Backing rings must be employed or other precautions taken to prevent the welding material from entering the pipe and reducing the cross-section area. This work must be performed by a qualified welder in compliance with local codes. Steel having a thickness of ½ in. or less and containing neither carbon in excess of 0.35 per cent nor molybdenum need not be preheated before welding or stress-relieved after welding (see Sec. 2.5 for a discussion of stress relieving of welded joints).

Socket-welded fittings are employed on pipe less than 1½ or 2 in.

in diameter in all high-temperature water systems and other applications where tight joints are necessary. Socket welding involves a special steel fitting into which the full-sized end of the pipe may be inserted. The pipe is welded in place to hold against the pressure and seal against leakage.

Various types of valves, including gate, globe, angle, quick-opening, ball, butterfly, plug-type, check, reducing, etc., are used in pipelines to stop or regulate the flow. Valves are available for all pressures and are arranged for use with screwed-, flanged-, or welded-joint piping.

Cast- or ductile-iron[1] pipe is used underground for water services and drainage systems and in other places to resist corrosion. The sections are connected together, and fittings are attached either by use of bell and spigot or by mechanical joints. A bell-and-spigot joint has one end enlarged (bell) to receive the neck (spigot) of the other. The space between the bell and spigot is calked with lead or provided with a rubber gasket to form a tight joint.

The mechanical joint likewise has one end enlarged to receive the other, but packing compressed by bolts is used to make a tight joint. Mechanical joints are more costly than bell-and-spigot joints but are preferred especially when the line is under a structure or roadway where a leak would result in great damage and be difficult and costly to repair. Valves and fittings are available for use in cast-iron piping systems employing either mechanical or bell-and-spigot joints. These joints do not restrain the pipe from longitudinal movement, and adequate anchors or ties must be provided when there is a change in direction or other cause for longitudinal thrust.

As the temperature of a pipeline changes, its length also changes, owing to expansion and contraction characteristics of the metal. This expansion results in a surprisingly large change in the length of the pipe and, if not taken care of, creates strains which might result in leaks or rupture. The amount of expansion can be calculated from the following:

$$E \text{ (expansion, in.)} = (Ti - t) \times L \times CoE \times 12$$

where Ti = final temperature of steam header, °F
   t = initial temperature of steam header, °F
   L = length of pipe, ft
   CoE = coefficient of expansion for 1°F (for steel, 0.00000734)

---

[1] Ductile iron is made by adding an alloy in the molten gray iron; the resulting change in the shape of the graphite particles results in a stronger, tougher, and more ductile material.

Example Assume a steel pipe carrying steam at a pressure of 100 psia and a total temperature of 500°F. The header is 200 ft long and room temperature 100°F. Find the amount of expansion when heated from room to steam temperature.

Solution

$$E = (500 - 100) \times 200 \times 0.00000734 \times 12 = 7.0464 \text{ in.}$$

In order to reduce the pressure drop in a pipeline to the minimum, the length of run should be as short as possible and bends and fittings kept to a minimum. However, this practice frequently leads to a pipeline which is too rigid to provide for expansion. Arranging the piping with several bends will frequently provide adequate flexibility, but when this method fails, special devices to provide for expansion must be used.

In high-pressure installations the U bend is universally used as an expansion joint. This consists of a section of pipe formed with a long-radius bend to allow for considerable variation in the length of the pipe without imposing undue strain (Fig. 13.8a). The size of the pipe and the radius of the bend determine the maximum permissible pipe movement that the pipe bend will accommodate. These U bends have the disadvantage of requiring considerable space, but when correctly installed they require no maintenance.

For low and medium pressures, the packed expansion joint is used to advantage (Fig. 13.8b). This consists of a sleeve which can move inside a larger sleeve with packing to prevent leakage. Joints of this type require less space than U bends and may, therefore, be installed in pipe tunnels. They can be made to adjust for a considerable amount of expansion. Stops are provided to prevent overtravel in both directions. The joints must be inspected and the packing tightened and occasionally replaced. Increased life of the packing may be obtained by applying a heavy heat-resistant grease to the packing gland with a grease gun.

Another compact expansion joint consists of a corrugated-metal section installed in the pipeline (Fig. 13.8c). The circular corrugations provide for a limited amount of endwise movement of the pipe. The number of corrugations can be varied to meet the maximum expansion requirements. When this type of joint is used for high pressure, metal support bands are clamped to the outside of the corrugations to prevent distortion. The side-bar arrangement on these bands equalizes the movement of the corrugations, thus preventing some from being overstressed. Care must be exercised not to allow these joints to move beyond the manufacturer's specifications.

The support of piping systems presents a difficult problem because the pipeline must be held in place and at the same time allowance

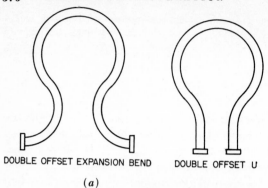

DOUBLE OFFSET EXPANSION BEND   DOUBLE OFFSET U

(a)

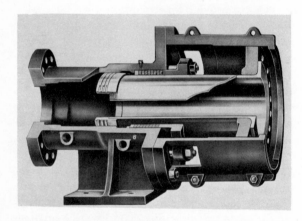

(b)

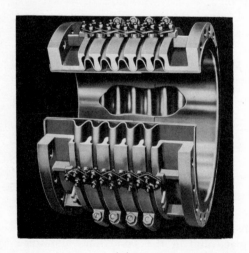

(c)

FIG. 13.8 (a) Expansion bends for pipelines. (b) Sliding sleeve—internally guided. (American District Steam Co.) (c) Corrugated expansion joint. (Zallea Brothers.)

must be made for movement resulting from expansion. The pipeline must be so anchored that the expansion will move the pipe toward the expansion joint. This is accomplished by securely anchoring the pipeline between the expansion joints and supporting it at the other points on rollers or spring hangers. In this way the pipe will be held in line, and as the temperature increases, the movement will be toward the expansion joint, allowing it to function as intended. The magnitude of the end thrust on the fixed anchors can be shown by an example. A 12-in. pipe carrying steam at 150 psig pressure is to be anchored at an ell so that the expansion will extend toward an expansion joint. The thrust upon the anchor will be

$$150 \times 12 \times 12 \times 0.7854 = 16{,}964.64 \text{ lb}$$

This shows that the thrust on the anchors will be approximately $8\frac{1}{2}$ tons. If the anchors should fail, the expansion joint would be expanded to the limit of its travel and the pipeline subjected to as much strain as if the expansion joint had not been installed.

Piping systems must be designed to limit the stress in the piping and the thrust imposed upon the equipment, to acceptable values. Overstressed pipe will result in failure. When piping exerts too much thrust or torque upon the nozzles of a pump, or a turbine, it will be forced out of alignment. Manufacturers of this type of equipment state the allowable forces that can be imposed upon the nozzles by the piping. After initial design of a piping system, the stress and forces at all major points are calculated by a computer. If the pipe is overstressed or if the forces are too great on the nozzle of the equipment to which the piping is attached, additional flexibility must be provided and the stresses and forces recalculated by the computer. It is essential that fixed hangers, spring supports, and anchors be maintained to prevent pipe failures and misalignment of equipment. When repairs or modifications are made to a piping system, consideration must be given to possible change in the flexibility.

Vibration is sometimes encountered in piping connected to air compressors and reciprocating pumps. This difficulty can be overcome by reducing the pulsating effects and by supporting the pipe more securely. In addition to the conventional air chambers for pumps, patented snubbers are available for reducing the pulsating effect produced by reciprocating machinery.

Steam-piping systems must be installed to prevent an accumulation of water while steam is flowing and to make it easy to drain the line when it is being placed in service. To accomplish this, the horizontal runs must slope in the direction of steam flow with steam traps at all points where water may accumulate. This precaution lessens the possibility of water collecting in the line and being carried along with the steam to produce waterhammer.

The temperature of the substance in the pipe is frequently different from that of the surrounding air, making it advisable to insulate the line. When the temperature of the substance in the pipe is higher than that of the surrounding air, there is a heat loss from the pipe and a rise in the surrounding-air temperature. This not only constitutes a loss of heat but frequently results in a high room temperature objectionable to personnel and injurious to equipment. When the temperature of the substance in the pipe is lower than that of the surrounding air, moisture will condense, resulting in corrosion and dripping of water from the line. This condition is especially prevalent when the air is warm and moist (high relative humidity).

High-temperature lines are effectively insulated by the use of 85 per cent magnesium blocks in varying thickness, depending upon the temperature. Actually the economic thickness of insulation should be based upon a balance of heat loss against the cost of insulation, with consideration to the comfort and safety of personnel. Heat loss is given in handbooks as Btu per hour per linear foot of pipe for various steam temperatures and thicknesses of insulation.

Example Consider 8-in. pipe carrying steam at 160 psig, with loss per linear foot of bare pipe 2140 Btu per hr. Loss per linear foot of 2-in. high-grade insulation is 190 Btu per hour. Savings per linear foot of pipe resulting from insulation is 1950 Btu per hr.[1] With 13,500 Btu per lb of coal and 70 per cent boiler efficiency, the insulation on 1,000 ft of 8-in. pipe would save

$$\frac{1950 \times 1,000 \times 30 \times 24}{13,500 \times 0.70 \times 2,000} = 74.3 \text{ tons of coal per month as compared with bare pipe}$$

Underground steam lines and return condensate lines are available in common insulated ducts. These lines are insulated and covered with a waterproof metal shield. Water must not come into contact with underground steam lines, as it will produce steam and give a false indication of leakage, as well as create a nuisance.

To prevent the accumulation of moisture, hair, felt, fiber glass, and cork are used to insulate lines that carry substances at a temperature lower than the surrounding atmosphere. These insulations must be applied with a vapor barrier to exclude the air, which, if allowed to enter the voids, will deposit moisture. When this occurs, the insulation will become saturated with water and mold will form, creating an unsightly, unsanitary condition and causing rapid deterioration of the insulation.

The operator must recognize the piping system as an essential part of the steam plant. Steam and compressed-air lines must be operated

[1] Kent's "Mechanical Engineers' Handbook" ("Power" volume).

to avoid waterhammer by preventing steam and water or air and water from mixing in the pipelines. Steam lines must be placed in service slowly to allow the water to drain out and expansion to take place without creating strain in isolated sections. Large steam-header valves should be provided with small bypass valves for this purpose. It is best to bypass the traps to ensure complete drainage during the time the steam line is being placed in service.

The cause of leakage should be determined immediately and repairs made as soon as possible. Leakage may be a warning of weakness that will cause serious rupture. Continual leakage of flanged joints results in cutting of the metal, making it necessary to replace or reface the flanges. All leaks cause waste, which is usually greater than would at first be estimated.

In time insulation deteriorates as a result of maintenance, leakage, and mechanical injury, and sections of the pipe and fittings become exposed. This condition should not be tolerated, as it exposes the personnel to burns, creates a source of heat loss, and makes the plant unsightly.

13.7   STEAM TRAPS. A steam trap is a device attached to the lower portion of a steam-filled line or vessel which will pass condensate but will not allow the escape of steam. Traps are used on steam mains, headers, separators, and purifiers, where they remove the water formed as the result of unavoidable condensation or carry-over from the boilers. They are also used on all kinds of steam-heating equipment in which the steam gives up heat and is converted into condensate. Coils used in heating buildings, in water heaters, and in a wide range of industrial processing equipment are included in this classification.

Whether a trap is used to keep condensate from accumulating in a steam line or to discharge water from a steam-heated machine, its operation is important. If it leaks, steam will be wasted; if it fails to operate, water will accumulate. A satisfactory trap installation must pass all the water that flows to it without discharging steam, must not be rendered inoperative by particles of dirt or by an accumulation of air, and must be rugged in construction with few moving parts so that it will remain operative with a minimum of attention. Several principles are employed in the operation of traps.

Figure 13.9 shows the operation of

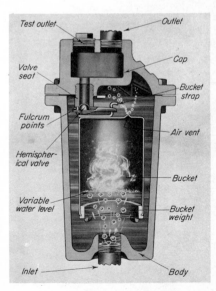

FIG.   13.9   *Inverted-bucket trap.* (*Armstrong Machine Works.*)

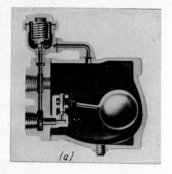
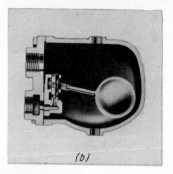

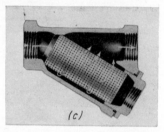

FIG. 13.10 *Float-actuated traps and pipeline strainers.*
(*A*) *Steam trap with thermostatic air vent.* (*B*) *Trap for removing water from compressed-air lines.*
(*C*) *Pipeline strainers.* (*Sarco Co. Inc.*)

an inverted-bucket-type trap.   The water and steam enter at the bottom and flow upward into the inverted bucket.   As long as the bucket contains steam, it is buoyed up in the same way that an inverted empty bucket is buoyed up in water.   While in this position, the valve is closed and there is no discharge of water or steam from the trap.   As water enters the bucket, it displaces the steam and the bucket loses its buoyancy and drops, causing the valve to open.   After the water has been discharged, the bucket again fills with steam, the buoyancy is restored, and the valve closes.   A small vent in the top of the bucket allows air to escape, thus preventing it from interfering with trap operation.   In some models, the release of air is controlled by a thermostatic vent which opens when the temperature decreases and allows the air to be discharged from the trap.   Under normal operating conditions with steam in the trap the vent remains closed.   The valve seats and disks are made of stainless steel or other alloys to resist corrosion and wear.   The moving parts, including the seats and disks, are easily replaced.

Float-actuated traps may be adapted to a wide range of operating conditions (Fig. 13.10).   When water fills the body of the trap, the float rises, opening the valve.   After the water has been discharged,

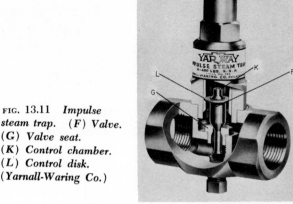

FIG. 13.11 *Impulse
steam trap.* (*F*) *Valve.*
(*G*) *Valve seat.*
(*K*) *Control chamber.*
(*L*) *Control disk.*
(*Yarnall-Waring Co.*)

the float drops, closing the valve and preventing the escape of steam.

The trap (Fig. 13.10*a*) is intended for use in removing condensate from steam systems. It has a thermovent valve which prevents air from interfering with the trap operation. When the temperature of the trap decreases this vent automatically opens, allowing the air to escape. The trap (Fig. 13.10*b*) is suitable for removing water from compressed-air systems. The strainer (Fig. 13.10*c*) is installed in the line, ahead of the trap, to prevent sediment from stopping up the trap orifice. When selected for the correct pressure and provided with strainers these traps will give satisfactory service. The valve assembly, including the float, float arm, and linkage, may be readily replaced.

The impulse steam trap (Fig. 13.11) operates on the principle that hot water under pressure tends to flash into vapor when the pressure is reduced. When hot water is flowing to the trap the pressure in the control chamber *K* is reduced, causing valve *F* to rise from its seat and discharge the water. As steam enters the trap, the pressure in chamber *K* increases, causing valve *F* to close, thus reducing the flow of steam to that which can pass through the small center orifice in valve *F*. It requires 3 to 5 per cent of full condensate capacity to prevent steam from being discharged through this small orifice. These traps are available for all pressures, they are compact, and the few parts subject to wear can be readily replaced. It is necessary to install a strainer, as small particles of foreign matter will interfere with the operation of valve *F*.

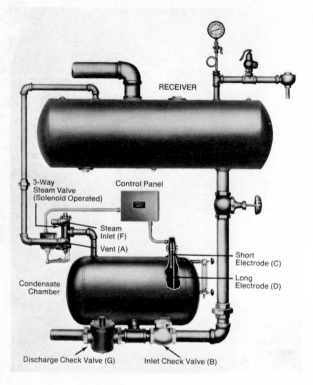

RECEIVER

3-Way
Steam Valve
(Solenoid Operated)

Control Panel

Steam
Inlet (F)

Vent (A)

Short
Electrode (C)

Long
Electrode (D)

Condensate
Chamber

Discharge Check Valve (G)   Inlet Check Valve (B)

FIG.   13.12 *Condensate-handling   system.   (The
Johnson Corporation.)*

Condensate is effectively returned from the point of steam usage to
the boiler plant by the trap system shown in Fig. 13.12.   The condensate flows into the receiver and then through the check valve into
the condensate chamber.   When the chamber fills to make contact with
the short electrode, a solenoid positions the three-way steam valve to
apply steam pressure to the chamber.   This pressure forces the condensate through the discharge check valve and back to the boiler plant.
When the level in the condensate chamber drops below the end of the
long electrode, the solenoid positions the three-way valve to shut off the
steam supply to the condensate chamber and at the same time vent this
chamber to the receiver.   Since the pressure is now equalized, the condensate will flow by gravity from the receiver to the chamber.   During
the condensate discharge cycle, when the chamber is pressurized, condensate accumulates in the receiver, providing an uninterrupted flow
from the steam-utilizing equipment.

This system delivers the condensate to the feedwater heater, under pressure. In this way the heat in the condensate, as discharged from the steam-utilizing equipment, is retained. This is in contrast to conventional condensate-pumping systems, in which the condensate tank must be vented to the atmosphere or cooling provided, to obtain satisfactory operation of the pumps.

Traps are termed "nonreturn" when the condensate is discharged into a receiver, or heater, rather than directly to the boiler. A "return" trap delivers the condensate directly to the boiler. Return traps are located above the boiler, and when filled with water a valve automatically opens and admits steam at boiler pressure. This equalizes the pressure, and the water flows into the boiler as a result of hydrostatic head caused by the elevation of the trap. These traps are sometimes used in connection with low-pressure heating boilers.

Figure 13.13 shows a trap piping arrangement on a vital piece of equipment which must remain in continuous service. The two-gate valves provide a means of shutting off the trap so that it can be removed for replacement or repairs. During this time the globe bypass valve is used to regulate the flow of condensate. Bypass on steam-header traps may be opened when the header is being placed in service. This procedure ensures against the possibility of water hammer if the trap should fail to drain the condensate from the steam line. The strainer is placed in the line ahead of the trap. A test valve is useful for checking trap operation by observing the discharge.

Traps correctly sized and installed should require very little attention or maintenance, but they should be inspected periodically. The frequency of inspection will depend upon the operating conditions. Gauge glasses provide an effective means of observing trap performance. When the trap is functioning satisfactorily, the water level will be near the center line of the body and will continue to show some fluctuations in level. If the water is above the glass (as observed by opening the gauge-glass drain), the trap is too small, the orifice is stopped up, the chamber contains air, there is an obstruction in the line, or the back pressure is too high. If the water is low or out of sight in the glass, the shutoff mechanism in the trap is not functioning and steam is blowing through and being wasted. The strainer should be blown out periodically to remove the accumulation of dirt. A test valve (Fig. 13.13) may be opened and the discharge of the trap observed to determine how often it is operating and how much condensate is being discharged. Do not be confused by the fact that a portion of the condensate discharged from the trap will appear as vapor; this is

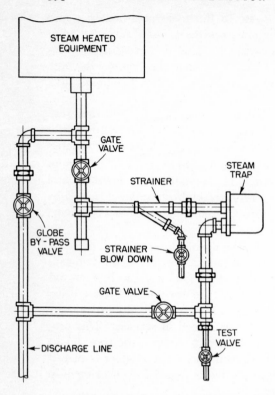

FIG. 13.13 *Typical steam-trap piping arrangement.*

the result of the water's flashing into steam as the pressure is reduced. The higher the pressure in the trap, the greater the amount of flashing. Traps can be checked by placing one end of a rod (sounding rod) to the ear and the other to the trap. In this way it is possible to detect by sound if the trap is blowing straight through or failing to operate and discharge the condensate.

13.8 PIPELINE SEPARATORS AND STRAINERS. The steam, air, and gas flowing in a pipeline often contain particles of moisture, oil, and other foreign matter. These impurities must be removed by the use of separators or strainers located in the pipeline. Although there are many types of separators, they all employ the same principle. The moving stream is directed against the baffle or obstruction, which suddenly changes the direction of flow of the steam, air, or gas. The particles of moisture or oil (being heavier than the steam, air, or gas) are thrown out of the main stream and are retained in the separator.

The separators shown in Fig. 13.14 have a ribbed baffle area extending outside the jet of steam or compressed air projected from the inlet pipe. Upon entering the separator, the flow is directed against this ribbed baffle. The particles of moisture or oil are first deposited upon the baffle and then drip into the collecting chamber. The ribs in the baffle retain the film of water or oil while the flow of steam or air passes through ports at the sides of the baffle. This arrangement prevents the flow from coming in contact with the water or oil drops as they fall into the chamber, thereby decreasing the tendency of water and oil to be picked up by the flow of steam or compressed air. The collect-

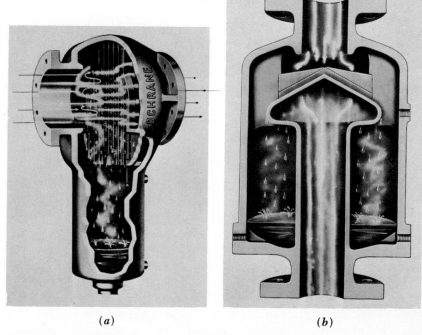

(a)   (b)

FIG. 13.14   *Steam and compressed-air separators. (a) For horizontal lines. (b) For downward flow in vertical lines. (Cochrane Division—Crane Co.)*

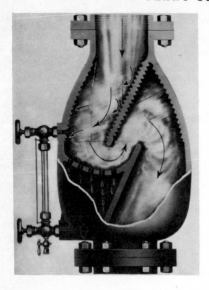

FIG. 13.15 *Live-steam separa-tor.* (*Wright-Austin Co.*)

ing chamber is provided with a trap for drainage and a gauge glass to check the liquid level.

A live-steam separator (Fig. 13.15) is suitable for use in a vertical steam line between the boiler and engine pump or turbine. This is the baffle-type separator in which both the baffle and the walls contain slanting corrugations. The moisture is collected by these corrugations, deposited in the well, and drained off by use of a steam trap. The complete reversal of flow at the lower edge of the baffle separates the steam from the moisture.

Centrifugal separators are used to remove oil, moisture, and solids from steam, compressed air, and gas. Figure 13.16 shows how these separators function to remove the entrainment.

The exhaust head used when steam is discharged to the atmosphere above the roof of a building is a special adaptation of this type of separator. When exhaust steam is discharged to the atmosphere, the exhaust head (separator) is installed on the discharge end of the pipe. In this way the oil and moisture are collected to prevent them from becoming a nuisance in the surrounding area.

Because oil separators are not effective in removing all the oil from steam, air, or gas, secondary filtering is often necessary. Condensate is often passed through a filter to remove the last trace of oil before it is returned to the boilers. Filters are available for cleaning compressed air when small amounts of oil are objectionable.

Strainers are available for removing foreign matter from liquids flowing in a pipeline. These strainers are made in various sizes for installation in the pipeline. The two-element arrangement provides for continuous service, as the shutoff valves place one element in operation while the other is being cleaned. The strainers or baskets may be made of any suitable material and mesh, depending upon the material in the line and the particle size of the impurities to be removed.

The difficulty most likely to be encountered with pipeline separators is failure of the trap to discharge the oil or water removed. Some separators have gauge glasses to observe the level of water and oil in the bowl, but it is difficult to keep these clean enough for them to be of any value. It is advisable to have a test valve on the outlet

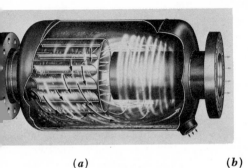

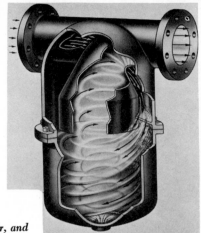

(a)   (b)

FIG. 13.16 *Centrifugal separators for steam, air, and gas.* (a) *Line-type for use in supply lines ahead of equipment.* (b) *T-type for compact in-line mounting.* (*Wright-Austin Co.*)

or bypass of the trap. By opening this valve and observing the discharge, it is possible to tell whether or not the trap is functioning.

13.9 LUBRICANTS AND LUBRICATING DEVICES. The primary purpose of a lubricant is to prevent two bearing surfaces from coming into direct contact. When one lubricated surface slides or rolls over another, the lubricant adheres to each surface and the motion takes place within the lubricant. The two metallic surfaces do not come into contact and, therefore, do not wear. Fluid friction occurs as one film of the lubricant moves over another.

In addition to this primary function, lubricants are sometimes required to carry away heat developed, as in a turbine bearing, to assist in sealing against leakage, as in a steam-engine cylinder, or to operate hydraulic cylinders and devices, as in turbine governors. In performing these functions, the oil must resist mixing with water (emulsifying), thinning out and carbonizing under the action of high temperature, oxidizing and becoming acid on exposure to air, and other contamination. It is evident from this that care should be exercised in selecting lubricants.

Progress has been made in the manufacture of lubricants to meet the needs of industry. Detergents are added to oil to prevent formation of dirt and sludge in crankcases and gearboxes. Wetting agents cause oil to adhere to the surfaces being lubricated. The compounding of mineral oils with vegetable oils produces lubricants that resist the washing action of wet steam in engine cylinders. Lubricants are available which contain additives that provide protection for gears operating with extreme pressure between the teeth.

Oils are thickened by the addition of soap to form plastic lubricants known as "greases." These greases are used as lubricants where oil would leak out of the bearings, water and dirt cannot be excluded, it is difficult to give the bearings frequent attention, or the operating speed is low and the bearings heavily loaded.

The plant operator can secure the benefits of advanced developments by presenting his problems to the engineer representative of the company from which he secures lubricants. Each application should receive separate consideration. The request presented by the operator should include a description of the machine, the size and speed of the shaft, the type of bearings, the possibility of leaking oil or grease, whether water and grit are excluded from the bearings, the method of applying the lubricant, whether service is continuous or intermittent, and the manufacturers' lubrication recommendations. From this information, the type of lubricant and frequency of application can be determined. For simplicity and economy the number of types of lubricant used in the plant should be kept at a minimum.

Viscosity is one of the qualifications mentioned in equipment manufacturers' specifications of lubricants. It may be defined as a measure of the resistance to flow. A high-viscosity lubricant is one that has high internal fluid friction. This means that a high-viscosity lubricant has a tendency to develop heat, especially when applied to high-speed machinery. It follows that, in general, light or low-viscosity lubricants are adaptable to high-speed machines and heavy or high-viscosity lubricants to low-speed machines.

The viscosity of oil can be measured by a Standard Saybolt Universal viscosimeter. This test consists of noting the time required for 60 ml (milliliters) of an oil at standard temperatures of 70, 100, 130, and 210°F to flow through the orifice. Thus the Saybolt Universal seconds (SSU) of 80 at 210°F would mean that it required 80 sec for the 60 ml of oil at 210°F to pass the orifice. The familiar Society of Automotive Engineers (SAE) viscosity number takes into account the SSU viscosity range of an oil. For example, an SAE 30 oil has a SSU viscosity of no less than 185 and no more than 255 at 130°F. Additives are available which reduce the effect of temperature changes upon the viscosity of lubricating oil. They are widely used where the machinery operates under varying temperatures. The automobile engine is a typical example.

Other lubrication specifications include *specific gravity*, which can be expressed either as the weight of oil compared with an equal volume of water or in API degrees (see Chap. 3); *pour point*, or lowest temperature at which the oil can be poured; *flash* or *fire points*, which are, respectively, the temperature at which the vapors given off will flash or burn continuously when exposed to an open flame; *demulsibility*, or ability of the oil to separate from water; and *acidity*, as measured

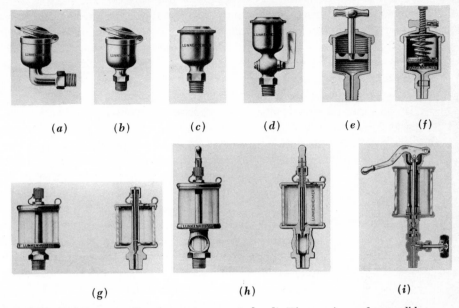

FIG. 13.17 *Lubricators—oil and grease cups.* (*a,b,c,d*) *Hinge-spring and screw-lid cups.* (*e*) *Plunger screw-feed grease cup.* (*f*) *Automatic-feed grease cup.* (*g*) *Oil cup without sight feed.* (*h*) *Oil cup with sight feed.* (*i*) *Hand-operated oil pump.* (*The Lunkenheimer Co.*)

by the amount of caustic of known strength required to neutralize a given quantity of oil.

The method of applying lubricants varies from frequent hand operation to automatic methods that require little attention. The method of application and the type of bearing must be considered when determining the frequency of application. The oil cups shown in Figs. 13.17a, b, c, and d are used in various ways to apply oil to bearings. The ring oil bearing is self-lubricating, but the level in the reservoir must be checked (Fig. 4.40). An oil cup as shown in Fig. 13.17a may be attached to the side of these bearings to provide a means of observing the oil level. Wool waste is sometimes placed in the oil cups to act as a feed and to decrease the necessary frequency of oiling. The oil cups in Figs. 13.17g and h have needle valves for regulating the flow of oil to the bearings. When the machine is stopped, the valve can be shut off without changing the needle-valve adjustment. The oilers shown in Fig. 13.17i include a hand pump for injecting oil into a system that is under pressure. Figure 13.17e shows a grease cup that can be filled and the grease applied as required by turning the handle. This method of grease application has been almost completely replaced by the high-pressure fitting and detachable grease gun. After the grease cup shown in Fig. 13.17f has been filled, the spring exerts a pressure which slowly feeds the grease into the bearing.

Multipoint lubricators are available for supplying grease to a large number of points automatically. Each lubrication point is provided with a lubricator. These lubricators are connected in series (line from one to the other) and supplied by one grease pump. The pump may be operated by hand or by power with a timer set for the desired interval. When pressure is applied to the system by the pump, grease is delivered to each point. The individual lubricators are adjustable so that the required amount of grease will be supplied.

The hydrostatic lubricator (Fig. 13.18) is used to introduce steam-engine and pump-cylinder lubricants into the steam line. It is called "hydrostatic" because it operates by means of a head of water in the following manner: The upper and lower connections to the lubricator are piped to the steam main leading to the engine. The distance from the top connection to the bottom of the lubricator is approximately 1½ to 2 ft. The steam enters on top, passing through line 3 to the condenser 4. Steam condenses in 4, water passing through valve 5 to line 6. The vessel 7 is filled with oil. As the water enters this vessel it displaces the oil, the oil passing through 8 and down through fitting 9 up through 10 to 11 to the steam line. Ordinarily, the pressures in the lower and upper steam connections are equal. The head of water, however, unbalances the lower pressure and forces oil into the line. The needle valve 9 is for the purpose of regulating the flow of oil to the engine, the rate of feed being determined from the sight glass 10. A gauge glass 13 is used to determine the water level in the reservoir 7. Filler plug 15 is used when refilling, and the cock 14 is the drain. In order to refill the reservoir, close valves 5 and 9. Open the drain 14 and then remove the filling plug 15. Close drain valve 14 and refill with oil. Replace filling plug 15 and open valve 5. Now adjust the needle valve 9 for the flow of oil desired.

This lubricator is not automatic and must be stopped and started by hand. It is somewhat difficult to maintain a uniform feed, as the viscosity of the oil changes with the temperature. Occasionally, owing to impurities in the line, feeding becomes irregular and sometimes stops altogether.

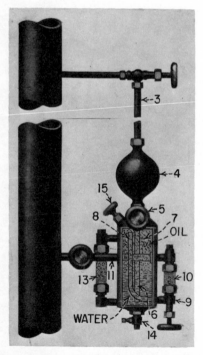

FIG. 13.18   *Hydrostatic lubricator.*   (*Socony-Vacuum Oil Co.*)

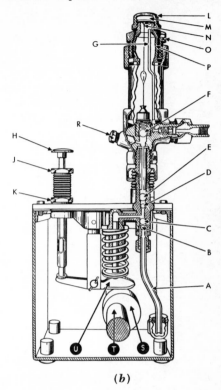

(a)   (b)

FIG. 13.19 *Forced-feed lubricator.* (a) *Exterior view.* (b) *Diagrammatic arrangement of working parts.* (*McCord Corp.*)

The forced-feed lubricator shown in Fig. 13.19 provides an automatic positive means of supplying lubricant to the steam line of engines and pumps and to bearings under pressure. It consists of a desired number of plunger pumps actuated by a cam which is driven from the reciprocating motion of some part of the machine it lubricates. The rate of feed may be individually adjusted and observed in the glycerine-filled sight glass. The cross section (Fig. 13.19b) shows the working parts of one of the lubricators. On the suction stroke, oil is drawn through tube A and suction valves B and C. On the delivery stroke, suction valves B and C close automatically, and oil is forced through delivery valves D and E and oil nozzle F into the liquid chamber, hence into the sight feed. The oil rises to the top owing to its having a lower specific gravity than that of the glycerine in the glass. From here the oil is forced through the delivery tube to the point of application. The rate of feed is adjusted by turning the threaded sleeve J, which can be held in position by locknut K. By alternately depressing and releasing the button H, the feeder can be hand-operated.   Figure 10.1 shows how this forced-feed type of lubricator can be applied to a steam-driven air compressor.

In pressure systems, the oil is circulated by pumps to the different points to be lubricated.   This oil accumulates impurities which must be removed, or serious injury to the equipment will result.   These impurities cause undue wear of the bearings, clogging of pipes in the lubrication system, coating of cooler tubes, and subsequent higher oil temperatures as a result of reduced heat transfer.   When oil comes into contact with water, there is a tendency to form emulsions, usually referred to as "sludge."   By removing these impurities from the oil, it is rendered satisfactory for service and the plant lubrication costs are thereby reduced.

Oil is conditioned by separators, filters, or a combination of both.   In some instances, the entire batch of oil is removed from the machine and reconditioned; in others, a portion of the oil is continually circulated through the reconditioning apparatus.

If contaminated oil is allowed to stand for a long period of time, the water and sludge, being heavier than the oil, will settle to the bottom of the container.   This process may be accelerated by passing the oil through a mechanical centrifuge which subjects it to a centrifugal force several thousand times that of gravity.   The centrifuge shown in Fig. 13.20 consists of a cast-iron frame which supports a bowl on a vertical shaft.   This shaft and bowl constitute the rotating elements, which are power-driven at high speed.   The dirty oil is fed into the top center of the separator, flows down through the hollow shaft, and is discharged into the bottom of the bowl by the centrifugal force.   The water and sediment, being heavier than the oil, are forced to the outside of the bowl and travel up its sides and overflow through the lower discharge tube.   A number of conical plates or baffles direct the movement of the oil toward the center of the bowl, where it flows upward and is discharged through the center spout.   The top spout acts as an emergency overflow.   Sediment collects on the conical plates and in the bowl, making it necessary to clean them frequently.

Conditioners which employ a combination of settling and stages of filtration are useful in maintaining the lubricating properties of low-viscosity oil.   These units remove small particles of solids and water from the oil, restoring its lubricating value.   Care must be exercised in selecting this equipment, as some filter mediums remove additives and inhibitors, thus changing the characteristics of the oil.

The oil reclaimer shown in Fig. 13.21 is capable of removing contamination from a lubricating system and thus of restoring the lubricating

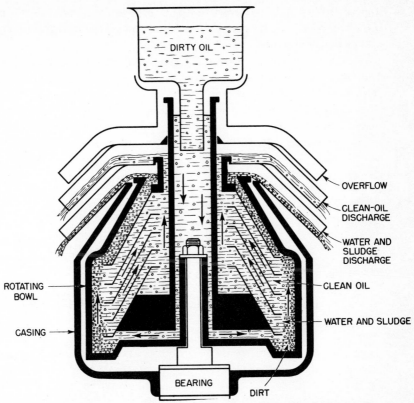

FIG. 13.20   *Cross section of centrifugal oil-separator bowl.*

qualities of the oil.   Figure 13.21*a* shows how the components are assembled to form a package unit.

The operation can be explained by reference to the flow diagram in Fig. 13.21*b*.   The dirty-oil pump forces the incoming oil through the heat exchanger, where it receives heat from the clean oil.   The dirty oil next passes through the electric preheater.   Then it is drawn by vacuum through a filter which removes carbon, sludge, abrasives, and other solids.   This filter also neutralizes the acidity and restores the color.   From the filter the oil enters the vaporizer.   Here moisture, solvents, dissolved gases, and other volatile impurities are removed by the combined action of the elevated temperature and the vacuum.   The volatile substances removed from the oil may be discharged to the atmosphere or condensed and reclaimed.   The clean-oil pump removes the oil from the vaporizer and forces it through the heat exchanger into a clean-oil storage tank or back into service.

The type of filter mediums is varied, depending upon the characteristics of the oil and the severity of the service.   The filter mediums must be changed when the gauges show that the pressure drop has exceeded the established limits.

The expendable or throwaway type of cartridge filter is effective in removing small particles of sediment from low-viscosity oil (Fig. 13.22). Filters of this type consist of a case designed to hold a cartridge which has a large filter area.   The cartridge is made of cellulose material and can be designed to remove particles as small as 0.00004 in.   When the filter becomes clogged, as indicated by an increase in pressure drop,

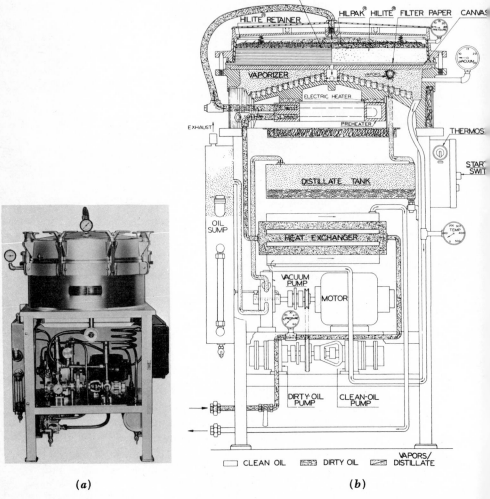

(a)   (b)

FIG. 13.21   Oil reclaimer.   (a) Package unit.   (b) Flow diagram.   (The Hilliard Corporation.)

TABLE 13.2   Lubrication Schedules

Form 10

BOILER-PLANT LUBRICATION SCHEDULE

Frequency: daily

Equipment and parts	Lubricant to be used	Work to be done	Work done by						
			M	T	W	T	F	S	S
Spreader stoker: Motor sheave pulley..	Light grease	1 fitting with gun							
Motor slide..........	Oil	Apply oil with can							
Drive shaft..........	Heavy grease	4 fittings with gun							
Coal-feed regulator...	Heavy grease	6 fittings with gun							
Stoker-feed mechanism	Heavy grease	5 fittings with gun							
Boiler feed pump and turbine: Pump shaft..........	Light turbine oil	Check oil level; 2 wells							
Turbine shaft..........	Light turbine oil	Check oil level; 2 wells							
Governor shaft.........	Light grease	1 grease cup							

Form 15

BOILER-PLANT LUBRICATION SCHEDULE

Frequency: 120 days

Equipment and parts	Lubricant to be used	Work to be done	Work done by
Spreader stoker: Motors...............	Light grease	2 motors; 2 fittings each	
Boiler feed pump: Pump shaft............	Light turbine oil	Flush and re-fill bearing	
Turbine shaft..........	Light turbine oil	Flush and re-fill bearing	

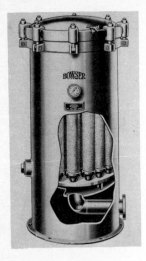

FIG. 13.22 *Cartridge oil filter.* (*Bowser, Inc.*)

the cartridge is replaced with a new one to restore the unit to its original performance.

After the operator has determined the correct type of lubricant and the frequency of application and made a study of the method of application, the next step is to supply these findings in the plant. Written procedures will serve as instructions and reduce the possibility of misunderstandings and neglect. Table 13.2 is a sample of a lubrication-schedule sheet for a small boiler plant. Such sheets are issued to the persons responsible for lubricating the equipment at the various specified intervals. Form 10 is issued each week and informs the oiler of the equipment involved, the points of application, and the lubrication to be used and provides a space for his initial indicating that the work has been done. The type of lubricant can be referred to by the manufacturers' specification numbers. At the end of the week the sheet is returned to the supervisor with comments on any difficulties that the oiler has encountered in carrying out the instructions. In like manner Form 15 is issued every 120 days and covers the lubrication work that is to be done at that frequency. Adequate lubrication is essential to continued operation and low maintenance costs.

QUESTIONS AND PROBLEMS

13.1. What is an open heater? A closed heater?

13.2. What functions are performed by open feedwater heaters?

13.3. Steam leaving the engine is exhausted to a heater. This steam heats the feedwater from 96 to 196°F. Plant steam, 125 psig dry and saturated. What saving in fuel would result by this preheating?

13.4. What kind of feedwater heater is best suited for hard water?

13.5. How would you test for leaking coils or tubes in a closed feedwater heater?

13.6. What determines the maximum temperature of the water leaving an open heater?

13.7. Where should the open feedwater heater be placed with respect to the boiler feed pump?

13.8. Discuss the function of a hot-process water softener supplying makeup water.

13.9. What effect does the accumulation of air and noncondensable gases have on the performance of feedwater heaters?

13.10. State the advantages of a sodium zeolite water softener over a hot-process lime–soda softener.

13.11. Explain the procedure for regenerating a sodium zeolite water softener.

13.12. What difficulties are likely to be encountered when using a sodium zeolite softener to condition boiler makeup water?

13.13. Under what condition would a demineralizer be used to condition boiler makeup water?

13.14. What is the function of an evaporator?

13.15. List the precautions to be taken when operating evaporators.

13.16. What is insulation? What are the advantages of using it? Name several materials which are used for insulating purposes.

13.17. Steam at 90 psia and 150°F superheat has a volume of 6.04 cu ft per lb. If 10,000 lb per hr flowed through the pipe to an engine, what size of pipe would be required?

13.18. Saturated steam at 90 psia has a volume of 4.89 cu ft per lb. Assuming the same flow as in Prob. 13.17, calculate the pipe size.

13.19. Upon steam being admitted to a steel header, the temperature changes from 80 to 650°F. The line being 135 ft long, what is its expansion?

13.20. A cast-iron pipe 300 ft long is heated from 100 to 350°F. How much does the line expand?

13.21. What are two methods of taking care of the expansion in steam lines?

13.22. Discuss the reasons for insulating hot pipes? Cold pipes?

13.23. What is a steam trap, and how does it operate?

13.24. Discuss three different types of steam traps.

13.25. To what dimension does the size of pipe refer?

13.26. Explain the correct procedure in checking the operation of steam traps.

13.27. Where is a return trap used? How does it operate?

13.28. What is a steam separator, and how does it work?

13.29. What is an oil separator, and how does it work?

13.30. What is meant by lubrication? Mention the qualities essential to a good lubricant.

13.31. Does a separator take care of all the oil?

13.32. What is grease? When is it used in preference to oil?

13.33. What is a hydrostatic lubricator, how does it operate, and how can it be refilled?

13.34. Explain the two methods of reconditioning low-viscosity lubricating oils.

13.35. What is meant by the viscosity of oil? When would you use a low- and when a high-viscosity oil?

13.36. What are the advantages of a forced-feed lubricator as compared with a hydrostatic?

13.37. How would you proceed to establish a lubrication program for a steam plant?

USEFUL
DATA

Measurements

Length:

        12 inches (in. or ″) = 1 foot (ft or ′)
          3 feet = 1 yard (yd)
    5,280 feet = 1 mile

Area:

        144 square inches (sq in.) = 1 square foot (sq ft)
            9 square feet = 1 square yard (sq yd)

Volume:

    1,728 cubic inches (cu in.) = 1 cubic foot (cu ft)
          27 cubic feet = 1 cubic yard (cu yd)

Liquid:

            2 pints (pt) = 1 quart (qt)
            4 quarts = 1 gallon (gal)
        231 cubic inches = 1 gallon

Weight:

    7,000 grains = 1 pound (lb)
          16 ounces (oz) = 1 pound
    2,000 pounds = 1 ton
    2,240 pounds = 1 long ton

Pressure:

        0.0361 pounds per square inch = column of water at 62°F
                                        1 inch high
        0.433 pound per square inch = column of water at 62°F
                                       1 foot high
        0.49 pound per square inch = column of mercury 1 inch high
        14.7 pounds per square inch = 1 atmosphere

Weight of Water at 62°F

    0.0361 pound   = 1 cubic inch
    62.355   pounds = 1 cubic foot
    8.3391 pounds* = 1 gallon of water
      * May be assumed as 8⅓ lb.

Weights of Common Substances

Substance	Lb/cu ft	Substance	Lb/cu ft
Air*	0.0763	Ice	57.5
Brass	525	Cast iron	450
Brick	125	Steel	490
Coal, bituminous	47–56	Zinc	438

* At 60°F, barometer 29.92 in. Hg.

Miscellaneous Data

3,413 British thermal units (Btu) = 1 kilowatthour (kwhr)

1,000 watts = 1 kilowatt (kw)

1.341 horsepower (hp) = 1 kilowatt

2,545 British thermal units = 1 horsepower-hour (hp-hr)

0.746 kilowatt = 1 horsepower

33,000 foot-pounds per minute = 1 horsepower

778 foot-pounds (ft-lb) = 1 British thermal unit

34.5 pounds of water evaporated per hour from and at 212°F
= 1 boiler horsepower (boiler hp)

10 square feet of heating surface = 1 rated boiler horsepower, water-tube boiler

12 square feet of heating surface = 1 rated boiler horsepower, fire-tube boiler

Linear Coefficient of Expansion per Degree Fahrenheit

Substance	Coefficient of expansion	Substance	Coefficient of expansion
Bronze	0.00001024	Cast iron,	0.00000589
Copper	0.00000926	Steel	0.00000734
Glass (flint)	0.00000438	Zinc	0.00001653

Specific Heat of Various Substances

Substance	Specific Heat
Air (constant pressure)	0.24
Dry flue gases (constant pressure)	0.24
Water vapor (atmospheric pressure)	0.48
Ice (0–32°F)	0.50
Steel	0.117

The Metric System

Terms

meter = unit of length
liter = unit of volume
are = unit of area for land
gram = unit of weight
kilo = denotes 1000
deca = denotes 10
deci = denotes 0.10
centi = denotes 0.01
micron = 0.000001 m

Measures of Length

10 millimeters (mm) = 1 centimeter (cm) = 0.01 m
10 centimeters = 1 decimeter (dm) = 0.1 m
10 decimeters = 1 meter (m)
1000 meters = 1 kilometer (km)

Equivalents for Reference

1 meter (m) = 39.37 in. (established by law)
1 gm = 15.432 grains
1 lb avoirdupois = 7000 grains
1 in. = 2.54 cm approx
1 kg = 2.2 lb approx

Measurements

Length:

1 in. = 2.54001 cm
1 ft = 30.4801 cm
1 km = 3280.83 ft = 0.62137 mile
1 mile = 1.60935 km

Area:

1 sq in. = 6.45163 cm^2
1 sq ft = 0.0929 m^2
1 sq yd = 0.836 m^2
1 cm^2 = 0.155 sq in.
1 m^2 = 10.76387 sq ft = 1.196 sq yd
1 are = 119.5985 sq yd
1 acre = 40.4687 ares

Volumes, Capacities:

$$1 \text{ cu in.} = 16.38716 \text{ cc}$$
$$1 \text{ cu ft} = 28.317 \text{ cm}^3$$
$$1 \text{ pt (liquid)} = 473.179 \text{ cc}$$
$$1 \text{ pt (dry)} = 550.6 \text{ cc}$$
$$1 \text{ qt (liquid)} = 946.358 \text{ cc}$$
$$1 \text{ qt (dry)} = 1101.228 \text{ cc}$$
$$1 \text{ cm}^3 = 0.061 \text{ cu in.}$$
$$1 \text{ liter} = 61.0234 \text{ cu in.}$$

Weights, Mass:

$$1 \text{ grain} = 0.0648 \text{ gm}$$
$$1 \text{ oz (avoirdupois)} = 28.35 \text{ gm}$$
$$1 \text{ lb (avoirdupois)} = 453.6 \text{ gm}$$
$$1 \text{ ton (short)} = 907.185 \text{ kg}$$
$$1 \text{ g} = 15.43 \text{ grains}$$
$$1 \text{ kg} = 2,205 \text{ lb}$$
$$1 \text{ metric ton} = 2204.6 \text{ lb}$$

ARITHMETIC

1. Multiplication Tables

1	2	3	4	5	6	7	8	9	10	11	12
2	4	6	8	10	12	14	16	18	20	22	24
3	6	9	12	15	18	21	24	27	30	33	36
4	8	12	16	20	24	28	32	36	40	44	48
5	10	15	20	25	30	35	40	45	50	55	60
6	12	18	24	30	36	42	48	54	60	66	72
7	14	21	28	35	42	49	56	63	70	77	84
8	16	24	32	40	48	56	64	72	80	88	96
9	18	27	36	45	54	63	72	81	90	99	108
10	20	30	40	50	60	70	80	90	100	110	120
11	22	33	44	55	66	77	88	99	110	121	132
12	24	36	48	60	72	84	96	108	120	132	144

Explanation. Locate one of the numbers to be multiplied in the top line and the other in the left-hand vertical column. Follow down the column from the number at the top and across to the right from the number in the left-hand column. The number at the intersection of this line and column is the required product.

2. Fractional Equivalents

Fourths
1 = 0.250
2 = 0.500
3 = 0.750

Eighths
1 = 0.125
3 = 0.375
5 = 0.625
7 = 0.875

Sixteenths
1 = 0.0625
3 = 0.1875
5 = 0.3125
7 = 0.4375
9 = 0.5625
11 = 0.6875
13 = 0.8125
15 = 0.9375

3. To Reduce a Common Fraction to a Decimal. Add decimal ciphers to the numerator. Divide by the denominator. Point off as many decimal places in the quotient as there are decimal ciphers added to the numerator.

Example Reduce $^{15}\!/_{16}$ to a decimal fraction.

$$16)15.0000(0.9375$$
$$14.4$$
$$60$$
$$48$$
$$120$$
$$112$$
$$80$$
$$80$$

4. Multiplication of Decimals. Multiply as in whole numbers. Point off decimal places in the product equal to the sum of the number of decimal places in the factors.

Examples

1. Multiply 6.82 by 0.042

$$
\begin{array}{r}
6.82 \\
0.042 \\
\hline
1364 \\
2728 \\
\hline
0.28644
\end{array}
$$

2. Multiply 56 by 0.7854

$$
\begin{array}{r}
0.7854 \\
56 \\
\hline
47124 \\
39270 \\
\hline
43.9824
\end{array}
$$

5. Division of Decimals. Add ciphers to the dividend until it has as many decimal places as, or more than, there are in the divisor. Divide as in whole numbers. Subtract the number of decimal places in the divisor from the number in the dividend. Point off decimal places in the quotient equal to this difference. To obtain more decimal places in the quotient (answer), add more ciphers to the dividend and proceed with the division.

Examples

1. Divide 0.6825 by 0.375

$$0.375)0.68250(1.82$$
$$375$$
$$3075$$
$$3000$$
$$750$$
$$750$$

2. Divide 685 by 0.025

$$0.025)685.000(27,400$$
$$50$$
$$185$$
$$175$$
$$100$$
$$100$$

6. Extracting the Square Root

Example   Extract the square root of 67,858.1 correct to the third decimal place.

```
         6 78 58. 10 00 00(260.4958)
         4
     46  278
         276
   5204    25810            Answer correct to the third deci-
           20816               mal place, 260.496
  52089    499400
           468801
 520985    3059900
           2604925
5209908    45497500
           41679264
```

Explanation.   Separate the number into periods of two figures each, beginning at the decimal point and proceeding both to the right and to the left. The places in the root (answer), both whole and decimal, will be the same as the number of periods. Add sufficient ciphers to the right to give the required number of decimal places in the root. Two ciphers are required for each decimal place in the root. (In the example, 58 is the first period to the left of the decimal point; 78 is the second; and 6, the third. Five ciphers are added to give the three decimal periods required for three decimal places in the root.)

Find the largest square that will be contained in the left-hand period. (In the example, 6 is the left-hand period, and the largest square is 4.)   Place the largest square (4) under the left-hand period (6) and subtract; the difference is 2. Place the square root (2) of the largest square (4) at the right of the number. This square root (2) is the left, or first, figure of the desired root.

Annex the next period (78) to the difference found by subtracting (2), and call the resulting number (278) the remainder. Multiply the root already found (2) by 2, and place the product (4) at the left of the remainder (278) to be used as the trial divisor.

Determine the number of times that the trial divisor (4) is contained an even number of times in the remainder, disregarding the right-hand number. (27 ÷ 4 = 6.)   Place this quotient (6) at the right of the trial divisor (4) to form the complete divisor (46) and also at the right of the root already found (2). This number (6) becomes the second number of the root.

Now, multiply the complete divisor (46) by the last number of the root (6), place the product (276) under the remainder (278), and subtract. To the difference (2) annex the next period (58), and proceed as before.

If the trial divisor fails to be contained in the remainder, omitting the right-hand figure (note in the example that 25 will not contain 52), another period (10) must be annexed to the remainder. In this case a cipher must be added to the root and the trial divisor before attempting to determine another place in the root.

After all the periods, including the ciphers, have been used in this manner, the root is pointed off, allowing one decimal place for each two places on the right of the decimal point.

In the example the fourth decimal place is found to be 8; since the answer was to be correct to the third place, the 8 is omitted and the 5 is increased to 6.

7. Rectangular Solid-surface Area and Volume. To find the *area* of a rectangle, multiply the two dimensions.

To find the *surface area* of a rectangular solid (a rectangular solid has six faces): find the area of each rectangular face; the total area is the sum of these areas.

To find the *volume* of a rectangular solid, multiply the three dimensions together.

8. Circumference and Area of a Circle. The circumference (distance around) a circle is 3.1416 times the diameter d (greatest distance across the circle).

The value 3.1416 is denoted by the symbol π (pi).

π divided by 4 ($\pi/4$) equals 0.7854.

The radius of a circle equals one-half of the diameter.

To find the *circumference* of a circle, multiply the diameter by 3.1416:

Circumference of circle $= d \times \pi$.

To find the *area* of a circle, multiply the square of the diameter by 0.7854:

Area of circle $= d^2 \times \pi/4$.

To find the *diameter* of a circle when the area is given, divide the area by 0.7854 and extract the square root of the quotient.

9. Surface and Volume of a Cylinder. To find the *surface* of a cylinder, multiply the circumference of the base by the height, and add to this product twice the area of the base.

To find the *volume* of a cylinder, multiply the area of the base by the height.

10. To reduce a temperature given in centigrade degrees to one in Fahrenheit degrees, multiply the centigrade reading by $\%$, and add 32 to the product.

To reduce the temperature given in Fahrenheit degrees to one given in centigrade degrees, subtract 32 from the Fahrenheit reading, and multiply the remainder by $\%$.

11. In solving problems using decimal fractions, carry as many decimal places as the accuracy of the original data makes advisable. In solving a problem in which the measurements are accurate to only 0.11 in., it is useless to carry the work to 0.001. The results of a problem solved by decimals should be carried one place farther than required. If this last place is less than 5, it should be dropped. If the last place is 5 or more, it should also be dropped and then the last remaining place increased by 1.

DEFINITIONS

Absolute Pressure. The common gauge expresses pressure in pounds per square inch, called "gauge pressure." When the gauge is open to the atmosphere, it reads zero. The gauge pressure plus atmospheric pressure is known as "absolute pressure." Atmospheric pressure is frequently assumed as 14.7 psi. Atmospheric pressure varies with location and atmospheric conditions. It is accurately indicated by a barometer.

Absolute Temperature. The temperature as read on the Fahrenheit scale plus 460 is the absolute temperature.

Absolute Zero of Pressure. The starting point of the absolute-pressure scale is absolute zero; it is lower than a "zero gauge" by an amount equal to the atmospheric pressure.

Absolute Zero of Temperature. The temperature 460° below zero Fahrenheit is "absolute zero." At absolute zero there is a complete absence of heat. Absolute zero has never been attained, but it has been approached within a few degrees.

Acidity. Water is a chemical combination of hydrogen (H) and oxygen (O). This may be written H_2O or as hydrogen ions H and hydroxyl ions OH. If there is a greater number of hydrogen ions than hydroxyl ions, the solution is acid. A greater number of hydroxyl ions results in an alkaline solution. The degree of acidity or alkalinity of a substance is known as the hydrogen-ion concentration and is called the pH value. A pH value of 7.0 indicates neutral water; a value less than 7.0, acidity; and a value greater than 7.0, alkalinity.

Adiabatic Expansion. When steam is expanded in such a manner that there is no heat flow into or away from the steam and all the heat energy lost by the steam is converted into work, the process is called "adiabatic expansion." Steam expanding behind the piston of a steam engine after the point of cutoff approaches adiabatic expansion.

Alkalinity. See *Acidity*.

Boiling Point of Water. Water at atmospheric pressure boils at 212°F, 100°C.

Boyle's Law of Gases. When the temperature of a gas remains constant, the volume will be reduced to one-half if the absolute pressure is doubled; the absolute pressure will be reduced to one-half if the volume is doubled. (The volume of a gas varies inversely as the pressure.)

British Thermal Unit (Btu). A British thermal unit is used to measure heat energy. It is defined as the quantity of heat required to raise the temperature of 1 lb of water 1°F.

600

Caustic Embrittlement. This refers to the cracking of boiler steel occurring while the boiler is in service but not subjected to excessive pressure or temperature. Such failures are attributed to the boiler water's being too caustic, that is, too alkaline; hence the term "caustic embrittlement."

Condensation. When steam or any other vapor is subjected to a change of state which reduces it to a liquid, it is said to be "condensed." Steam is condensed in a condenser or heater by extracting heat. The water formed is called "condensate."

Conduction. When heat is transmitted through a substance or from one substance in contact with another but without the bodies' themselves moving, the transfer is by conduction. Heat is conducted through the metal in the shell and tubes of a boiler. Substances differ widely in their ability to conduct heat. Metal is a good conductor; soot and boiler scale are very poor.

Convection. When heat is carried by means of the movement of currents within a body, it is said to be transmitted by "convection." The change in density of the substance, due to the heating, causes the movement. The circulation of water in a boiler carries heat from the tubes near the fire to the boiler drum.

Conversion of Heat Energy and Mechanical Energy. Heat energy and mechanical energy are convertible. There is a direct relation between heat energy and mechanical energy; 778 ft-lb is equivalent to 1 Btu.

Density. The density of a substance is the number of units of weight that it contains per unit of volume. Water has a density of 62.5 lb per cu ft.

Dew Point. The temperature at which a vapor liquefies is called the "dew point."

Efficiency. The efficiency of any contrivance is the output divided by the input, sometimes stated as the useful energy divided by the energy expended. The input and output may be expressed in any energy units. They must, however, be in the same units.

Energy. Energy is the ability to do work. Mechanical energy is expressed in foot-pounds or horsepower-hours; electric energy, in kilowatthours; and heat energy, in British thermal units.

Enthalpy. Enthalpy is the number of Btu which a substance contains above a specific datum. In the case of water and steam the reference condition is water at 32°F. The enthalpy values given in the steam tables are the Btu required to raise water or steam from 32°F to the specific condition of temperature and pressure. These values are also referred to as "total heat."

Evaporation. The process of changing a liquid into a vapor or a gas is known as "vaporization." This is usually accomplished by the application of heat.

Excess Air. The combination of fuel is primarily the combining of combustible substances of the fuel with oxygen of the air. A fuel requires a definite amount of oxygen, therefore air, to effect complete combustion. The amount of air used in excess of this amount is known as "excess air." Excess air is necessary to effect complete combustion, but too much causes a decrease in efficiency.

Expansion. A change in temperature produces a change in the size of practically all substances. Each material changes a different amount for a given change in temperature. The change in length per degree-temperature change is known as the "linear coefficient of expansion." Such coefficients for common materials may be found in handbooks. If the pressure of a gas is kept constant, the volume will change in proportion to the absolute temperature.

Factor of Evaporation. If 970.3 Btu is added to 1 lb of water which is at atmospheric pressure and 212°F, it will be converted into steam and the steam will be at atmospheric pressure and 212°F. This is termed "evaporation from and at 212°F," or, briefly, "from and at." The heat added to a pound of water by the boiler (from the time at which it enters until it leaves as steam) divided by 970.3 is the factor of evaporation.

Force. Force is that which produces, or tends to produce, motion. The force on the piston of a steam engine produces motion. The force exerted on the head of a steam boiler does not produce motion, but it tends to; both are examples of force.

Freezing Point of Water. Water freezes at 32 on the Fahrenheit scale of temperature measurement. This corresponds to 0 on the centigrade scale. When water freezes, its volume increases by about 9 per cent.

Furnace Heat Release. This is the number of Btu developed per hour in each cubic foot of furnace volume. It is usually assumed that all the heat available in the fuel burned is transformed into heat in the combustion process. Therefore the furnace heat release equals the total fuel burned per hour times the Btu content divided by the furnace volume.

Fusion Temperature. Fusion is the act or process of melting by heat or the state of being fused or melted. For coal, the fusion temperature is reported as initial-deformation (I.D.T.), ash-softening (A.S.T.), and ash-fusion (A.F.T.), the test being made under a reducing atmosphere.

Head. Head is the energy per pound of fluid.

Potential Head. This refers to energy of position, measured by work possible in dropping a vertical distance.

Static Pressure. This refers to energy per pound due to pressure; it is the height to which liquid can be raised by a given pressure.

Velocity. This refers to kinetic energy per pound; it is the vertical distance a liquid would have to fall to acquire the velocity V.

Total. This refers to the net difference between total suction and discharge heads.

Net Positive Suction (N.P.S.H.). This is the amount of energy in the liquid at the pump datum.

Heat. Heat is a form of energy.

Heat Content of Steam, or Total Heat. This refers to the Btu that must be added to produce steam at the condition in question. Water at 32°F is usually taken as the starting point.

Horsepower. Horsepower is a unit of power that tells the rate at which work is being performed, namely, 33,000 ft-lb per min.

Horsepower-Hour. A horsepower-hour is 1 hp of energy expended continuously for 1 hr.

Ideal Engine. An ideal engine is a theoretical engine which is assumed to take full advantage of the heat available between a given set of throttle and exhaust conditions. The efficiency obtained with the ideal engine cannot be obtained with an actual engine owing to friction, wiredrawing, leakage, cylinder condensation, and radiation. However, the efficiency of the ideal engine is useful in rating and comparing actual engines.

Kilowatt. A kilowatt is a unit of electric energy and is equal to 1,000 watts. For direct current, watts equals amperes times volts; for alternating current, watts equals amperes times volts times power factor.

Kilowatthour. A kilowatthour is 1 kw of energy expended continuously for 1 hr.

Kinetic Energy. Kinetic energy is energy that a body has due to its motion. The flywheel has kinetic energy due to the motion of its heavy rim. Doubling the velocity of a body makes its kinetic energy 4 times as great.

Latent Heat of Evaporation. When a liquid is vaporized, a large amount of heat must be added to produce the change. This heat does not increase the temperature and is therefore called "latent heat." The latent heat of vaporization of water at amospheric pressure is 970.3 Btu.

Law of Conservation of Energy. The amount of energy in existence is constant. The machines that we build and operate do not produce energy; they merely change it from one form to another.

Mechanical Equivalent of Heat. This is sometimes called Joule's equivalent: 778 ft-lb of mechanical energy is equivalent to 1 Btu of heat energy.

Power. Power is the rate at which work is done. Foot-pounds express work, but the rate or time required determines the power: 33,000 ft-lb per min is a horsepower.

Radiation. Radiation is the transmission of heat without the use of a material carrier. The earth receives heat from the sun, and most of that distance the heat travels through a vacuum. When a furnace door is open, you can feel the heat even though air is being pulled into the furnace through the door. The lower rows of boiler tubes receive much heat by radiation. Radiated heat is very similar to visible light; they both travel at the same speed, namely, 186,000 miles per sec.

Saturated Steam. Saturated steam is steam that contains no moisture; and yet if heat were added, its temperature would be increased. It is saturated with heat.

Smoke. Smoke refers to flue gases that contain enough unburned carbon and hydrocarbons to cause discoloration. The degree of coloration depends upon the carbon present.

Specific heat. Different substances have different heat capacities. In fact the heat capacity of some substances changes as the temperature changes. By the definition of a Btu the heat content of a pound of water per degree Fahrenheit is 1. The specific heat of any substance is the heat required to raise 1 lb of it 1°F.

Stoker Combustion Rate. This rate is expressed in terms of the number of pounds of fuel or the Btu developed per square foot of stoker grate area per hour.

Superheated Steam. Steam that has been raised to a temperature higher than the boiling temperature corresponding to the boiling pressure is said to be "superheated."

Temperature. The temperature of a substance must be carefully distinguished from the heat content. Temperature is thermal pressure and is a measure of the ability of a substance to give or receive heat from another.

Thermal Efficiency. Thermal efficiency refers to heat engines and is the output expressed in Btu divided by the input expressed in Btu.

Torque. Torque is a force which tends to produce rotation. It is measured by the product of the force and the applied distance from the center of rotation. A force of 100 lb applied at a distance of 2 ft will produce 200 lb-ft of torque.

Vacuum. The word "vacuum" refers to pressures below atmospheric, units expressed in inches of mercury (14.7 psi atmospheric equals 30 in. of mercury).

Viscosity. Viscosity is the resistance that a fluid offers to flow. That of lubricating oil is an important characteristic. Viscosity varies with the temperature.

Wet Steam. When steam contains particles of water that have not been evaporated, it is said to be "wet."

Wiredrawing. When steam forces its way through a small opening with a corresponding loss in pressure, the process is called "wiredrawing," or "throttling." This occurs to some extent during the admission of steam to a steam engine.

Work. Work is force exerted through a distance. No mention is made of the time required. Work is conveniently expressed in foot-pounds.

STEAM TABLES
AND
CHARTS

Steam Tables. The properties of steam, including pressure, temperature, specific volume, total heat (enthalpy), entropy, and superheat, are given in steam tables for use in solving problems (see Sec. 2.10). Table D.1 gives these properties of saturated steam with reference to the absolute pressure in pounds per square inch, which is shown in the left column. Table D.2 gives the same data except that the data correspond to the temperature which appears in the left column. When the pressure of saturated steam is known, use Table D.1 to find the other properties. When the temperature of saturated steam is known, use Table D.2 to find the other properties.

The properties of wet steam are not given directly in the steam tables but may be calculated from the data given in these tables when the moisture content is known. (See Sec. 2.10 for the method used in making wet-steam calculations.)

When steam is heated above the saturation temperature, it is said to be "superheated"; it contains more heat per pound and has a greater volume than shown in the saturated-steam tables. Table D.3 gives the volume in cubic feet per pound of steam (v), the total heat (enthalpy) in Btu per lb (h), and entropy (s) for superheated steam at various absolute pressures and temperatures in degrees Fahrenheit. The saturated temperatures for the respective absolute pressures are given by the numbers in parentheses directly under the pressures in the left column. The temperature above saturation (amount of superheat) is found by subtracting the saturation temperature from the total temperature. For example, at 400 psia and 800°F total temperature, the superheat is

$$800 - 444.59 = 355.41°F$$

since the saturated temperature at 400 psia is 444.59°F.

TABLE D.1   Dry Saturated Steam Pressure*

Abs press., psi	Temp, °F	Specific volume, cu ft/lb		Enthalpy, Btu/lb			Entropy		
		Sat. liquid	Sat. vapor	Sat. liquid	Evap.	Sat. vapor	Sat. liquid	Evap.	Sat. vapor
p	t	v_f	v_g	h_f	h_{fg}	h_g	s_f	s_{fg}	s_g
1.0	101.74	0.01614	333.6	69.70	1036.3	1106.0	0.1326	1.8456	1.9782
2.0	126.08	0.01623	173.73	93.99	1022.2	1116.2	0.1749	1.7451	1.9200
3.0	141.48	0.01630	118.71	109.37	1013.2	1122.6	0.2008	1.6855	1.8863
4.0	152.97	0.01636	90.63	120.86	1006.4	1127.3	0.2198	1.6427	1.8625
5.0	162.24	0.01640	73.52	130.13	1001.0	1131.1	0.2347	1.6094	1.8441
6.0	170.06	0.01645	61.98	137.96	996.2	1134.2	0.2472	1.5820	1.8292
7.0	176.85	0.01649	53.64	144.76	992.1	1136.9	0.2581	1.5586	1.8167
8.0	182.86	0.01653	47.34	150.79	988.5	1139.3	0.2674	1.5383	1.8057
9.0	188.28	0.01656	42.40	156.22	985.2	1141.4	0.2759	1.5203	1.7962
10	193.21	0.01659	38.42	161.17	982.1	1143.3	0.2835	1.5041	1.7876
14.696	212.00	0.01672	26.80	180.07	970.3	1150.4	0.3120	1.4446	1.7566
15	213.03	0.01672	26.29	181.11	969.7	1150.8	0.3135	1.4415	1.7549
20	227.96	0.01683	20.089	196.16	960.1	1156.3	0.3356	1.3962	1.7319
25	240.07	0.01692	16.303	208.42	952.1	1160.6	0.3533	1.3606	1.7139
30	250.33	0.01701	13.746	218.82	945.3	1164.1	0.3680	1.3313	1.6993
35	259.28	0.01708	11.898	227.91	939.2	1167.1	0.3807	1.3063	1.6870
40	267.25	0.01715	10.498	236.03	933.7	1169.7	0.3919	1.2844	1.6763
45	274.44	0.01721	9.401	243.36	928.6	1172.0	0.4019	1.2650	1.6669
50	281.01	0.01727	8.515	250.09	924.0	1174.1	0.4110	1.2474	1.6585
55	287.07	0.01732	7.787	256.30	919.6	1175.9	0.4193	1.2316	1.6509
60	292.71	0.01738	7.175	262.09	915.5	1177.6	0.4270	1.2168	1.6438
65	297.97	0.01743	6.655	267.50	911.6	1179.1	0.4342	1.2032	1.6374
70	302.92	0.01748	6.206	272.61	907.9	1180.6	0.4409	1.1906	1.6315
75	307.60	0.01753	5.816	277.43	904.5	1181.9	0.4472	1.1787	1.6259
80	312.03	0.01757	5.472	282.02	901.1	1183.1	0.4531	1.1676	1.6207
85	316.25	0.01761	5.168	286.39	897.8	1184.2	0.4587	1.1571	1.6158
90	320.27	0.01766	4.896	290.56	894.7	1185.3	0.4641	1.1471	1.6112
95	324.12	0.01770	4.652	294.56	891.7	1186.2	0.4692	1.1376	1.6068
100	327.81	0.01774	4.432	298.40	888.8	1187.2	0.4740	1.1286	1.6026
110	334.77	0.01782	4.049	305.66	883.2	1188.9	0.4832	1.1117	1.5948
120	341.25	0.01789	3.728	312.44	877.9	1190.4	0.4916	1.0962	1.5878

* Abridged from "Thermodynamic Properties of Steam," by Joseph H. Keenan and Frederick G. Keyes, John Wiley & Sons, Inc., New York, 1937.

TABLE D.1   (Continued)

Abs press., psi p	Temp, °F t	Specific volume, cu ft/lb		Enthalpy, Btu/lb			Entropy		
		Sat. liquid v_f	Sat. vapor v_g	Sat. liquid h_f	Evap. h_{fg}	Sat. vapor h_g	Sat. liquid s_f	Evap. s_{fg}	Sat. vapor s_g
130	347.32	0.01796	3.455	318.81	872.9	1191.7	0.4995	1.0817	1.5812
140	353.02	0.01802	3.220	324.82	868.2	1193.0	0.5069	1.0682	1.5751
150	358.42	0.01809	3.015	330.51	863.6	1194.1	0.5138	1.0556	1.5694
160	363.53	0.01815	2.834	335.93	859.2	1195.1	0.5204	1.0436	1.5640
170	368.41	0.01822	2.675	341.09	854.9	1196.0	0.5266	1.0324	1.5590
180	373.06	0.01827	2.532	346.03	850.8	1196.9	0.5325	1.0217	1.5542
190	377.51	0.01833	2.404	350.79	846.8	1197.6	0.5381	1.0116	1.5497
200	381.79	0.01839	2.288	355.36	843.0	1198.4	0.5435	1.0018	1.5453
250	400.95	0.01865	1.8438	376.00	825.1	1201.1	0.5675	0.9588	1.5263
300	417.33	0.01890	1.5433	393.84	809.0	1202.8	0.5879	0.9225	1.5104
350	431.72	0.01913	1.3260	409.69	794.2	1203.9	0.6056	0.8910	1.4966
400	444.59	0.0193	1.1613	424.0	780.5	1204.5	0.6214	0.8630	1.4844
450	456.28	0.0195	1.0320	437.2	767.4	1204.6	0.6356	0.8378	1.4734
500	467.01	0.0197	0.9278	449.4	755.0	1204.4	0.6487	0.8147	1.4634
550	476.94	0.0199	0.8424	460.8	743.1	1203.9	0.6608	0.7934	1.4542
600	486.21	0.0201	0.7698	471.6	731.6	1203.2	0.6720	0.7734	1.4454
650	494.90	0.0203	0.7083	481.8	720.5	1202.3	0.6826	0.7548	1.4374
700	503.10	0.0205	0.6554	491.5	709.7	1201.2	0.6925	0.7371	1.4296
750	510.86	0.0207	0.6092	500.8	699.2	1200.0	0.7019	0.7204	1.4223
800	518.23	0.0209	0.5687	509.7	688.9	1198.6	0.7108	0.7045	1.4153
850	525.26	0.0210	0.5327	518.3	678.8	1197.1	0.7194	0.6891	1.4085
900	531.98	0.0212	0.5006	526.6	668.8	1195.4	0.7275	0.6744	1.4020
950	538.43	0.0214	0.4717	534.6	659.1	1193.7	0.7355	0.6602	1.3957
1000	544.61	0.0216	0.4456	542.4	649.4	1191.8	0.7430	0.6467	1.3897
1100	556.31	0.0220	0.4001	557.4	630.4	1187.8	0.7575	0.6205	1.3780
1200	567.22	0.0223	0.3619	571.7	611.7	1183.4	0.7711	0.5956	1.3667
1300	577.46	0.0227	0.3293	585.4	593.2	1178.6	0.7840	0.5719	1.3559
1400	587.10	0.0231	0.3012	598.7	574.7	1173.4	0.7963	0.5491	1.3454
1500	596.23	0.0235	0.2765	611.6	556.3	1167.9	0.8082	0.5269	1.3351
2000	635.82	0.0257	0.1878	671.7	463.4	1135.1	0.8619	0.4230	1.2849
2500	668.13	0.0287	0.1307	730.6	360.5	1091.1	0.9126	0.3197	1.2322
3000	695.36	0.0346	0.0858	802.5	217.8	1020.3	0.9731	0.1885	1.1615
3206.2	705.40	0.0503	0.0503	902.7	0	902.7	1.0580	0	1.0580

TABLE D.2   Dry Saturated Steam Temperature*

Temp, °F	Abs press., psi	Specific volume, cu ft/lb			Enthalpy, Btu/lb			Entropy		
		Sat. liquid	Evap.	Sat. vapor	Sat. liquid	Evap.	Sat. vapor	Sat. liquid	Evap.	Sat. vapor
t	p	v_f	v_{fg}	v_g	h_f	h_{fg}	h_g	s_f	s_{fg}	s_g
32	0.08854	0.01602	3306	3306	0.00	1075.8	1075.8	0.0000	2.1877	2.1877
35	0.09995	0.01602	2947	2947	3.02	1074.1	1077.1	0.0061	2.1709	2.1770
40	0.12170	0.01602	2444	2444	8.05	1071.3	1079.3	0.0162	2.1435	2.1597
45	0.14752	0.01602	2036.4	2036.4	13.06	1068.4	1081.5	0.0262	2.1167	2.1429
50	0.17811	0.01603	1703.2	1703.2	18.07	1065.6	1083.7	0.0361	2.0903	2.1264
60	0.2563	0.01604	1206.6	1206.7	28.06	1059.9	1088.0	0.0555	2.0393	2.0948
70	0.3631	0.01606	867.8	867.9	38.04	1054.3	1092.3	0.0745	1.9902	2.0647
80	0.5069	0.01608	633.1	633.1	48.02	1048.6	1096.6	0.0932	1.9428	2.0360
90	0.6982	0.01610	468.0	468.0	57.99	1042.9	1100.9	0.1115	1.8972	2.0087
100	0.9492	0.01613	350.3	350.4	67.97	1037.2	1105.2	0.1295	1.8531	1.9826
110	1.2748	0.01617	265.3	265.4	77.94	1031.6	1109.5	0 1471	1.8106	1.9577
120	1.6924	0.01620	203.25	203.27	87.92	1025.8	1113.7	0.1645	1.7694	1.9339
130	2.2225	0.01625	157.32	157.34	97.90	1020.0	1117.9	0.1816	1.7296	1.9112
140	2.8886	0.01629	122.99	123.01	107.89	1014.1	1122.0	0.1984	1.6910	1.8894
150	3.718	0.01634	97.06	97.07	117.89	1008.2	1126.1	0.2149	1.6537	1.8685
160	4.741	0.01639	77.27	77.29	127.89	1002.3	1130.2	0.2311	1.6174	1.8485
170	5.992	0.01645	62.04	62.06	137.90	996.3	1134.2	0.2472	1.5822	1.8293
180	7.510	0.01651	50.21	50.23	147.92	990.2	1138.1	0.2630	1.5480	1.8109
190	9.339	0.01657	40.94	40.96	157.95	984.1	1142.0	0.2785	1.5147	1.7932
200	11.526	0.01663	33.62	33.64	167.99	977.9	1145.9	0.2938	1.4824	1.7762
210	14.123	0.01670	27.80	27.82	178.05	971.6	1149.7	0.3090	1.4508	1.7598
212	14.696	0.01672	26.78	26.80	180.07	970.3	1150.4	0.3120	1.4446	1.7566
220	17.186	0.01677	23.13	23.15	188.13	965.2	1153.4	0.3239	1.4201	1.7440
230	20.780	0.01684	19.365	19.382	198.23	958.8	1157.0	0.3387	1.3901	1.7288
240	24.969	0.01692	16.306	16.323	208.34	952.2	1160.5	0.3531	1.3609	1.7140
250	29.825	0.01700	13.804	13.821	216.48	945.5	1164.0	0.3675	1.3323	1.6998
260	35.429	0.01709	11.746	11.763	228.64	938.7	1167.3	0.3817	1.3043	1.6860
270	41.858	0.01717	10.044	10.061	238.84	931.8	1170.6	0.3958	1.2769	1.6727
280	49.203	0.01726	8.628	8.645	249.06	924.7	1173.8	0.4096	1.2501	1.6597
290	57.556	0.01735	7.444	7.461	259.31	917.5	1176.8	0.4234	1.2238	1.6472

* Abridged from "Thermodynamic Properties of Steam," by Joseph H. Keenan and Frederick G. Keyes, John Wiley & Sons, Inc., New York, 1937.

TABLE D.2   (Continued)

Temp, °F	Abs press., psi	Specific volume, cu ft/lb			Enthalpy, Btu/lb			Entropy		
		Sat. liquid	Evap.	Sat. vapor	Sat. liquid	Evap.	Sat. vapor	Sat. liquid	Evap.	Sat. vapor
t	p	v_f	v_{fg}	v_g	h_f	h_{fg}	h_g	s_f	s_{fg}	s_g
300	67.013	0.01745	6.449	6.466	269.59	910.1	1179.7	0.4369	1.1980	1.6350
310	77.68	0.01755	5.609	5.626	279.92	902.6	1182.5	0.4504	1.1727	1.6231
320	89.66	0.01765	4.896	4.914	290.28	894.9	1185.2	0.4637	1.1478	1.6115
330	103.06	0.01776	4.289	4.307	300.68	887.0	1187.7	0.4769	1.1233	1.6002
340	118.01	0.01787	3.770	3.788	311.13	879.0	1190.1	0.4900	1.0992	1.5891
350	134.63	0.01799	3.324	3.342	321.63	870.7	1192.3	0.5029	1.0754	1.5783
360	153.04	0.01811	2.939	2.957	332.18	862.2	1194.4	0.5158	1.0519	1.5677
370	173.37	0.01823	2.606	2.625	342.79	853.5	1196.3	0.5286	1.0287	1.5573
380	195.77	0.01836	2.317	2.335	353.45	844.6	1198.1	0.5413	1.0059	1.5471
390	220.37	0.01850	2.0651	2.0836	364.17	835.4	1199.6	0.5539	0.9832	1.5371
400	247.31	0.01864	1.8447	1.8633	374.97	826.0	1201.0	0.5664	0.9608	1.5272
410	276.75	0.01878	1.6512	1.6700	385.83	816.3	1202.1	0.5788	0.9386	1.5174
420	308.83	0.01894	1.4811	1.5000	396.77	806.3	1203.1	0.5912	0.9166	1.5078
430	343.72	0.01910	1.3308	1.3499	407.79	796.0	1203.8	0.6035	0.8947	1.4982
440	381.59	0.01926	1.1979	1.2171	418.90	785.4	1204.3	0.6158	0.8730	1.4887
450	422.6	0.0194	1.0799	1.0993	430.1	774.5	1204.6	0.6280	0.8513	1.4793
460	466.9	0.0196	0.9748	0.9944	441.4	763.2	1204.6	0.6402	0.8298	1.4700
470	514.7	0.0198	0.8811	0.9009	452.8	751.5	1204.3	0.6523	0.8083	1.4606
480	566.1	0.0200	0.7972	0.8172	464.4	739.4	1203.7	0.6645	0.7868	1.4513
490	621.4	0.0202	0.7221	0.7423	476.0	726.8	1202.8	0.6766	0.7653	1.4419
500	680.8	0.0204	0.6545	0.6749	487.8	713.9	1201.7	0.6887	0.7438	1.4325
520	812.4	0.0209	0.5385	0.5594	511.9	686.4	1198.2	0.7130	0.7006	1.4136
540	962.5	0.0215	0.4434	0.4649	536.6	656.6	1193.2	0.7374	0.6568	1.3942
560	1133.1	0.0221	0.3647	0.3868	562.2	624.2	1186.4	0.7621	0.6121	1.3742
580	1325.8	0.0228	0.2989	0.3217	588.9	588.4	1177.3	0.7872	0.5659	1.3532
600	1542.9	0.0236	0.2432	0.2668	617.0	548.5	1165.5	0.8131	0.5176	1.3307
620	1786.6	0.0247	0.1955	0.2201	646.7	503.6	1150.3	0.8398	0.4664	1.3062
640	2059.7	0.0260	0.1538	0.1798	678.6	452.0	1130.5	0.8679	0.4110	1.2789
660	2365.4	0.0278	0.1165	0.1442	714.2	390.2	1104.4	0.8987	0.3485	1.2472
680	2708.1	0.0305	0.0810	0.1115	757.3	309.9	1067.2	0.9351	0.2719	1.2071
700	3093.7	0.0369	0.0392	0.0761	823.3	172.1	995.4	0.9905	0.1484	1.1389
705.4	3206.2	0.0503	0	0.0503	902.7	0	902.7	1.0580	0	1.0580

TABLE D.3 Properties of Superheated Steam*

v = specific volume, cu ft/lb; h = enthalpy, Btu/lb; s = entropy

Abs press., psi (sat. temp)		Temperature, °F									
		400	500	600	700	800	900	1000	1100	1200	1400
1 (101.74)	v	512.0	571.6	631.2	690.8	750.4	809.9	869.5	929.1	988.7	1107.8
	h	1241.7	1288.3	1335.7	1383.8	1432.8	1482.7	1533.5	1585.2	1637.7	1745.7
	s	2.1720	2.2233	2.2702	2.3137	2.3542	2.3923	2.4283	2.4625	2.4952	2.5566
5 (162.24)	v	102.26	114.22	126.16	138.10	150.03	161.95	173.87	185.79	197.71	221.6
	h	1241.2	1288.0	1335.4	1383.6	1432.7	1482.6	1533.4	1585.1	1637.7	1745.7
	s	1.9942	2.0456	2.0927	2.1361	2.1767	2.2148	2.2509	2.2851	2.3178	2.3792
10 (193.21)	v	51.04	57.05	63.03	69.01	74.98	80.95	86.92	92.88	98.84	110.77
	h	1240.6	1287.5	1335.1	1383.4	1432.5	1482.4	1533.2	1585.0	1637.6	1745.6
	s	1.9172	1.9689	2.0160	2.0596	2.1002	2.1383	2.1744	2.2086	2.2413	2.3028
14.696 (212.00)	v	34.68	38.78	42.86	46.94	51.00	55.07	59.13	63.19	67.25	75.37
	h	1239.9	1287.1	1334.8	1383.2	1432.3	1482.3	1533.1	1584.8	1637.5	1745.5
	s	1.8743	1.9261	1.9734	2.0170	2.0576	2.0958	213.19	2.1662	2.1989	2.2603
20 (227.96)	v	25.43	28.46	31.47	34.47	37.46	40.45	43.44	46.42	49.41	55.37
	h	1239.2	1286.6	1334.4	1382.9	1432.1	1482.1	1533.0	1584.7	1637.4	1745.4
	s	1.8396	1.8918	1.9392	1.9829	2.0235	2.0618	2.0978	2.1321	2.1648	2.2263
40 (267.25)	v	12.628	14.168	15.688	17.198	18.702	20.20	21.70	23.20	24.69	27.68
	h	1236.5	1284.8	1333.1	1381.9	1431.3	1481.4	1532.4	1584.3	1637.0	1745.1
	s	1.7608	1.8140	1.8619	1.9058	1.9467	1.9850	2.0212	2.0555	2.0883	2.1498
60 (292.71)	v	8.357	9.403	10.427	11.441	12.449	13.452	14.454	15.453	16.451	18.446
	h	1233.6	1283.0	1331.8	1380.9	1430.5	1480.8	1531.9	1583.8	1636.6	1744.8
	s	1.7135	1.7678	1.8162	1.8605	1.9015	1.9400	1.9762	2.0106	2.0434	2.1049
80 (312.03)	v	6.220	7.020	7.797	8.562	9.322	10.077	10.830	11.582	12.332	13.830
	h	1230.7	1281.1	1330.5	1379.9	1429.7	1480.1	1531.3	1583.4	1636.2	1744.5
	s	1.6791	1.7346	1.7836	1.8281	1.8694	1.9079	1.9442	1.9787	2.0115	2.0731
100 (327.81)	v	4.937	5.589	6.218	6.835	7.446	8.052	8.656	9.259	9.860	11.060
	h	1227.6	1279.1	1329.1	1378.9	1428.9	1479.5	1530.8	1582.9	1635.7	1744.2
	s	1.6518	1.7085	1.7581	1.8029	1.8443	1.8829	1.9193	1.9538	1.9867	2.0484
120 (341.25)	v	4.081	4.636	5.165	5.683	6.195	6.702	7.207	7.710	8.212	9.214
	h	1224.4	1277.2	1327.7	1377.8	1428.1	1478.8	1530.2	1582.4	1635.3	1743.9
	s	1.6287	1.6869	1.7370	1.7822	1.8237	1.8625	1.8990	1.9335	1.9664	2.0281
140 (353.02)	v	3.468	3.954	4.413	4.861	5.301	5.738	6.172	6.604	7.035	7.895
	h	1221.1	1275.2	1326.4	1376.8	1427.3	1478.2	1529.7	1581.9	1634.9	1743.5
	s	1.6087	1.6683	1.7190	1.7645	1.8063	1.8451	1.8817	1.9163	1.9493	2.0110
160 (363.53)	v	3.008	3.443	3.849	4.244	4.631	5.015	5.396	5.775	6.152	6.906
	h	1217.6	1273.1	1325.0	1375.7	1426.4	1477.5	1529.1	1581.4	1634.5	1743.2
	s	1.5908	1.6519	1.7033	1.7491	1.7911	1.8301	1.8667	1.9014	1.9344	1.9962
180 (373.06)	v	2.649	3.044	3.411	3.764	4.110	4.452	4.792	5.129	5.466	6.136
	h	1214.0	1271.0	1323.5	1374.7	1425.6	1476.8	1528.6	1581.0	1634.1	1742.9
	s	1.5745	1.6373	1.6894	1.7355	1.7776	1.8167	1.8534	1.8882	1.9212	1.9831
200 (381.79)	v	2.361	2.726	3.060	3.380	3.693	4.002	4.309	4.613	4.917	5.521
	h	1210.3	1268.9	1322.1	1373.6	1424.8	1476.2	1528.0	1580.5	1633.7	1742.6
	s	1.5594	1.6240	1.6767	1.7232	1.7655	1.8048	1.8415	1.8763	1.9094	1.9713

TABLE D.3 (Continued)

i = specific volume, cu ft/lb; h = enthalpy, Btu/lb; s = entropy

Abs press., psi (sat. temp)		Temperature, °F									
		400	500	600	700	800	900	1000	1100	1200	1400
220 (389.86)	v.......	2.125	2.465	2.772	3.066	3.352	3.634	3.913	4.191	4.467	5.017
	h......	1206.5	1266.7	1320.7	1372.6	1424.0	1475.5	1527.5	1580.0	1633.3	1742.3
	s.......	1.5453	1.6117	1.6652	1.7120	1.7545	1.7939	1.8308	1.8656	1.8987	1.9607
240 (397.37)	v.......	1.9276	2.247	2.533	2.804	3.068	3.327	3.584	3.839	4.093	4.597
	h......	1202.5	1264.5	1319.2	1371.5	1423.2	1474.8	1526.9	1579.6	1632.9	1742.0
	s.......	1.5319	1.6003	1.6546	1.7017	1.7444	1.7839	1.8209	1.8558	1.8889	1.9510
260 (404.42)	v.......	......	2.063	2.330	2.582	2.827	3.067	3.305	3.541	3.776	4.242
	h......	......	1262.3	1317.7	1370.4	1422.3	1474.2	1526.3	1579.1	1632.5	1741.7
	s.......	......	1.5897	1.6447	1.6922	1.7352	1.7748	1.8118	1.8467	1.8799	1.9420
280 (411.05)	v.......	......	1.9047	2.156	2.392	2.621	2.845	3.066	3.286	3.504	3.938
	h......	......	1260.0	1316.2	1369.4	1421.5	1473.5	1525.8	1578.6	1632.1	1741.4
	s.......	......	1.5796	1.6354	1.6834	1.7264	1.7662	1.8033	1.8383	1.8716	1.9337
300 (417.33)	v.......	......	1.7675	2.005	2.227	2.442	2.652	2.859	3.065	3.269	3.674
	h......	......	1257.6	1314.7	1368.3	1420.6	1472.8	1525.2	1578.1	1631.7	1741.0
	s.......	......	1.5701	1.6268	1.6751	1.7184	1.7582	1.7954	1.8305	1.8638	1.9260
350 (431.72)	v.......	......	1.4923	1.7036	1.8980	2.084	2.266	2.445	2.622	2.798	3.147
	h......	......	1251.5	1310.9	1365.5	1418.5	1471.1	1523.8	1577.0	1630.7	1740.3
	s.......	......	1.5481	1.6070	1.6563	1.7002	1.7403	1.7777	1.8130	1.8463	1.9086
400 (444.59)	v.......	......	1.2851	1.4770	1.6508	1.8161	1.9767	2.134	2.290	2.445	2.751
	h......	......	1245.1	1306.9	1362.7	1416.4	1469.4	1522.4	1575.8	1629.6	1739.5
	s.......	......	1.5281	1.5894	1.6398	1.6842	1.7247	1.7623	1.7977	1.8311	1.8936

TABLE D.3   (Continued)

v = specific volume, cu ft/lb; h = enthalpy, Btu/lb; s = entropy

Abs press., psi (sat. temp)		Temperature, °F											
		500	600	620	640	660	680	700	800	900	1000	1200	1400
450 (456.28)	v	1.1231	1.3005	1.3332	1.3652	1.3967	1.4278	1.4584	1.6074	1.7516	1.8928	2.170	2.443
	h	1238.4	1302.8	1314.6	1326.2	1337.5	1348.8	1359.9	1414.3	1467.7	1521.0	1628.6	1738.7
	s	1.5095	1.5735	1.5845	1.5951	1.6054	1.6153	1.6250	1.6699	1.7108	1.7486	1.8177	1.8803
500 (467.01)	v	0.9927	1.1591	1.1893	1.2188	1.2478	1.2763	1.3044	1.4405	1.5715	1.6996	1.9504	2.197
	h	1231.3	1298.6	1310.7	1322.6	1334.2	1345.7	1357.0	1412.1	1466.0	1519.6	1627.6	1737.9
	s	1.4919	1.5588	1.5701	1.5810	1.5915	1.6016	1.6115	1.6571	1.6982	1.7363	1.8056	1.8683
550 (476.94)	v	0.8852	1.0431	1.0714	1.0989	1.1259	1.1523	1.1783	1.3038	1.4241	1.5414	1.7706	1.9957
	h	1223.7	1294.3	1306.8	1318.9	1330.8	1342.5	1354.0	1409.9	1464.3	1518.2	1626.6	1737.1
	s	1.4751	1.5451	1.5568	1.5680	1.5787	1.5890	1.5991	1.6452	1.6868	1.7250	1.7946	1.8575
600 (486.21)	v	0.7947	0.9463	0.9729	0.9988	1.0241	1.0489	1.0732	1.1899	1.3013	1.4096	1.6208	1.8279
	h	1215.7	1289.9	1302.7	1315.2	1327.4	1339.3	1351.1	1407.7	1462.5	1516.7	1625.5	1736.3
	s	1.4586	1.5323	1.5443	1.5558	1.5667	1.5773	1.5875	1.6343	1.6762	1.7147	1.7846	1.8476
700 (503.10)	v		0.7934	0.8177	0.8411	0.8639	0.8860	0.9077	1.0108	1.1082	1.2024	1.3853	1.5641
	h		1280.6	1294.3	1307.5	1320.3	1332.8	1345.0	1403.2	1459.0	1513.9	1623.5	1734.8
	s		1.5084	1.5212	1.5333	1.5449	1.5559	1.5665	1.6147	1.6573	1.6963	1.7666	1.8299
800 (518.23)	v		0.6779	0.7006	0.7223	0.7433	0.7635	0.7833	0.8763	0.9633	1.0470	1.2088	1.3662
	h		1270.7	1285.4	1299.4	1312.9	1325.9	1338.6	1398.6	1455.4	1511.0	1621.4	1733.2
	s		1.4863	1.5000	1.5129	1.5250	1.5366	1.5476	1.5972	1.6407	1.6801	1.7510	1.8146
900 (531.98)	v		0.5873	0.6089	0.6294	0.6491	0.6680	0.6863	0.7716	0.8506	0.9262	1.0714	1.2124
	h		1260.1	1275.9	1290.9	1305.1	1318.8	1332.1	1393.9	1451.8	1508.1	1619.3	1731.6
	s		1.4653	1.4800	1.4938	1.5066	1.5187	1.5303	1.5814	1.6257	1.6656	1.7371	1.8009
1000 (544.61)	v		0.5140	0.5350	0.5546	0.5733	0.5912	0.6084	0.6878	0.7604	0.8294	0.9615	1.0893
	h		1248.8	1265.9	1281.9	1297.0	1311.4	1325.3	1389.2	1448.2	1505.1	1617.3	1730.0
	s		1.4450	1.4610	1.4757	1.4893	1.5021	1.5141	1.5670	1.6121	1.6525	1.7245	1.7886
1100 (556.31)	v		0.4532	0.4738	0.4929	0.5110	0.5281	0.5445	0.6191	0.6866	0.7503	0.8716	0.9885
	h		1236.7	1255.3	1272.4	1288.5	1303.7	1318.3	1384.3	1444.5	1502.2	1615.2	1728.4
	s		1.4251	1.4425	1.4583	1.4728	1.4862	1.4989	1.5535	1.5995	1.6405	1.7130	1.7775
1200 (567.22)	v		0.4016	0.4222	0.4410	0.4586	0.4752	0.4909	0.5617	0.6250	0.6843	0.7967	0.9046
	h		1223.5	1243.9	1262.4	1279.6	1295.7	1311.0	1379.3	1440.7	1499.2	1613.1	1726.9
	s		1.4052	1.4243	1.4413	1.4568	1.4710	1.4843	1.5409	1.5879	1.6293	1.7025	1.7672
1400 (587.10)	v		0.3174	0.3390	0.3580	0.3753	0.3912	0.4062	0.4714	0.5281	0.5805	0.6789	0.7727
	h		1193.0	1218.4	1240.4	1260.3	1278.5	1295.5	1369.1	1433.1	1493.2	1608.9	1723.7
	s		1.3639	1.3877	1.4079	1.4258	1.4419	1.4567	1.5177	1.5666	1.6093	1.6836	1.7489
1600 (604.90)	v			0.2733	0.2936	0.3112	0.3271	0.3417	0.4034	0.4553	0.5027	0.5906	0.6738
	h			1187.8	1215.2	1238.7	1259.6	1278.7	1358.4	1425.3	1487.0	1604.6	1720.5
	s			1.3489	1.3741	1.3952	1.4137	1.4303	1.4964	1.5476	1.5914	1.6669	1.7328
1800 (621.03)	v				0.2407	0.2597	0.2760	0.2907	0.3502	0.3986	0.4421	0.5218	0.5968
	h				1185.1	1214.0	1238.5	1260.3	1347.2	1417.4	1480.8	1600.4	1717.3
	s				1.3377	1.3638	1.3855	1.4044	1.4765	1.5301	1.5752	1.6520	1.7185
2000 (635.82)	v				0.1936	0.2161	0.2337	0.2489	0.3074	0.3532	0.3935	0.4668	0.5352
	h				1145.6	1184.9	1214.8	1240.0	1335.5	1409.2	1474.5	1596.1	1714.1
	s				1.2945	1.3300	1.3564	1.3783	1.4576	1.5139	1.5603	1.6384	1.7055

TABLE D.3 **(Continued)**

v = specific volume, cu ft/lb; h = enthalpy, Btu/lb; s = entropy

Abs press., psi (sat. temp)		Temperature, °F											
		500	600	620	640	660	680	700	800	900	1000	1200	1400
2500 (668.13)	v						0.1484	0.1686	0.2294	0.2710	0.3061	0.3678	0.4244
	h						1132.3	1176.8	1303.6	1387.8	1458.4	1585.3	1706.1
	s						1.2687	1.3073	1.4127	1.4772	1.5273	1.6088	1.6775
3000 (695.36)	v							0.0984	0.1760	0.2159	0.2476	0.3018	0.3505
	h							1060.7	1267.2	1365.0	1441.8	1574.3	1698.0
	s							1.1966	1.3690	1.4439	1.4984	1.5837	1.6540
3206.2 (705.40)	v								0.1583	0.1981	0.2288	0.2806	0.3267
	h								1250.5	1355.2	1434.7	1569.8	1694.6
	s								1.3508	1.4309	1.4874	1.5742	1.6452

* Abridged from "Thermodynamic Properties of Steam," by Joseph H. Keenan and Frederick G. Keyes, John Wiley & Sons, Inc., New York, 1937.

Steam Charts. The properties of steam may be arranged graphically in the form of charts for convenience in solving problems. The Enthalpy-Entropy Diagram for Steam (Chart D.1), known as the "Mollier chart," has a wide application. The vertical axis represents the total heat (enthalpy) in Btu per pound; the horizontal axis, the entropy per pound. The curved lines plotted on the chart represent the pressure in pounds per square inch absolute, the steam temperature and superheat in degrees Fahrenheit, and the percentage of moisture in the wet-steam range. When two properties of steam are known, the others may be read directly from the chart. This applies to wet, saturated, and superheated steam.

The chart shows that steam at 100 psia and 600°F will have a heat content of 1329 Btu as compared with 1329.1 as given in Table D.3.

Steam having an absolute pressure of 50 psia and containing 1100 Btu per lb is found from the chart to contain 8.0 per cent moisture. Many problems involving the use of steam may be solved directly by use of the Mollier chart.

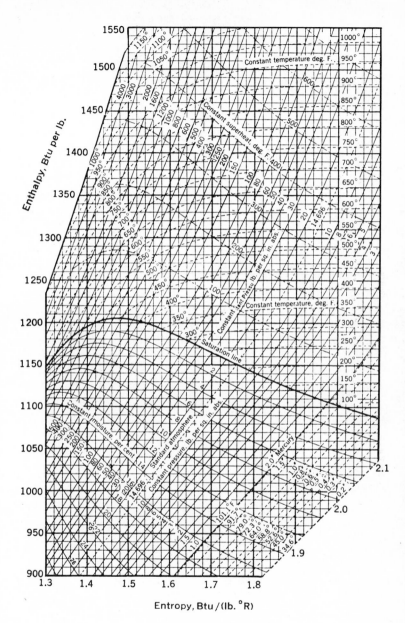

CHART **D.1** *A Mollier chart for steam. (Abstracted by permission from "Thermodynamic Properties of Steam," by J. H. Keenan and F. G. Keyes, published by John Wiley & Sons, Inc.)*

CHART D.1 (*Continued*) Saturation pressures and temperature of steam.

Abs press., in. Hg.	Sat. temp., °F	Abs. press., lb. per sq. in.	Sat. temp., °F	Abs press., lb. per sq. in.	Sat. temp., °F
0.20	34.56	1.0	101.74	120	341.25
0.25	40.23	2	126.08	140	353.02
0.30	44.96	3	141.48	160	363.53
0.35	49.06	4	152.97	180	373.06
0.40	52.04	5	162.24	200	381.79
0.45	55.87	6	170.06	220	389.86
0.50	58.80	7	176.85	240	397.37
0.55	61.48	8	182.86	260	404.42
0.60	63.95	9	188.28	280	411.05
0.65	66.26	10	193.21	300	417.33
0.70	68.40	12	201.96	400	444.59
0.75	70.43	14	209.56	500	467.01
0.80	72.32	14.696	212.00	600	486.21
0.85	74.13	16	216.32	700	503.10
0.90	75.84	18	222.41	800	518.23
0.95	77.47	20	227.96	1000	544.61
1.00	79.03	25	240.07	1200	567.22
1.10	81.95	30	250.33	1400	587.10
1.20	84.65	35	259.28	1600	604.90
1.30	87.17	40	267.25	1800	621.03
1.40	89.51	45	274.44	2000	635.82
.50	91.72	50	281.01	2200	649.46
1.60	93.80	60	292.71	2400	662.12
1.70	95.77	70	302.92	2600	673.94
1.80	97.65	80	312.03	2800	684.99
1.90	99.43	90	320.27	3000	695.36
2.00	101.14	100	327.81	3206.2	705.40

CALCULATION OF
THE STAYED AREA
OF THE HEADS
OF HORIZONTAL-RETURN
TUBULAR BOILERS

The head of a return tubular boiler must be stayed. The tubes are rolled into the sheets, and the ends beaded over so that they act as stays for part of the area. Additional stays must be provided for the remaining area. The area that must be stayed can be approximated by the following method: A 2-in. strip above the tubes is usually considered as supported by them. The width of the strip d (Fig. E.1) supported by the shell is found from the following formula:

$$d = \frac{5 \times T}{\sqrt{P}}$$

where d = width of strip supported by shell
T = thickness of tube sheet, expressed in sixteenths of an inch
P = maximum allowable working pressure, lb per sq in.

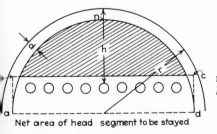

Net area of head segment to be stayed

FIG. E.1 *Net area of head segment to be stayed.*

The area of the segment of the tube sheet that must be stayed may be determined approximately as follows:

$$(D - 2d)^2 \frac{\pi}{8} = \text{area of semicircle}$$

where D = diameter of drum

d = width of strip supported by shell

The area of the semicircle is larger than the required segment by the area *abcd*.

$$\text{Area } abcd = \left(\frac{D}{2} - h + 2\right) \times (D - 2d)$$

where h = distance from top tubes to shell

Then the area of the segment to be stayed would be

$$(D - 2d)^2 \frac{\pi}{8} - \left(\frac{D}{2} - h + 2\right)(D - 2d)$$

This formula assumes that *abcd* is a rectangle and disregards the two small triangles at the ends. This makes the final area of the segment too small. However, as long as the height h is large with respect to the radius, the formula is sufficiently accurate for practical purposes.

Example   A boiler shell is 60 in. in diameter, and the top tubes are 20 in. from the top of the shell. The tube sheets are $\frac{7}{16}$ in. thick. Pressure is 100 psi. Calculate the area to be stayed. If stay bolts have a strength of 8,000 lb, how many must be used?

Solution
Width of the tube sheet supported by the shell:

$$d = \frac{5 \times T}{\sqrt{P}} = \frac{5 \times 7}{\sqrt{100}} = \frac{35}{10} = 3\frac{1}{2} \text{ in.}$$

Area of the segment to be supported:

$$(D - 2d)^2 \frac{\pi}{8} - \left(\frac{D}{2} - h + 2\right)(D - 2d)$$

$$D = 60 \qquad d = 3\frac{1}{2} \qquad h = 20$$

Then $\qquad (60 - 2 \times 3.5)^2 \dfrac{3.1416}{8} \left(\dfrac{60}{2} - 20 + 2\right)(60 - 2 \times 3.5) =$

1,103.09 − 636 = 467.09 sq in which must be stayed
467.09 × 100 = 46,709, pressure stays must hold
46,709 ÷ 8,000 = 5.8 stays required (6 stays would be used)

Problem   A return tubular boiler has a shell 60 in. in diameter; the head is made of $\frac{5}{8}$-in. plate. The tops of the tubes are 22 in. from the top of the shell. The boiler is to operate at 100 lb pressure. Crowfoot stays having a strength of 8,000 lb each are used. How many are required?

CALCULATION OF
THE MEAN EFFECTIVE
PRESSURE OF A
STEAM ENGINE BY
APPLYING THE
ORDINATE METHOD TO
THE INDICATOR DIAGRAM

When a planimeter is not available, the average height of the indicator diagram may be determined by use of the ordinate method. This consists of dividing the diagram into a number of strips and measuring the height of each strip. Although an approximation, this method is sufficiently accurate for most practical work. The diagram must be divided into at least 10 strips, but a greater number increases the accuracy.

The procedure is explained by reference to the typical indicator diagram shown in Fig. F.1. The indicator diagram is first divided by vertical lines into a number of strips of equal width. In the example a minimum of 10 strips has been used. The strips, although equal in width, are unequal in height. The dotted lines, *aa'*, *bb'*, etc., drawn midway between the full lines, represent approximately the average height of the corresponding strips. Since the diagram is composed of these strips, their average height is the average height of the diagram.

The average height of the diagram can be determined by measuring the length of each of the dotted lines, *aa'*, *bb'*, etc. The sum of these measurements expressed in inches and divided by the number of strips into which the diagram is divided (in this case 10) gives the average height of it in inches. (It is advisable to reduce the fractions to decimals before adding.) This average height in inches multiplied by the spring scale of the indicator equals

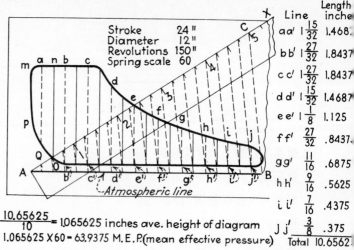

Line		Length inches
a a'	$1\frac{15}{32}$	1.468
b b'	$1\frac{27}{32}$	1.8437
c c'	$1\frac{27}{32}$	1.8437
d d'	$1\frac{15}{32}$	1.4687
e e'	$1\frac{1}{8}$	1.125
f f'	$\frac{27}{32}$	.8437
g g'	$\frac{11}{16}$	.6875
h h'	$\frac{9}{16}$	.5625
i i'	$\frac{7}{16}$	.4375
j j'	$\frac{3}{8}$	.375
Total		10.6562

Stroke 24"
Diameter 12"
Revolutions 150"
Spring scale 60

$$\frac{10.65625}{10} = 1.065625 \text{ inches ave. height of diagram}$$

$$1.065625 \times 60 = 63.9375 \text{ M.E.P.(mean effective pressure)}$$

FIG. F.1 *Calculating mean effective pressure from indicator diagram.*

the mean effective pressure in pounds per square inch as represented by the diagram. The ordinary rule graduated in inches may be replaced by a special scale graduated in pounds corresponding to the scale of the spring used in the indicator. When these scales are used to measure the height of the lines *aa', bb', cc'*, etc., the readings are in pounds per square inch. It is then necessary only to average the readings to obtain the mean effective pressure. The area of the diagram, representing the work done in the cylinder, is found by multiplying the average height by the length.

The following is a convenient method to use in dividing the diagram into strips of equal width and drawing the lines which represent the average height of these strips. In the example shown in Fig. F.1 we shall continue to use 10 strips, but the same principle can be applied to any number of divisions.

1. Draw the line *AX* at *any* convenient angle to the atmospheric line *AB*.
2. By use of a ruler start at *A* and place dots ¼ in. apart along *AX*.
3. Denote the 5-in. division on *AX* as *C* and draw a line from *B* to *C*.
4. With the aid of a square or triangle draw lines parallel to *BC* from each of the dots to the line *AC* until they intersect the line *AB*. The line *AB* is now divided into 20 equal parts.
5. From these intersection points draw lines at right angles to *AB*. Make the lines projected from the ½- and 1-in. divisions of the ruler full lines and those from the ¼- and ¾-in. divisions dotted lines. The full lines divide the diagram into strips of equal width, and the dotted lines represent the average height of these strips. They are used in finding the average height of the diagram and the mean effective pressure.

Answers
To
Problems

Chapter 2

2.1. 0.79 sq in.
2.2. 573 sq in. 3.98 sq ft
2.3. 101,800 lb
2.4. 1⅝ or 1.625, in.
2.23. 27,200 psi elastic limit
56,600 psi ultimate strength
2.24. 610,700 lb
2.25. 7,776,000 lb
2.26. 9,167 psi bursting pressure
1,528 psi working pressure
2.27. 1,354 psi working pressure
2.28. 740 psi bursting pressure
123 psi working pressure
2.30. 0.366 in. (F.S. = 6) use ⅜-in.
plate

2.31. 0.45 in., use ½-in. plate
2.32. 3,841 psi tension on plate
6,787 psi shearing stress on rivets
2.33. 31.8 per cent
2.35. 1,381 sq ft
2.36. 115 boiler hp
2.37. 308 sq ft of heating surface
2.38. 106 boiler hp
2.39. 287 boiler hp developed; 153
per cent rating
2.40. 17,518, 13,841, 3,200 Btu per hr
per sq ft; 88,543 Btu per hr
per cu ft

Chapter 3

3.7. 2.09 lb of oxygen
9.028 lb of air
3.10. 28.71 per cent

3.12. 13.54 lb
3.13. 12,336 Btu
3.23. 22.3° API

Chapter 4

4.39. 1.135 in. of water theoretical
(0.908 available)

Chapter 5

5.11. 115.8 lb

5.12. 6.5 lb low

5.24 3,976 lb

Chapter 6

6.20. 25,250 lb per hr

Chapter 7

7.63. 43.3 lb

7.64. 288.7 ft

7.65. 1,451,520 cu in.; 6,283.6 gal;
52,400 lb; 840 cu ft

7.66. 48.3 gpm; 2,898 gph

7.67. 130 min

7.68. 9 hr approx

7.69. 256 psi; 591 ft

7.70. 3,745 gal

7.71. 33 ft

7.72. 112.5 psi

7.73. 3.53 in. diameter

7.74. 23.56 cu in

7.75. 102 strokes per min

7.76. 15.15 hp

7.77. 485 hp

Chapter 8

8.8. 12.57, 50.27, 95.03, 143.14,
176.72 sq in. area; 1,257,
5,027, 9,503, 14,314, 17,672
lb pressure

8.9. 1,237, 4,948, 9,327, 14,073,
17,357 lb pressure

8.10. 100.5 cu in.; 603.2 cu in.

8.11. 7,540 lb head end; 7,422 lb
crank end

8.21. 2,827 fpm

8.22. 318 rpm

8.26. 6 in.

8.38. 667 rpm

8.39. 111.1 rpm

8.40. 3.5 in.

8.42. 83.3 fpm

8.43. 583.3 fpm

8.44. 9.5 ft

8.45. 250 fpm

Chapter 10

10.27. 109.2 ihp

10.28. 53.6 ihp

10.29. 65.0 ihp

10.30. 58.5 brake hp

10.31. 23.6 brake hp

10.32. 27.5 lb; 9.16
% thermal efficiency

Chapter 12

12.15. 1.83 psi

12.16. 0.53 psi

Chapter 13

13.3. 8.86 per cent

13.17. 5.55 in. use 6 in. pipe

13.18. 4.99 in. use 6 in. pipe

13.19. 6.78 in.

13.20. 5.30 in.

INDEX

Absolute pressure, 89, 100, 600
Absolute temperature, 100, 101, 600
Absolute zero of pressure, 89, 600
Adiabatic expansion, 455, 600
Admission in steam engines, 405–408
Air:
  for combustion, 96, 100, 111–117
  excess, 100, 108, 109, 298, 601
  infiltration of, 159
  nitrogen in, 106
  over-fire, 19, 48, 98, 148–149, 154,
    158, 168, 173, 222–223
  oxygen in, 106, 107
  pollution control, 223–230
  preheaters for, 29–32
  primary, 98, 186, 193
  properties of, 106
  specific heat of, 593
  supply of, 96
  weight of: per pound of fuel, 106
    at 60° F, barometer 29.92 in. Hg.,
    593
Air-admitting grates, 154
Air chambers for pumps, 335
Air pollution control, 223–230
Air preheaters, 34–37
Air space in grate bars, 146
Alarms, high-low water, 233
Analysis:
  of coal: proximate, 118, 121
    ultimate, 118, 121
  of feedwater, 291–294
  of flue gases, 110–112, 136–140
Angle:
  of advance, 412
  of lap, 405–415
  of lead, 405–415
Answers to problems, 619, 620
Anthracite coal, 120, 121
Arches:
  for furnaces, 143, 144
  ignition, 144
Arithmetic, 596–599
Ash of coal:
  fusion of, 119, 120
  influence of, on operation, 36, 124
  minimum required for combustion, 124
Atmospheric line on indicator card, 456
Atmospheric pressure, 100
Atmospheric relief valve, 494, 511
Atomic energy, 56–58
Atomic weights, 56–60
Automatic combustion control, 210–214,
  275–280
Automatic engine, 393
Automatic nonreturn valves, 259–262

Back pressure, 407, 518
Baffles, boiler, 7, 13–15, 25, 34, 319–321
Bag (protrusion on boiler), 289, 314, 317
Balanced valve, 242, 274, 391, 415–417
Ball pulverizer mill, 182–184
Barometric condensers, 496, 497, 498,
  514, 531–532

Bearings:
  engine, 381, 385–386, 449
  Kingsbury thrust, 355, 480, 481
  thrust, 351, 355–356
    in pumps, 351, 355–356
    in turbines, 474, 480–482
Bessemer converter, 64
Bituminous coal, 120, 121
  burning characteristics of, 120, 121
Blister on boiler plate, 290, 314, 317
Blowers, soot, 263–269
Blowoff:
  operation of, 254–259
  surface, 254
Blowoff apparatus, 254–259
Blowoff piping, 6
Blowoff tank, 259
Blowoff valves, 256–259
Boiler assembly, 1–59, 82–85
Boiler baffles, 7, 13–15, 25, 34, 319–321
Boiler drums, 65–71
  construction of, 65, 71–73, 76
  longitudinal, 5, 6
  mud, 34
Boiler joints, 73–75
Boiler plate, ultimate strength of, 70
Boiler-rated capacity, 85–93
Boiler settings, 142–145
Boiler shells, 65–71
Boiler stays, 66, 79–82
Boilers, 1–59
  application of, 1–59
  ash removal from, 124, 287
  assembly of, 1–59, 82–85
  automatic combustion control, 275–280
  automatic operation of, 210–214
  baffles, 7, 13–15, 25, 34, 319–321
  bag on, 289, 314, 317
  blister on plate, 290, 314, 317
  blowdown in, 254–259, 562–564
  blowoff pipe in, 4, 13, 82, 254–256
  blowoff tank for, 259
  bursting pressure of, 69, 70
  calculations for, 86–88
  capacity of, 85–93
  calking of, 75, 313, 318
  carry-over in, 72
  caustic embrittlement of, 4, 75, 93,
    223, 293, 317, 601
  comparison of, fire-tube vs. water-tube,
    20, 21
  construction of, 2, 24, 25, 61–65, 71–81
  corrosion of, 289, 290, 293
  cross-drum, 13
  crown sheet in, 3–5
  definition of, 1
  draft in, 111, 112, 214–222
  dry pipe in, 6, 8, 21, 72, 73
  dry sheet in, 143
  economic loading of, 297
  efficiency of, 296, 297
  emergency operation of, 299–305
  evaporation rate of, 91
  expansion in, 84
  externally fired, 2, 3

Boilers (*Cont.*):
feed piping in, 5, 232–241
feedwater: analysis of, 289–295
   connections for, 232–237
   regulators for, 241–246
fire cracks in, 313
fire-tube, 2–9
   exposed-type, 3
   submerged-type, 4
firebox in, 3–6, 13
firing of (*see* Firing of boilers)
flues in, 7
foaming in, 300, 301
   and priming in, 300, 301
furnace volume of, 88
fusible plugs in, 237–239
gas firing of, 130–132, 206–210
gauge cocks in, 232, 233
gauge connections to (*see* Gauges
   connected to boilers)
hand firing of, 97, 98, 146
handholes for, 81, 82
heat loss of, 135
heat-recovery equipment in, 26–32
heating surface of, 85–87, 93
high-low water alarms for, 233
high-temperature water, 19–20
high water level in, 289
horizontal-return tubular, 5–7
   calculation of stayed area of heads,
   615–616
horsepower of, 86, 369
hydrostatic test for, 283, 284, 314
ideal, 1
idle, care of, 303–305
industrial and utility, 39–57
inspection of, 312–314
instruments used with, 275–280
internally fired, 2
laying up, 303–305
low water level in, 295, 296
maintenance of, 305–311, 314–322
material used in construction of, 61–64
mechanical tube cleaners for, 307, 308
nonreturn valves, 259–262
oil firing of, 126–130
operation of, 283–286
   in emergency, 299–303
   normal, 286–299
packaged units, 16–18
pressure of, safe, 68–71
preventive maintenance of, 311
priming in, 300, 301
pulverized fuel firing of, 190–193
ratings for, 91
repairs to, 314–322
return-tubular, 6–7
safe working pressure of, 68–71
safety in, factor of, 69
safety precautions for, 283–323
safety valves in, 246–254
scale formation in, 292
   chemical cleaning of, 292, 293
   effect on operation, 289
   mechanical cleaners for, 307, 308
   prevention of, 291–294, 305–309
   removal of, from boiler, 307–309
' Scotch marine, 8

Boilers (*Cont.*):
settings of, 1–59, 142–145
shells of, 65–67
soot blowers in, 263–269
starting up, 283–286
stays of, 61, 79–81
steam gauges, 239–241
steam headers for, 262–263
steam separators in, 21–23
Stirling, 13–15, 35
stoker fired, 147–176
stress relieving in, 77
superheaters in, 24–26
suspended setting of, 6
trouble shooting of, 299–303
try cocks in, 232
tube cleaners for, 307, 308
tube renewal of, 306–308, 315–317
tube sheets of, 72
tube size of, 7
valves in, 269–275
vertical, 2–5
vertical-tubular, 4, 12
water column on, 232–237
water conditioning of, 289–295
water glass on, 232, 237
water legs on, 3–4
water level in, 235, 236
water-tube, 9–21
welded construction of (*see* Welded
   construction of boilers)
welding of, 76–79
Bolts, stay, 66, 80, 193
Bourdon pressure gauge, 239, 240
Boyle's Law of Gases, 600
Braces, 66, 67
Brake, Prony, 460
Brake horsepower, 453, 459–462
Brass, weight of, 593
Breakers, vacuum, 495
Brick, weight of, 593
Bridge walls, 6, 143
British thermal unit (BTU), 600
Buckets for steam turbines, 471
Burners, oil, 196, 198–203
   pressure-type, 200–202
   rotary-type, 198–200
   steam-atomizing, 195, 198, 199
Butt strap, 74, 75
By-product fuels, 132–133

Calking of boilers, 75, 317, 318
Calorific value of coal, 118, 119, 121
Calorimeter, 111, 118, 119
Capacity, boiler-rated, 85–93
Carbon:
   combustion of, 121
   fixed, 121
Carbon packing, 474, 483, 484, 527
Carry-over, 21
Cast iron, weight of, 593
Caustic embrittlement, 4, 75, 93, 223,
   293, 317, 601
Centrifugal force, 390–396
Centrifugal governors, 390–396
Centrifugal oil purifiers, 586–590
Centrifugal pumps, 346–360
   classification of, 324, 325

Centrifugal pumps (*Cont.*):
    diffusion rings in, 347, 349–352
    impellers for, 347–353
    multistage, 349–354
    performance of, 362–368
    volute, 347, 348, 355, 356
Chain-grate stokers, 147–149
Check valves, 273, 274
Chemical analysis:
    of feedwater, 291–294
    of fuels, 117–132
Chimneys, 214–216
    draft, 111–113
Cinder reinjection, 168
Circle, area and circumference of, 599
Circumferential joints, 6
Cleaners, mechanical tube, 307, 308
Cleaning plugs, 234
Clearance:
    engine piston, 386–387
    methods of finding, 383–387
    volume of, 384, 445
Clinker grinder, 157
Closed feedwater heaters, 554–557
Coal, 117–126
    air required for, 96, 106, 111, 117
    analysis of, 119, 121
        proximate, 118, 121
        ultimate, 118, 121
    anthracite, 120, 121
    ash in, effect on, 36, 124
    ash fusion temperature, 119, 120
    bituminous, 120, 121
        weight of, 593
    burning characteristics of, 119–122
    calorific value of, 118, 119, 121
    classification of, 120
    coking characteristics of, 120
    combustion of, 18, 97, 98, 117–126
    fixed carbon in, 121
    free-swelling index, 120
    fusion temperature of, 602
    grindability of, 123
    heat of combustion of, 121
    heat balance for, 118, 134, 135
    heat value of, 121
    lignite, 121
    moisture in, 118, 121
    peat, 121
    proximate analysis of, 121
    semianthracite, 121
    semibituminous, 121
    size of, 123
    storage of, 125
    tempering of, 122
    ultimate analysis of, 121
    volatile matter in, 121
Cocks, try or gauge, 232
Coefficient of expansion per degree
    fahrenheit, 593
Collectors, fly ash, 225, 228
Combining weights, laws of, 102–104
Combustion, 6, 100
    air requirements for, 96, 100, 111–117
        actual, 107, 111, 112
    of by-product fuels, 132–133
    chamber for, 142, 143
    chemistry of, 100–111

Combustion (*Cont.*):
    of coal, 117–126
    combustion process, 96–100
    complete, 108
    control of, 133–140, 210–213
        automatic, 210, 214, 275–280
    excess air in, 100, 108, 109
    fluidized-bed, 194–195
    of fuel oil, 126–130
    of gas, 130–132
    ignition temperature, 97
    incomplete, 108
    oxygen required for, 102, 103 106, 107
    perfect, 106
    process of, 96–100
    products of, 106–108
    spontaneous, 125, 126
    temperature required for, 97
    theoretical, 107
    theory of, 100–111
    time required for, 97
Combustion equipment and heating sur-
    face, settings of, 142–230
Compound engines (*see* Engines,
    compound)
Compression in engines, 407, 408
    point of, 407
Condensate pumps, 511–513, 533, 543
Condensation in engine cylinders, 429,
    452
Condenser auxiliaries, 511–515
Condensers, 497–504, 530–536
    advantage of, 530, 531
    air leakage in, 501, 515, 535
    air pumps for, 511
    atmospheric-relief valves for, 515
    barometric, 496, 497, 498, 514, 531–
        532
    barometric jet, 496, 530, 531
    circulating pumps for, 511, 512
    condensation in, 601
    cooling water for, 497, 499–504
    jet, 497–499, 513, 517, 530, 531, 534
    maintenance of, 534–536
    operation of, 532, 533
    packing for, 483, 484
    pumps for, 511–513
    surface, 499–504, 531–534
    tubes for, 501–503
    water required for, 533, 534
Condensing steam engine, 400–401, 429
Condensing steam turbines, 484, 497–
    504, 522
Conduction, 1, 2, 601
Conduction heat, 2, 601
Connecting rods, 380–387
Contact pulverizer mill, 177–182
Control:
    automatic, 133–140, 210–214, 276–280
    combustion, 275–280
Convection heat, 2, 24, 25, 601
Conversion of units, 589–593
Converter, bessemer, 64
Coolers, oil, 529
Cooling ponds, 504, 505
Cooling towers, 505–511, 536–541
    forced-draft, 507, 508, 538–541
    induced-draft, 506, 507, 510, 538–541

Cooling towers (*Cont.*):
  maintenance of, 541
  natural-draft, 505–508
  operation of, 508, 538–541
Corliss engines, 419–425
Corrosion of boilers, 289, 290, 293
Counterbore, 381
Covering pipe, 6, 572, 573
Cross-drum boilers, 13
Crown sheet, 3–5
Cushion valves, 332
Cutoff governor, 390–396
  point of cutoff, 395, 399, 400, 406–408
Cyclone furnace, 52–56
Cylinder calculations, 452
  in engines, 400, 401

D slide valve, 405–417
  setting of, 412–415
Dampers, 146, 151, 154
Dashpots:
  for Corliss valve gear, 420, 421, 423
  for governors, 391, 423, 488
Dead center of engine, 383, 405
Dead plates, 145, 146
Decimal fractions, 596–599
Definition of terms, 600–603
Density, 601
Desuperheaters, 48
Dew point, 32, 105, 125, 129, 130, 601
Diffusion rings, 347–352
Direct-connected valve gear, 412
Disc, rupture, 494, 495, 555
Displacement volume, 371, 384
Draft, 111, 214–216
  balanced, 114, 159, 216
  in chimneys, 111–113
  forced, 113, 151, 154, 216–221
  induced, 116, 216–221
  mechanical, 113, 159, 216–222
  natural, 112, 113
  for various fuels, 114
Drums, boiler (*see* Boiler drums)
Dry bulb temperature, 537
Dry pipe, 6, 8, 21, 22
Dulong's formula, 110, 111
Dummy piston, 480
Duplex pumps, steam, 328–337, 342
Dust collectors:
  bag filters, 228
  electrostatic, 226, 227
  mechanical, 225, 226
  scrubbers, 227, 228

Eccentricity, 390
Eccentrics on engines, 381, 387–390
Economizers, 26–29
Ejectors, steam-air, 222, 513, 532, 534, 542–544
Elastic limit, 62, 69
Electric boiler, 12
Elongation, 62
Embrittlement, caustic, 4, 75, 93, 223, 293, 317, 601
Energy, 601

Engines:
  angle of advance in, 412–414
  automatic, 393
  back pressure in, 455
  bearings of, 381, 385–386, 449
  bull rings of, 382, 449
  clearance of, 383, 384, 386, 387
  clearance volume of, 384, 417, 452, 457
  cocks of, 382
  compound, 382
    cross, 398–400
    receivers for, 399
    tandem, 398–400
  compression in, 407, 408
    point of, 407
  condensation of, 400, 401, 429, 452
  condensing, economy of, 400–401, 429
  connecting rods of, 380, 381, 384, 386
    effect of angularity, 388
  Corliss, 419–425
  counterbore, 381, 447
  cranks of, 380, 381, 387–390
  crossheads of, 380, 384–386
  cycle of events in, 405–408
  cyclinder calculations in, 400, 401
  cylinders of, 381
    condensation in, 429, 452
    pressure cocks for, 382, 414
  dashpots, 423
  dead center of, 383, 405
  eccentrics on, 381, 387–390
  economy of, 451–453, 454, 463
  efficiency of, 445, 458–463
  flywheels in, 380, 387–390
  foundation of, 390
  frames of, 390
  governors of, 390–396
  high-speed, 397
  horsepower of, 453–459, 461, 462
  indicators for, 453–458
  inspection of, 435–437
  lap of, 408, 409
    effect of, 405–415
  lead of: for compound engines, 399
    effect of, 405–415
  low-speed, 397
  lubrication of, 437–440
  maintenance and alignment of, 447–451
  mean effective pressure, 455–456, 459
  mean effective pressure calculations, 460, 617–618
  noncondensing, 400, 401
  operation of, 444–447
    emergency in, 445, 446
    precaution for, 444, 445
    starting up, 444–447
  packing of, 380–382
  pistons for, 380, 381, 384–387
    rings of, 382
    rods of, 382, 384
  poppet-valve, 425–428
  radiation loss in, 384
  ratings for, 458–463
  reciprocating, 380–402
  running over of, 389
  running under of, 389

Engines (*Cont.*):
  setting valves on D slide valve, 412, 413
  slide valve (*see* Slide valve engines)
  speed of, 397–398
    piston, 397
    rotative, 389
  steam:
    admission in engine, 405–408
    effect of lap on, 416
    effect of lead on, 410, 411
    pistons for, 384–387
    release in, 407
  striking points of, 376, 436
  stuffing boxes in, 382, 385
  tandem-compound, 398–400, 518
  uniflow, 401–402, 428–433, 592, 593
  valves of, 387–390
    balanced, 415, 416
    Corliss, 419–425
    D slide, 405–417
    multiported, 415, 416
    piston, 417–419
    poppet, 425–429
    riding cutoff, 415
    uniflow, 428–433
  water rate of, 461, 462
  wire drawing (throttling) in, 452
  wristplate on, 421
Equivalents, tables of, 593–595
Evaporation:
  equivalent, 87
  factor of, 87, 91, 601
  latent heat of, 575
  rate of, in boilers, 85, 87
Evaporators, 560–562
Excess air, 100, 108, 109, 298, 601
Exhaust heads, 580
Expansion:
  adiabatic, 455, 600
  in boilers, 84
  in piping, 568–571
Externally fired boilers, 2, 3
Extraction turbines, 487, 488, 518, 521

Factor settings, 142–145
Fan shaft horsepower, 218–222
Fans, 216
Fatigue of metals, 63
Feed piping in boilers, 5, 232–241
Feedwater:
  chemical analysis of, 291–294
  hardness of, 289, 554
  heaters for, 547–560
  importance of treatment of, 289–294
  impurities in, 289–291
  treatment of, 281, 282, 553, 557–560
Feedwater heaters:
  closed, 554–557
  open, 547–554
Feedwater regulators, 241–246, 273, 274
Filters, oil, 528
Fire cracks in boilers, 313
Fire-tube boilers (*see* Boilers, fire-tube)
Firebox in boilers, 3–6, 13

Firing of boilers:
  with gas, 131, 132
  by hand, 97–99, 145–147
  with oil, 128, 129, 195–206
  with pulverized coal, 121, 124
    cyclone fired, 52–56
  with stokers (*see* Stokers)
Fittings, 81–82
Fixed carbon, 121
Flanged pipe fittings, 564–573
Flue gas:
  analysis of, 110–112, 136–140
  dry, specific heat of, 593
  and heat loss, 135
Flues, 7
Fluidized-bed combustion, 194–195
Fly ash collectors, 225, 228
Fly wheels, 387–390, 394
Foaming in boilers, 300, 301
  and priming, 300, 301
Foot valves, 336, 361
Forced draft, 113, 151, 154, 216–221
Forced feed lubrication, 437–444, 585
Free-swelling index, 119, 160
Fuel oils, 126–130, 195–206
  advantages of, 128
  analysis of, 129
  atomizing burners for, 202, 203
  for burners, 129, 198–203
    pressure-type, 200, 201
    rotary, 198
    steam-atomizing, 198, 199
  classification of, 129
  combustion of, 126–130
  operating boiler with, 203–206
  viscosity of, 127, 583, 603
  (*See also* Oils)
Fuels:
  by-product, 132, 133
  calorific value of, 121
    calorimeter in determing, 111, 118, 119
  chemical analysis of, 117–132
  combustion of, 117–133
  heating value of, 121, 129, 131–133
  pulverized (*see* Pulverized fuel)
Furnace construction, 32–39
Furnaces:
  arches for, 143, 144
  cyclone, 52–56
  dry bottom, 17, 18, 144
  externally fired, 2
  hand fired, 2–6, 42, 98, 145–147
  heat release in, 36, 93, 601
  internally fired, 2
  open-hearth, 64
  pressurized, 116, 117, 216
  slag-tapped, 49, 188
    cyclone fired 52–56
  volume of, 88
  water cooled, 36–38
  wet bottom, 48, 144
Fusible plugs, 237–239

Gas, 206–210
  analysis of, 131
  blast-furnace, 133

Gas (*Cont.*):
  by-product, 207–210
  chimney (*see* Flue gas)
  coke oven, 133
  combustion of, 130–132
  flue (*see* Flue gas)
  as fuel, 130–132, 206–209
  natural, 130, 207
  producer, 125, 130
Gas laws, 100–102
Gate value, 270–273
Gauges connected to boilers, 232–237
  glass, 232–237
  pressure, 239–241
  water column correction for, 240, 241
Globe values, 269–271
Governors:
  automatic, 390–396
  balanced, 391
  centrifugal, 390–396
  cutoff, 390–396
    point of cutoff, 395, 399, 400,
      406–408
  D slide value for, 391–392
  dashpots for, 391, 423, 488
  flyball, 392–393
  hunting in, 391, 488
  inertia (*see* Inertia governors)
  oil-relay, 486, 487
  overspeed trip of, 486, 487
  shaft or flywheel, 390–396
  speed regulation of turbines, 394, 488,
    489
  throttling, 390–396
  turbine (*see* Steam turbines)
Grate bars, 145, 148
  air space in, 146
Grate surface, ratio of, to heating surface,
  146
Grates:
  air-admitting, 154
  traveling, 147, 148
Gravity, specific, 126, 127, 582
Grease cups, 444, 583
Grinder for clinkers, 157

Hand fired furnaces, 2–6, 42, 98, 145–147
Handholes for boilers, 81–82
Heads:
  exhaust, 600, 601
  net positive suction (N.P.S.H.), 343,
    345, 356, 358
Heat:
  of combustion of coal, 121
  conduction, 2, 601
  convection, 2, 14, 34, 601
  latent, of evaporation, 602
  liberated per pound of carbon, 102
  loss of, in chimney gases, 134
  radiant, 1, 24, 25
  specific, 91, 593, 603
Heat-release volume in furnace, 88, 89
Heat value of coal, 121
Heaters:
  closed feedwater, 554–557
  deaerators, 548–550
  feedwater, 547–560

Heaters (*Cont.*):
  fuel savings with, 556
  open, 547–554
Heating surface of boilers, 85–87, 93
Heating surface calculations, 86–88
Heating value of fuels, 121, 129, 131–133
High-low alarms, water column, 233
High-speed engines, 397
Horizontal-return tubular boilers, 5–7
  calculation of stayed area of heads,
    615–616
Horsepower:
  of boiler, 86, 369
  brake, 453, 459–462
  definition of, 86, 575
  of engine, 453–459, 461, 462
  fan shaft, 218–222
  indicated, 453, 459
Horsepower hour, 463, 575
Hot process softener, 551–554
Humidity, relative, 537
Hydrostatic lubricator, 437, 438, 584, 585
Hydrostatic test for boilers, 283, 284, 314

Ice:
  specific heat of, 593
  weight of, 593
Ignition arches, 144
Impact pulverizer mill, 184, 185
Impellers for centrifugal pumps, 347–371
Impulse turbines (*see* Steam turbines)
Incomplete combustion, 108
Indicated horsepower, 453, 459
Indicator card:
  atmospheric line on, 456
  calculation of horsepower from, 459,
    460
    mean effective pressure shown on,
      455–456, 459
  method of taking, 454, 455
Indicator diagram:
  theoretical, 456
  valve setting by, 455–458
Indicators:
  explanation of, 453–458
  operation of, 454, 455
  reducing motion for, 454
  for steam engines, 453–458
Indirect-connected valve gear, 411, 412,
  415
Inertia governors, 395, 396
  automatic, 390–396
  performance of, 395, 396
Injectors, 325–328
Inside-packed plunger pump, 336
Instruments, 275–280
Internally fired boiler, 2

Jacket, steam, 452
Jet condensers, 497–499, 513, 517, 530,
  531, 534
Jets:
  over-fire air, 98, 148, 154, 155, 158,
    162, 168, 173, 222, 530, 531
  steam and air, 98, 222–223
Joints:
  boiler, 73–75

Joints (*Cont.*):
  calculated strength of, 68–71
  circumferential, 6
  efficiency of, 69
  lap and butt, 74
  longitudinal, 4, 5
  riveted (*see* Riveted joints)
Joule's equivalent, 603

Kilowatt, 575
Kilowatt hour, 575
Kingsbury thrust bearing, 355, 480, 481

Labyrinth packing, 477, 483, 484
Lap, effect of, on steam engine, 405–415
Latent heat of evaporation, 575
Lead, effect of, on steam engine,
  405–415
Lignite, 120, 121
Lime-soda softener, 291, 293, 553, 559
Linear coefficient of expansion, 593
Longitudinal drum, 5, 6
Longitudinal joint, 4, 5
Low-speed engines, 397
Lubricants, 581–590
  physical characteristics of, 581–590
  suitable for engines, 437–444
  viscosity of, 444, 582
Lubrication, forced feed, 437–444, 585
Lubricators, 437–444, 583–586
  grease cups, 444, 583
  hydrostatic, 437, 438, 584, 585
  mechanical, 438, 439
  oilers, 438, 583, 584

Manholes, 81–82
Matter, forms of, 100
Mean effective pressure of engines,
  455–456, 459
  calculation of, 617–618
Measurements, tables of, 592–595
Mechanical draft, 113, 159, 216–222
Mechanical stokers (*see* Stokers)
Mechanical tube cleaners for boilers, 307,
  308
Metallic packing, 385, 484, 527
Metals:
  fatique of, 63
  shearing strength of, 61–63, 75
Metric system, 594, 595
Mixed-pressure turbines, 517, 521
Moisture:
  in coal, 118, 121
  in steam, 87, 90, 521
Mud drums, 84
Multiported valves, 415, 416
Multistage centrifugal pumps, 349–354

Natural draft, 112, 113
Net positive suction head (N.P.S.H.),
  343, 345, 356, 358
Nitrogen:
  in air, 106
  oxides of, 229, 230
Noncondensing engines (*see* specific
  engine)

Noncondensing turbine, 493, 494, 518,
  520–522
Nonreturn valves, 259–262
Nuclear steam, 55–57

Oil:
  purification of, 586–590
  relay governors for, 486, 487
  separators for 578–581, 587
  (*See also* Fuel oils)
Oil burners, 196, 198–203
Oil coolers, 529
Oil filters, 528
Oil purifiers, centrifugal, 586–590
Oilers, sight feed for, 434, 583, 584
Oiling systems for turbines, 474, 489–491,
  528, 581–590
Open-hearth furnace, 64
Open heaters, 547–554
Orsat gas analyzer, operation of, 136–140
Outside-packed plunger pumps, 335, 336
Over-fire air, 19, 48, 98, 148–149, 154,
  158, 168, 173, 222–223
Over-fire air jet, 98, 148, 154, 155, 158
  162, 168, 173, 222, 530, 531
Overfeed stokers, 147–152, 155–176
Overspeed trip device for turbines, 472,
  486, 487
Oxygen:
  in air, 106, 107
  in feedwater, 293
  used in combustion, 102, 103, 106, 107

Packing:
  carbon, 474, 483, 484, 527
  labyrinth, 477, 483, 484
  metallic, 385, 484, 527
Patches, 317
Peat, 121
Phosphate, use of, 292
Pipe:
  covering, 6, 572, 573
  dry, 6, 8, 21, 22
  extra strong, 565
Pipeline separators and strainers, 578–581
Piping, 564–573
  blowoff, 6
  capacity of, 564–573
  covering of, 6, 572–573
  expansion of, 568–571
  feedwater (*see* Boilers, feedwater)
  fittings for, 564–573
  joints of, 564–568
  size of, 564–567
  specifications for, 564–566
  steam, 262, 263
  steam-gauge, 239–241
Piston valves, 417–419
Pistons:
  dummy, 480
  for pumps, 328–333
  for steam engines, 380–382, 384–387
Pitting, 313
Planimeter, 458, 459, 615, 616
Plugs:
  cleaning, 234
  fusible, 237–239

Plungers for pumps, 328, 329, 335–336
Ponds, spray (*see* Spray Ponds)
Pop-type safety valve (*see Safety* valves)
Poppet-valve engine, 425–428
Power pumps, 339–340, 365, 366
Preheaters, air, 29–32
Pressure:
    absolute, 89, 100, 600
    absolute zero of, 89, 600
    atmospheric, 100
    back, 407, 518
    measurement of, 100, 240, 241
Pressure gauge, Bourdon, 239, 240
Pressure staging in turbines, 467, 469
Pressure systems, bearings for lubricating
    turbines, 482, 489–491
Pressure volume diagram, 400, 499
Pressurized furnace, 116, 117, 216
Primary air, 98, 186, 193
Priming:
    in boilers, 300, 301
    of pumps, 362, 364, 543
Prony brake, 460
Proximate analysis of coal, 121
Pulverized fuel, 176–194
    advantages of, 189–192
    burners for, 186–188
    central system of pulverization, 176
    cyclone firing of, 51–55
    firing procedures for, 189, 190
    maintenance of equipment for, 192–194
    mills for pulverizing: ball, 182–184
        contact, 177–182
        impact, 184, 185
    unit system of pulverization, 176
Pumps, 324–376
    advantages of (*see* specific pump)
    air chambers for, 335
    by-pass valves for, 359
    calculations for, 364–370
    centrifugal (*see* Centrifugal pumps)
    classification of, 324, 325
    clearance volume of, 332
    condensate, 511–513, 533, 543
    cushion valves for, 332
    diffusion rings for, 347, 349, 351, 352
    direct acting, 328–335
    double-stage, 347, 348
    double-suction, 347, 348, 349, 355
    duplex steam, 328–337, 342
    foot valves for, 336, 361
    head on, 359, 360, 364, 370
    impellers for, 348–351
    injectors, 325–328
    inside-packed plunger, 336
    installation of, 360–364
    lost motion of, 332–334
    lubrication of, 356, 364, 372–376
    maintenance of, 370–376
    mechanical seals, 345, 346, 349, 354,
        355
    multistage centrifugal, 347, 348
    net positive suction head of, 343, 345,
        356, 358
    operation of, 360–364
    outside-packed plunger, 335, 336
    packing in, 335, 345, 349, 353,

Pumps (*Cont.*):
    354, 373, 374
    pistons for, 328, 329, 335
    plungers for, 328, 329, 335–336
    power, 339–340, 365, 366
    priming of, 362, 364, 543
    rated capacity of, 369, 370
    relief valves in, 345
    rotary, 343–346
    setting valves for duplex, 333
    short stroking, 334
    simplex steam, 338–339
    single-stage, 347, 348
    single-suction, 347, 348
    slip in, 335, 345
    steam (*see* Steam pumps)
    steam-bound, 359
    strainers for, 345, 361, 486, 493, 543
    suction lift of, 336, 356–359, 361
    testing and calculations for, 364–370
    thrust bearing in, 351, 355, 356
    turbine, 347
    vacuum, 340–343, 362, 543, 552
    vacuum chambers on, 340
    volute centrifugal, 347, 348, 355, 356

Radiation, 1, 24, 25, 603
Rating:
    for boilers, 91
    for engines, 458–463
Reaction turbines, 467–470
Receivers for compound engines, 399
Reciprocating engines, 380–397
Reducing motion for indicators, 454
Reducing valves, 274, 548
Regulators:
    feedwater, 241–246
    maintenance of, 245, 246
Reheated steam, 518–522
Relative humidity, 537
Release in steam engines, 407
Relief valves:
    atmospheric, 494, 511
    in pumps, 345
    in turbines, 515
Return-tubular boilers, 6–7
Riding cutoff valves, 416
Rings, diffusion, 347–352
Riveted joints, 4, 5, 73–75
    double, 74
    single, 74
Rivets, shearing strength of, 73
Rods, connecting, 380–387
Rotary burners, 198–200
Rupture discs, 494, 495, 555

Safety, factor of, in boilers, 69
Safety valves, 246–254
    accumulation test for, 247
    adjustment of, 248–250, 253, 254
    blowback in, 250, 254
    blowdown in, 250
    capacity of, 246, 248, 250, 251
    installation of, 246–248
    setting of, 246, 248–250
Scotch marine boiler, 8
Screen, traveling, 515
Semianthracite coal, 121

Semibituminous coal, 121
Separators:
  oil, 578–581, 587
  steam, 578–581
Setting valves on D slide valve, 412–415
Settings:
  of boilers, 1–59, 142–145
  of combustion equipment and heating
    surface, 142–230
Shaft or flywheel governors, 390–396
Shearing strength of metal, 61–63, 75
Shell construction, 71–73
  stress in, 65–71
Sight-feed oilers, 434, 583
Simplex steam pumps, 338–339
Slide valve engines, 390, 405–428
  advantages of (see specific engine)
  disadvantages of (see specific engine)
  events in cycle of, 405–408
  setting D slide valve, 412, 413
Slip in pumps, 335, 345
Smoke, 143, 224, 603
Softener:
  hot process, 551–554
  lime-soda, 291, 293, 553, 559
  zeolite, 291, 421, 475, 559, 560
Soot blowers, 263–269
Specific gravity, 126, 127, 582
Specific heat, 91, 593, 603
  of air, 593
Spray ponds, 504, 514, 536–539
  operation of, 505, 506, 537
Square root, 598
Stay bolts, 66, 80, 193
Stays:
  boiler, 66, 79–81
  crowfoot or diagonal, 80
  gusset, 80
  through, 80
Steam:
  dry, 89
  moisture in, 87, 90, 521
  nuclear, 55–57
  reheated, 518–522
  saturated, 89, 603
  superheated, 90, 603
  wet, 90, 583, 603
  wiredrawing of, 452, 453, 603
Steam-air ejectors, 222, 513, 532, 534,
  542–544
Steam-air jets, 98, 222–223
Steam-atomizing burners, 195, 198, 199
Steam boilers (see Boilers)
Steam-bound pump, 359
Steam domes, 72
Steam engines (see Engines, steam)
Steam-gauge piping, 239–241
Steam gauges, 239–241
Steam headers, 269
Steam indicators, 453–458
Steam jacket, 452
Steam piping, 262, 263
Steam pumps, 324–376
  duplex, 328–337, 342
  simplex, 338, 339
Steam separators, 578–581
Steam tables, 604–612
Steam traps, 573–578

Steam turbines, 466–545
  auxiliaries, 541–545
  bearings, in, 472, 474, 477–482, 493,
    527
  blades for, 470, 471, 477
  buckets for, 471
  carbon packing in, 474, 483, 484, 527
  classification of, 517–519
  condensing, 488, 497–504, 522,
    530–536
    auxiliaries for, 511–515
  cooling system, 504–511, 536–541
  design and construction, 471–480
  diaphragm, 474, 477
  double-flow principle in, 480
  elementary, theory of, 466–470
  erection of, 491–497
  extraction of steam, 487, 488, 518, 521
  governors in, 471–489, 528
  high-pressure, 517, 518
  impulse, 466–469
  inspection of, 526–530
  labyrinth packing in, 477, 483, 484
  low-pressure, 517, 518
  lubrication of, 474, 489–491, 525, 528
  maintenance of, 526–530
  metallic packing in, 484, 527
  mixed-pressure, 517, 521
  noncondensing, 493, 494, 518, 520–522
  nonextraction, 494, 519
  nozzles for, 467–469, 477
  oil-relay systems for, 486, 487
  oiling systems for, 474, 489–491, 528,
    581–590
  operation of, 522–526
  overspeed trip in, 472, 486, 487
  packing in, 483–484
  pressure staging in, 467, 469
  pressure system of lubrication, 482,
    489–491
  pumps for, 543–545
  reaction, 467–470
  release in, 407
  relief valves in, 515
  shroud ring, 471, 474
  speed regulation of, 391, 472, 474, 488,
    489
  tandem compound, 398–400, 518
  Terry, 485
  thrust bearings in, 477, 480–483, 527
  vacuum breakers in, 495
  velocity compounding, 468–470
Steel:
  specific heat of, 593
  strength of, 62, 63
  weight of, 593
Stirling boilers, 13–15, 35
Stokers:
  advantages of (see specific stoker)
  chain-grate, 147–149
  maintenance of (see specific stoker)
  multiple-retort, 155–159
  operation of (see specific stoker)
  overfeed, 147–152, 155–176
  rate of combustion, 150
  single-retort, 155
  spreader, 148, 149, 162–176

Stokers (*Cont.*):
traveling-grate, 147–152
turndown ratio, 163
underfeed, 152–162
vibrating-grate, 149–150
Strainers for pumps, 345, 361, 486, 493, 543
Stress relieving in boilers, 77
Stresses in boiler drums and shells, 65–71
Substances, weight of, 589–593
Suction lift of pumps, 336, 356–359, 361
Sulfur oxides, 229
Superheaters, 24–26
advantage of, 24
convection, 24
radiant, 24
Surface blowoff, 254
Surface condensers, 494–504, 531–534
Suspended boiler setting, 6

Tables of measurements, 592–595
Tandem-compound engines, 398–400, 518
Tank, blowoff, 259
Temperature, 603
absolute, 100, 101, 600
dry-bulb, 537
wet-bulb, 537
Tempering of coal, 122
Tensile strength, 61, 62
Terry turbine, 485
Throttling governors, 390–396
Thrust bearings:
in pumps, 351, 355–356
in turbines, 474, 480–482
Traps, steam, 573–578
installation of, 573–578
Traveling-grate stokers, 147–152
Traveling screens, 515, 541, 543
Try or gauge cocks, 232
Tube sheets, 72
Tubes:
cleaners for, 307, 308
expanders for, 307, 308
installation of, 314–317
mechanical cleaners for, 307, 308
replacement of, 314–317
stress in, 65–71
Turbines (*see* Steam turbines)
Tuyères, 154

Ultimate analysis of coal, 121
Ultimate strength of boiler plates, 70
Underfeed stokers, 152–162
Uniflow engines, 401–402, 428–433, 592, 593
Units, conversion of, 592–595

Vacuum breakers, 495
Vacuum chambers on pumps, 340
Vacuum pumps, 340–343, 362, 543, 552
Valve gears:
direct-connected, 412
indirect-connected, 411, 412, 415
Valves, 269–275, 405–433
atmospheric-relief, 494, 511
automatic nonreturn, 259–262
balanced, 242, 274, 391, 415–417

Valves (*Cont.*):
blowoff, 256–259
by-pass on, 359
check, 273, 274
Corliss, 419–425
dashpots for, 420, 421, 423
cushion, 332
cycle of events, 405–408
D slide, 405–417
setting valves on, 412–415
foot, 336, 361
function of, 408–415
gate, 270–273
globe, 269–271
multiported, 415–416
nonreturn, 259–262
piston, 417–419
plug, 275
pop-type safety (*see* Safety valves)
poppet, 425–428
pumps (*see* specific pump)
reducing, 274, 548
relief: in pumps, 345
in turbines, 493
riding cutoff, 415–416
safety or pop (*see* Safety valves)
setting duplex pump, 333
slide, 415–417
in uniflow engine, 428–433
Vertical-tubular boilers, 4, 12
Viscosity of oils, 127, 583, 603
Volatile matter in coal, 121
Volute centrifugal pumps, 347, 348, 355, 356

Water:
analysis of, 290–294
automatic boiler, injectors for, 325–328
level of, in boilers, 235, 236
pH valve of, 540
weight of, 592
Water column on boilers, 232–237
Water conditioning, 557–560
external treatment of, 291
internal, treatment of, 291–294
softening, 290–294, 551, 553
of tube boilers, 289–295
walls of, 36–39, 321
Water glass on boilers, 232, 237
Water hammer, 300, 547, 551
Water legs on boilers, 3–4
Water vapor, specific heat of, 593
Waterwalls, 36–39, 321
Weight of substances, 593
Weights, laws of combining, 100–111
Welded construction of boilers, 76–79
and steam drum, 76–79
tensile testing of, 77
x-ray testing of, 78
Wet-bulb temperature, 537
Wet steam, 90, 583, 603
Wiredrawing of steam, 452, 453, 603
Work, 603
Wristplate, 421, 445

Zeolite softeners, 291, 421, 475, 559, 560
Zinc, weight of, 593